Acta Numerica 2008

Acta Numerica

Volume 17   2008

CAMBRIDGE
UNIVERSITY PRESS

CAMBRIDGE UNIVERSITY PRESS
Cambridge, New York, Melbourne, Madrid, Cape Town,
Singapore, São Paulo, Delhi, Tokyo, Mexico City

Cambridge University Press
The Edinburgh Building, Cambridge CB2 8RU, UK

Published in the United States of America by Cambridge University Press, New York

www.cambridge.org
Information on this title: www.cambridge.org/9780521174350

First published 2008
First paperback edition 2011

A catalogue record for this publication is available from the British Library

ISBN 978-0-521-51642-6 Hardback
ISBN 978-0-521-17435-0 Paperback

Contents

Contents

Acta Numerica (2008), pp. 1–86
doi: 10.1017/S0962492906340019

Linear algebra algorithms as dynamical systems

Moody T. Chu*

Department of Mathematics,

North Carolina State University,

Raleigh, North Carolina, 27695-8205, USA

E-mail: chu@math.ncsu.edu

In memory of Gene Golub, a good friend and mentor

Any logical procedure that is used to reason or to infer either deductively or inductively, so as to draw conclusions or make decisions, can be called, in a broad sense, a realization process. A realization process usually assumes the recursive form that one state develops into another state by following a certain specific rule. Such an action is generally formalized as a dynamical system. In mathematics, especially for existence questions, a realization process often appears in the form of an iterative procedure or a differential equation. For years researchers have taken great effort to describe, analyse, and modify realization processes for various applications.

The thrust in this exposition is to exploit the notion of dynamical systems as a special realization process for problems arising from the field of linear algebra. Several differential equations whose solutions evolve in submanifolds of matrices are cast in fairly general frameworks, of which special cases have been found to afford unified and fundamental insights into the structure and behaviour of existing discrete methods and, now and then, suggest new and improved numerical methods. In some cases, there are remarkable connections between smooth flows and discrete numerical algorithms. In other cases, the flow approach seems advantageous in tackling very difficult open problems. Various aspects of the recent development and application in this direction are discussed in this paper.

* This research was supported in part by the National Science Foundation under grants DMS-0505880 and CCF-0732299.

CONTENTS

1. Introduction

At the risk of oversimplifying an extremely complex mechanism of thinking, we begin with a large and loose metaphor to delineate the characteristics of a realization process. A realization process usually comprises three components. First, we have two abstract problems, of which one is an artificial problem whose solution is easy to find, while the other is the real problem whose solution is hard to attain. Secondly, we need to design a bridge or a path that connects the easy problem to the difficult problem. The basic idea is to utilize the bridge to set the rule for a certain dynamical system that evolves from the solution of the easy problem to the solution of the difficult problem. Once the blueprint for the bridge construction is in place, we finally need a practical method allowing us to move along the path so that the desirable solution is reached at the end of the process.

The steps taken for the realization, that is, the changes from one state to the next state along the bridge, can be discrete or continuous. Given the limitations of current computing technology, however, it is generally accepted that the most common and effective way to execute a computation is by means of floating-point arithmetic (Goldberg 1991). As such, it is almost a mandate that a continuous realization process must be discretized first before it can be put into operation numerically (Allgower and Georg 2003). For this reason, and perhaps more so for convenience, we have observed that a majority of numerical algorithms in practice are iterative in nature. It could very well be the case that an iterative scheme was initially devised without the notion of a 'connecting bridge' in mind. Its convergence and hence the appearance of a bridge connecting the starting point to the limit point are often not immediately evident, but are rather the result of hard analysis. In hindsight, we now recognize that most iterative methods can be categorically classified as realization processes.

Our principal goal in this exposition is to characterize the relationship between the dynamics of classical iterative methods and that of certain differential systems. We note that in certain cases the continuous model 'interpolates' exactly the iterates of the corresponding discrete method, or that the discrete model 'samples' the solution flow of the corresponding differential equation at integer times, while in other case we can only suggest a straightforward continuous extension or an obvious discretization. In all cases, we think that the interplay between dynamical systems and computational methods is not only of theoretical interest but also has important consequences, as will be made manifest in the subsequent discussion.

Needless to say, the success of a realization process depends on how the bridge is extended from the trivial solution to the desirable solution. Sometimes we have specific guidelines in building the bridge. Bridges underlying the projected gradient method (Chu and Driessel 1990), the interior point method (Karmarkar 1984, Wright 1997, Potra and Wright 2000, Wright

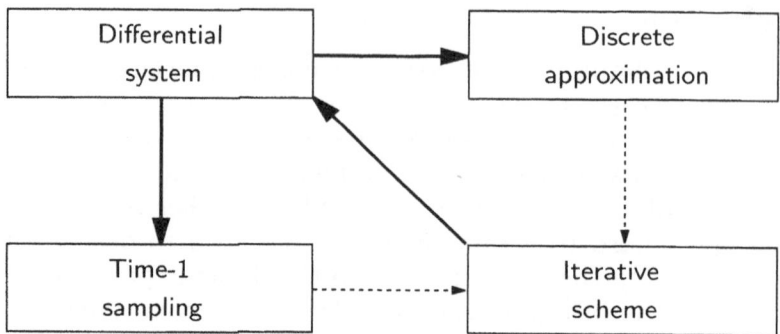

Figure 1.1. Possible links between continuous
and discrete dynamical systems.

2005) or the conjugate gradient method (Hestenes and Stiefel 1952, Greenbaum 1997, Meurant 2006), for example, are based on the principle of systematically optimizing the values of certain objective functions. Sometimes
the bridge is developed more or less on the basis of 'innate inclination',
where we can only hope that the bridge will connect to the other end.
The continuous Newton method (Smale 1977), or the homotopy method
(Allgower and Georg 1980, García and Gould 1980, Morgan 1987), for example, requires extra efforts to make sure that the bridge actually makes
the desirable connection. In other situations, such as for the QR algorithm
(Francis 1961/1962, Watkins 1982) or the Rayleigh quotient iteration (Parlett 1974), it appears that the bridge comes into existence in an anomalous
way. But the fact is that a much deeper mathematical or physical cause is
often involved. When the theory is unveiled, we are often amazed to see
that these seemingly aberrant bridges do exist by themselves naturally.

Figure 1.1 serves as a reminder of the possible links between continuous
and discrete dynamical systems. The dotted lines indicate that an iterative
scheme might be generated or regenerated from a differential system. Going
from a continuous system to a discrete system is usually regarded as 'natural'
since most numerical ODE techniques are doing precisely that task, but
one major thrust of this paper is to illustrate some non-traditional ways of
discretization that are not as straightforward as an ordinary ODE scheme,
but could lead to new and effective algorithms. On the other hand, going
from an iterative scheme to a differential system is not always as obvious
as merely considering the discrete scheme as an Euler step of a differential
system. Other mechanisms, such as control, acceleration, optimization, or
structure preservation, can also induce continuous dynamical systems. Our
presentation in this paper centres around describing, case by case, each
direction in the flowchart of Figure 1.1, with applications arising from linear
algebra algorithms.

2. Numerical analysis versus dynamical systems

Most of the iterative methods developed for practical purposes assume the format of an *m-step sequential process* (Ortega and Rheinboldt 2000),

$$\mathbf{x}_{k+1} = G_k(\mathbf{x}_k, \ldots, \mathbf{x}_{k-m+1}), \quad k = 0, 1, \ldots, \tag{2.1}$$

where

$$G_k : D_k \subset V^m \to V \tag{2.2}$$

are some predetermined maps, V is a designated vector space and m is a fixed integer. Obviously, to start up an m-step iteration, initial values $\mathbf{x}_0, \mathbf{x}_{-1}, \ldots, \mathbf{x}_{-m+1}$ must be specified first. An m-step process is said to be *stationary* if all iteration maps G_k together with the domains D_k are independent of k.

Conventional numerical integrators such as the Runge–Kutta methods and the Adams methods for an initial value problem,

$$\frac{d\mathbf{x}}{dt} = \mathbf{f}(t, \mathbf{x}), \quad \mathbf{x}(0) = \mathbf{x}_0, \tag{2.3}$$

are typical one-step and multi-step sequential processes, respectively. The corresponding iterative maps G_k have evident definitions for explicit methods, but their construction is more devious for implicit methods. Discussions on issues of stability and convergence for discrete methods in this context are abundant in the literature. We shall not review any numerical ODE techniques in this paper, but would recommend the seminal books by Hairer, Nørsett and Wanner (1993) and Hairer and Wanner (1996) as general references on this subject. Our focus in this paper is concentrated primarily on a few very specific iterative processes that were developed originally for problems from fields other than ODEs. It will become apparent that the differential systems associated with the applications to be discussed are of a distinct character and that special numerical techniques might be needed. It is perhaps fitting to echo what Gear (1981) has suggested: that there are more things to do with ODE techniques.

It should be stressed that the subject of discrete dynamical systems has its own distinguished role in nonlinear analysis, providing models for many natural phenomena, and is itself a discipline of extensive and deep research activity. For example, there is Sarkovskii's theorem, remarkable for its lack of hypotheses and for its qualitative universality, asserting that if the discrete dynamical system formed by iterating a continuous function $f :$ $\mathbb{R} \to \mathbb{R}$ has a point of period 3, then it has points of all periods. This topic is beyond the scope of our current discussion, but we find the introductory textbooks by Devaney (1992) and Elaydi (2005), as well as the extended article by Galor (2005), very accessible. The book by Kulenović and Merino (2002) is interesting in that it contains ready-to-use software for computer simulation. For more rigorous theoretical development and a rich collection of applications, we recommend the monograph by Sedaghat (2003). Of course, the fundamental textbook by Wimp (1984) remains the absolute reference for computational issues associated with finite difference equations.

2.1. Dynamics of iterative maps

A subtle line must be drawn in that the classical convergence analysis and stability theory of numerical analysis consider only systems with trivial asymptotic behaviour, namely convergence to a unique equilibrium point, whereas most dynamical systems show more complicated behaviour, with limit cycles or even strange attractors (Stuart and Humphries 1996). From a numerical analysis point of view, the discretization of a differential equation is primarily meant to trace the solution flow with reliable and reasonable accuracy. From a dynamical systems perspective, however, the analysis of a sequential process seeks to differentiate the intrinsic geometric structure. There is considerable overlap between these two disciplines, but there are also significant differences, as Guckenheimer (2002) explains:

'The tension between geometric and more traditional analysis of numerical integration algorithms can be caricatured as the interchange between two limits. The object of study is systems of ordinary differential equations and their flows. Numerical solution of initial value problems for systems of ordinary differential equations discretizes the equations in time and produces sequences of points that approximate solutions over time intervals. Dynamical systems theory concentrates on questions about long-time behavior of the solution trajectories, often investigating intricate geometry in structures formed by the trajectories. The two limits of (1) discretizing the equations with finer and finer resolution in time and (2) letting time tend to infinity do not commute. Classical theories of numerical analysis give little information about the limit behavior of numerical trajectories with increasing time. Extending these theories to do so is feasible only by making the analysis specific to classes of systems with restricted geometric properties. The blend of geometry and numerical analysis that is taking place in current research has begun to produce a subject with lots of detail and richness.'

Perhaps a simple example can best demonstrate the above points. Consider the task of solving the logistic equation,

$$\frac{\mathrm{d}x}{\mathrm{d}t} = x(1-x), \quad x(0) = x_0, \tag{2.4}$$

by the Euler method,

$$x_{k+1} = x_k + \epsilon x_k(1 - x_k), \tag{2.5}$$

with step size ϵ. The exact solution of (2.4) is given by

$$x(t) = \frac{x_0}{x_0 + \mathrm{e}^{-t}(1 - x_0)}, \tag{2.6}$$

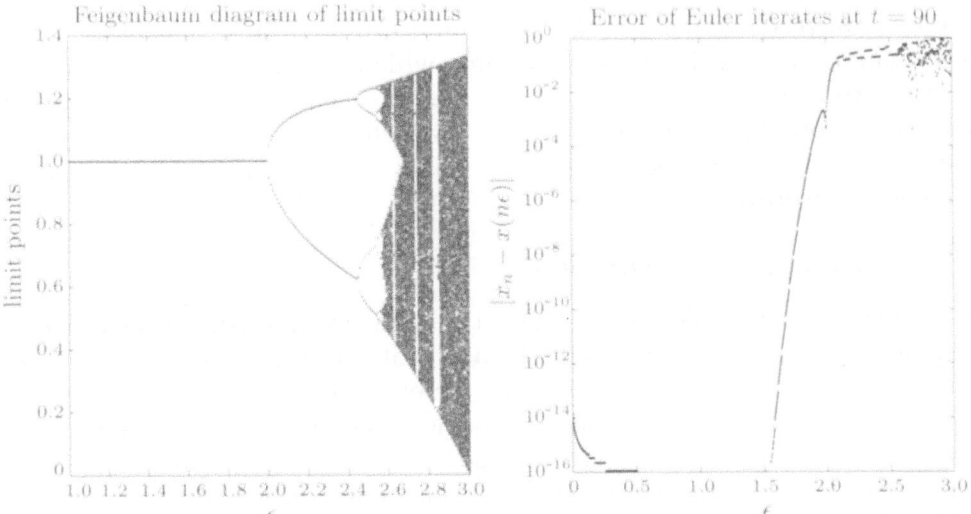

Figure 2.1. Euler iterations for the logistic equation.

which converges to the equilibrium $x(\infty) = 1$ for any initial value $x_0 \neq 0$. Traditional numerical analysis concerns and proves the convergence of x_n to $x(t)$ at each fixed t in the sense that $n \to \infty$ but $t = n\epsilon$. With $n = \lceil \frac{90}{\epsilon} \rceil$ and $0 < \epsilon \leq 3$, we plot the absolute error $|x_n - x(n\epsilon)|$ in the right-hand graph of Figure 2.1. The graph for ϵ between approximately 0.5 and 1.5 is omitted because of the logarithm at machine zero. Note that even at ridiculously large step sizes the errors follow the theoretic estimate $\mathcal{O}(\epsilon)$. On the other hand, with each fixed ϵ, if we iterate the Euler steps 5000 times, then the sequence $\{x_k\}$ exhibits period doubling when ϵ is larger than 2. The left-hand graph in Figure 2.1 shows the limit points, as a function of ϵ, of the corresponding sequence $\{x_k\}$. The so-called Feigenbaum diagram clearly indicates a cascade of period doubling as ϵ increases, which eventually leads to numerical chaos. Note in particular that the equilibrium $x(\infty) = 1$ for (2.4) is no longer an attractor to the discrete dynamical system (2.5) when ϵ is sufficiently large. This equilibrium of the original differential equation does not even appear in the Feigenbaum diagram for large ϵ values. In contrast, implicit schemes such as

$$x_{k+1} = x_k + \epsilon x_k (1 - x_{k+1}) \qquad (2.7)$$

or

$$x_{k+1} = x_k + \epsilon x_{k+1}(1 - x_{k+1}) \qquad (2.8)$$

converge to the equilibrium $x(\infty) = 1$ for any step size ϵ.

With this lesson in mind, we must be careful in distinguishing between the limiting behaviour of an iterative algorithm, which is designed originally by a numerical practitioner to solve a specific problem, and that of a discrete approximation of a differential system, which is formulated to mimic an existing iterative algorithm. Likewise, we must also distinguish the asymptotic behaviour of a differential system, which is developed originally from a specific realization process, and that of its discrete approximation, which becomes an iterative scheme.

2.2. Pseudo-transient continuation

It might be worthwhile to illustrate a general mechanism for advancing a specific continuous system. This idea is not the only way to discretize a continuous system and does not work for every kind of differential system, but it illustrates an interesting view of how the trajectory of a continuous system can be approximately tracked so as to find the equilibrium point, by using numerical ODE techniques in a somewhat non-traditional way.

We shall see in Section 7.3 that it is often the case in many applications that the solution $\mathbf{x}^*$ is realized as the limit point

$$\mathbf{x}^* = \lim_{t \to \infty} \mathbf{x}(t), \tag{2.9}$$

where $\mathbf{x}(t)$ is the solution to the gradient flow

$$\frac{d\mathbf{x}}{dt} = -\nabla F(\mathbf{x}), \quad \mathbf{x}(0) = \mathbf{x}_0, \tag{2.10}$$

with respect to a specified smooth objective function $F : \mathbb{R}^n \to \mathbb{R}$. At first glance, we should be able to find $\mathbf{x}^*$ by solving the first-order optimality condition

$$\nabla F(\mathbf{x}) = 0,$$

with some general-purpose Newton-like iterative methods. Such an approach, however, ignores the gradient property of ∇F and may locate a solution which is different from $\mathbf{x}^*$, and might even be dynamically unstable. Employing some existing ODE integrators to carefully trace the trajectory $\mathbf{x}(t)$ is another way of finding $\mathbf{x}^*$. As reliable as this approach might be, it requires expensive computation at the transient states which is not needed for computing $\mathbf{x}^*$.

One feasible discretization of (2.10) is as follows. Assuming that an approximate solution $\mathbf{x}_k$ has already been computed, one implicit Euler step with step size ϵ_k to (2.10) yields a nonlinear equation,

$$\mathbf{x}_{k+1} = \mathbf{x}_k - \epsilon_k \nabla F(\mathbf{x}_{k+1}), \tag{2.11}$$

for the next step $\mathbf{x}_{k+1}$. Instead of solving (2.11) to high precision as an ODE integrator would normally do, we perform the correction using only one Newton iteration starting at $\mathbf{x}_k$ and accept the outcome as $\mathbf{x}_{k+1}$. The idea is to stay near the true trajectory, but not to strive for accuracy. It is not difficult to see that one Newton step for (2.11) leads to the iterative scheme

$$\mathbf{x}_{k+1} = \mathbf{x}_k - \left(\frac{1}{\epsilon_k} I_n + \nabla^2 F(\mathbf{x}_k)\right)^{-1} \nabla F(\mathbf{x}_k). \qquad (2.12)$$

This scheme is a special implicit upwind method which has been applied successfully for computing steady-state solutions in the PDE community (Mulder and van Leer 1985). Note that for small values of ϵ_k the scheme (2.12) behaves like a steepest descent method, whereas for large values of ϵ_k it behaves like a Newton iteration. Taking into account the fact that $\nabla F(\mathbf{x})$ should have small norm near the optimal point $\mathbf{x}^*$, the so-called 'switched evolution relaxation' strategy for selecting the step sizes, namely,

$$\epsilon_{k+1} = \epsilon_k \frac{\|\nabla F(\mathbf{x}_k)\|}{\|\nabla F(\mathbf{x}_{k+1})\|}, \qquad (2.13)$$

seems to be able to capture the characteristics of being relatively large in the initial phase, and small in the terminal phase of the iteration. The method described above is referred to as *pseudo-transient continuation* in Kelley and Keyes (1998), where convergence theory and implementation issues are also discussed. For a review of its applications, see the recent paper by Kelley, Liao, Qi, Chu, Reese and Winton (2007).

In the subsequent sections of this paper, we shall review various kinds of numerical algorithms, especially those related to linear algebra problems, and explore the possibility of recasting them as dynamical systems. Not only do we want to establish the relationship for theoretical interest, but we also wish to gain some insights via this interpretation, and to develop some new algorithms. A few of these ideas have already been reported in an earlier review by Chu (1988). It is hoped that this paper will bring up to date some more recent developments advanced in the past two decades and point out some new areas for research.

3. Dynamical systems for linear equations

Iterative methods for linear systems have a significant role in history and in applications. This class of methods has come a long way with a dazzling array of developments. See, for example, the various 'templates' discussed in the book by Barrett *et al.* (1994). Research is still evolving even now. Current techniques range from the ingenious acceleration of classical iterative schemes (Hageman and Young 1981) to effective Krylov subspace approximation (van der Vorst 2003), to the more geometrically motivated multi-grid (Briggs 1987, Bramble 1993) or domain decomposition approaches (Toselli and Widlund 2005). Some favourites of practitioners include the preconditioned conjugate gradient (PCG) method (Hestenes and Stiefel 1952), the generalized minimum residual (GMRES) method (Saad and Schultz 1986), the quasi-minimal residual (QMR) method (Freund and Nachtigal 1991), and so on. It is impossible to discuss the dynamics of these methods one by one in this presentation. We outline briefly only two principal ideas in this section.

3.1. Stationary iteration

Most classical iterative methods, such as the Jacobi, the Gauss–Seidel, or the SOR methods, for the linear system

$$A\mathbf{x} = \mathbf{b}, \tag{3.1}$$

where $A \in \mathbb{R}^{n \times n}$ is non-singular and $\mathbf{b} \in \mathbb{R}^n$, are one-step stationary sequential processes of the form

$$\mathbf{x}_{k+1} = G\mathbf{x}_k + \mathbf{c}, \quad k = 0, 1, 2, \ldots. \tag{3.2}$$

The *iteration matrix* $G \in \mathbb{R}^{n \times n}$ plays a crucial role in the convergence of $\{\mathbf{x}_k\}$ in this scheme. Indeed, a necessary and sufficient condition for the convergence of (3.2) from any given starting value $\mathbf{x}_0$ to the unique solution $\mathbf{x}^*$ of (3.1) is that the spectral radius $\rho(G)$ is strictly less than one (Varga 2000). Extensive efforts have been made to construct G to ensure convergence. This is usually done as follows. At the fixed point $\mathbf{x}^*$, we see the relationship

$$G = I - K^{-1}A, \tag{3.3}$$
$$\mathbf{c} = K^{-1}\mathbf{b},$$

for some non-singular matrix K. Because A is 'split' by K in the sense that

$$A = K - KG,$$

K is called a *splitting matrix* of A. Thus, in designing an effective iterative method, attention turns to the selection of a splitting matrix K of A, such that $\rho(I - K^{-1}A) < 1$, for which K^{-1} is relatively easy to compute. The mathematical theory developed for this traditional approach can be found in the seminal book by Varga (2000).

It is trivially seen that the iterative scheme (3.2) is equivalent to an Euler step with unit step size applied to the differential system

$$\frac{d\mathbf{x}}{dt} = \mathbf{f}(\mathbf{x}; K) := -K^{-1}(A\mathbf{x} - \mathbf{b}), \tag{3.4}$$

whose analytic solution is given by

$$\mathbf{x}(t) = e^{-K^{-1}At}(\mathbf{x}_0 - A^{-1}\mathbf{b}) + A^{-1}\mathbf{b}. \tag{3.5}$$

For convergence, however, there is a fundamental difference between the difference equation (3.2) and the differential equation (3.4) in the condition to be imposed on the splitting matrix K. The concern in (3.2) is to make $\rho(I - K^{-1}A)$ as small as possible. Indeed, an ideal K would be one for which the eigenvalues of $K^{-1}A$ are clustered around the real value $\lambda = 1$. Of course, the obvious choice $K = A$ is not practical, because computing A^{-1} is precisely the task we want to circumvent by doing iteration. In contrast, the concern in (3.4) is to make the real part of eigenvalues of $K^{-1}A$ positive and large for fast convergence to the limit point $\mathbf{x}^*$. It might also be desirable to keep the eigenvalues of $K^{-1}A$ clustered to avoid stiffness or high oscillation.

All of these requirements imposed on eigenvalues of $K^{-1}A$ in either case can be met by employing techniques for multiplicative inverse eigenvalue problems, which are discussed in the book by Chu and Golub (2005). For specific applications, finding the most suitable preconditioner has been a major research effort, since it can significantly improve the efficiency of an iterative method. In practice, however, preconditioning is an inexact science because different preconditioners work better for different kinds of problems. To stay within the theme of this article, we shall not elaborate on the choice of K, but assume that it has been constructed in some fashion.

The question now is how to integrate (3.4) so as to reach its equilibrium point quickly. Certainly there are various ways to discretize the differential system (3.4), including the pseudo-transient continuation method described earlier. There are also many different choices of the splitting matrix K, including an obvious choice $K^{-1} = A^\top$ which leads to a gradient flow

$$\frac{d\mathbf{x}}{dt} = -A^\top(A\mathbf{x} - \mathbf{b}), \tag{3.6}$$

for the objective function $f(\mathbf{x}) = \frac{1}{2}\|A\mathbf{x} - \mathbf{b}\|_2^2$, which works even when A is a rectangular matrix. Once a decision is made, what is the dynamics of the resulting iterative map?

We shall describe in the next section how the discretization of (3.4) can be related to the Krylov subspace method. At present, it might be appropriate to recall two scenarios already described in (Chu 1988) that demonstrate the 'tension' referred to by Guckenheimer (2002) between geometric and more traditional analysis of numerical integration algorithms.

First, suppose that the trapezoidal rule with step size ϵ is applied to (3.4). We obtain an iterative scheme,

$$\mathbf{x}_{k+1} = \left(I + \frac{\epsilon}{2}K^{-1}A\right)^{-1}\left(I - \frac{\epsilon}{2}K^{-1}A\right)\mathbf{x}_k + \epsilon\left(I + \frac{\epsilon}{2}K^{-1}A\right)^{-1}K^{-1}\mathbf{b}, \tag{3.7}$$

which makes an interesting comparison with the analytic solution,

$$\mathbf{x}(t + \epsilon) = e^{-\epsilon K^{-1}A}\mathbf{x}(t) + \int_t^{t+\epsilon} e^{(t+\epsilon-u)A}(K^{-1}\mathbf{b})\,du. \tag{3.8}$$

Specifically, the iteration matrix $(I + \frac{\epsilon}{2}K^{-1}A)^{-1}(I - \frac{\epsilon}{2}K^{-1}A)$, being the $(1,1)$-pair Padé approximation, agrees with the exponential matrix $e^{-\epsilon K^{-1}A}$ up to the ϵ^2 term in the series expansion. Likewise, the second term in (3.7) agrees with the integral in (3.8) to the same order of accuracy. Though it might not be practical for real computation, the iterative scheme (3.7), using the trapezoidal rule, on one hand tracks the solution curve closely for small ϵ, and on the other hand converges to $\mathbf{x}^*$ for any step size ϵ.

Secondly, recall that the well-known polynomial acceleration methods applied to (3.2) usually assume a three-term recursive relationship,

$$\mathbf{x}_1 = \epsilon_1(G\mathbf{x}_0 + \mathbf{c}) + (1 - \epsilon_1)\mathbf{x}_0,$$

$$\mathbf{x}_{k+1} = \alpha_{k+1}\big[\epsilon_{k+1}(G\mathbf{x}_k + \mathbf{c}) + (1 - \epsilon_{k+1})\mathbf{x}_k\big] + (1 - \alpha_{k+1})\mathbf{x}_{k-1}, \qquad (3.9)$$

with some properly defined real numbers α_k and ϵ_k (Hageman and Young 1981, Chapters 4–6). Note that the scheme (3.9) amounts to a two-step sequential process. It is not difficult to rewrite the recursive relationship as

$$\mathbf{x}_1 = \mathbf{x}_0 + \epsilon_1\mathbf{f}_0,$$

$$\mathbf{x}_{k+1} = \alpha_{k+1}\mathbf{x}_k + (1 - \alpha_{k+1})\mathbf{x}_{k-1} + \epsilon_{k+1}\alpha_{k+1}\mathbf{f}_k, \qquad (3.10)$$

with $\mathbf{f}_k := \mathbf{f}(\mathbf{x}_k; K)$, which is the vector field in (3.4). This identification offers an interesting interpretation, that is, the polynomial acceleration procedure (3.9) can be regarded as the application of a sequence of explicit two-step methods (3.10) to the differential system (3.4) with step size ϵ_{k+1}. Beware, however, of the subtle distinction that the two-step method (3.10) has a low order of accuracy (of order one, indeed) if regarded as an ODE method, but has a faster rate of convergence (with appropriately selected step size ϵ_k) to the equilibrium $\mathbf{x}^*$ if regarded as an iterative scheme.

3.2. Krylov subspace methods

We have seen how a basic iterative system (3.2) motivates the continuous system (3.4), which we now rewrite as

$$\frac{\mathrm{d}\mathbf{x}}{\mathrm{d}t} = K^{-1}\mathbf{r}, \qquad (3.11)$$

with $\mathbf{r} := \mathbf{b} - A\mathbf{x}$ denoting the residual vector. Instead of considering the iterative scheme,

$$\mathbf{x}_{k+1} = \mathbf{x}_k + \epsilon_k K^{-1}\mathbf{r}_k, \qquad (3.12)$$

as one Euler step with variable step size ϵ_k, we interpret (3.12) as a line search in the $K^{-1}\mathbf{r}_k$ direction for a given K^{-1}. In this context, we can even put aside the concern of requiring eigenvalues of $K^{-1}A$ to reside in the right half of the complex plane. If the search is intended to minimize the size of the residual vector, say, $\mathbf{r}_{k+1}^{\top}\mathbf{r}_{k+1}$, then the optimal step size is given by

$$\epsilon_k = \frac{\langle AK^{-1}\mathbf{r}_k, \mathbf{r}_k\rangle}{\langle AK^{-1}\mathbf{r}_k, AK^{-1}\mathbf{r}_k\rangle}, \tag{3.13}$$

where $\langle \mathbf{u}, \mathbf{v}\rangle := \mathbf{u}^{\top}\mathbf{v}$ stands for the inner product. If A is symmetric and positive definite and $\mathbf{r}_{k+1}A^{-1}\mathbf{r}_{k+1}$ is to be minimized, then the optimal step size is given by

$$\epsilon_k = \frac{\langle K^{-1}\mathbf{r}_k, \mathbf{r}_k\rangle}{\langle AK^{-1}\mathbf{r}_k, K^{-1}\mathbf{r}_k\rangle}. \tag{3.14}$$

In the special case $K = I$, the two step size selection strategies (3.13) and (3.14) correspond precisely to the ORTHOMIN(1) and steepest descent methods (Greenbaum 1997), respectively.

We can also adopt a two-step sequential process similar to the accelerator (3.10), except that conventionally we prefer to write the scheme as

$$\mathbf{x}_{k+1} = \mathbf{x}_k + \epsilon_k\left[K^{-1}\mathbf{r}_k + \gamma_k(\mathbf{x}_k - \mathbf{x}_{k-1})\right], \tag{3.15}$$

with step size ϵ_k. Such a scheme, if regarded as an ODE method for the differential system (3.11), would have low order of accuracy. However, by defining $\mathbf{p}_0 = K^{-1}\mathbf{r}_0$ and

$$\mathbf{p}_k := K^{-1}\mathbf{r}_k + \gamma_k(\mathbf{x}_k - \mathbf{x}_{k-1}) = K^{-1}\mathbf{r}_k + \beta_k\mathbf{p}_{k-1}, \tag{3.16}$$

with $\beta_k := \epsilon_{k-1}\gamma_k$, we see an interesting non-stationary iteration embedded in (3.15), that is,

$$\mathbf{x}_{k+1} = \mathbf{x}_k + \epsilon_k\mathbf{p}_k,$$

$$\mathbf{r}_{k+1} = \mathbf{r}_k - \epsilon_k A\mathbf{p}_k,$$

which has profound consequences. In particular, under the assumption that A is symmetric and positive definite and K is symmetric, it can be verified that the iterative scheme (3.15) with the specially selected scalars

$$\epsilon_k = \frac{\langle \mathbf{p}_k, \mathbf{r}_k\rangle}{\langle A\mathbf{p}_k, \mathbf{p}_k\rangle}, \tag{3.17}$$

$$\beta_{k+1} = -\frac{\langle K^{-1}\mathbf{r}_{k+1}, A\mathbf{p}_k\rangle}{\langle A\mathbf{p}_k, \mathbf{p}_k\rangle}, \quad k = 0, 1, \ldots, \tag{3.18}$$

corresponds precisely to the well-known preconditioned conjugate gradient method with K^{-1} as the preconditioner (Greenbaum 1997). Among the many nice properties of the conjugate gradient method, the most significant one is that the sequence $\{\mathbf{x}_k\}$ converges in exact arithmetic to the equilibrium point $\mathbf{x}_*$ in at most n iterations. Such a phenomenon of reaching convergence in only a finite number of steps (by a somewhat laughably inaccurate method as far as solving (3.11) is concerned) is perhaps unexpected from a numerical ODE point of view.

There is a variety of different formulations of the Krylov subspace methods (van der Vorst 2003). We remark that quite a few of them can be derived in a similar spirit, but space limitation prohibits us from giving the details here. Referring to the diagram in Figure 1.1, the lesson we have learned is that from a very basic discrete dynamical system such as (3.2) we can arrive at a very general continuous dynamical system such as (3.4). Instead of tracing the continuous dynamics by some very refined numerical ODE methods, we could use the system as a guide to draw up some general procedures such as (3.10) or (3.15). These discrete procedures roughly solve the continuous system, but not with great accuracy. However, upon aptly tuning the parameters which masquerade as the step sizes in the procedures, we can often achieve fast convergence to the equilibrium point of the continuous system, eventually accomplishing the goal of the original basic discrete dynamical system.

4. Control systems for nonlinear equations

The dynamical system (3.11) for linear equations $Ax = b$, where K is interpreted as a splitting matrix or a preconditioner of A, is merely a special case of a much more general setting. The following approach sets forth a framework from which many new algorithms can be derived.

The notion that many important numerical algorithms can be interpreted via systems and control theory has long been in the minds of researchers. In the seminal book by Tsypkin (1971) and the follow-up volume (Tsypkin 1973), for example, it was advocated that the gradient dynamical systems 'cover many iterative formulas of numerical analysis'. Following the ideas suggested by Bhaya and Kaszkurewicz (2006), we cast the various numerical techniques for finding zero(s) of a given differentiable function

$$\mathbf{g} : \mathbb{R}^n \to \mathbb{R}^n$$

in an input–output control framework with different control strategies. Our point is, again, a comparison of similarities between continuous and discrete dynamical systems.

4.1. Continuous control

Consider the basic model

$$\frac{\mathrm{d}\mathbf{x}(t)}{\mathrm{d}t} = \mathbf{u}(t), \tag{4.1}$$

$$\mathbf{y}(t) = -\mathbf{r}(t),$$

where the state variable $\mathbf{x}(t)$ is controlled by $\mathbf{u}(t)$ while the output variable $\mathbf{y}(t)$ is observed from the residue function

$$\mathbf{r}(t) = -\mathbf{g}(\mathbf{x}(t)).$$

Table 4.1. Control strategies and the associated dynamical systems.

$\phi(\mathbf{x}, \mathbf{r})$	$\frac{\mathrm{d}V}{\mathrm{d}t}$	$\frac{\mathrm{d}\mathbf{x}}{\mathrm{d}t}$
$\mathbf{g}'(\mathbf{x})^{-1}\mathbf{r}$	$-\|\mathbf{r}\|_2^2$	$-\mathbf{g}'(\mathbf{x})^{-1}\mathbf{g}(\mathbf{x})$
$\mathbf{g}'(\mathbf{x})^{\top}\mathbf{r}$	$-\|\mathbf{g}'(\mathbf{x})^{\top}\mathbf{r}\|_2^2$	$-\mathbf{g}'(\mathbf{x})^{\top}\mathbf{g}(\mathbf{x})$
$\mathbf{g}'(\mathbf{x})^{-1}\mathrm{sgn}(\mathbf{r})$	$-\|\mathbf{r}\|_1$	$-\mathbf{g}'(\mathbf{x})^{-1}\mathrm{sgn}(\mathbf{g}(\mathbf{x}))$
$\mathrm{sgn}(\mathbf{g}'(\mathbf{x})^{\top}\mathbf{r})$	$-\|\mathbf{g}'(\mathbf{x})^{\top}\mathbf{r}\|_1$	$-\mathrm{sgn}(\mathbf{g}'(\mathbf{x})^{\top}\mathbf{g}(\mathbf{r}))$
$\mathbf{g}'(\mathbf{x})^{\top}\mathrm{sgn}(\mathbf{r})$	$-\|\mathbf{g}'(\mathbf{x})^{\top}\mathrm{sgn}(\mathbf{r})\|_2^2$	$-\mathbf{g}'(\mathbf{x})^{\top}\mathrm{sgn}(\mathbf{g}(\mathbf{x}))$

One obvious approach is to employ both the state and the output as a feedback to estimate the control strategy, that is,

$$\mathbf{u} = \phi(\mathbf{x}, \mathbf{r}), \qquad (4.2)$$

based on some properly selected ϕ. Different choices of ϕ can be used to design the control and, hence, lead to various algorithms. Of course, it is often that case that the choice of the control strategy ϕ depends on what cost function $V(\mathbf{x}(t), \mathbf{u}(t))$ is to be optimized. In turn, the cost function often plays the role as a Lyapunov function for the dynamical system. Table 4.1 summarizes just a few possible choices for the control $\mathbf{u}$ and the derivatives of the associated cost functions (Bhaya and Kaszkurewicz 2006). Notably, the first case in the table is the well-known continuous Newton method (Hirsch and Smale 1979, Smale 1977).

It is not difficult to verify that the cost functions are $V(t) = \frac{1}{2}\|\mathbf{r}(t)\|_2^2$ in the first four cases and $V(t) = \|\mathbf{r}(t)\|_1$ in the last case, respectively. Be aware of the fact that the vector fields for $\mathbf{x}(t)$ are only piecewise continuous in the last three cases. A discretization of the differential system may not be trivial, which we will draw a distinct line from the discrete control in the next section. trivial, which we will make a clear distinction from the discrete control in the next section. Regardless of the possible non-smoothness in the trajectory $\mathbf{x}(t)$, it is evident that the choice of the control $\mathbf{u}(t)$ always causes the cost function $V(t)$ to decrease in t and, if $\mathbf{g}'(\mathbf{x}(t))$ is always non-singular, the residual function $\mathbf{r}(t)$ converges to zero.

4.2. Discrete control

An Euler analogue of (4.1) is the discrete input-output control system,

$$\mathbf{x}_{k+1} = \mathbf{x}_k + \mathbf{u}_k, \qquad (4.3)$$

where the control $\mathbf{u}_k$ follows the feedback law,

$$\mathbf{u}_k = \epsilon_k \phi(\mathbf{x}_k, \mathbf{r}_k), \tag{4.4}$$

with $\mathbf{r}_k = -\mathbf{g}(\mathbf{x}_k)$. To estimate the step size ϵ_k, observe the *informal* Taylor series expansion,

$$\mathbf{r}_{k+1} \approx \mathbf{r}_k - \epsilon_k \mathbf{g}'(\mathbf{x}_k) \phi(\mathbf{x}_k, \mathbf{r}_k). \tag{4.5}$$

The step size that best reduce the Euclidean norm of the vector on the right side of (4.5) is given by the expression,

$$\epsilon_k = \frac{\langle \mathbf{g}'(\mathbf{x}_k)\phi(\mathbf{x}_k, \mathbf{r}_k), \mathbf{r}_k \rangle}{\langle \mathbf{g}'(\mathbf{x}_k)\phi(\mathbf{x}_k, \mathbf{r}_k), \mathbf{g}'(\mathbf{x}_k)\phi(\mathbf{x}_k, \mathbf{r}_k) \rangle}. \tag{4.6}$$

We have already seen a special case of (4.6) in (3.13) when the equation $\mathbf{g}(\mathbf{x}) = A\mathbf{x} - \mathbf{b}$ is linear and the control $\phi(\mathbf{x}, \mathbf{r}) = K^{-1}\mathbf{r}$ is employed, which is the ORTHOMIN(1) method. Another special case corresponding to the choice of control $\phi(\mathbf{x}, \mathbf{r}) = \mathbf{g}'(\mathbf{x})^{-1}\mathbf{r}$ leads to $\epsilon_k = 1$, which of course is the classical Newton iteration. Interestingly enough, the various choices of $\phi(\mathbf{x}, \mathbf{r})$ described in Table 4.1 together with the associated ϵ_k defined in (4.6) set forth different zero-finding iterative schemes, some of which are perhaps new. We do not think that all convergence properties of these schemes have been well understood.

Be aware that the approximation in (4.5) is not necessarily true in general. The increment $\mathbf{u}_k$ from $\mathbf{x}_k$ to $\mathbf{x}_{k+1}$, for instance, may not be small enough to warrant the expansion of $\mathbf{g}(\mathbf{x}_{k+1})$ at $\mathbf{x}_k$: the approximation is in jeopardy. The step size ϵ_k defined in (4.6) therefore does not necessarily decrease the magnitude of the residual function $\mathbf{r}(\mathbf{x})$. This is precisely the dividing line between a discrete dynamical system which often converges only locally and the continuous dynamical system which converges globally. The well-known convergence behaviour of the classical Newton iteration and the continuous Newton algorithm serves well to exemplify our points: the classical Newton iteration with $\epsilon_k = 1$ does not necessarily give rise to a descent step for the residual function $\mathbf{r}(\mathbf{x})$, whereas the continuous Newton flow always does. The relationship between the convergence rates of iterative and continous processes has recently been studied in Hauser and Nedić (2007).

It is certainly possible to adopt models more sophisticated than (4.1) or (4.3). For example, the two-step scheme

$$\mathbf{x}_{k+1} = \mathbf{x}_k + \epsilon_k \left[\phi(\mathbf{x}_k, \mathbf{r}_k) + \gamma_k (\mathbf{x}_k - \mathbf{x}_{k-1}) \right] \tag{4.7}$$

is analogous to (3.15) and can be converted into a nonlinear conjugate gradient method (Daniel 1967, Savinov 1983, Yabe and Takano 2004). We shall not elaborate on zero-finding algorithms here, but we hope the above discussion has shed some light on how a realization process, either continuous or discrete, can be developed either from or for a dynamical system in the way suggested in Figure 1.1. There seems to be a rich interpretation of the analogy between a discrete scheme and its continuous counterpart. It would be interesting to see whether further consideration along these lines, such as higher-order or multiple-step processes, can develop into new numerical algorithms. Indeed, such a notion has been known as 'higher-order controllers' for given plants in the community of control systems. Some of the theory developed in that discipline might be useful in this regard, and *vice versa* (Bhaya and Kaszkurewicz 2006).

5. Lax dynamical systems and isospectrality

One classical problem of fundamental importance in many critical applications is to find the spectral decomposition,

$$A_0 = U_0 \Lambda_0 U_0^\top, \tag{5.1}$$

of a given real-valued symmetric matrix A_0. In the factorization, U_0 is an orthogonal matrix composed of eigenvectors of A_0 and Λ is the diagonal matrix of the corresponding eigenvalues. Currently, one of the most effective techniques for eigenvalue computation is by an iterative process called the QR algorithm (Golub and Van Loan 1996). The algorithm performs well due to the cooperation of several ingenious components, one of which is the employment of suitable shift strategies that greatly improve the convergence behaviour. Viewing the shifts as feedback control variables, some studies have been made by Helmke and Wirth (2000, 2001) to analyse the controllability of the inverse power method. As far as we know, however, modelling the shift strategies used in a practical QR algorithm by a dynamical system is still an open question. For simplicity, we demonstrate only the basic QR algorithm with no shift.

Recall the fact that any matrix A enjoys the QR decomposition

$$A = QR,$$

where Q is orthogonal and R is upper triangular. The basic QR scheme defines a sequence of matrices $\{A_k\}$ via the recursion (Francis 1961/1962)

$$\begin{cases} A_k &= Q_k R_k, \\ A_{k+1} &= R_k Q_k. \end{cases} \tag{5.2}$$

The iteration implies that

$$A_{k+1} = Q_k^T A_k Q_k, \tag{5.3}$$

showing not only the isospectrality of A_k to A_0, but also the mechanism of orthogonal congruence transformations applied to A_0. It can be proved that the sequence $\{A_k\}$ converges to a diagonal matrix and, hence, the decomposition (5.1) is realized through the iterative scheme (5.2). One is immediately curious why the swapping of Q_k and R_k works in (5.2). Indeed, there is a much deeper theory involved. Referring to the diagram in Figure 1.1, we now identify a differential system to which the QR algorithm corresponds, not as a discrete approximation but rather as a time-1 sampling.

5.1. Isospectral flow

Consider the initial value problem,

$$\frac{\mathrm{d}X(t)}{\mathrm{d}t} := [X(t), k_1(X(t))], \quad X(0) := X_0, \tag{5.4}$$

where $k_1 : \mathbb{R}^{n \times n} \to \mathbb{R}^{n \times n}$ is some selected matrix-valued function to be specified later, and

$$[A, B] := AB - BA \tag{5.5}$$

denotes the Lie commutator (bracket) operation between matrices A and B. We shall refer to (5.4) as a general *Lax dynamical system* with the Lax pair (X, k_1). Associated with (5.4), we define two *parameter dynamical systems*:

$$\frac{\mathrm{d}g_1(t)}{\mathrm{d}t} := g_1(t)k_1(X(t)), \quad g_1(0) := I, \tag{5.6}$$

and

$$\frac{\mathrm{d}g_2(t)}{\mathrm{d}t} := k_2(X(t))g_2(t), \quad g_2(0) := I, \tag{5.7}$$

with the property that

$$k_1(X) + k_2(X) = X. \tag{5.8}$$

The following facts are useful but easy to prove, and have been established in an early paper by Chu and Norris (1988).

Theorem 5.1. For any t within the interval of existence, the solutions $X(t)$, $g_1(t)$, and $g_2(t)$ of the systems (5.4), (5.6), and (5.7), respectively, are related to each other by the following three properties.

(1) Similarity property:

$$X(t) = g_1(t)^{-1} X_0 g_1(t) = g_2(t) X_0 g_2(t)^{-1}. \tag{5.9}$$

(2) Decomposition property:

$$\exp(t X_0) = g_1(t) g_2(t). \tag{5.10}$$

(3) Reversal property:

$$\exp(t X(t)) = g_2(t) g_1(t). \tag{5.11}$$

The implication of Theorem 5.1 is quite remarkable. First, it shows that eigenvalues are invariant. For this reason, $X(t)$ is called an isospectral flow. Secondly, let the product $g_1(t)g_2(t)$ in (5.10) be called the *abstract g_1g_2 decomposition* of $\exp(tX_0)$ because at present we do not know the individual structure, if there is any, of the parameter matrices $g_1(t)$ or $g_2(t)$. By setting $t = 1$ in both (5.10) and (5.11), we see the relationship

$$\begin{cases} \exp(X(0)) = g_1(1)g_2(1), \\ \exp(X(1)) = g_2(1)g_1(1). \end{cases} \qquad (5.12)$$

Since the dynamical system for $X(t)$ is autonomous, it follows that the phenomenon characterized by (5.12) will occur at every integer time within the interval of existence for these initial value problems. Corresponding to the abstract g_1g_2 decomposition, the above iterative process (5.12) for all feasible integers will be called the *abstract g_1g_2 algorithm*. It is thus seen that the curious iteration in the QR algorithm is completely generalized and abstracted via the mere splitting (5.8) of the identity map. Choosing a different splitting leads to a different algorithm.

In particular, let any given matrix X be decomposed as

$$X = X^o + X^- + X^+,$$

where X^o, X^-, and X^+ denote the diagonal, the strictly lower triangular, and the strictly upper triangular parts of X, respectively. Define

$$k_1(X) = \Pi_0(X) := X^- - X^{-\top}. \qquad (5.13)$$

The resulting Lax dynamical system,

$$\frac{\mathrm{d}X(t)}{\mathrm{d}t} = [X(t), \Pi_0(X(t))], \quad X(0) = X_0, \qquad (5.14)$$

is known as the *Toda lattice* (though initially the lattice is referred to only in the case when X_0 is symmetric and tridiagonal). It is important to note that the matrix $k_1(X(t))$ in the Toda lattice is skew-symmetric and thus $g_1(X(t))$ is orthogonal for all t. Furthermore, $k_2(X(t))$ is upper triangular and thus so is $g_2(X(t))$. In other words, the abstract g_1g_2 decomposition of $\exp(X)$ is precisely the QR decomposition of $\exp(X)$. It follows that the sequence $\{X(k)\}$ by sampling the solution of the Toda flow (5.14) at integer times gives rise to exactly the same iterates as the QR algorithm (5.2) applied to the matrix $A_0 = \exp(X_0)$.

The connection between the QR algorithm and the Toda lattice was first discovered by Symes (1981/82) when studying the asymptotic behaviour of momenta of particles in a non-periodic Toda lattice. The same relationship was found later to be also closely related to the quotient-difference algorithm developed much earlier by Rutishauser (1954).

In contrast to the association between a discrete system and a continuous system described earlier in Sections 3 and 4, which perhaps can be best characterized as 'mimicry', the correspondence between the QR algorithm and the Toda lattice exhibits a new type of involvement, namely, the result of an iterative scheme is entirely 'embedded' in the solution curve of a continuous dynamical system or, equivalently, the solution curve of a differential equation smoothly 'interpolates' all points generated by a discrete dynamical system. Because of this close relationship, the evolution of $X(t)$, which starts from a symmetric initial value X_0 and converges isospectrally to a limit point which is a diagonal matrix, can almost be expected without the need for any extra inculcation in the classical theory of the QR algorithm, and *vice versa* (Deift, Nanda and Tomei 1983).

It is important to point out that, strictly speaking, the QR algorithm applied to a non-symmetric matrix A_0 with complex eigenvalues does not converge to any fixed limit point at all in the conventional mathematical sense. The iterates from the QR algorithm only *pseudo-converge* to a block upper triangular form with at most 1×1 or 2×2 blocks along the main diagonal. Such a structure is a necessity when dealing with complex-conjugate eigenvalues of a real-valued matrix by real arithmetic. For later reference, we shall refer to any matrix with this kind of structure as an *upper quasi-triangular matrix*. We stress again that the QR algorithm (and many other algorithms) produces only this 'form', but not any fixed matrix, in its limiting behaviour.

Likewise, the Toda flow applied to a non-symmetric matrix X_0 does not have any asymptotically stable equilibrium point in general. Rather, the flow converges to an upper quasi-triangular form where each of the 2×2 blocks actually represents an ω-limit cycle. Now that we know the Toda flow interpolates the iterates of the QR algorithm, the limit cycle behaviour of the Toda flow offers a nice theoretical explanation of the pseudo-convergence behaviour of the QR algorithm. Without causing ambiguity, we shall henceforth refer to such limiting behaviour as 'convergence to an upper quasi-triangular matrix'.

5.2. Complete integrability

The Lax dynamical system (5.4) actually arises in a much broader area of applications. Consider the one-dimensional Korteweg–de Vries (KdV) equation,

$$\frac{\partial u}{\partial t} + 6u\frac{\partial u}{\partial x} + \frac{\partial^3 u}{\partial x^3} = 0, \tag{5.15}$$

for $u = u(x, t)$. It is a classical result that the KdV equation is completely integrable in the sense there are infinitely many conserved quantities or constants of motion. Lax (1968) proved that the KdV equation is precisely the *compatibility condition*

$$\frac{\mathrm{d}L}{\mathrm{d}t} = [B, L], \tag{5.16}$$

for the pair of differential operators

$$L\psi := \frac{\partial^2 \psi}{\partial x^2} + u\psi, \tag{5.17}$$

$$B\psi := -4\frac{\partial^3 \psi}{\partial x^3} - 6u\frac{\partial \psi}{\partial x} - 3\frac{\partial u}{\partial x}\psi. \tag{5.18}$$

In other words, by recognizing the fact that

$$\left[\frac{\partial}{\partial x}, x\right]\psi = \frac{\partial(x\psi)}{\partial x} - x\frac{\partial \psi}{\partial x} = \psi$$

as the identity of differential operator,

$$\left[\frac{\partial}{\partial x}, x\right] = \mathrm{id},$$

the equation (5.16) holds if and only if u satisfies (5.15). The eigenvalues $\lambda \in \mathbb{R}$ of the one-dimensional Schrödinger equation,

$$L\psi = \lambda\psi, \tag{5.19}$$

$$\frac{\partial \psi}{\partial t} = B\psi, \tag{5.20}$$

for the wave function $\psi = \psi(x, t; \lambda)$ with $u(x, t)$ as the potential constitute precisely the integrals of the KdV equation. The second equation, (5.20), characterizes how the wave function evolves in time. The pair of operators (L, B) is referred to as a *Lax pair*.

Under the assumption that λ is invariant over t, note that the two equations (5.19) and (5.20) are sufficient to imply the compatibility condition (5.16) when acting on the eigenfunction ψ of the operator L. This is true regardless of how the operators L and B are defined. In terms of the notation adopted in our preceding section, we may interpret the Lax pair as $(X, k_2(X))$, where

$$\frac{\mathrm{d}X}{\mathrm{d}t} = [k_2(X), X], \tag{5.21}$$

$$\frac{\mathrm{d}\psi}{\mathrm{d}t} = k_2(X)\psi, \tag{5.22}$$

and $\psi(t)$ tells us how the eigenvector corresponding to the invariant eigenvalue λ varies in time.

We have seen that sampling the solution flow $X(t)$ at integer times gives rise to an iterative scheme, such as the QR algorithm. The question now is whether an effective discretization can be derived to handle the integration of equation (5.21) directly.

It has to be pointed out that a central theme in the game of engaging dynamical systems such as (5.21) is to maintain isospectrality. Nonetheless, Calvo, Iserles and Zanna (1997) proved that most of the conventional numerical ODE methods, in particular the multi-step and the Runge–Kutta schemes, simply cannot preserve isospectral flows. One remedy is to perform numerical integration over one of the parameter dynamical systems (5.6) or (5.7) and then employ the similarity property (5.9) to reclaim $X(t)$. Solving the parameter dynamical system still requires the preservation of some structures, but can be handled more easily. In the case of (5.14), for example, the flow $g_1(X(t))$ of orthogonal matrices can be tracked by orthogonal integrators developed by Dieci, Russell and Van Vleck (1994). Approaches such as this follow the paradigm of discretization from the numerical analysis perspective. We want to emphasize that there is more beyond this traditional way of thinking. The Toda lattice itself has more structure, so that a completely different perspective of discretization could be, and should be, taken into account.

Two separate but related approaches that suggest integrable discretization of the Toda lattice (for symmetric and tridiagonal matrices) are outlined in Sections 5.3 and 5.4. We shall present the theory in these two sections, but refrain from discussing the actual implementation, since eigenvalue computation is a well-developed subject. Even so, the facts we are about to introduce, namely, that the solution to the Toda lattice and, hence, the iterates generated by the QR algorithm can be represented in 'closed form', strongly suggest that an appropriate discretization can make the computation very effective. In Section 6, we will have a chance to exploit these ideas further, and describe in detail an integrable discretization for the more complicated singular value decomposition.

5.3. Orthogonal polynomials, moments and measure deformation

The first approach makes an interesting connection between orthogonal polynomials and the solution of (5.14) when X_0 is tridiagonal, which sheds light on the notion of integrable discretization. In particular, we shall represent the solution to the Toda lattice in terms of moments associated with a specific measure.

Recall that a set of orthogonal polynomials $\{p_k(x)\}$ defined by a positive measure $\mu(x)$ over $\mathbb{R}$, that is,

$$\int p_k(x)p_\ell(x)\,\mathrm{d}\mu(x) = \delta_{k,\ell}, \quad k, \ell = 0, 1, \ldots,$$

always satisfies a three-term recurrence relationship,

$$xp_k(x) = a_k p_{k+1}(x) + b_k p_k(x) + a_{k-1}p_{k-1}(x), \quad k = 1, 2, \ldots, \qquad (5.23)$$

with $p_{-1}(x) \equiv 0$ and $p_0(x) \equiv 1$. This recurrence can be neatly written in a semi-infinite matrix form:

$$\underbrace{\begin{bmatrix} b_0 & a_0 & 0 & & \\ a_0 & b_1 & a_1 & 0 & \\ 0 & a_1 & b_2 & a_2 & 0 \\ & \ddots & \ddots & \ddots & \ddots & \ddots \end{bmatrix}}_{J} \begin{bmatrix} p_0(x) \\ p_1(x) \\ p_2(x) \\ \vdots \end{bmatrix} = x \begin{bmatrix} p_0(x) \\ p_1(x) \\ p_2(x) \\ \vdots \end{bmatrix}. \tag{5.24}$$

Indeed, there is a one-to-one correspondence between the measure μ and the coefficient matrix J defined above (Akhiezer 1965, Aptekarev, Branquinho and Marcellán 1997). This is closely related to the classical moment problem. Let the moments corresponding to μ be denoted by

$$s_j := \int x^j \, d\mu(x), \quad j = 0, 1, \ldots. \tag{5.25}$$

Define further the so-called *Hankel determinants*,

$$H_k := \det \begin{bmatrix} s_0 & s_1 & \cdots & s_{k-1} \\ s_1 & s_2 & & s_k \\ \vdots & & & \vdots \\ s_{k-1} & s_k & \cdots & s_{2k-2} \end{bmatrix}. \tag{5.26}$$

It is known that the monic orthogonal polynomials $\{\tilde{p}_k(x)\}$ associated with $\{p_k(x)\}$ are given by (Akhiezer 1965, Szegő 1975)

$$\tilde{p}_k(x) = \frac{1}{H_k} \det \begin{bmatrix} s_0 & s_1 & \cdots & s_k \\ s_1 & s_2 & & s_{k+1} \\ \vdots & & & \vdots \\ s_{k-1} & s_k & \cdots & s_{2k-1} \\ 1 & x & \cdots & x^k \end{bmatrix}. \tag{5.27}$$

If we write $\tilde{p}_k(x)$ as

$$\tilde{p}_k(x) = x^k + c_1^{(k)} x^{k-1} + \ldots + c_{k-1}^{(k)} x + c_k^{(k)},$$

then its coefficients are given by

$$c_j^{(k)} = \frac{(-1)^j}{H_k} \det \begin{bmatrix} s_0 & \cdots & s_{k-j-1} & s_{k-j+1} & \cdots & s_k \\ s_1 & & & & & s_{k+1} \\ \vdots & & \vdots & \vdots & & \vdots \\ s_{k-1} & \cdots & s_{2k-j-2} & s_{2k-j} & \cdots & s_{2k-1} \end{bmatrix}. \tag{5.28}$$

Corresponding to (5.23), the recurrence relation for $\{\tilde{p}_k(x)\}$ becomes

$$x\tilde{p}_k(x) = \tilde{p}_{k+1} + b_k\tilde{p}_k(x) + a_{k-1}^2 \tilde{p}_{k-1}(x). \tag{5.29}$$

By comparing the corresponding coefficients, we conclude that

$$a_k^2 = \frac{H_k H_{k+2}}{H_{k+1}^2}, \tag{5.30}$$

$$b_k = c_1^{(k)} - c_1^{(k+1)}. \tag{5.31}$$

This is a classical result connecting the measure $\mu(x)$, the moments $s_j(x)$ and the orthogonal polynomials $p_k(x)$.

Suppose now that the coefficients in J are time-dependent. Then the corresponding measure μ is also time-dependent. Finding the relationship

$$J(t) \leftrightarrow \mu(x;t)$$

allows us to write the coefficients of the orthogonal polynomials $\{p_k(x;t)\}$ in terms of the corresponding moments $s_j(t)$. In general, this is a fairly difficult task. Only very few cases are known to have exact solutions, among which one is the $J(t)$ associated with the Toda lattice (Aptekarev *et al.* 1997).

The relationship is most conspicuous in the semi-infinite Toda lattice. By identifying (the symmetric and tridiagonal matrix) $X(t)$ with the tridiagonal matrix J in (5.24), the entries in the differential system (5.14) can be expressed as the system

$$\frac{\mathrm{d}a_k}{\mathrm{d}t} = a_k(b_{k+1} - b_k), \tag{5.32}$$

$$\frac{\mathrm{d}b_k}{\mathrm{d}t} = 2(a_k^2 - a_{k-1}^2), \tag{5.33}$$

with $a_{-1} \equiv 0$. This differential system characterizes how $X(t)$ or, equivalently, the family of polynomials varies in time. We just need a measure that can ensure the orthogonality of these polynomials. It turns out that the corresponding one-parameter deformation of the measure that can introduce the desirable orthogonality has been shown by Moser (1975) to be

$$\mathrm{d}\mu(x;t) := \mathrm{e}^{tx}\,\mathrm{d}\mu(x;0). \tag{5.34}$$

Equipped with this measure, we can easily calculate the solution to the Toda lattice. That is, the entries $a_k(t)$ and $b_k(t)$ of $X(t)$ can be calculated via (5.30) and (5.31), once the moments given by the integrals (5.25) are computed. In fact, note that with this measure (5.34) we even enjoy the recursion

$$\frac{\mathrm{d}s_\ell}{\mathrm{d}t} = s_{\ell+1}, \quad \ell = 0, 1, \ldots. \tag{5.35}$$

Since these moments are computable in analytic form, we may say that the solution to the Toda lattice (of symmetric and tridiagonal matrices) and hence the iterates by the QR algorithm are now characterized in closed form.

In this sense, we have obtained a discretization while maintaining complete integrability.

It is informative to depict the relationship just described for the Toda lattice as solid lines in Figure 5.1. We stress that the commuting diagram composed of the top four boxes holds in general. That is, the coefficients of the orthogonal polynomials corresponding to a given measure can be

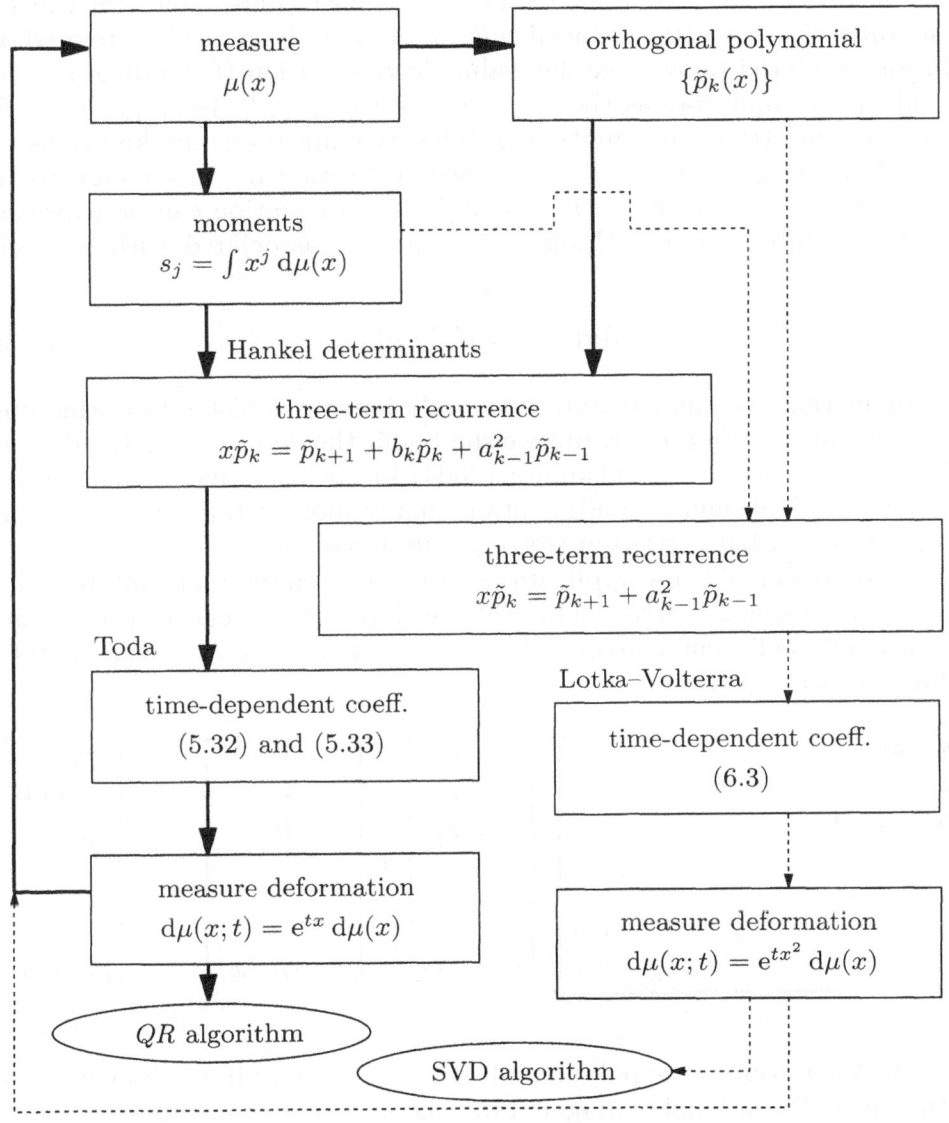

Figure 5.1. Integrable discretization of Toda lattice (solid line) and Lotka–Volterra equation (dashed line) via Hankel determinants.

expressed in terms of Hankel determinants of the corresponding moments. For the Toda lattice where the coefficients $\{a_k\}$ and $\{b_k\}$ are governed by the differential system (5.32) and (5.33), respectively, the commuting diagram comprised of the left five boxes indicates how the inverse problem is solved. An efficient calculation of the Hankel determinants is all we need for an effective eigenvalue computation. This *modus operandi* is very different from the orthogonal integrator approach mentioned earlier.

We mention in passing that a similar relationship also holds for the singular value decomposition. Specifically, there is a dynamical system whose solution is related to the singular value decomposition (for bidiagonal matrices) in the same way as the Toda lattice to the QR decomposition (for symmetric and tridiagonal matrices). This dynamical system, known as the Lotka–Volterra equation, will be defined in Section 6. In analogy to the Toda lattice, the solution to the Lotka–Volterra equation can be expressed in terms of moments and Hankel determinants associated with a special measure,

$$\mathrm{d}\mu(x;t) = \mathrm{e}^{tx^2}\,\mathrm{d}\mu(x). \tag{5.36}$$

For completion and comparison, such a relationship depicted as dotted lines is also included in Figure 5.1, but we shall omit the details here. Readers are referred to the paper by Nakamura (2004) for an overview of this subject. The book by Nakamura (2006) contains many more details and interesting historical notes, but is written (for now) in Japanese.

In most linear algebra applications, we are perhaps more interested in a finite-dimensional matrix. This can be done by truncating the infinite-dimensional coefficient matrix J into an $n \times n$ matrix L. Then (5.24) is reduced to the equation

$$\underbrace{\begin{bmatrix} b_0 & a_0 & 0 \\ a_0 & b_1 & a_1 & 0 \\ 0 & a_1 & b_2 & a_2 & 0 \\ & \ddots & \ddots & \ddots \\ & & & & a_{n-2} \\ & & 0 & a_{n-2} & b_{n-1} \end{bmatrix}}_{L} \begin{bmatrix} p_0(x) \\ p_1(x) \\ p_2(x) \\ \vdots \\ p_{n-1}(x) \end{bmatrix} + \begin{bmatrix} 0 \\ 0 \\ 0 \\ \vdots \\ a_{n-1}p_n(x) \end{bmatrix} = x \begin{bmatrix} p_0(x) \\ p_1(x) \\ p_2(x) \\ \vdots \\ p_{n-1}(x) \end{bmatrix}.$$

Clearly, λ is a root of the polynomial $p_n(x)$ if and only if λ is an eigenvalue of the finite-dimensional tridiagonal matrix L. Other than this requirement of special values for λ, this finite-dimensional eigenvalue problem remains a segment of the semi-infinite system (5.24). As far as the evolution of the entries of L is concerned, it is the same as those of J as long as λ is time-invariant. This isospectrality is precisely what is entailed in the one-dimensional Schrödinger equation (5.19). Under the condition of isospectrality throughout the evolution, the theory developed above for the semi-infinite Toda lattice remains applicable to the finite-dimensional eigenvalue problem. In particular, the solution to the finite-dimensional Toda lattice can still be represented in terms of moments.

5.4. Tau functions and determinantal solution

The second approach utilizes the notion of τ functions originally introduced by the 'Kyoto school' as a central element in the description of the soliton theory for the Kadomtsev–Petviashvili or Hirota–Miwa hierarchies (Date, Kashiwara, Jimbo and Miwa 1983, Hirota, Tsujimoto and Imai 1993, Pöppe 1989). We limit our attention in this section to the basic idea applied to the Toda lattice only.

With the change of variable,

$$c_k(t) := a_k^2\left(\frac{t}{2}\right), \qquad (5.37)$$

the off-diagonal entries in the Toda lattice (5.32) can be expressed as a second-order but self-contained equation,

$$\frac{d^2 \ln c_k}{dt^2} = c_{k+1} - 2c_k + c_{k-1}. \qquad (5.38)$$

If we impose another sequence of new variables $\{\tau_k(t)\}$ implicitly via the relationship

$$c_k = \frac{\tau_{k+1}\tau_{k-1}}{\tau_k^2}, \qquad (5.39)$$

then naturally we have

$$\ln c_k = \ln \tau_{k+1} - 2 \ln \tau_k + \ln \tau_{k-1}. \qquad (5.40)$$

Upon comparison of (5.38) and (5.40), a compatibility condition is that

$$c_k = \frac{d^2 \ln \tau_k}{dt^2} \qquad (5.41)$$

or, equivalently, that $\{\tau_k\}$ must satisfy the Hirota bilinear form

$$\tau_k \frac{d^2 \tau_k}{dt^2} - \left(\frac{d\tau_k}{dt}\right)^2 = \tau_{k-1}\tau_{k+1}, \qquad (5.42)$$

with $\tau_0 \equiv 1$. The bilinear form (5.42) is sufficient for generating a sequence $\{\tau_k(t)\}$ of solution recursively. For example, starting with an arbitrary initial value $\tau_1(t) = \phi(t)$ that is infinitely differentiable, we obtain

$$\tau_2(t) = \phi \frac{\mathrm{d}^2\phi}{\mathrm{d}t^2} - \left(\frac{\mathrm{d}\phi}{\mathrm{d}t}\right)^2,$$

$$\tau_3(t) = -\left(\frac{\mathrm{d}^2\phi}{\mathrm{d}t^2}\right)^3 + \phi\left(\frac{\mathrm{d}^2\phi}{\mathrm{d}t^2}\right)\frac{\mathrm{d}^4\phi}{\mathrm{d}t^4} - \left(\frac{\mathrm{d}\phi}{\mathrm{d}t}\right)^2\frac{\mathrm{d}^4\phi}{\mathrm{d}t^4}$$

$$+ 2\left(\frac{\mathrm{d}\phi}{\mathrm{d}t}\right)\left(\frac{\mathrm{d}^2\phi}{\mathrm{d}t^2}\right)\frac{\mathrm{d}^3\phi}{\mathrm{d}t^3} - \phi\left(\frac{\mathrm{d}^3\phi}{\mathrm{d}t^3}\right)^2,$$

and so on. Obviously, the expression for $\tau_k(t)$ becomes more and more involved when k gets higher. The beauty of the τ functions is that there is a much better representation for $\tau_k(t)$ in general.

From a given $\phi(t)$, define the Hankel determinant $\hat{H}_k(t)$ by

$$\hat{H}_k(t) := \det \begin{bmatrix} \phi & \phi^{(1)} & \cdots & \phi^{(k-1)} \\ \phi^{(1)} & \phi^{(2)} & & \phi^{(k)} \\ \vdots & & & \vdots \\ \phi^{(k-1)} & \phi^{(k)} & \cdots & \phi^{(2k-2)} \end{bmatrix}, \tag{5.43}$$

where for simplicity we adopt the abbreviation

$$\phi^{(\ell)} = \frac{\mathrm{d}^\ell\phi}{\mathrm{d}t^\ell}, \quad \ell = 1, 2, \ldots.$$

Let $\hat{H}_k\begin{bmatrix} i \\ j \end{bmatrix}$ denote the determinant of the submatrix by deleting the ith row and the jth column from the matrix defining $\hat{H}_k$. Observe that

$$\frac{\mathrm{d}\hat{H}_k}{\mathrm{d}t} = \hat{H}_{k+1}\begin{bmatrix} k+1 \\ k \end{bmatrix}, \tag{5.44}$$

$$\frac{\mathrm{d}^2\hat{H}_k}{\mathrm{d}t^2} = \hat{H}_{k+1}\begin{bmatrix} k \\ k \end{bmatrix}. \tag{5.45}$$

On the other hand, recall the Sylvester determinant identity (Horn and Johnson 1990)

$$\hat{H}_{k+1}\hat{H}_{k-1} = \det \begin{bmatrix} \hat{H}_{k+1}\begin{bmatrix} k+1 \\ k+1 \end{bmatrix} & \hat{H}_{k+1}\begin{bmatrix} k+1 \\ k \end{bmatrix} \\ \hat{H}_{k+1}\begin{bmatrix} k \\ k+1 \end{bmatrix} & \hat{H}_{k+1}\begin{bmatrix} k \\ k \end{bmatrix} \end{bmatrix}. \tag{5.46}$$

In conclusion, we see that $\hat{H}_k(t)$ satisfies precisely the differential equation (5.42). As a consequence, we have obtained a closed form solution for $c_k(t)$

via (5.39), where $\tau_k(t)$ is given by

$$
\tau_k(t) = \det \begin{bmatrix} \phi & \phi^{(1)} & \cdots & \phi^{(k-1)} \\ \phi^{(1)} & \phi^{(2)} & & \phi^{(k)} \\ \vdots & & & \vdots \\ \phi^{(k-1)} & \phi^{(k)} & \cdots & \phi^{(2k-2)} \end{bmatrix}. \tag{5.47}
$$

The existence of a determinantal solution to the Toda lattice provides insightful information for the discretization of integrable systems (Iwasaki and Nakamura 2006). With appropriate discretization, for example, it can be shown that the above formula leads to the Rutishauser qd algorithm (Nakamura 2004, Rutishauser 1954). Instead of detailing here how this can be done for the eigenvalue computation, which is a well-studied subject, we shall demonstrate a similar application to the much more sophisticated singular value decomposition in the next section.

6. Lotka–Volterra equation and singular values

Given a rectangular matrix $A_0 \in \mathbb{R}^{m \times n}$ with $m \geq n$, the singular value decomposition (SVD) of A_0 is a factorization of the form

$$
A_0 = U_0 \Sigma_0 V_0^\top, \tag{6.1}
$$

where $U_0 \in \mathbb{R}^{m \times m}$ and $V_0 \in \mathbb{R}^{n \times n}$ are unitary matrices and $\Sigma_0 \in \mathbb{R}^{m \times n}$ is a diagonal matrix with non-negative diagonal entries. The notion of SVD has been a powerful tool for matrix analysis and has been a centrepiece in many areas of applications (Golub and Van Loan 1996, Horn and Johnson 1990).

The use of the SVD and associated ideas has a rich history. In the interesting treatise of Stewart (1993), the early history of the SVD was traced back to Beltrami in 1873 and Jordan in 1874. Before high-speed digital computers became available, the SVD could only be approximated (Chu and Funderlic 2002, Horst 1965). Today, there are a number of highly efficient ways to compute the SVD (Demmel, Gu, Eisenstat, Slapničar, Veselić and Drmač 1999). Some are perhaps more polished and possibly more accurate than others (Demmel and Kahan 1990). In this section, we consider only the basic and conventional approach proposed by Golub and Kahan (1965).

A standard practice in the SVD computation consists of two phases. First, two orthogonal matrices P_1 and Q_1 are found such that $B_0 = P_1^\top A_0 Q_1$ is in bidiagonal form. This step of reduction can be done directly. Then an iterative procedure is employed to compute the SVD of B_0. This main step of iteration is mathematically equivalent to the QR algorithm applied to the tridiagonal matrix $B_0^\top B_0$, except that the product $B_0^\top B_0$ is never formed explicitly. Needless to say, extra tactics, such as implicit-shift, could be added to the iterative process to increase efficiency in computation.

6.1. SVD flow

In view of how the Toda lattice is related to the QR algorithm, Chu (1986b) proposed a peculiar continuous dynamical system of the form

$$\frac{dB}{dt} = B\Pi_0(B^\top B) - \Pi_0(BB^\top)B, \quad B(0) = B_0, \qquad (6.2)$$

where Π_0 is the operator defined in (5.13), and proved that the sequence $\{B(\ell)\}$ produced by $B(t)$ corresponds to the iterates produced by the Golub–Kahan SVD algorithm. One special feature of (6.2) is that $B(t)$ stays bidiagonal for all t. What other properties of this SVD flow can we exploit for applications?

Without loss of generality, we shall assume henceforth that B_0 is an $n \times n$ matrix. By denoting

$$B(t) := \mathrm{diag} \left\{ \begin{matrix} & b_2(t) & & \cdots & & b_{2n-2}(t) & \\ b_1(t) & & b_3(t) & & \cdots & & b_{2n-1}(t) \end{matrix} \right\},$$

and defining

$$u_{2k-1}(t) := b_{2k-1}^2\left(\frac{t}{2}\right),$$

$$u_{2k}(t) := b_{2k}^2\left(\frac{t}{2}\right),$$

the differential system (6.2) can be condensed into the expression

$$\frac{du_k}{dt} = u_k(u_{k+1} - u_{k-1}), \quad k = 1, 2, \ldots, 2n - 1, \qquad (6.3)$$

with $u_0(t) \equiv 0$ and $u_{2n}(t) \equiv 0$, which is known as the *continuous-time finite Lotka–Volterra equation*.

The dynamical system (6.3) is Hamiltonian, that is, it can be written in the form of Hamilton's equations (Deift, Demmel, Li and Tomei 1991). The system is also integrable and enjoys a determinantal solution which can be derived from the theory of τ functions as follows.

Define a change of variable by

$$u_k = \frac{\tau_{k+2}\tau_{k-1}}{\tau_{k+1}\tau_k}. \qquad (6.4)$$

Clearly, we have

$$\frac{d\ln u_k}{dt} = \frac{d}{dt}\ln\frac{\tau_{k+2}}{\tau_{k+1}} - \frac{d}{dt}\ln\frac{\tau_k}{\tau_{k-1}}. \qquad (6.5)$$

A comparison between (6.3) and (6.5) suggests that a compatibility condition could be

$$\frac{\tau_{k+2}\tau_{k-1}}{\tau_{k+1}\tau_k} = \frac{d}{dt}\ln\frac{\tau_{k+1}}{\tau_k}, \qquad (6.6)$$

which is equivalent to

$$\frac{\mathrm{d}\tau_k}{\mathrm{d}t}\tau_{k+1} - \tau_k\frac{\mathrm{d}\tau_{k+1}}{\mathrm{d}t} + \tau_{k-1}\tau_{k+2} = 0. \tag{6.7}$$

The differential equation (6.7) can be used to generate $\tau_k(t)$ recursively. Assuming starting values $\tau_{-1} \equiv 0$, $\tau_0 \equiv 1$, $\tau_1(t) = 1$ and $\tau_2(t) = \psi(t)$, we obtain from (6.7)

$$\tau_3 = \frac{\mathrm{d}\psi}{\mathrm{d}t},$$

$$\tau_4 = \det\begin{bmatrix} \psi & \psi^{(1)} \\ \psi^{(1)} & \psi^{(2)} \end{bmatrix},$$

and in general it can be proved that (Tsujimoto 1995)

$$\tau_{2k-1} = \overline{H}_{k-1,1}, \tag{6.8}$$

$$\tau_{2k} = \overline{H}_{k,0}, \tag{6.9}$$

where

$$\overline{H}_{k,j}(t) := \det\begin{bmatrix} \psi^{(j)} & \psi^{(j+1)} & \cdots & \psi^{(j+k-1)} \\ \psi^{(j+1)} & \psi^{(j+2)} & \cdots & \psi^{(j+k)} \\ \vdots & \vdots & & \vdots \\ \psi^{(j+k-1)} & \psi^{(j+k)} & & \psi^{(j+2k-2)} \end{bmatrix}, \quad j = 0 \text{ or } 1, \tag{6.10}$$

is the determinant of a $k \times k$ Hankel matrix and

$$\overline{H}_{-1,j}(t) \equiv 0, \quad \overline{H}_{0,j}(t) \equiv 1, \quad \overline{H}_{n+1,j}(t) \equiv 0. \tag{6.11}$$

The general solution to the Lotka–Volterra equation, therefore, is given by the formula (Tsujimoto, Nakamura and Iwasaki 2001)

$$u_{2k-1}(t) = \frac{\overline{H}_{k,1}(t)\overline{H}_{k-1,0}(t)}{\overline{H}_{k,0}(t)\overline{H}_{k-1,1}(t)}, \tag{6.12}$$

$$u_{2k}(t) = \frac{\overline{H}_{k+1,0}(t)\overline{H}_{k-1,1}(t)}{\overline{H}_{k,1}(t)\overline{H}_{k,0}(t)}, \quad k = 1, 2, \ldots, n, \tag{6.13}$$

By assuming that all the derivatives of ψ are obtainable from elementary calculus, it is true in principle that all these Hankel determinants can be calculated algebraically. Since all quantities involved in (6.12) and (6.13) are now in the analytic form, we may say that the SVD flow and, hence, the iterates from the SVD algorithm are representable in closed form.

This determinantal solution for the continuous Lotka–Volterra equation can be utilized to effectuate numerical computation. Indeed, it motivates the notion of integrable discretization of (6.3), which we consider in the next section.

6.2. Integrable discretization

A key step in the integrable discretization of the Lotka–Volterra equation
(6.3) is a particular Euler-type scheme of the form (Hirota *et al.* 1993)

$$u_k^{[\ell+1]} = u_k^{[\ell]} + \delta\left(u_k^{[\ell]}u_{k+1}^{[\ell]} - u_k^{[\ell+1]}u_{k-1}^{[\ell+1]}\right), \qquad (6.14)$$

where $u_k^{[\ell]}$ represents the approximation solution of $u_k(t)$ at $t = \ell\delta$ with
boundary conditions $u_0^{[\ell]} \equiv 0$ and $u_{2n}^{[\ell]} \equiv 0$ for all ℓ. Be aware of the notation
that the superscript $^{[\ell+1]}$ in brackets indicates the advance in time by a step
of size δ whereas the subscript $_{k+1}$ indicates the $(k+1)$th bidiagonal entry
of the matrix $B(t)$.

In hindsight, the scheme (6.14) appears to be simply a mixture of both
explicit and implicit Euler methods. The fact of the matter is that it takes
considerable insight to get the right combination so that, as in the continuous case, the discrete Lotka–Volterra equation (6.14) still enjoys a determinantal solution. Specifically, we claim without proof that the solution to the
finite difference equation (6.14) is given by (Iwasaki and Nakamura 2002)

$$u_{2k-1}^{[\ell]} = \frac{\widetilde{H}_{k,1}^{[\ell]}\,\widetilde{H}_{k-1,0}^{[\ell+1]}}{\widetilde{H}_{k,0}^{[\ell]}\,\widetilde{H}_{k-1,1}^{[\ell+1]}}, \qquad (6.15)$$

$$u_{2k}^{[\ell]} = \frac{\widetilde{H}_{k+1,0}^{[\ell]}\,\widetilde{H}_{k-1,1}^{[\ell+1]}}{\widetilde{H}_{k,1}^{[\ell]}\,\widetilde{H}_{k,0}^{[\ell+1]}}, \qquad k = 1, 2, \ldots, n, \qquad (6.16)$$

where $\widetilde{H}_{k,j}^{[\ell]}$ is the Hankel determinant defined by

$$\widetilde{H}_{k,j}^{[\ell]} = \det\begin{bmatrix} \widetilde{\psi}_j^{[\ell]} & \widetilde{\psi}_j^{[\ell+1]} & \cdots & \widetilde{\psi}_j^{[\ell+k-1]} \\ \widetilde{\psi}_j^{[\ell+1]} & \widetilde{\psi}_j^{[\ell+2]} & \cdots & \widetilde{\psi}_j^{[\ell+k]} \\ \vdots & \vdots & & \vdots \\ \widetilde{\psi}_j^{[\ell+k-1]} & \widetilde{\psi}_j^{[\ell+k]} & & \widetilde{\psi}_j^{[\ell+2k-2]} \end{bmatrix}, \qquad j = 0 \text{ or } 1, \qquad (6.17)$$

with boundary conditions

$$\widetilde{H}_{-1,j}^{[\ell]} \equiv 0, \quad \widetilde{H}_{0,j}^{[\ell]} \equiv 1, \quad \widetilde{H}_{n+1,j}^{[\ell]} \equiv 0, \qquad (6.18)$$

in which $\{\widetilde{\psi}_0^{[\ell]}\}$ is a given initial sequence and $\widetilde{\psi}_1^{[\ell]}$ is the quotient difference
defined by

$$\widetilde{\psi}_1^{[\ell]} := \frac{\widetilde{\psi}_0^{[\ell+1]} - \widetilde{\psi}_0^{[\ell]}}{\delta}. \qquad (6.19)$$

The knowledge of a solution $u_k^{[\ell]}$ in the form of (6.15) and (6.16) enables
us to gain considerable insight into its asymptotic behaviour as ℓ goes to

infinity. We shall skip that part of discussion in this paper, but rather pay more attention to a possible numerical implementation for the remainder of this section.

We modify (6.14) to the more general variable-step scheme

$$u_k^{[\ell+1]}\left(1 + \delta^{[\ell+1]}u_{k-1}^{[\ell+1]}\right) = u_k^{[\ell]}\left(1 + \delta^{[\ell]}u_{k+1}^{[\ell]}\right), \qquad (6.20)$$

referred to hereafter as the *vdLV scheme*. In a series of extensive studies (Tsujimoto *et al.* 2001, Iwasaki and Nakamura 2002, 2004, 2006), the *vdLV* scheme has been implemented as an alternative means for the SVD computation. Numerical experiments show its strong competitiveness with existing SVD software packages. We briefly outline the ideas below, which also provides another example of Figure 1.1 on how a differential system might be carefully discretized and implemented to become an effective algorithm.

It will be most convenient if we present the interrelationships in matrix form, even though the actual computation should involve only a few scalars. For each ℓ, define two sequences of scalars,

$$q_i^{[\ell]} := \frac{1}{\delta^{[\ell]}}\left(1 + \delta^{[\ell]}u_{2i-2}^{[\ell]}\right)\left(1 + \delta^{[\ell]}u_{2i-1}^{[\ell]}\right), \quad i = 1,\ldots,n, \qquad (6.21)$$

$$e_j^{[\ell]} := \delta^{[\ell]}u_{2j-1}^{[\ell]}u_{2j}^{[\ell]}, \quad j = 1,\ldots n-1, \qquad (6.22)$$

and assemble them into two $n \times n$ bidiagonal matrices,

$$L^{[\ell]} := \begin{bmatrix} q_1^{[\ell]} & 0 & & & 0 \\ 1 & q_2^{[\ell]} & & & \\ & & \ddots & & \\ & & & \ddots & \\ & & & 1 & q_n^{[\ell]} \end{bmatrix}, \qquad (6.23)$$

$$R^{[\ell]} := \begin{bmatrix} 1 & e_1^{[\ell]} & & \\ 0 & 1 & & \\ & & \ddots & \ddots & \\ & & & & e_{n-1}^{[\ell]} \\ & & & & 1 \end{bmatrix}. \qquad (6.24)$$

From the relationship (6.20), it is readily verifiable that the matrix equation

$$L^{[\ell+1]}R^{[\ell+1]} = R^{[\ell]}L^{[\ell]} - \left(\frac{1}{\delta^{[\ell]}} - \frac{1}{\delta^{[\ell+1]}}\right)I_n \qquad (6.25)$$

holds for all ℓ. It should not be a surprise to discover that the above formulation corresponds to the so-called *progressive qd algorithm* already described by Rutishauser (1954, 1960).

As a matter of fact, equation (6.25) is even more closely related to the so-called *differential quotient-difference algorithm with shift* (*dqds*) proposed by Fernando and Parlett (1994) and implemented in Parlett and Marques (2000). More specifically, if we abbreviate the left-hand side of the *vdLV* scheme in (6.20) as

$$w_k^{[\ell]} := u_k^{[\ell]}\big(1 + \delta^{[\ell]} u_{k-1}^{[\ell]}\big), \tag{6.26}$$

and introduce the tridiagonal matrix $Y^{[\ell]}$ defined by

$$Y^{[\ell]} := L^{[\ell]} R^{[\ell]} - \frac{1}{\delta^{[\ell]}} I_n, \tag{6.27}$$

then we find from (6.20) that $Y^{[\ell]}$ can be expressed in the form

$$Y^{[\ell]} = \begin{bmatrix} w_1^{[\ell]} & w_1^{[\ell]} w_2^{[\ell]} & 0 & & & 0 \\ 1 & w_2^{[\ell]} + w_3^{[\ell]} & w_3^{[\ell]} w_4^{[\ell]} & & & \\ & & \ddots & & & \\ & & & \ddots & & \\ & & & & \ddots & \\ 0 & & & & & w_{2n-3}^{[\ell]} w_{2n-2}^{[\ell]} \\ 0 & & & & 1 & w_{2n-2}^{[\ell]} + w_{2n-1}^{[\ell]} \end{bmatrix}, \tag{6.28}$$

and that the relationship

$$Y^{[\ell+1]} = R^{[\ell]} Y^{[\ell]} R^{[\ell]-1} \tag{6.29}$$

holds for all ℓ. Clearly, all matrices in the sequence $\{Y^{[\ell]}\}$ are isospectral. To connect back to our original goal of computing the singular values, observe that $w_k^{[\ell]} > 0$ as long as $u_k^{[0]} > 0$ and $\delta^{[\ell]} > 0$, which can easily be achieved. We thus can symmetrize the tridiagonal matrix $Y^{[\ell]}$ by a diagonal similarity transformation,

$$Y_S^{[\ell]} := D^{[\ell]-1} Y^{[\ell]} D^{[\ell]}, \tag{6.30}$$

with

$$D^{[\ell]} := \mathrm{diag}\bigg\{ \prod_{i=1}^{n-1} \sqrt{w_{2i-1}^{[\ell]} w_{2i}^{[\ell]}}, \ \prod_{i=2}^{n-1} \sqrt{w_{2i-1}^{[\ell]} w_{2i}^{[\ell]}}, \dots, \sqrt{w_{2n-3}^{[\ell]} w_{2n-2}^{[\ell]}}, 1 \bigg\}.$$

Again, it is easy to check that the positivity of $w_k^{[\ell]}$ guarantees that $Y_S^{[\ell]}$ enjoys a Cholesky decomposition

$$Y_S^{[\ell]} = B^{[\ell]\top} B^{[\ell]}, \tag{6.31}$$

with

$$
B^{[\ell]} := \begin{bmatrix}
\sqrt{w_1^{[\ell]}} & \sqrt{w_2^{[\ell]}} & & & & \\
0 & \sqrt{w_3^{[\ell]}} & \sqrt{w_4^{[\ell]}} & & & \\
& & \ddots & & & \\
& & & \ddots & & \\
& & & & \sqrt{w_{2n-3}^{[\ell]}} & \sqrt{w_{2n-2}^{[\ell]}} \\
& & & & & \sqrt{w_{2n-1}^{[\ell]}}
\end{bmatrix}. \tag{6.32}
$$

The above recurrence relationships, all derived from an integrable discretization (6.20) of the Lotka–Volterra equation (6.3), have useful application to the SVD computation. We summarize the discussion thus far in the following theorem.

Theorem 6.1. Given the boundary conditions $u_0^{[\ell]} \equiv 0$ and $u_{2n}^{[\ell]} \equiv 0$, let the sequence $\{u_k^{[\ell]}\}$ be generated by the scheme (6.20). Then the singular values of the bidiagonal matrices $\{B^{[\ell]}\}$ which is defined in (6.32) with its entries $\{w_k^{[\ell]}\}$ given by (6.26) are invariant in ℓ.

For our application, we are interested in computing the singular values of a given matrix B_0. Thus, we need to make sure that the initial values for the iterative scheme (6.20) should be

$$
u_k^{[0]} := \frac{b_k(0)^2}{1 + \delta^{[0]} u_{k-1}^{[0]}}, \quad k = 1, 2, \ldots, 2n - 1. \tag{6.33}
$$

The calculation of $u_k^{[\ell+1]}$ proceeds in the fashion depicted in Figure 6.1, where the quantity

$$
v_k^{[\ell]} := u_k^{[\ell]}\left(1 + \delta^{[\ell]} u_{k+1}^{[\ell]}\right) \tag{6.34}
$$

is an intermediate value listed for convenience, but is also used later. The bold-faced arrows point to the input and output in one step of the calculation. The shaded region indicates the array of initial values and progresses downward as ℓ increases. In the meantime, it is important to note that the boundary conditions from the two vertical boxes in Figure 6.1 help to make the computation explicit in ℓ.

Convergence theory and stability analysis of the *vdLV* scheme are well established in the series of papers referred to earlier and, in particular, the book by Nakamura (2006). It has been proved, for example, that with initial values (6.33) and any step sizes $\delta^{[\ell]} > 0$, the sequence $\{u_1^{[\ell]}, u_3^{[\ell]}, \ldots, u_{2n-1}^{[\ell]}\}$ converges to the squares of singular values of B_0 in descending order while

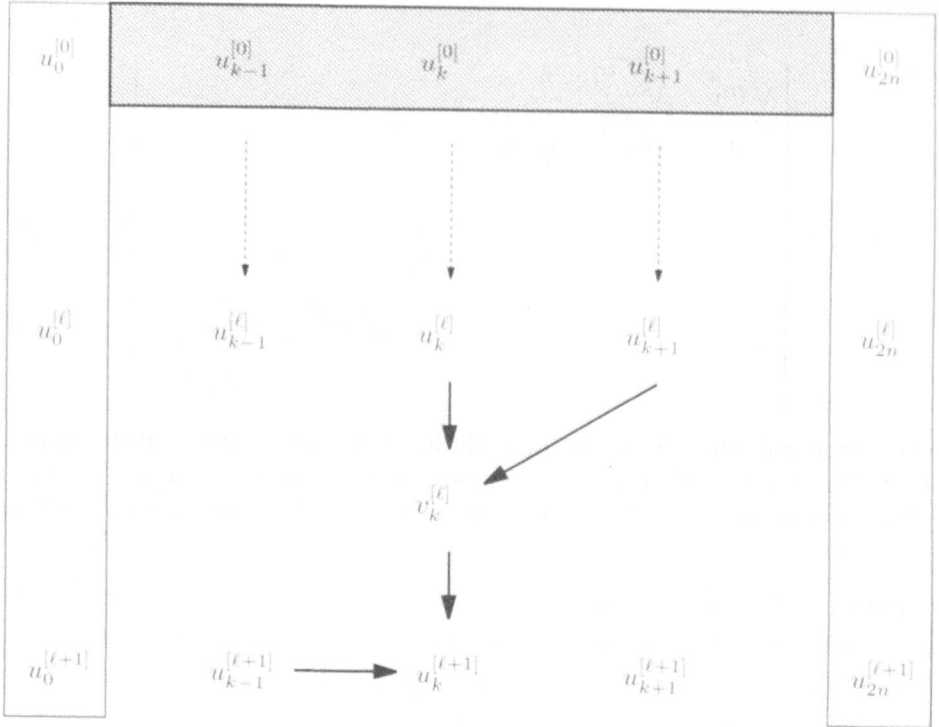

Figure 6.1. Computing $u_k^{[\ell+1]}$ via $vdLV$.

$u_{2k}^{[\ell]}$ converges to 0 for all k as ℓ goes to infinity. The $vdLV$ scheme (6.20) enjoys additional nice features: no subtraction is involved and all quantities are bounded by $\|B_0\|$, implying its numerical stability.

What we have shown thus far is that the Lotka–Volterra equation gives rise to, on one hand, the iterates of the standard SVD algorithm when its solution is sampled at integer times and, on the other hand, an entirely different iterative scheme when the differential system is discretized under some proper conditions. The relationship (6.25) indicates that the $vdLU$ scheme is algebraically equivalent to the $dqds$ with the shift

$$s := \frac{1}{\delta^{[\ell]}} - \frac{1}{\delta^{[\ell+1]}}. \tag{6.35}$$

However, up to this point, we have not given any clear strategy on how the step size $\delta^{[\ell]}$ should be selected in the $vdLV$ scheme. In the case of constant step size $\delta^{[\ell]} \equiv \delta$, Iwasaki and Nakamura (2002) have shown that the convergence is linear, with asymptotic convergence factor given by

$$\alpha = \max_{k=1,\ldots,n-1} \frac{\sigma_{k+1} + \frac{1}{\delta}}{\sigma_k + \frac{1}{\delta}},$$

where $\sigma_1 > \sigma_2 > \cdots > \sigma_n$ are the singular values of B_0. It implies that larger step sizes might reduce the value of α to a certain extent. Linear convergence with the built-in shift (6.35) certainly cannot make the $vdLV$ algorithm efficient enough.

Strictly speaking, the shift (6.35) has never entered into the matrix $B^{[\ell]}$ effectually. In the case of constant step size, $s = 0$. In the case of variable step size, the effect of s is diminished as δ^{ℓ} is increased. The true shift that is really needed should be of the form (Iwasaki and Nakamura 2006)

$$\overline{B}^{[\ell]\top}\overline{B}^{[\ell]} = B^{[\ell]\top}B^{[\ell]} - \theta^{[\ell]2}, \tag{6.36}$$

while we keep the bidiagonal form

$$\overline{B}^{[\ell]} := \begin{bmatrix} \sqrt{\overline{w}_1^{[\ell]}} & \sqrt{\overline{w}_2^{[\ell]}} & & & & \\ 0 & \sqrt{\overline{w}_3^{[\ell]}} & \sqrt{\overline{w}_4^{[\ell]}} & & & \\ & & \ddots & & \\ & & & \ddots & & \\ & & & & \sqrt{\overline{w}_{2n-3}^{[\ell]}} & \sqrt{\overline{w}_{2n-2}^{[\ell]}} \\ & & & & & \sqrt{\overline{w}_{2n-1}^{[\ell]}} \end{bmatrix}. \tag{6.37}$$

Upon comparing the entries, we find the nonlinear relationship that

$$\overline{w}_{2k}^{[\ell]} + \overline{w}_{2k+1}^{[\ell]} = w_{2k}^{[\ell]} + w_{2k+1}^{[\ell]} - \theta^{[\ell]2}, \tag{6.38}$$

$$\overline{w}_{2k-1}^{[\ell]}\overline{w}_{2k}^{[\ell]} = w_{2k-1}^{[\ell]}w_{2k}^{[\ell]}, \quad k = 0, \ldots, n-1, \tag{6.39}$$

with $\overline{w}_0^{[\ell]} = w_0^{[\ell])} \equiv 0$. Though nonlinear, this relationship is a bijection correspondence between $(w_1^{[\ell]}, \ldots, w_{2n-1}^{[\ell]})$ and $(\overline{w}_1^{[\ell]}, \ldots, \overline{w}_{2n-1}^{[\ell]})$. The non-linear map in (6.38) and (6.39) can easily be carried out by recurrence for computation, starting at the vertical box on the left with zero boundary conditions and progressing to the right, as indicated in Figure 6.2.

Recall that

$$v_k^{[\ell]} = w_k^{[\ell+1]},$$

by definitions in (6.26) and (6.34), and that

$$u_k^{[\ell+1]} = \frac{w_k^{[\ell+1]}}{1 + \delta^{[\ell+1]}u_{k-1}^{[\ell+1]}},$$

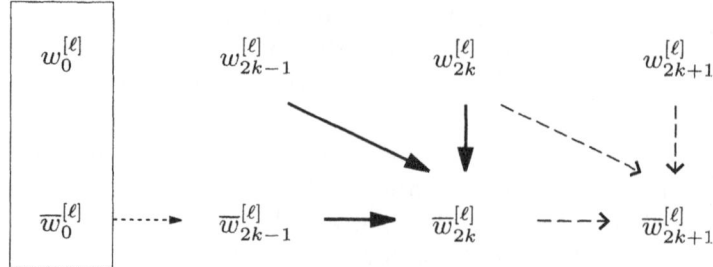

Figure 6.2. Updating $w_k^{[\ell]}$ to $\overline{w}_k^{[\ell]}$ with shift
(dashed line (6.38); solid line (6.39)).

by the *vdLV* scheme (6.20). The modified scheme with shift becomes

$$u_k^{[\ell+1]} = \frac{\overline{w}_k^{[\ell+1]}}{1 + \delta^{[\ell+1]} u_{k-1}^{[\ell+1]}}. \tag{6.40}$$

This variant, called the *mdLVs*, has been studied thoroughly in Iwasaki and Nakamura (2006). The diagram in Figure 6.1 is therefore modified to become Figure 6.3. Be aware of the possible 'psychological illusion' perceived in Figure 6.3. It does appear that the emphasis is on the computation of $u_k^{[\ell+1]}$. However, the diagram can also be interpreted as a path to advance $w_k^{[\ell]}$ and $\overline{w}_k^{[\ell]}$ to $w_k^{[\ell+1]}$ and $\overline{w}_k^{[\ell+1]}$, respectively, whereas $u_k^{[\ell]}$ should be regarded as an intermediate value for convenience.

Many more research results and interesting properties could have been described. Singular vector computation by taking advantage of the *mdLVs* scheme, for example, is another important topic. However, to stay within the theme of this article, we shall stop short of giving more detailed shift strategies and convergence analysis which are available in the literature (Nakamura 2006). Suffice it to say that numerical experiments reported in Takata, Iwasaki, Kimura and Nakamura (2005, 2006) seem to suggest strongly that the resulting algorithm is competitive in both speed and accuracy with existing SVD packages.

The discourse presented in Sections 5 and 6 appears verbose. However, these two sections manifest a successful story about viewing numerical linear algebra algorithms as dynamical systems. We hope to have accomplished two goals through this important deliberation.

First, powerful discrete dynamical systems such as the *QR* algorithm and the SVD algorithm do have their continuous counterparts, namely, the Toda lattice and the Lotka–Volterra equation, which often arise from seemingly rather distinct fields of disciplines. We think it is truly remarkable that

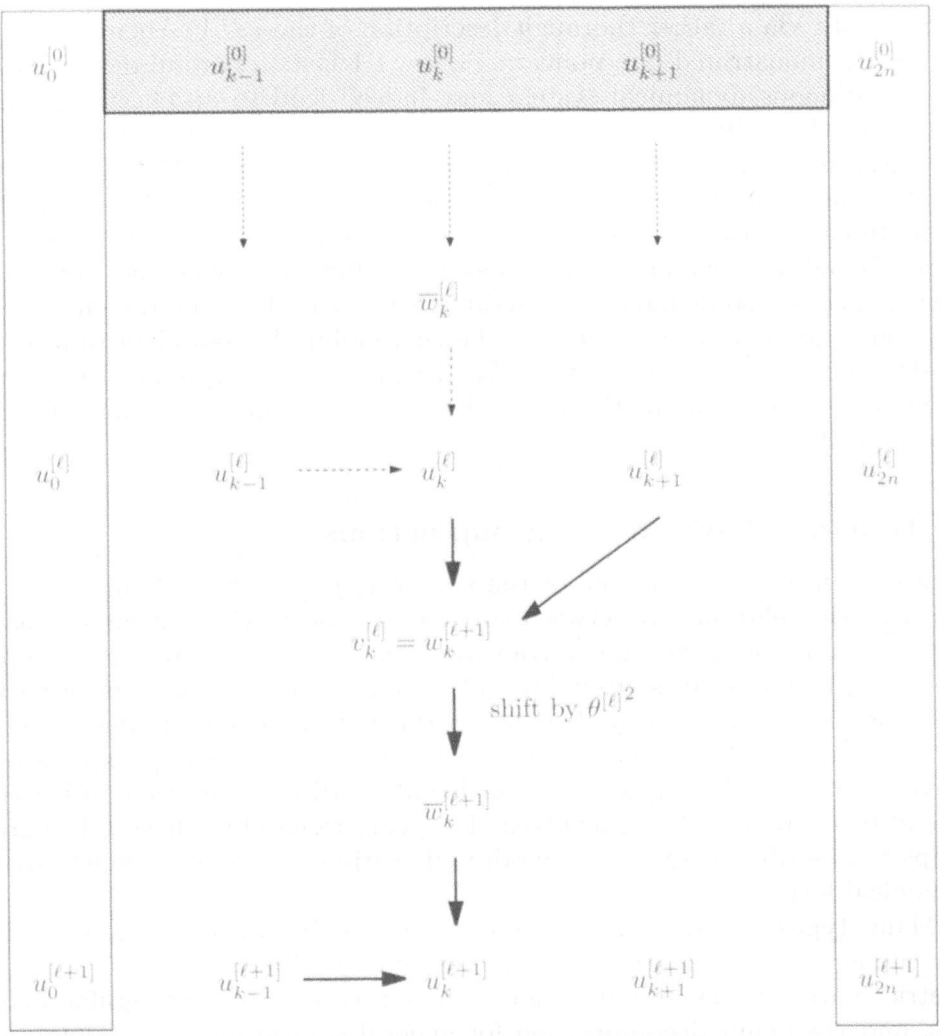

Figure 6.3. Computing $u_k^{[\ell+1]}$ via $mdLVs$.

diverse topics, such as soliton theory, integrable systems, continuous fractions, τ functions, orthogonal polynomials, the Sylvester identity, moments, and Hankel determinants, can all play together, intertwine, and eventually lead to the fact abstractly, but literally, that the eigenvalues and the singular values of a given matrix can be expressed as the limit of some closed-form formulas! We hope to have offered complete particulars on the determinantal solutions to the Toda lattice and the Lotka–Volterra equation, which we know are tied to the eigenvalues and singular values of the underlying matrix.

Secondly, via a rather thorough description of the *vdLV* scheme, we hope to have demonstrated our point in Figure 1.1 that a careful discretization of a continuous dynamical system may indeed lead to an effective numerical algorithm. By a 'careful discretization', it is important to note that the discrete scheme (6.20) maintains its complete integrability, which in the limit is the same integrability as that of the original Lotka–Volterra equation. Integrability-preserving discretization seems to be the key to success here, though a great many details such as shift strategies and implementation tactics also demand considerable attention. It is interesting to note the route we have taken, from the classical Golub–Kahan algorithm to the Lotka–Volterra equation, to the *dLV* scheme, to the *dqds* algorithm, and then to a brand new method, *mdLVs*, for computing the singular value decomposition.

7. Dynamical systems as group actions

Linear transformation is one of the simplest, yet most profound, ways to describe the relationship between two vector spaces. Over linear subspaces with a countable basis, linear transformations can be conveniently represented by matrices. It is often desirable to represent a linear transformation in some characteristic way, leading to the notion of identifying a matrix by its *canonical form*. The canonical form, most frequently expressed in terms of matrix decomposition, can facilitate discussions that, otherwise, would be complicated and involved. For years researchers have taken great steps to describe, analyse, and modify algorithms to reduce a matrix to its canonical form.

Many types of canonical forms exist in the literature. Those feasible for numerical computation include the spectral decomposition for symmetric matrices, the singular value decomposition for rectangular matrices, and the Schur decomposition for general square matrices (Golub and Van Loan 1996, Horn and Johnson 1990). The Jordan canonical form, on one hand, is perhaps the most fundamental and classical in matrix theory. On the other hand, the Jordan decomposition is generally considered a 'taboo' for numerical computation because it is hard to distinguish between eigenvalues that are repeated exactly and eigenvalues that are clustered closely together (Beelen and Van Dooren 1990, Golub and Wilkinson 1976). Nonetheless, it might be worthwhile to mention the notion of *pejorative manifold* proposed in an unpublished paper by Kahan (1972), who argues that multiple roots are well behaved under perturbation when the multiplicity structure is preserved. Loosely speaking, it is suggested that problems that are sensitive to arbitrary perturbations might be less sensitive to structured perturbations. Exploiting this idea, Zeng and Li (2007) have recently proposed an interesting approach to tackle the Jordan decomposition.

Thus far, most matrix decompositions are processed through iterative procedures whose success is made evident by the many available discrete methods. Our goal in this section is to recast some of those iterative schemes as dynamical systems via group actions.

We need to adjust our mind-set before continuing: the meaning of a canonical form should be understood with a much broader field of view than just matrix factorizations. Arnold (1988) asks a question in a similar spirit:

'What is *the simplest form* to which a family of matrices depending smoothly on the parameters can be reduced by *a change of coordinates* depending smoothly on the parameters?'

Obviously, an essential component to any answer is a qualification of the simplest form to which, and a mechanism by which, the coordinates are continuously changed. Before being specific about the qualification and the mechanism, we may categorically characterize the proposed procedure, whether discrete or continuous in nature, as a realization process. The canonical form, or the simplest form, that the process intends to realize ultimately should be interpreted broadly as any 'mode' from which we gain the agility to think and draw conclusions. Some useful modes, as well as the mechanisms to realize these modes, will be exemplified in the subsequent discussion.

The precise meaning of our points above will become clear later, but for now we hastily point out that, as a whole, the procedure to find the simplest form in most applications appears to follow the orbit of a certain matrix group action on the underlying matrix. This connection should not come as a surprise because the representation of a group by its homomorphisms into bijective linear maps over a certain vector space is well known (Curtis 1984, Shaw 1982, Smirnov 1970). For groups whose elements depend on continuously varying parameters, so do the corresponding matrix representations. The obvious advantage of this tie is that we have the group structure on one side and the matrix structure on the other side. A *matrix group*, that is, a subset of non-singular matrices which are closed under matrix multiplication and inversion, does form a Lie group (Howe 1983). The well-developed Lie theory therefore lends us greater advantages over iterations lacking this structure.

The question then becomes: What canonical form can a matrix, or a family of matrices, be linked to by the orbit of a group action? The choice of the group, the definition of the action, and the intended targets will constrain the various paths of transitions and, thus, the algorithms. Earlier work along these lines can be found in Della-Dora (1975). We will try to expound the various aspects of the recent development and applications in this direction. Some newly developed dynamical systems seem able to offer promising channels to tackle some linear algebra problems that, otherwise, are difficult to solve by iterative means.

7.1. Group actions and canonical forms

In a dynamical system, the state variable gets evolved in accordance with a certain rule. How the rule of transition is defined determines the dynamical behaviour. The emphasis of this section is on a specific rule characterized by group actions.

Given a group G and a set $\mathbb{V}$, a *group action* of G on $\mathbb{V}$ is a map $\mu : G \times \mathbb{V} \longrightarrow \mathbb{V}$ satisfying the associative law

$$\mu(gh, \mathbf{x}) = \mu(g, \mu(h, \mathbf{x})), \quad g, h \in G, \tag{7.1}$$

Table 7.1. Examples of classical matrix groups over $\mathbb{R}$.

Group	Subgroup	Notation	Characteristics
general linear		$\mathcal{Gl}(n)$	$\{A \in \mathbb{R}^{n \times n} \mid \det(A) \neq 0\}$
	special linear	$\mathcal{Sl}(n)$	$\{A \in \mathcal{Gl}(n) \mid \det(A) = 1\}$
upper triangular		$\mathcal{U}(n)$	$\{A \in \mathcal{Gl}(n) \mid A \text{ is upper triangular}\}$
	unipotent	$\mathcal{Unip}(n)$	$\{A \in \mathcal{U}(n) \mid a_{ii} = 1 \text{ for all } i\}$
orthogonal		$\mathcal{O}(n)$	$\{Q \in \mathcal{Gl}(n) \mid Q^{\top}Q = I\}$
generalized orthogonal		$\mathcal{O}_S(n)$	$\{Q \in \mathcal{Gl}(n) \mid Q^{\top}SQ = S\}$, S is a fixed symmetric matrix
	symplectic	$\mathcal{Sp}(2n)$	$\mathcal{O}_J(2n), \quad J := \begin{bmatrix} 0 & I \\ -I & 0 \end{bmatrix}$
	Lorentz	$\mathcal{Lor}(n, k)$	$\mathcal{O}_L(n + k)$, $L := \mathrm{diag}\{\underbrace{1, \ldots, 1}_{n}, \underbrace{-1, \ldots -1}_{k}\}$
affine		$\mathcal{Aff}(n)$	$\left\{ \begin{bmatrix} A & \mathbf{t} \\ \mathbf{0} & 1 \end{bmatrix} \mid A \in \mathcal{Gl}(n), \mathbf{t} \in \mathbb{R}^n \right\}$
	translation	$\mathcal{Trans}(n)$	$\left\{ \begin{bmatrix} I & \mathbf{t} \\ \mathbf{0} & 1 \end{bmatrix} \mid \mathbf{t} \in \mathbb{R}^n \right\}$
	isometry	$\mathcal{Isom}(n)$	$\left\{ \begin{bmatrix} Q & \mathbf{t} \\ \mathbf{0} & 1 \end{bmatrix} \mid Q \in \mathcal{O}(n), \mathbf{t} \in \mathbb{R}^n \right\}$
product of G_1 and G_2		$G_1 \times G_2$	$\{(g_1, g_2) \mid g_1 \in G_1, g_2 \in G_2\}$, $(g_1, g_2) * (h_1, h_2) := (g_1 h_1, g_2 h_2)$, G_1 and G_2 are given groups
automorphism		$\mathbb{G}_M$	$\{A \in \mathcal{Gl}(n) \mid \langle A\mathbf{x}, A\mathbf{y} \rangle_M = \langle \mathbf{x}, \mathbf{y} \rangle_M\}$, $\langle \mathbf{x}, \mathbf{y} \rangle_M = \mathbf{x}^{\top} M \mathbf{y}$, M is a given matrix

and the identity property

$$\mu(e, \mathbf{x}) = \mathbf{x}, \tag{7.2}$$

where e is the identity element in G, for all $\mathbf{x} \in \mathbb{V}$. Given a fixed $\mathbf{x} \in \mathbb{V}$, the *orbit of* $\mathbf{x}$ associated to an action μ of G is defined to be the set

$$\text{Orb}_G(\mathbf{x}) := \{\mu(g, \mathbf{x}) | g \in G\}. \tag{7.3}$$

For our applications, we are interested in using matrix groups and various actions to help transform a given matrix into an appropriate canonical form. The transformation is to take place along the associated orbit of the given matrix. To get this idea going, we need four components working together: a group that characterizes the coordinates to be used, an action that constrains the transformations to be allowed, a canonical form that sets the goal to be reached, and a rule that delineates the path to be followed. Each of these four components affects the final result.

For demonstration, Table 7.1 is a short list of matrix groups compiled from the books of Baker (2002), Chu and Golub (2005) and Curtis (1984). We remark that the automorphism group $\mathbb{G}_M$ associated with a non-degenerate bilinear form $\langle \mathbf{x}, \mathbf{y} \rangle_M = \mathbf{x}^\top M \mathbf{y}$ contains as special cases the orthogonal group and the symplectic group (Mackey, Mackey and Tisseur 2003).

Table 7.2 typifies some group actions that have been commonly used in numerical linear algorithm algorithms. Traditionally, numerical analysts prefer to use the orthogonal group for actions because of its cost efficiency and numerical stability. Such a restriction, however, could have limited the canonical forms that we otherwise would be able to reach by different groups.

Table 7.2. Examples of group actions and their applications.

Set $\mathbb{V}$	Group G	Action $\mu(g, A)$	Application
$\mathbb{R}^{n \times n}$	any subgroup	$g^{-1}Ag$	conjugation
$\mathbb{R}^{n \times n}$	$\mathcal{O}(n)$	$g^\top Ag$	orthogonal similarity
$\underbrace{\mathbb{R}^{n \times n} \times \ldots \times \mathbb{R}^{n \times n}}_{k}$	any subgroup	$(g^{-1}A_1g, \ldots, g^{-1}A_kg)$	simultaneous reduction
$\mathbb{S}(n) \times \mathbb{S}_{PD}(n)$	any subgroup	$(g^\top Ag, g^\top Bg)$	symm. positive definite pencil reduction
$\mathbb{R}^{n \times n} \times \mathbb{R}^{n \times n}$	$\mathcal{O}(n) \times \mathcal{O}(n)$	$(g_1^\top Ag_2, g_1^\top Bg_2)$	QZ decomposition
$\mathbb{R}^{m \times n}$	$\mathcal{O}(m) \times \mathcal{O}(n)$	$g_1^\top Ag_2$	singular value decomp.
$\mathbb{R}^{m \times n} \times \mathbb{R}^{p \times n}$	$\mathcal{O}(m) \times \mathcal{O}(p) \times \mathcal{Gl}(n)$	$(g_1^\top Ag_3, g_2^\top Bg_3)$	generalized singular value decomp.

Table 7.3 makes evident the wide scope of canonical forms that group actions can (or be desired to) accomplish, ranging from a typical structure with a specified pattern of zeros, such as a diagonal, tridiagonal, or triangular matrix, to a matrix with a specified construct, such as Toeplitz, Hamiltonian, stochastic, or other linear varieties, to a matrix with a specified algebraic constraint, such as low rank or non-negativity.

With the group, action and orbit in place, we finally need a properly defined dynamical system, either continuous or discrete, so that its integral curves or iterates stay on the specified orbit and connect one state to the next state. The Toda lattice and the Lotka–Volterra equation discussed earlier serve as typical examples in this regard, although in both cases the group actions are built into the dynamical systems and are not exploited explicitly. We shall develop a general framework of the projected gradient

Table 7.3. Examples of canonical forms used in practice.

Canonical form	Also known as	Action
bidiagonal J	quasi-Jordan decomp., $A \in \mathbb{R}^{n \times n}$	$P^{-1}AP = J$, $P \in \mathcal{Gl}(n)$
diagonal Σ	sing. value decomp., $A \in \mathbb{R}^{m \times n}$	$U^\top AV = \Sigma$, $(U, V) \in \mathcal{O}(m) \times \mathcal{O}(n)$
diagonal pair (Σ_1, Σ_2)	gen. sing. value decomp., $(A, B) \in \mathbb{R}^{m \times n} \times \mathbb{R}^{p \times n}$	$(U^\top AX, V^\top BX) = (\Sigma_1, \Sigma_2)$, $(U, V, X) \in \mathcal{O}(m) \times \mathcal{O}(p) \times \mathcal{Gl}(n)$
upper quasi-triangular H	real Schur decomp., $A \in \mathbb{R}^{n \times n}$	$Q^\top AQ = H$, $Q \in \mathcal{O}(n)$
upper quasi-triangular H upper triangular U	gen. real Schur decomp., $A, B \in \mathbb{R}^{n \times n}$	$(Q^\top AZ, Q^\top BZ) = (H, U)$, $Q, Z \in \mathcal{O}(n)$
symmetric Toeplitz T	Toeplitz inv. eigenv. prob., $\{\lambda_1, \ldots, \lambda_n\} \subset \mathbb{R}$ is given	$Q^\top \operatorname{diag}\{\lambda_1, \ldots, \lambda_n\}Q = T$, $Q \in \mathcal{O}(n)$
non-negative $N \geq 0$	non-neg. inv. eigenv. prob., $\{\lambda_1, \ldots, \lambda_n\} \subset \mathbb{C}$ is given	$P^{-1} \operatorname{diag}\{\lambda_1, \ldots, \lambda_n\}P = N$, $P \in \mathcal{Gl}(n)$
linear variety X with fixed entries at fixed locations	matrix completion prob., $\{\lambda_1, \ldots, \lambda_n\} \subset \mathbb{C}$ is given $X_{i_\nu, j_\nu} = a_\nu, \nu = 1, \ldots, \ell$	$P^{-1}\{\lambda_1, \ldots, \lambda_n\}P = X$, $P \in \mathcal{Gl}(n)$
nonlinear variety with fixed singular values and eigenvalues	test matrix construction, $\Lambda = \operatorname{diag}\{\lambda_1, \ldots, \lambda_n\}$ and $\Sigma = \operatorname{diag}\{\sigma_1, \ldots \sigma_n\}$ are given	$P^{-1}\Lambda P = U^\top \Sigma V$ $P \in \mathcal{Gl}(n), \quad U, V \in \mathcal{O}(n)$
maximal fidelity	structured low-rank approx. $A \in \mathbb{R}^{m \times n}$	$(\operatorname{diag}(USS^\top U^\top))^{-1/2}USV^\top$, $(U, S, V) \in \mathcal{O}(m) \times \mathbb{R}_\times^k \times \mathcal{O}(n)$

approach in the next section to help to construct other useful dynamical systems. The projected gradient flows from continuous group actions are often easy to formulate and analyse, and are sometimes able to tackle problems that are seemingly impossible to resolve by conventional discrete methods.

An area that has been active for research, and remains open for further work, is to develop numerical algorithms that can effectively trace dynamical systems arising from various group actions. We note that there are many new techniques developed recently for dynamical systems on Lie groups, including the RK-MK methods (Engø 2003, Munthe-Kaas 1998), Magnus and Fer expansions (Blanes, Casas, Oteo and Ros 1998, Zhang and Deng 2005), and so on. A good collection of Lie structure-preserving algorithms and pertaining references can be found in the seminal review paper by Iserles, Munthe-Kaas, Nørsett and Zanna (2000) and the book by Hairer, Lubich and Wanner (2006). These new geometric integration techniques certainly can benefit the computations needed for the projected gradient flow, but still we are seeking a method that also takes into account the descent property of a gradient flow. For a gradient flow where finding its stable equilibrium point is the ultimate goal of computation, recall that the pseudo-transient continuation described in Section 2.2 has been suggested as a possible numerical method.

7.2. Projected gradient flows

The idea of projected gradient flows stems from the constrained least-squares approximation to a desirable canonical form. From a given matrix A in a subset $\mathbb{V}$ of matrices of fixed sizes, the constraint on the variable is that the transformation of A must be limited to the orbit $\mathrm{Orb}_G(A)$ determined by a prescribed continuous matrix group G and a group action $\mu : G \times \mathbb{V} \longrightarrow \mathbb{V}$. The objective function itself is built with two additional limitations. One is a differentiable map $f : \mathbb{V} \longrightarrow \mathbb{V}$ designed to regulate certain 'inherent' properties such as symmetry, diagonal, isospectrality, low rank, or other algebraic conditions. The other is a projection map $P : \mathbb{V} \longrightarrow \mathbb{P}$, where $\mathbb{P}$ denotes the subset of matrices in $\mathbb{V}$ carrying a certain desirable structure, that is, the canonical form. The set $\mathbb{P}$ could be a singleton, an affine subspace, or a cone, or other geometric entities. Consider the functional $F : G \longrightarrow \mathbb{R}$ where

$$F(Q) := \frac{1}{2} \| f(\mu(Q, A)) - P(\mu(Q, A)) \|_F^2. \tag{7.4}$$

The goal is to minimize F over the group G. The meaning of this constrained minimization is that, while staying in the orbit of A under the action of μ and maintaining the inherent property guaranteed by the function f, we look for the element $Q \in G$ so that the matrix $f(\mu(Q, A))$ best realizes the desired canonical structure in the sense of least squares.

In principle, the functional (7.4) can be minimized by conventional optimization techniques which mostly are iterative in nature. However, we find that the projected gradient flow approach can conveniently be formulated as a dynamical system,

$$\frac{dQ}{dt} = -\mathrm{Proj}_{\mathcal{T}_Q G} \nabla F(Q), \tag{7.5}$$

where $\mathcal{T}_Q G$ and $\nabla F(Q)$ stand for the tangent space of the group G and the gradient of the objective functional F at Q, respectively.

One advantage of working with a matrix group is that its tangent spaces at every element g have the same structure as the tangent space $\mathfrak{g} = \mathcal{T}_e G$ at the identity element e of G. More specifically, the tangent space at any element Q in G is a translation of $\mathfrak{g}$ via the relationship

$$\mathcal{T}_Q G = Q\mathfrak{g}. \tag{7.6}$$

Thus the projection in (7.5) is fairly easy to do once the tangent space $\mathfrak{g}$ is identified.

It might be instructive to illustrate the idea of projection by the following calculation (Chu and Driessel 1990). By (7.6), the tangent space of $\mathcal{O}(n)$ at any orthogonal matrix Q is

$$\mathcal{T}_Q \mathcal{O}(n) = Qo(n),$$

where $o(n)$ denotes the subspace of all skew-symmetric matrices in $\mathbb{R}^{n \times n}$. It can easily be argued that the normal space of $\mathcal{O}(n)$ at any orthogonal matrix Q is

$$\mathcal{N}_Q \mathcal{O}(n) = Qo(n)^\perp,$$

where the orthogonal complement $o(n)^\perp$ is precisely the subspace of all symmetric matrices. The space $\mathbb{R}^{n \times n}$ can be split as the direct sum of

$$\mathbb{R}^{n \times n} = Qo(n) \oplus Qo(n)^\perp.$$

Any $X \in \mathbb{R}^{n \times n}$ therefore has the unique orthogonal splitting

$$X = Q(Q^\top X) = Q\left\{ \frac{1}{2}(Q^\top X - X^\top Q) \right\} + Q\left\{ \frac{1}{2}(Q^\top X + X^\top Q) \right\}.$$

The projection of X onto the tangent space $\mathcal{T}_Q \mathcal{O}(n)$ is therefore given by the formula

$$\mathrm{Proj}_{\mathcal{T}_Q \mathcal{O}(n)} X = Q\left\{ \frac{1}{2}(Q^\top X - X^\top Q) \right\}. \tag{7.7}$$

For other groups, the projection can be done in a similar way.

We briefly touch upon the realm of differential geometry with two remarks. First, the notion of 'projected' gradient described above is indeed the 'ordinary' gradient with respect to the Killing form or the normal metric on the tangent space $\mathfrak{g}$ (Edelman, Arias and Smith 1999, Tam 2004).

Secondly, the set $\mathfrak{g}$ is a *Lie subalgebra*, that is, its elements are closed under the Lie bracket operation. The Lie subalgebra $\mathfrak{g}$ can be characterized as the logarithm of G in the sense that

$$\mathfrak{g} = \{M \in \mathbb{R}^{n \times n} \mid \exp(tM) \in G, \text{ for all } t \in \mathbb{R}\}. \qquad (7.8)$$

The exponential map $\exp : \mathfrak{g} \to G$, as we have seen in Theorem 5.1, is a central step from a Lie algebra $\mathfrak{g}$ to the associated Lie group G (Celledoni and Iserles 2000, Howe 1983). Since exp is a local homeomorphism which maps a neighbourhood of the zero O in the algebra $\mathfrak{g}$ onto a neighbourhood of the identity e in the group G, any dynamical system in G, in the neighbourhood of e, would therefore have a corresponding dynamical system in $\mathfrak{g}$, in the neighbourhood of O. Because of this, the decomposition we have observed in Section 5.1 can be interpreted as follows. It is known that the Lie group $\mathcal{G}l(n)$ can be decomposed as the product of two Lie subgroups in the neighbourhood of the identity matrix I if and only if the corresponding tangent space $gl(n)$ of real-valued $n \times n$ matrices can be decomposed into the sum of two Lie subalgebras. By the decomposition property and the reversal property in Theorem 5.1, the Lie structure is apparently not needed for isospectral flows. A subspace decomposition of $gl(n)$ as is indicated in (5.8) suffices to guarantee a factorization of a *one-parameter semigroup* in the neighbourhood of I as the product of two non-singular matrices, that is, the decomposition indicated in (5.10).

Before we talk about specific applications, a misconception about the gradient flow (2.10) in general and the projected gradient flow (7.5) in particular must be clarified. It is true that the objective value $F(\mathbf{x}(t))$ is non-increasing in t if $\mathbf{x}(t)$ follows the gradient flow (2.10). If F is further known to be bounded below, the $F(\mathbf{x}(t))$ converges to a limit value. However, the flow $\mathbf{x}(t)$ itself might not converge at all. Examples can be constructed to show the case that a local minimum of an infinitely differentiable objective function F may not be an equilibrium point of the differential system (2.10). Likewise, a stable equilibrium point of (2.10) may not be a local minimum of F at all (Absil and Kurdyka 2006). A cone-shaped minaret with outside spiral ramp, or a helicoid, can be modified to serve as examples where a gradient flow converges to a limit cycle. The important message we want to convey is that infinite smoothness of the gradient vector field is not sufficient to guarantee the convergence of a gradient trajectory. A sufficient condition that happens to fit our applications is the analyticity of the objective function. More specifically, the Łojasiewicz–Simon theorem asserts that if the objective function F is real analytic, then the trajectory of a gradient flow cannot have more than one limit point (Chill 2003, Łojasiewicz 1963, Simon 1983). Furthermore, under the analyticity assumption, a stable equilibrium point of the differential system (2.10) is a local minimum of F, and *vice versa* (Absil, Mahony and

Andrews 2005, Absil and Kurdyka 2006). In our applications, group actions, linear projections and squares of the Frobenius norm are naturally analytic. Our gradient flows are defined by an analytic vector field, so convergence is ensured.

7.3. Applications

From the framework outlined above, projected gradient dynamical systems can be tailored to meet the need arising from various circumstances. We shall demonstrate four interesting designs in this section. Many additional applications and the associated dynamical systems can be found in the literature. See, for instance, the problems discussed in the paper by Brockett (1993) and the book by Helmke and Moore (1994). Our intention in this section is to demonstrate the versatility of projected gradient flows. Some applications can be solved more efficiently by other means, but there are problems where the continuous dynamical systems approach is particularly easy to formulate and compute.

Example 7.1. Given a symmetric matrix Λ and a desirable structure $\mathbb{P}$, suppose we want to find a symmetric matrix that is closest to $\mathbb{P}$ and has the same spectrum as Λ (Chu and Driessel 1990). By defining the isospectral matrix $X := Q^\top \Lambda Q$ with $Q \in \mathcal{O}(n)$, the objective functional $F : \mathcal{O}(n) \to \mathbb{R}$ is taken to be

$$F(Q) := \frac{1}{2} \|Q^\top \Lambda Q - P(Q^\top \Lambda Q)\|_F^2, \tag{7.9}$$

where the Frobenius norm of a real matrix M is, as usual, defined by

$$\|M\|_F = \sqrt{\operatorname{trace}(MM^\top)}.$$

It can be verified that the projected gradient flow (7.5) on the group $\mathcal{O}(n)$ is equivalent to the isospectral flow,

$$\frac{\mathrm{d}X}{\mathrm{d}t} = [X, [X, P(X)]], \tag{7.10}$$

on the orbit $\operatorname{Orb}_{\mathcal{O}(n)}(\Lambda)$.

With different choices of Λ and $\mathbb{P}$, the dynamical system (7.10) enjoys different interpretation of applications. For example, if $P(X) = \operatorname{diag}(X)$, then $X(t)$ stands for a continuous Jacobi-type flow that gradually reduces the off-diagonal elements of X while maintaining isospectrality. As another example, by specifying the structure retained in $\mathbb{P}$, the flow (7.10) offers an avenue to tackle various kinds of very difficult structured inverse eigenvalue problems (Chu and Golub 2002).

The so-called double bracket flow by Brockett (1991) corresponds to the special case where $\mathbb{P} = \{N\}$ contains a single constant symmetric matrix N

and hence $P(Q^\top \Lambda Q) \equiv N$. The resulting qualitative behaviour is relatively easier to analyse, but this seemingly ingenuous nearest matrix approximation to a fixed matrix has the following sorting property, which appears universal in a wide spectrum of applications, including the interior-point algorithm (Faybusovich 1991), the QR algorithm (Deift *et al.* 1983), moment maps (Bloch, Brockett and Ratiu 1992) and many others (Helmke and Moore 1994).

Theorem 7.2. Suppose that both $\Lambda = \mathrm{diag}\{\lambda_1, \dots, \lambda_n\}$ and the spectrum of N have distinct elements. Then $X = Q^\top \Lambda Q$ is the unique nearest matrix to N on the isospectral orbit of Λ if and only if the columns of $Q^\top$ are the orthonormal eigenvectors of N, corresponding to eigenvalues arranged in the same order as $\{\lambda_1, \dots, \lambda_n\}$.

We have to mention one remarkable connection. If Λ is a tridiagonal matrix to begin with and if $N = \mathrm{diag}\{n, n - 1, \dots, 2, 1\}$, then the double bracket flow becomes exactly the Toda lattice that has been discussed in great length in Section 5.1. The sorting property asserted in Theorem 7.2 therefore explains the sorting property of the QR algorithm. It is interesting that 'the same set of equations is thus Hamiltonian and a gradient flow on the isospectral set' (Bloch *et al.* 1992).

Given the wide range of applications, an effective way of integrating either the isospectral dynamical system (7.10) for $X(t)$ over the orbit or the associated parameter dynamical system for $Q(t)$ over the group therefore would be extremely useful and desirable. We think that an efficient discretization would probably not come from the traditional numerical ODE approaches, but rather could be more in line with the *vdLV* approach, where a certain structure is preserved. On the other hand, it is worth noting that the versatile double bracket flow $\frac{\mathrm{d}X}{\mathrm{d}t} = [X, [X, N]]$ might be handled differently. By representing the isospectral solution in the form $X(t) = e^{\Omega(t)} X_0 E^{-\Omega(t)}$, Iserles (2002) has developed an interesting approach to the discretization of $X(t)$. Specifically, each term in the Taylor series expansion of $\Omega(t)$ can be constructed explicitly and recursively by means of rooted trees with bicolour leaves.

Example 7.3. In analogy to Example 7.1, we could also consider the nearest approximation by iso-singular-value matrices. Given a rectangular matrix Σ of size $m \times n$ and a desirable structure $\mathbb{P}$ over $\mathbb{R}^{m \times n}$, all matrices on the orbit $\mathrm{Orb}_{\mathcal{O}(m) \times \mathcal{O}(n)}(\Lambda) := \{X = U^\top \Sigma V \,|\, U \in \mathcal{O}(m), V \in \mathcal{O}(n)\}$ have the same singular values as Σ. The objective functional $F : \mathcal{O}(m) \times \mathcal{O}(n) \to \mathbb{R}$ defined by

$$F(U, V) := \|U^\top \Sigma V - P(U^\top \Sigma V)\|_F^2 \qquad (7.11)$$

is meant to best approach the structure $\mathbb{P}$ while maintaining the singular values. A continuous transformation $X := U^\top \Sigma V$ is governed by the dynamical system

$$\frac{\mathrm{d}X}{\mathrm{d}t} = \{X(X^\top P(X) - P(X)^\top X) - (XP(X)^\top - P(X)X^\top)X\}, \quad (7.12)$$

which, at first glance, is not exactly in the double bracket form. However, upon recasting the original action of equivalence $U^\top \Sigma V$ by the product group $\mathcal{O}(m) \times \mathcal{O}(n)$ as a new action of conjugation,

$$\begin{bmatrix} U^\top & 0 \\ 0 & V^\top \end{bmatrix} \begin{bmatrix} 0 & \Sigma \\ \Sigma^\top & 0 \end{bmatrix} \begin{bmatrix} U & 0 \\ 0 & V \end{bmatrix},$$

by a subgroup of $\mathcal{O}(m+n)$, Tam (2004) has observed that (7.12) can indeed be written in a double bracket form,

$$\frac{\mathrm{d}\mathfrak{X}}{\mathrm{d}t} = [\mathfrak{X}, [\mathfrak{X}, \mathfrak{P}(\mathfrak{X})]], \quad (7.13)$$

with the definition

$$\mathfrak{X} := \begin{bmatrix} 0 & X \\ X^\top & 0 \end{bmatrix},$$

$$\mathfrak{P}(\mathfrak{X}) := \begin{bmatrix} 0 & P(X) \\ P(X)^\top & 0 \end{bmatrix}.$$

Some applications of the gradient flow (7.12) include a sorting property similar to Theorem 7.2 if $\mathbb{P}$ consists of a single constant matrix (Chu and Driessel 1990, Smith 1991), structured inverse singular value problems, and a Jacobi-type algorithm if $P(X) = \mathrm{diag}(X)$. In the last case, the corresponding dynamical system is

$$\frac{\mathrm{d}X}{\mathrm{d}t} = \{X(X^\top \mathrm{diag}(X) - \mathrm{diag}(X)^\top X) - (X \mathrm{diag}(X)^\top - \mathrm{diag}(X)X^\top)X\}.$$

It is interesting to note that by merely a change of sign in the above equation, we obtain the system

$$\frac{\mathrm{d}X}{\mathrm{d}t} = \{X(X^\top \mathrm{diag}(X) - \mathrm{diag}(X)^\top X) + (X \mathrm{diag}(X)^\top - \mathrm{diag}(X)X^\top)X\}$$

$$= XX^\top \mathrm{diag}(X) - \mathrm{diag}(X)X^\top X, \quad (7.14)$$

which is precisely the SVD flow (6.2). Recall that the SVD flow was originally formulated with the intention to preserve the bidiagonal structure if Σ is bidiagonal to begin with. The fact that the SVD flow can be expressed differently as in (7.14) is interesting. At present, whether (7.14) is just an algebraic coincidence or is a result of a deeper theory is not clear to us.

Example 7.4.   Consider the classical matrix nearness problem of finding the closest normal matrix to a given matrix $A \in \mathbb{C}^{n \times n}$ (Higham 1989, Ruhe

1987). This problem is equivalent to minimizing the functional

$$F(U) = \frac{1}{2}\|U^*AU - \operatorname{diag}(U^*AU)\|_F^2, \qquad (7.15)$$

subject to the constraint that $U \in \mathbb{C}^{n \times n}$ is unitary. Once the minimizer $\tilde{U}$ of (7.15) is found, the nearest normal matrix to A is given by $\tilde{U} \operatorname{diag}(\tilde{U}^*A\tilde{U})\tilde{U}^*$.

The objective function (7.15) is similar to (7.9) except that we are dealing with complex-valued matrices. A projected gradient flow,

$$\frac{dZ}{dt} = \left[Z, \frac{[Z, \operatorname{diag}(Z^*)] - [Z, \operatorname{diag}(Z^*)]^*}{2} \right], \qquad (7.16)$$

for the complex matrix $Z = U^*AU$ can be derived as the action of the unitary group over $\mathbb{C}^{n \times n}$. One advantage of this differential equation approach is that many theoretical results concerning the nearest normal matrix approximation which have been challenging to matrix theorists can be obtained naturally from analysing the equilibrium point of the dynamical system (Chu 1991, Ruhe 1987).

Example 7.5. We now illustrate how the 'regulator' of f in (7.4) comes into play in some applications. Given two vectors $\mathbf{a}, \boldsymbol{\lambda} \in \mathbb{R}^n$, the Schur–Horn theorem states that there exists a Hermitian matrix H with eigenvalues $\boldsymbol{\lambda}$ and diagonal entries $\mathbf{a}$ if and only if $\boldsymbol{\lambda}$ is majorized by $\mathbf{a}$ (Horn and Johnson 1990). The harder part of this classical result is the inverse problem of construct a symmetric matrix with prescribed diagonal entries $\mathbf{a}$ and spectrum $\{\lambda_1, \ldots, \lambda_n\}$. We recast the inverse problem as the problem of minimizing the functional

$$F(Q) := \frac{1}{2}\| \operatorname{diag}(Q^\top \Lambda Q) - \operatorname{diag}(\mathbf{a})\|_F^2, \qquad (7.17)$$

subject to $Q \in \mathcal{O}(n)$. Note that we have taken $f(X) = \operatorname{diag}(X)$ for the isospectral matrices $X := Q^\top \Lambda Q$. It can be shown that the projected gradient flow becomes a double bracket equation (Chu 1995):

$$\frac{dX}{dt} = [X, [X, \operatorname{diag}(\mathbf{a}) - \operatorname{diag}(X)]]. \qquad (7.18)$$

Stability analysis at the equilibrium yields an easy existence proof of the Schur–Horn theorem.

We should re-emphasize that, unless special care is given to the discretization and implementation, the differential equation approach generally is not necessarily the most effective numerical means for solving problems. For the Schur–Horn problem, a finite-step recursive algorithm is computationally more efficient (Zha and Zhang 1995).

7.4. Generalization beyond group actions

The primary purpose of employing group actions in linear transformations is to keep eigenvalues or singular values invariant under the change of coordinates. It sometimes becomes desirable to keep other properties invariant. In many cases, the notion of gradient flows can be generalized to other geometric entities that do not hold any group structure. Examples of applications include the Stiefel manifold for the orthonormal Procrustes problem, or the more general Penrose regression problem (Chu and Trendafilov 2001), the convex set of positive definite real symmetric matrices for the balanced realization (Helmke, Moore and Perkins 1994), the Grassmann manifold for the geometric optimization methods (Edelman *et al.* 1999), the manifold of oblique matrices for the multi-dimensional scaling (Cox and Cox 1994, Del Buono and Lopez 2002) or the data fitting on the unit sphere (Chu, Del Buono, Lopez and Politi 2005), the cone of non-negative matrices for inverse eigenvalue problem (Chu and Guo 1998, Orsi 2006), and so on.

We wrap up this section by demonstrating one of these generalizations. At first glance, no group structure is involved in the formulation of the dynamical system. We then modify the coordinate systems to bring in group actions.

Example 7.6. The non-negative inverse eigenvalue problem concerns the construction of a entry-wise non-negative matrix $A \in \mathbb{R}^{n \times n}$ with a prescribed set $\{\lambda_1, \ldots, \lambda_n\} \subset \mathbb{C}$, closed under conjugation, as its spectrum. This has been a classical but hard problem, long investigated by many matrix theorists. The inadequacy of the current development is evidenced by the fact that the necessary condition for solvability is usually too general, while the sufficient condition is too specific (Chu and Golub 2005).

Recently it has been proved that, given an arbitrary $(n-1)$-tuple

$$\Omega = (\lambda_2, \ldots, \lambda_n) \in \mathbb{C}^{n-1},$$

whose components are closed under complex conjugation, there exists a unique positive real number $\mathcal{R}(\Omega)$, called the *minimal realizable spectral radius of* Ω, such that the set $\{\lambda_1, \ldots, \lambda_n\}$ is precisely the spectrum of a certain $n \times n$ non-negative matrix with λ_1 as its spectral radius if and only if $\lambda_1 \geq \mathcal{R}(\Omega)$. Employing any existing necessary conditions as a mode of checking criteria, Chu and Xu (2005) have proposed a simple bisection procedure to approximate the location of $\mathcal{R}(\Omega)$. As an immediate application, it offers a quick numerical way to check whether a given n-tuple could be the spectrum of a certain non-negative matrix. However, even after a potential spectrum is identified as feasible, very few general numerical procedures are available for the actual construction of non-negative matrices. Generalizing the above ideas and taking the advantage of its easy formulation, a gradient flow can come to serve this purpose (Chu and Guo 1998).

Since the spectrum is closed under complex conjugation, we may assume a real-valued matrix J to carry the prescribed spectrum. We cast the inverse problem as a constrained minimization problem by working with two matrix parameters (g, R),

$$\text{minimize} \quad F(g, R) := \frac{1}{2} \|gJg^{-1} - R \circ R\|_F^2,$$

$$\text{subject to} \quad g \in \mathcal{Gl}(n), \; R \in \mathfrak{gl}(n),$$

where $\circ$ denotes the component-to-component Hadamard product. The idea behind $F(g, R)$ is similar to that in (7.4), except that this time we want to minimize the distance between the orbit $\text{Orb}_{\mathcal{Gl}(n)}(J)$ and the cone of non-negative matrices. The constraints literally do not exist because both $\mathcal{Gl}(n)$ and $\mathfrak{gl}(n)$ are open sets. No projection onto the constraints is needed. The steepest descent flow for $F(g, R)$ is given by straightforward calculation,

$$\frac{\mathrm{d}g}{\mathrm{d}t} = \left[(gJg^{-1})^\top, \alpha(g, R)\right]g^{-\top}, \tag{7.19}$$

$$\frac{\mathrm{d}R}{\mathrm{d}t} = 2\alpha(g, R) \circ R, \tag{7.20}$$

with $\alpha(g, R) := gJg^{-1} - R \circ R$.

The requirement of computing g^{-1} in the gradient flow is worrisome. We can diminish concern at the cost of re-parametrizing g by its analytic singular value decomposition (Bunse-Gerstner, Byers, Mehrmann and Nichols 1991, Wright 1992). Suppose $g(t) = X(t)S(t)Y(t)^\top$ is the singular value decomposition of $g(t)$, where $S(t)$ is a diagonal matrix with elements from the multiplicative group $\mathbb{R}_\times$ of non-zero real numbers and $X(t)$ and $Y(t)$ are elements from the orthogonal group $\mathcal{O}(n)$. From the relationship of derivatives,

$$X^\top \frac{\mathrm{d}g}{\mathrm{d}t} Y = \underbrace{X^\top \frac{\mathrm{d}X}{\mathrm{d}t}}_{Z} S + \frac{\mathrm{d}S}{\mathrm{d}t} + S \underbrace{\frac{\mathrm{d}Y^\top}{\mathrm{d}t} Y}_{W}, \tag{7.21}$$

we can specify the dynamics of evolution for the parameters (X, S, Y). In particular, let $\Upsilon := X^\top \frac{\mathrm{d}g}{\mathrm{d}t} Y$, where $\frac{\mathrm{d}g}{\mathrm{d}t}$ is given by (7.20). Given initial values $(X(0), S(0), Y(0))$, we see that the equation for $S(t)$ is readily available,

$$\frac{\mathrm{d}S}{\mathrm{d}t} = \text{diag}(\Upsilon), \tag{7.22}$$

whereas the two equations

$$\frac{\mathrm{d}X}{\mathrm{d}t} = XZ, \tag{7.23}$$

$$\frac{\mathrm{d}Y}{\mathrm{d}t} = YW \tag{7.24}$$

can also be defined, since the skew-symmetric matrices Z and W can be retrieved from off-diagonal elements of Υ and S. In total, we have constructed a gradient flow for the objective function F in terms of the four matrix parameters (X, S, Y, R) that evolve on the manifold $\mathcal{O}(n) \times \mathbb{R}^n_\times \times \mathcal{O}(n) \times gl(n)$.

8. Structure-preserving dynamical systems

The notion of structure preservation has been put into practice in numerical linear algebra since its very early stage of development. The upper Hessenberg form has been used in the QR algorithm, the upper Hessenberg/triangular form in the QZ algorithm, and the bidiagonal form in the SVD algorithm (Golub and Van Loan 1996), to mention a few. These structures are not only preserved throughout the iterative processes, but also play a fundamental role in making the algorithms effective for computation.

Each of the three above-mentioned iterative schemes has a corresponding continuous analogue. It is well known that the generalized Toda flow preserves the tridiagonal form for symmetric matrices and the upper Hessenberg form for general matrices (Chu 1988, Watkins and Elsner 1988). The QZ flow and the SVD flow, on the other hand, were designed specifically to preserve the upper Hessenberg/triangular and the bidiagonal structures, respectively. Recall that the Lotka–Volterra equation discussed extensively in Section 6 is precisely the SVD flow applied to bidiagonal matrices.

As before, the meaning of structure should be interpreted broadly to include any invariant properties under the flow. The Toda flow, therefore, preserves at least two structures: the spectrum and the upper Hessenberg form. Likewise, the SVD flow preserves the singular values and the bidiagonal form. It then becomes interesting to ask whether there are other structures invariant under these flows. To distinguish these special matrix forms from other invariant properties to be discussed later, we shall use the term *zero structure* to refer collectively to any specific zero pattern of a matrix. The flip side of the question is equally interesting and perhaps more important: Given a set of structures related to a fixed matrix, can a dynamical system, continuous or discrete, be designed to preserve the specified structures?

The importance of structure preservation goes far beyond the realm of linear algebra alone. There are properties other than zero structures that we want to maintain. Stability and passivity preservation, for example, are highly desirable in model reduction (Antoulas 2005). Standard simplicity preservation allows a doubling algorithm to effectively separate stable and unstable eigenvalues when solving the discrete algebraic Riccati equation (Lin and Xu 2006). See also an interesting discussion by Mackey *et al.* (2003) for structured matrices arising in the context of a bilinear or sesquilinear form. A quick search for the key phrase 'structure preserving' over the internet brings up a wide range of applications across multiple disciplines. We will not and are unable to review the various situations in the literature where structure preservation is essential. However, it might be safe to state that structure preservation is essential in applications because it often makes possible more efficient computation, improves physical feasibility or interpretability, and is more robust.

In this section, we shall explore dynamical systems that preserve some interesting structures arising from linear algebra. We intend to disclose some of the structures that are elusive from consideration of the dynamical systems. Be warned that we have to pose several observations as conjectures because no mathematical proofs are available at present. Even so, numerical experiments strongly suggest that these conjectures should be true.

8.1. Staircase structure

The upper Hessenberg form is actually a special case of the more general form known as the staircase structure. Given a matrix $A = [a_{ij}] \in \mathbb{R}^{m \times n}$, define the step index for each column by

$$t_k(A) := \max\Big\{k, \max_{k < i \leq m} \{i \,|\, a_{ik} \neq 0\}\Big\}, \quad k = 1, \ldots, n. \qquad (8.1)$$

We say that A is in *staircase form* if and only if

$$t_k(A) \leq t_{k+1}(A), \quad k = 1, \ldots, n - 1. \qquad (8.2)$$

Both of the following matrices, for example,

$$
\begin{bmatrix}
\times & \times & \times & \times & \times \\
0 & \times & \times & \times & \times \\
0 & \times & 0 & \times & \times \\
0 & 0 & \times & \times & \times \\
0 & 0 & 0 & 0 & \times
\end{bmatrix}, \quad
\underbrace{\begin{bmatrix}
\times & \times & \times & \times & \times \\
0 & \times & \times & \times & \times \\
0 & \times & \times & \times & \times \\
0 & 0 & \times & \times & \times \\
0 & 0 & 0 & 0 & \times
\end{bmatrix}}_{\text{full staircase}}
$$

are staircase matrices with step indices $\{1, 3, 4, 4, 5\}$. When there are no zero elements above the stairs, we say that the matrix is of *full staircase*.

Recall that the QR algorithm is the most efficient method for eigenvalue computation due to its stability and isospectrality. The following result by Arbenz and Golub (1995) identifies the zero structure that is preserved under the QR algorithm when applied to symmetric matrices.

Theorem 8.1. Assume that A_0 is symmetric. Let $\{A_k\}$ be the iterates generated by the QR algorithm (5.2). Then the following are true.

(1) If A_0 is reducible by some permutation matrix P, that is,

$$PA_0P^\top = \begin{bmatrix} A_{01} & A_{02} \\ 0 & A_{03} \end{bmatrix},$$

then each A_k is also reducible by means of the same permutation P.

(2) If A_0 is irreducible, then the zero pattern of A_0 is preserved throughout $\{A_k\}$ if and only if A_0 is a full staircase matrix.

Consider the zero structure of the following two 7×7 symmetric matrices,

$$\begin{bmatrix}
\times & 0 & \times & 0 & \times & 0 & \times \\
0 & \times & 0 & \times & 0 & \times & 0 \\
\times & 0 & \times & 0 & \times & 0 & \times \\
0 & \times & 0 & \times & 0 & \times & 0 \\
\times & 0 & \times & 0 & \times & 0 & \times \\
0 & \times & 0 & \times & 0 & \times & 0 \\
\times & 0 & \times & 0 & \times & 0 & \times
\end{bmatrix},
\begin{bmatrix}
\times & 0 & \times & 0 & \times & \times & \times \\
0 & \times & 0 & \times & 0 & \times & 0 \\
\times & 0 & \times & 0 & \times & 0 & \times \\
0 & \times & 0 & \times & 0 & \times & 0 \\
\times & 0 & \times & 0 & \times & 0 & \times \\
\times & \times & 0 & \times & 0 & \times & 0 \\
\times & 0 & \times & 0 & \times & 0 & \times
\end{bmatrix}, \qquad (8.3)$$

which differ only at the $(1, 6)$ and $(6, 1)$ positions. The QR algorithm using these two matrices as the initial values produces very different behaviour. Theorem 8.1 asserts that the zero pattern for the left matrix is preserved because it is reducible, but the zero pattern for the right matrix is totally destroyed even after one iteration.

For non-symmetric matrices, the reducibility is not guaranteed to be preserved. However, the staircase form remains a sufficient, but not necessary, condition for shape preservation under the QR algorithm. Given the close relationship between the QR algorithm and the Toda flow, it should not be surprising that if X_0 is a staircase matrix, then so is $X(t)$ under the dynamical system (5.14) (Ashlock, Driessel and Hentzel 1997, Chu and Norris 1988).

For the generalized eigenvalue problem,

$$A_0\mathbf{x} = \lambda B_0\mathbf{x}, \qquad (8.4)$$

a typical iterative scheme is the QZ algorithm. For practical purposes, the matrix A_0 is usually first reduced to an upper Hessenberg form and B_0 to an upper triangular form by orthogonal equivalence transformations. The basic idea behind the QZ algorithm is to simulate the effect of the QR algorithm on the matrix $B_0^{-1}A_0$ (assuming B_0 is invertible) without explicitly forming the inverse or the product. Throughout the QZ iteration, a critical component in the algorithm is that the upper Hessenberg/triangular structure is preserved.

Suppose now that a smooth orthogonal equivalence transformation has been applied to the pencil $B_0\lambda - A_0$,

$$\mathscr{L}(t) = Q(t)(B_0\lambda - A_0)Z(t), \quad Q(t), Z(t) \in \mathcal{O}(n). \tag{8.5}$$

Upon differentiation, the isospectral flow $\mathscr{L}(t)$ is necessarily governed by a differential system of the form

$$\frac{\mathrm{d}\mathscr{L}}{\mathrm{d}t} = \mathscr{L}R - L\mathscr{L}, \quad \mathscr{L}(0) = B_0\lambda - A_0, \tag{8.6}$$

where the coordinate transformation must satisfy the system

$$\frac{\mathrm{d}Q}{\mathrm{d}t} = -LQ,$$

$$\frac{\mathrm{d}Z}{\mathrm{d}t} = ZR,$$

with some $L, R \in o(n)$. The choice of skew-symmetric matrix parameters $L(t)$ and $R(t)$ determines the dynamics. Write

$$X(t) = Q(t)A_0Z(t),$$

$$Y(t) = Q(t)B_0Z(t).$$

To mimic the QZ algorithm, we prefer to choose $L(t)$ and $R(t)$ so that the resulting vector fields $\frac{\mathrm{d}X}{\mathrm{d}t}$ and $\frac{\mathrm{d}Y}{\mathrm{d}t}$ remain upper Hessenberg/triangular whenever $X(t)$ and $Y(t)$ are, respectively. Among many possibilities, one selection out of naïveté but with proper symmetry is the choice

$$L = \Pi_0(XY^{-1}), \tag{8.7}$$

$$R = \Pi_0(Y^{-1}X), \tag{8.8}$$

where the operator Π_0 is given in (5.13). Define the QZ flow accordingly by

$$\frac{\mathrm{d}\mathscr{L}}{\mathrm{d}t} = \mathscr{L}\Pi_0(Y^{-1}X) - \Pi_0(XY^{-1})\mathscr{L}, \quad \mathscr{L}(0) = B_0\lambda - A_0. \tag{8.9}$$

Note that if $X(t)$ and $Y(t)$ are upper Hessenberg/triangular, then both $L(t)$ and $R(t)$ are tridiagonal. Note also that if we define

$$E(t) := X(t)Y^{-1}(t), \tag{8.10}$$

$$F(t) := Y^{-1}(t)X(t), \tag{8.11}$$

then it can readily be proved that

$$\frac{\mathrm{d}E}{\mathrm{d}t} = [E, \Pi_0(E)], \tag{8.12}$$

$$\frac{\mathrm{d}F}{\mathrm{d}t} = [F, \Pi_0(F)]. \tag{8.13}$$

In other words, the QZ flow (8.9) is related to the QZ algorithm in the same way as the Toda flow is related to the QR algorithm. The convergence of the QZ flow therefore follows naturally from the dynamics of the Toda flow (Chu 1986a).

Thus far, the peculiar right-hand sides of (8.9) are designed solely for the purpose of maintaining the upper Hessenberg/triangular form. However, one interesting phenomenon as a by-product is worth mentioning. It has been observed that the QZ flow and, consequently, the corresponding QZ algorithm preserve the staircase structure. A more precise description of our empirical observation is given in the following conjecture, of which a rigorous proof has not been established at present.

Conjecture 8.2. If both A_0 and B_0 are staircase matrices, not necessarily of the same pattern, then the structures of A_0 and B_0 are preserved by $X(t)$ and $Y(t)$ under the QZ flow defined by (8.9), respectively.

We elaborate on the implication of Conjecture 8.2 a little bit more. The distinct zero patterns of the two matrices,

$$
A_0 = \begin{bmatrix}
\times & \times & \times & \times & \times & \times & \times \\
\times & \times & \times & \times & \times & \times & \times \\
0 & \times & \times & \times & \times & \times & \times \\
0 & \times & \times & \times & \times & \times & \times \\
0 & 0 & 0 & \times & \times & \times & \times \\
0 & 0 & 0 & \times & \times & \times & \times \\
0 & 0 & 0 & 0 & 0 & 0 & \times
\end{bmatrix}, \quad
B_0 = \begin{bmatrix}
\times & \times & \times & \times & \times & \times & \times \\
\times & \times & \times & \times & \times & \times & \times \\
\times & \times & \times & \times & \times & \times & \times \\
\times & \times & \times & \times & \times & \times & \times \\
0 & 0 & 0 & \times & \times & \times & \times \\
0 & 0 & 0 & 0 & \times & \times & \times \\
0 & 0 & 0 & 0 & 0 & \times & \times
\end{bmatrix},
$$

for example, are preserved, respectively, in the QZ flow. It is not obvious why the separate stair structures are kept without interference. It is amazing that the procedure of 'mixing' Y^{-1}, which is usually full and dense with the structured X followed by the operations in the way specified in (8.9), will eventually separate and give back the original staircase structures of X and Y, respectively. Direct manipulation is hard to come by, because algebraic expression would be considerably complicated. Perhaps it is for this reason that the staircase structure has been reticent thus far. Although not necessarily of practical value, such a structure-preserving property of the QZ flow (and of the QZ algorithm) is mathematically intriguing.

An equally interesting structure-preserving property is also found in the SVD flow (6.2). Our original idea in deriving this particular matrix form of dynamical system was simply to maintain the bidiagonal structure (Chu 1986b). Because of this property, the SVD flow is reduced to the Lotka–Volterra equation (6.3) when B_0 is bidiagonal to begin with. Surprisingly, if we continue to use the SVD flow in its matrix form (6.2), then we have empirical evidence to support the following conjecture.

Conjecture 8.3. Suppose B_0 is a staircase matrix. Then the SVD flow $B(t)$ defined by (6.2) and the corresponding SVD algorithm maintains the same staircase structure.

For small size matrices, the validity of Conjecture 8.3 can be proved by an *ad hoc* calculation. We are curious whether there is a more elegant way to validate this conjecture in general.

Finally, we remark that the staircase form is only a sufficient condition for shape preservation under the SVD flow. There are other structures invariant under the dynamical system (6.2). The chessboard structure of the left matrix in (8.3), for example, is preserved under the SVD flow, but unlike the symmetric QR flow, the SVD flow does not preserve the reducibility.

8.2. Lancaster structure

The *Lancaster structure* of three given matrices M_0, C_0 and K_0 in $\mathbb{R}^{n \times n}$ refers to a linear pencil of the form (Gohberg, Lancaster and Rodman 1982)

$$\mathfrak{L}(\lambda) := \mathfrak{L}(\lambda; M_0, C_0, K_0) = \begin{bmatrix} C_0 & M_0 \\ M_0 & 0 \end{bmatrix} \lambda - \begin{bmatrix} -K_0 & 0 \\ 0 & M_0 \end{bmatrix}. \quad (8.14)$$

The matrices need not have any additional properties such as symmetry or positive definiteness. The Lancaster structure consists of more than just zero patterns. It also requires the matrix M_0 to appear at three specified locations. It is easy to see that the linear pencil (8.14) is equivalent to the quadratic pencil,

$$\mathfrak{Q}(\lambda) := \mathfrak{Q}(\lambda; M_0, C_0, K_0) = \lambda^2 M_0 + \lambda C_0 + K_0, \quad (8.15)$$

in the sense that

$$\left(\begin{bmatrix} C_0 & M_0 \\ M_0 & 0 \end{bmatrix} \lambda - \begin{bmatrix} -K_0 & 0 \\ 0 & M_0 \end{bmatrix} \right) \begin{bmatrix} \mathbf{u} \\ \mathbf{v} \end{bmatrix} = 0 \quad (8.16)$$

if and only if

$$\begin{cases} (\lambda C_0 + K_0)\mathbf{u} + \lambda M_0 \mathbf{v} = 0, \\ \lambda M_0 \mathbf{u} - M_0 \mathbf{v} = 0. \end{cases} \quad (8.17)$$

Indeed, if M_0 is non-singular, then we know further that $\mathbf{v} = \lambda \mathbf{u}$. Obviously, the Lancaster structure implies that if $\mathfrak{Q}(\lambda)$ is self-adjoint, then so is $\mathfrak{L}(\lambda)$. The eigen-information $(\lambda, \mathbf{u}) \in \mathbb{C} \times \mathbb{C}^n$ of the quadratic pencil $\mathfrak{Q}(\lambda)$ is critical to the understanding of the dynamical system

$$M_0 \ddot{\mathbf{x}} + C_0 \dot{\mathbf{x}} + K_0 \mathbf{x} = f(t), \quad (8.18)$$

which arises frequently in many important applications, including applied mechanics, electrical oscillations, vibro-acoustics, fluid mechanics, and signal processing (Tisseur and Meerbergen 2001).

We are interested in the Lancaster structure because, in contrast to the common knowledge that generally no three matrices can be diagonalized simultaneously by equivalence transformations, it has been shown that for almost all quadratic pencils there exist real-valued $2n \times 2n$ real matrices Π_ℓ and Π_r such that

$$\Pi_\ell^\top \mathfrak{L}(\lambda)\Pi_r = \mathfrak{L}(\lambda; M_D, C_D, K_D), \tag{8.19}$$

where M_D, C_D, K_D are all real-valued $n \times n$ diagonal matrices. In other words, almost all n-degree-of-freedom second-order systems can be reduced to n totally independent single-degree-of-freedom second-order subsystems by real-valued isospectral transformations (Chu and Del Buono 2008a, Garvey, Friswell and Prells 2002a, 2002b). Such an isospectral transformation is significant in that it links the dynamical behaviour of a multiple-degree-of-freedom system directly to that of a system consisting of n independent single-degree-of-freedom subsystems. It breaks down the interlocking connectivity in the original system into totally disconnected subsystems while preserving the entire spectral properties. Thus it will be of great value in practice if the transformations Π_ℓ and Π_r can be found from any given pencil. We may consider (8.19) as a special kind of canonical form for the linear pencil (8.14).

The current theory of existence expresses Π_ℓ and Π_r in terms of the complete spectral information of $\mathfrak{L}(\lambda)$. The need for spectral information in the construction of Π_ℓ and Π_r is certainly not practical. Employing the notion of structure-preserving isospectral flows, it is possible to construct Π_ℓ and Π_r numerically without knowing the spectral information.

We first explore the 'orbit' of $\mathfrak{L}(\Lambda)$ under (Lancaster) structure-preserving equivalence transformations. Denote

$$\Pi_\ell = \begin{bmatrix} \ell_{11} & \ell_{12} \\ \ell_{21} & \ell_{22} \end{bmatrix}, \quad \Pi_r = \begin{bmatrix} r_{11} & r_{12} \\ r_{21} & r_{22} \end{bmatrix}, \tag{8.20}$$

where each ℓ_{ij} or r_{ij} is an $n \times n$ matrix. In order to maintain the Lancaster structure in the transformation $\Pi_\ell^\top \mathfrak{L}(\Lambda)\Pi_r$, it is necessary that the following five equations hold:

$$-\ell_{11}^\top K_0 r_{12} + \ell_{21}^\top M_0 r_{22} = 0,$$

$$-\ell_{12}^\top K_0 r_{11} + \ell_{22}^\top M_0 r_{21} = 0,$$

$$\ell_{12}^\top C_0 r_{12} + \ell_{22}^\top M_0 r_{12} + \ell_{12}^\top M_0 r_{22} = 0, \tag{8.21}$$

$$\ell_{11}^\top C_0 r_{12} + \ell_{21}^\top M_0 r_{12} + \ell_{11}^\top M_0 r_{22} = \ell_{12}^\top C_0 r_{11} + \ell_{22}^\top M_0 r_{11} + \ell_{12}^\top M_0 r_{21}$$

$$= -\ell_{12}^\top K_0 r_{12} + \ell_{22}^\top M_0 r_{22}.$$

Ultimately, in order to produce the canonical form, the matrices Π_ℓ and Π_r

must be such that the left-hand sides of the following three expressions,

$$-\ell_{12}^\top K_0 r_{12} + \ell_{22}^\top M_0 r_{22} = M_D,$$
$$\ell_{11}^\top C_0 r_{11} + \ell_{21}^\top M_0 r_{11} + \ell_{11}^\top M_0 r_{21} = C_D, \qquad (8.22)$$
$$\ell_{11}^\top K_0 r_{11} - \ell_{21}^\top M_0 r_{21} = K_D,$$

are diagonal matrices. The conditions (8.21) and (8.22) together constitute a homogeneous second-degree polynomial system of $8n^2 - 3n$ equations in $8n^2$ unknowns. It is not obvious how the nonlinear algebraic system could be solved analytically, but the underdetermined system does imply that there is plenty of room to choose the transformation matrices Π_ℓ and Π_r. In particular, a smooth path connecting (M_0, C_0, K_0) to (M_D, C_D, K_D) can be defined.

To characterize the path, denote the Lancaster pair in (8.14) by (A_0, B_0), where

$$A_0 = \begin{bmatrix} -K_0 & 0 \\ 0 & M_0 \end{bmatrix}, \qquad B_0 = \begin{bmatrix} C_0 & M_0 \\ M_0 & 0 \end{bmatrix}. \qquad (8.23)$$

We now develop two one-parameter families $T_\ell(t)$ and $T_r(t)$ in $\mathbb{R}^{2n \times 2n}$ of structure-preserving transformations starting with $T_\ell(0) = T_r(0) = I_{2n}$. Assume that these families of transformations act on (A_0, B_0) via the form

$$A(t) = T_\ell^\top(t) A_0 T_r(t),$$
$$B(t) = T_\ell^\top(t) B_0 T_r(t),$$

respectively. Clearly, regardless of how $T_\ell(t)$ and $T_R(t)$ are defined, the transformed pencil $(A(t), B(t))$ is isospectral to (A_0, B_0) for any t. For simplicity, we limit ourselves to a special class of transformations where matrices $T_\ell(t)$ and $T_r(t)$ are governed by the dynamical systems

$$\frac{dT_\ell(t)}{dt} = T_\ell(t)\mathcal{L}(t) = T_\ell(t) \begin{bmatrix} L_{11}(t) & L_{12}(t) \\ L_{21}(t) & L_{22}(t) \end{bmatrix}, \qquad (8.24)$$

$$\frac{dT_r(t)}{dt} = T_r(t)\mathcal{R}(t) = T_r(t) \begin{bmatrix} R_{11}(t) & R_{12}(t) \\ R_{21}(t) & R_{22}(t) \end{bmatrix}, \qquad (8.25)$$

respectively, where each $L_{ij}(t)$ or $R_{ij}(t)$, $i, j = 1, 2$, is a $n \times n$ real one-parameter matrix yet to be defined. Upon substitution, we observe that the pencil

$$\mathcal{L}(t) = B(t)\lambda - A(t)$$

must satisfy the equation

$$\frac{d\mathcal{L}}{dt} = \mathcal{L}^\top \mathcal{L} + \mathcal{L}\mathcal{R}, \qquad \mathcal{L}(0) = \mathfrak{L}(\lambda).$$

It is interesting to note that these differential equations are similar to those discussed by Bloch and Iserles (2006), which led to a Lie–Poisson system. By insisting that $(A(t), B(t))$ maintains the Lancaster structure throughout the transformation, that is,

$$A(t) = \begin{bmatrix} K(t) & 0 \\ 0 & -M(t) \end{bmatrix}, \quad B(t) = \begin{bmatrix} C(t) & M(t) \\ M(t) & 0 \end{bmatrix}, \qquad (8.26)$$

we see that the entries of $\mathcal{L}(t)$ and $\mathcal{R}(t)$ should satisfy

$$R_{12} = -DM, \qquad (8.27)$$

$$R_{21} = DK, \qquad (8.28)$$

$$L_{12} = D^{\mathsf{T}} M^{\mathsf{T}}, \qquad (8.29)$$

$$L_{21} = -D^{\mathsf{T}} K^{\mathsf{T}}, \qquad (8.30)$$

$$L_{11} - L_{22} = D^{\mathsf{T}} C^{\mathsf{T}}, \qquad (8.31)$$

$$R_{11} - R_{22} = -DC, \qquad (8.32)$$

where $D \in \mathbb{R}^{n \times n}$ is an arbitrary matrix parameter. Note that hidden in (8.31) and (8.32) are two other free matrix parameters denoted by N_L and N_R, respectively.

There are several possible ways to choose the parameters and to arrange the diagonal blocks of $\mathcal{L}(t)$ and $\mathcal{R}(t)$. For instance, corresponding to the choice

$$\mathcal{L} = \begin{bmatrix} D^{\mathsf{T}} & 0 \\ 0 & D^{\mathsf{T}} \end{bmatrix} \begin{bmatrix} \frac{C^{\mathsf{T}}}{2} & M^{\mathsf{T}} \\ -K^{\mathsf{T}} & -\frac{C^{\mathsf{T}}}{2} \end{bmatrix} + \begin{bmatrix} N_L^{\mathsf{T}} & 0 \\ 0 & N_L^{\mathsf{T}} \end{bmatrix}, \qquad (8.33)$$

$$\mathcal{R} = \begin{bmatrix} D & 0 \\ 0 & D \end{bmatrix} \begin{bmatrix} -\frac{C}{2} & -M \\ K & \frac{C}{2} \end{bmatrix} + \begin{bmatrix} N_R & 0 \\ 0 & N_R \end{bmatrix}, \qquad (8.34)$$

an isospectral flow of the triplet $(M(t), C(t), K(t))$ can be defined by the autonomous system

$$\frac{\mathrm{d}K}{\mathrm{d}t} = \frac{1}{2}(CDK - KDC) + N_L^{\mathsf{T}} K + K N_R,$$

$$\frac{\mathrm{d}C}{\mathrm{d}t} = (MDK - KDM) + N_L^{\mathsf{T}} C + C N_R, \qquad (8.35)$$

$$\frac{\mathrm{d}M}{\mathrm{d}t} = \frac{1}{2}(MDC - CDM) + N_L^{\mathsf{T}} M + M N_R.$$

Furthermore, by assuming $N_R(t) = N_L(t)$, the symmetry retained in the matrix parameter D has the effect of preserving the symmetry for the flow $(M(t), K(t), C(t))$ defined by the dynamical system (8.35). The various symmetry-preserving properties are summarized in Table 8.1.

The remaining task is to 'control' the free matrix parameters in such a way that the structure-preserving isospectral flow $(A(t), B(t))$ converges to the

Table 8.1. Preserving symmetry of $(M(t), C(t), K(t))$ by $D(t)$, if $N_R(t) = N_L(t)$.

$D(t)$	$M(t)$	$C(t)$	$K(t)$
skew-symmetric	symmetric	symmetric	symmetric
symmetric	symmetric	skew-symmetric	symmetric
symmetric	skew-symmetric	skew-symmetric	skew-symmetric
skew-symmetric	skew-symmetric	symmetric	skew-symmetric

canonical form (8.19). Consider the idea of minimizing a given sufficiently smooth objection function $f : \mathbb{R}^n \to \mathbb{R}$ whose state variable $\mathbf{x} \in \mathbb{R}^n$ is constrained to the integral curve of

$$\frac{d\mathbf{x}}{dt} = g(\mathbf{x})\mathbf{u}, \quad \mathbf{x}(0) = \mathbf{x}_0, \tag{8.36}$$

where $\mathbf{g} : \mathbb{R}^n \longrightarrow \mathbb{R}^m$ is a fixed function and $\mathbf{u}(t) \in \mathbb{R}^m$ is the control. For minimization, one way to choose the control $\mathbf{u}$ is to make the vector $\dot{\mathbf{x}}$ as close to $-\nabla f(\mathbf{x})$ as possible. This amounts to the selection of the least squares solution $\mathbf{u}$ defined by

$$\mathbf{u}(t) = -g(\mathbf{x}(t))^{\dagger} \nabla f(\mathbf{x}(t)), \tag{8.37}$$

where $g(\mathbf{x})^{\dagger}$ stands for the Moore–Penrose generalized inverse of $g(\mathbf{x})$. It follows that the closed-loop[1] dynamical system,

$$\frac{d\mathbf{x}}{dt} = -g(\mathbf{x})g(\mathbf{x})^{\dagger} \nabla f(\mathbf{x}), \tag{8.38}$$

defines a descent flow $\mathbf{x}(t)$ for the objective function $f(\mathbf{x})$.

For our application, we wish the structure-preserving isospectral flow $(M(t), C(t), K(t))$ to be driven to diagonal matrices. However, unlike the isospectral flow by orthogonal transformations, our flow $(M(t), C(t), K(t))$ preserves only the Lancaster structure but not the norm. Thus, we seek matrix parameters N_R, N_L and D to minimize the function

$$f(K, C, M) := \frac{1}{2} \{ \|\text{offdiag}(M)\|_F^2 + \|\text{offdiag}(C)\|_F^2 + \|\text{offdiag}(K)\|_F^2 \}$$
$$+ \delta h(\text{diag}(M), \text{diag}(C), \text{diag}(K)), \tag{8.39}$$

subject to the condition that (M, C, K) is governed by the differential system (8.35). The crux of choosing this particular objective function is to minimize

---

[1] The system (8.36) is 'closed-loop' in the sense that it is now self-contained: the reference to $\mathbf{u}$ is no longer needed directly.

the off-diagonal entries of (M, C, K) while using the function h to regulate the behaviour of the diagonal entries by a factor of δ. Note that we may rewrite the dynamical system (8.35) in the same control scheme,

$$\frac{\mathrm{d}}{\mathrm{d}t} \begin{bmatrix} \mathbf{vec}(M) \\ \mathbf{vec}(C) \\ \mathbf{vec}(K) \end{bmatrix} = \begin{bmatrix} \frac{1}{2}(K \otimes C - C \otimes K) & K \otimes I & I \otimes K \\ K \otimes M - M \otimes K & C \otimes I & I \otimes C \\ \frac{1}{2}(C \otimes M - M \otimes C) & M \otimes I & I \otimes M \end{bmatrix} \begin{bmatrix} \mathbf{vec}(D) \\ \mathbf{vec}(N_L^\top) \\ \mathbf{vec}(N_R) \end{bmatrix},$$

as that of (8.36). The above-mentioned control strategy fits perfectly. In this way, we have developed a 'controlled' gradient flow which not only preserves both the Lancaster structure and the isospectrality, but also moves in the direction of total decoupling of a quadratic pencil. More detailed discussion can be found in Chu and Del Buono (2008b).

8.3. Hamiltonian structure

A matrix $\mathcal{H} \in \mathbb{R}^{2n \times 2n}$ is said to be *Hamiltonian* if it satisfies the relationship $(\mathcal{H}J)^\top = \mathcal{H}J$, where

$$J := \begin{bmatrix} 0 & I_n \\ -I_n & 0 \end{bmatrix}.$$

It is easy to see that a Hamiltonian matrix must have the structure

$$\mathcal{H} = \begin{bmatrix} M & P \\ Q & -M^\top \end{bmatrix}, \quad P \text{ and } Q \text{ are symmetric.} \tag{8.40}$$

Likewise, a *skew-Hamiltonian* matrix $\mathcal{W}$ satisfies $(\mathcal{W}J)^\top = -\mathcal{W}J$, and has the structure

$$\mathcal{W} = \begin{bmatrix} M & F \\ G & M^\top \end{bmatrix}, \quad F \text{ and } G \text{ are skew-symmetric.} \tag{8.41}$$

Without causing ambiguity, we shall refer to a form of either (8.40) or (8.41) collectively as a *Hamiltonian structure*. We shall call up the more specific reference to a Hamiltonian matrix or a skew-Hamiltonian matrix only when a clear distinction is necessary. The notation $\mathcal{H}$ and $\mathcal{W}$, specifically reserved for the Hamiltonian matrix and the skew-Hamiltonian matrix, respectively, should offer a clue as to which structure we are referring to in the context.

Matrices with Hamiltonian structure arise from a variety of applications, including systems and controls, algebraic Riccati equations, and quadratic eigenvalue problems (Benner, Kressner and Mehrmann 2005). Inherent in the Hamiltonian structure are many interesting properties. For example, the eigenvalues of $\mathcal{H}$ are symmetric with respect to the imaginary axis, and the eigenvalues of $\mathcal{W}$ have even algebraic and geometric multiplicities. These properties are often tied to the physical settings that lead to the underlying structure. For feasibility and interpretability, therefore, any transformation of $\mathcal{H}$ or $\mathcal{W}$ should respect the original Hamiltonian structure. Because conventional algorithms usually fail to meet this requirement, there has been considerable research effort to derive special methods for matrices with Hamiltonian structure. Some principal references will be given in the course of our presentation. Needless to say, special methods mean more delicate manipulations. The description of these methods are usually quite involved.

In this section, we are mainly interested in deriving continuous dynamical systems that mimic existing iterative schemes. In contrast to the iterative methods, most of our Hamiltonian structure-preserving dynamical systems can be characterized as a single line equation. Nonetheless, despite the fact that our extensive numerical experiments have given convincing evidence for the resulting dynamical behaviour, a major drawback in our current work is the lack of a complete asymptotic analysis of these differential systems. We have to leave these gaps as conjectures in this presentation.

To maintain the Hamiltonian structure, it is typical in practice that a similarity transformation of $\mathcal{H}$ or $\mathcal{W}$ should involve only symplectic matrices $S \in \mathbb{R}^{2n \times 2n}$. A symplectic matrix S must satisfy the condition

$$S^\top J S = J, \tag{8.42}$$

which naturally implies the symmetry $SJS^\top = J$ as well. Recall that we mentioned earlier in Table 7.1 that symplectic matrices form a group $\mathcal{S}p(2n)$. For numerical stability, it is often further required that the transformation matrix S be orthogonal symplectic.

The following three facts, leading to the particular structure called the *real Schur–Hamiltonian form* in the first two cases and the *URV form* in the third case, play fundamental roles in the computation of eigenvalues for matrices with Hamiltonian structure.

Theorem 8.4. Given $\mathcal{H}, \mathcal{W} \in \mathbb{R}^{2n \times 2n}$ which are Hamiltonian and skew-Hamiltonian matrices, respectively, then we have the following.

(1) (Paige and Van Loan 1981) If $\mathcal{H}$ has no purely imaginary eigenvalues, then there exists an orthogonal symplectic matrix $U \in \mathbb{R}^{2n \times 2n}$ such that $\widetilde{\mathcal{H}} = U^\top \mathcal{H} U$ is Hamiltonian and is of the form

$$\widetilde{\mathcal{H}} = \begin{bmatrix} R & P \\ 0 & -R^\top \end{bmatrix}, \tag{8.43}$$

where P is symmetric and R is upper quasi-triangular.

(2) (Van Loan 1984) There exists an orthogonal symplectic matrix $U \in \mathbb{R}^{2n \times 2n}$ such that $\widetilde{\mathcal{W}} = U^\top \mathcal{W} U$ is skew-Hamiltonian, and is of the form

$$\widetilde{\mathcal{W}} = \begin{bmatrix} R & F \\ 0 & R^\top \end{bmatrix}, \tag{8.44}$$

where F is skew-symmetric and R is upper quasi-triangular.

(3) (Benner *et al.* 2005) There exist orthogonal symplectic matrices $U, V \in \mathbb{R}^{2n \times 2n}$ such that $\widehat{\mathcal{H}} = U^\top \mathcal{H} V$ is of the form

$$\widehat{\mathcal{H}} = \begin{bmatrix} T & N \\ 0 & R^\top \end{bmatrix}, \tag{8.45}$$

where N has no particular structure, T is upper triangular and R is upper quasi-triangular.

Evidently, being able to reduce a matrix of Hamiltonian structure to its Schur–Hamiltonian form is sufficient for retrieving eigenvalue information. Most existing numerical methods for eigenvalue problems with Hamiltonian structure consist of two steps: first, endeavour to obtain the reduced form and, secondly, employ some classical iterative schemes to solve the reduced eigenproblem.

Currently, stable procedures for computing eigenvalues of skew-Hamiltonian matrices are well developed (Benner *et al.* 2005, Van Loan 1984). For Hamiltonian matrices, the task is much harder. The prevailing idea is to square a Hamiltonian $\mathcal{H}$ due to the fact that $\mathcal{H}^2$ is skew-Hamiltonian. Indeed, by (8.45), we see that $\mathcal{H}^2$ can be factorized as

$$U^\top \mathcal{H}^2 U = \begin{bmatrix} -TR & TN^\top - NT^\top \\ 0 & -R^\top T^\top \end{bmatrix}, \tag{8.46}$$

showing that the eigenvalues of $\mathcal{H}$ are the square roots of the eigenvalues from the matrix $-TR$. The $2n \times 2n$ eigenvalue problem is therefore effectively halved. A QZ-type algorithm can be applied to find the eigenvalues of the product TR without explicitly forming the product. Implementation details can be found in the paper by Benner and Kressner (2006). A similar idea but with improved invariant subspace computation is explored in Chu, Liu and Mehrmann (2007). We shall present in the following an interesting contrast that a continuous approach is easier to formulate for the Hamiltonian eigenproblem than for the skew-Hamiltonian eigenproblem.

In a spirit similar to that of the QR, the QZ or the SVD algorithms, we are interested in deriving dynamical systems that can realize the Schur–Hamiltonian form or its like. Towards that end, we need to understand how a smooth curve $S(t)$ moves on the manifold of symplectic group $Sp(2n)$. It suffices to know that the tangent space $\mathfrak{g} = T_{I_{2n}} Sp(2n)$ for $Sp(2n)$ at the identity is simply the collection of Hamiltonian matrices. The tangent vectors of $S(t)$ must be given by

$$\frac{\mathrm{d}S}{\mathrm{d}t} = S\mathfrak{K}, \quad (\text{or } \mathfrak{K}S), \tag{8.47}$$

where $\mathfrak{K}$ is Hamiltonian. If the symplectic $S(t)$ is also orthogonal, then the

Hamiltonian matrix $\mathfrak{K}$ must be of the special form

$$\mathfrak{K} = \begin{bmatrix} M & -Q \\ Q & M \end{bmatrix}, \tag{8.48}$$

where M is skew-symmetric and Q is symmetric.

We demonstrate a simple application of (8.48) to the Hamiltonian eigenproblem. Given a matrix $\mathcal{H}_0 \in \mathbb{R}^{2n \times 2n}$, consider a special kind of Lax dynamical system described in (5.4),

$$\frac{\mathrm{d}X}{\mathrm{d}t} = [X, \mathcal{P}_0(X)], \quad X(0) = \mathcal{H}_0, \tag{8.49}$$

where the operator $\mathcal{P}_0$ acting on X is defined to be the skew-symmetric matrix,

$$\mathcal{P}_0(X) := \begin{bmatrix} 0 & -X_{21}^\top \\ X_{21} & 0 \end{bmatrix}, \tag{8.50}$$

if X is partitioned into four blocks of size $n \times n$,

$$X = \begin{bmatrix} X_{11} & X_{12} \\ X_{21} & X_{22} \end{bmatrix}.$$

Following (5.6), define the parameter dynamical system

$$\frac{\mathrm{d}g}{\mathrm{d}t} = g\mathcal{P}_0(X), \quad g(0) = I_{2n}. \tag{8.51}$$

Note that if $\mathcal{P}_0(X)$ is Hamiltonian, then $g(t)$ is automatically orthogonal symplectic. In particular, if $\mathcal{H}_0$ is Hamiltonian to begin with, then we know by Theorem 5.1 that $X(t) = g^\top(t)\mathcal{H}_0 g(t)$ remains Hamiltonian for all t. Under some mild conditions, it can be proved that $X(t)$ converges to an upper block triangular form, that is, $X_{21}(t) \longrightarrow 0$ as $t \longrightarrow \infty$ (Chu and Norris 1988). Though the limit point of the isospectral flow (8.49) is not exactly of the Schur–Hamiltonian form, it suffices to halve the Hamiltonian eigenproblem. The flow approach is remarkably simple, given that in the literature the Hamiltonian eigenproblem is known to be notoriously hard to solve by iterative methods.

Unfortunately, the corresponding $\mathcal{P}_0(X)$ is not Hamiltonian if X is skew-Hamiltonian. The simple dynamical system (8.49) therefore cannot preserve the skew-Hamiltonian structure. Since the skew-Hamiltonian eigenproblem is supposed to be relatively easier to handle than the Hamiltonian eigenproblem by iterative methods, it becomes interesting to ask whether the Schur–Hamiltonian form of a skew-Hamiltonian matrix $\mathcal{W}_0$ can ever be realized continuously. We offer a partial answer that looks pleasingly neat in theory, but is probably of little use in practice.

It is known that every real skew-Hamiltonian matrix has a real Hamiltonian square root (Faßbender, Mackey, Mackey and Xu 1999). Thus, given a skew-Hamiltonian matrix $\mathcal{W}_0$, if we define $\mathcal{H}_0$ to be its real Hamiltonian square root and define $X(t)$ according to (8.49), then the corresponding $\mathcal{W}(t) = X^2(t)$ is skew-Hamiltonian and will converge to an upper block triangular form. In particular, the very same parameter $g(t)$ defined in (8.51) (in terms of the Hamiltonian square root $X(t)$) serves as the continuous coordinate transformation for $\mathcal{W}(t) = g^\top(t)\mathcal{W}_0 g(t)$ and leads to convergence. It is not difficult to verify that symbolically we can write the motion of $\mathcal{W}(t)$ via the dynamical system

$$\frac{d\mathcal{W}}{dt} = [\mathcal{W}, \mathcal{P}_0(\mathcal{W}^{1/2})], \quad \mathcal{W}(0) = \mathcal{W}_0, \tag{8.52}$$

where $\mathcal{W}^{1/2}$ represents the real Hamiltonian square root of $\mathcal{W}$. We hasten to point out that caution must be taken in the above expression because a skew-Hamiltonian matrix $\mathcal{W}$ has infinitely many Hamiltonian square roots (Faßbender *et al.* 1999).

In the Lax dynamical system (5.14), the operation $\Pi_0(X)$ provides the magic of convergence to the real Schur form for a general square matrix X_0. We seek a similar dynamical system that converges to the real Schur–Hamiltonian for a Hamiltonian matrix $\mathcal{H}_0$. The operator $\mathcal{P}_1$ applied to a Hamiltonian matrix X via the definition

$$\mathcal{P}_1(X) := \begin{bmatrix} \Pi_0(X_{11}) & -X_{21} \\ X_{21} & \Pi_0(X_{11}) \end{bmatrix} \tag{8.53}$$

appears to be a compromise between the overall $\Pi_0(X)$ required by the QR flow for reaching sensible convergence and the form (8.48) required by the orthogonal symplecticity for keeping the Hamiltonian structure. The two operators Π_0 and $\mathcal{P}_1$ for a Hamiltonian matrix X differ only in the $(2,2)$-block. We propose the dynamical system

$$\frac{d\mathcal{H}}{dt} = [\mathcal{H}, \mathcal{P}_1(\mathcal{H})], \quad \mathcal{H}(0) = \mathcal{H}_0, \tag{8.54}$$

for finding the real Schur–Hamiltonian form of a Hamiltonian matrix $\mathcal{H}_0$. The following conjecture characterizes the convergence behaviour we have observed numerically, but we cannot offer a theoretical proof for the present.

Conjecture 8.5. Suppose $\mathcal{H}_0$ is Hamiltonian with no purely imaginary eigenvalues. Then the solution flow $\mathcal{H}(t)$ of (8.54) remains Hamiltonian and converges to the real Schur–Hamiltonian form as is specified in (8.43).

If the square root is interpreted in the same way as in (8.52), then a similar conjecture can be made for the system

$$\frac{d\mathcal{W}}{dt} = [\mathcal{W}, \mathcal{P}_1(\mathcal{W}^{1/2})], \quad \mathcal{W}(0) = \mathcal{W}_0. \tag{8.55}$$

The solution flow $\mathcal{W}(t)$ preserves the skew-Hamiltonian structure of an initial matrix $\mathcal{W}_0$ and converges to the real Schur skew-Hamiltonian form as is characterized in (8.44).

Regarding the URV decomposition, it is necessary that a flow $X(t) = U^\top(t)X_0V(t)$ satisfies a differential equation of the form

$$\frac{dX}{dt} = XR - LX, \quad X(0) = X_0, \tag{8.56}$$

where the coordinate transformations are governed by

$$\frac{dU}{dt} = -UL^\top, \tag{8.57}$$

$$\frac{dV}{dt} = VR, \tag{8.58}$$

with L and R to be determined. The setting thus far is very similar to that of the SVD flow. Let the operator $\mathcal{P}_3$ denote a generalization of $\mathcal{P}_0$ in that the partition of X is not necessarily at the midpoint of its diagonal. In particular, the off-diagonal block X_{21} can be of size $(2n-k) \times k$ with $k \leq n$. Consider the dynamical system

$$\frac{dX}{dt} = X\mathcal{P}_3(X^\top X) - \mathcal{P}_3(XX^\top)X, \quad X(0) = X_0, \tag{8.59}$$

for a general $2n \times 2n$ matrix X_0, Note that (8.59) is analogous to the SVD flow (6.2) except that $\mathcal{P}_3$ is used in the place of Π_0. Clearly, $X(t)$ maintains the same singular values as X_0. Numerical experiments support the following conjecture, which seems new and interesting.

Conjecture 8.6. Given a general $2n \times 2n$ matrix X_0 with distinct singular values and an integer $k \leq n$, the solution flow $X(t)$ of (8.59) converges to a block diagonal matrix $\mathrm{diag}\{\widehat{X}_{11}, \widehat{X}_{22}\}$ of size $k \times k$ and $(2n-k) \times (2n-k)$, respectively. Furthermore, the singular values of $\widehat{X}_{11}$ are the first k largest singular values of X_0.

The coordinate transformations involved in Conjecture 8.6 are orthogonal similarity at most. To really achieve the URV decomposition specified in Theorem 8.4 part (3) for a Hamiltonian matrix $\mathcal{H}_0$, we have to employ orthogonal symplectic transformations. The clue comes at recognizing from (8.46) that the U transformation that does the URV decomposition for $\mathcal{H}_0$ should be the same U transformation that does the real Schur–Hamiltonian form for $\mathcal{H}_0$. That is, by Conjecture 8.5, $L = \mathcal{P}_1(U^\top \mathcal{H}_0 U)$. Similarly, the V matrix in the URV decomposition should be the same V matrix that transforms $\mathcal{H}_0^\top$ to *lower* quasi-triangular Schur–Hamiltonian form. That is, by defining the operator

$$\mathcal{P}_2(X) := \begin{bmatrix} -\Pi_0(X_{11}^\top) & X_{12} \\ -X_{12} & -\Pi_0(X_{11}^\top) \end{bmatrix} \tag{8.60}$$

for a given Hamiltonian matrix X, we take $R = \mathcal{P}_2(V^\top \mathcal{H}_0^\top V)$. We are interested in a *URV* flow $X(t) = U^\top(t)\mathcal{H}_0 V(t)$. From the relations

$$U^\top \mathcal{H}_0^2 U = XJX^\top J,$$
$$V^\top \mathcal{H}_0^2 V = X^\top JXJ,$$

we can express the *URV* flow symbolically through the autonomous dynamical system

$$\frac{\mathrm{d}X}{\mathrm{d}t} = X\mathcal{P}_2((X^\top JXJ)^{1/2}) - \mathcal{P}_1((XJX^\top J)^{1/2})X, \quad X(0) = \mathcal{H}_0, \quad (8.61)$$

where again $\mathcal{W}^{1/2}$ represents a proper real Hamiltonian square root of the skew-Hamiltonian matrix $\mathcal{W}$.

Hamiltonian structure-preserving differential systems like (8.49), (8.54), or even (8.61) might not be practically useful right away, but they neatly represent complicated dynamics that otherwise will be quite tedious, if not formidable, to describe by iterative procedures. Maybe, and only maybe, these flows could be suitably discretized and lead to effective numerical algorithms. One precedent is the realization of the *vdLV* algorithm for the Lotka–Volterra equation which, when first proposed two decades ago, was regarded as 'impractical' as well. These flows might be worth further investigation.

8.4. Hamiltonian pencils

We have already seen linear pencils with the Lancaster structure resulting from a special linearization of a quadratic pencil. There are also linear pencils with the Hamiltonian structure. To start off, two different definitions in the literature must be carefully differentiated from each other. First, a linear pencil $B\lambda - A$ is said to be *Hamiltonian* if and only if

$$BJA^\top = -AJB^\top. \quad (8.62)$$

This definition is equivalent to saying that the product $B^{-1}A$ is Hamiltonian, provided B^{-1} exists (Lin, Mehrmann and Xu 1999). If λ is an eigenvalue of a Hamiltonian pencil, then so are $-\lambda, \bar{\lambda}, -\bar{\lambda}$. Secondly, a linear pencil $B\lambda - A$ is said to be *skew-Hamiltonian/Hamiltonian* (sHH) if and only if B is skew-Hamiltonian and A is Hamiltonian (Mehl 1999). Pencils with the sHH structure appear in gyroscopic systems, structural mechanics, linear response theory, quadratic optimal control problems and many other applications (Benner, Byers, Mehrmann and Xu 2002, Mehrmann and Watkins 2000). Although it is a natural generalization in mathematics, we have rarely seen Hamiltonian/Hamiltonian (HH) pencils in applications. One indicator that an HH pencil is probably too general to deserve any special attention is the fact that the HH structure does not generally carry any additional symmetric properties in its spectrum. We note, for example, that any self-adjoint quadratic pencil (8.15) can be linearized as the pencil

$$\begin{bmatrix} M_0 & 0 \\ -C_0 & -M_0 \end{bmatrix} \lambda - \begin{bmatrix} 0 & M_0 \\ K_0 & 0 \end{bmatrix}, \qquad (8.63)$$

which is equivalent to the Lancaster pair (8.14), is of the HH structure, and can literally have arbitrary eigenvalues.

In analogy to (8.5), the one-parameter isospectral flow

$$\mathscr{L}(t) = Q(t)(B_0\lambda - A_0)Z(t)$$

should satisfy a differential equation of the form

$$\frac{d\mathscr{L}}{dt} = \mathscr{L}R - L\mathscr{L}, \quad \mathscr{L}(0) = B_0\lambda - A_0, \qquad (8.64)$$

where the coordinate transformations are governed by

$$\frac{dQ}{dt} = -LQ, \qquad (8.65)$$

$$\frac{dZ}{dt} = ZR. \qquad (8.66)$$

with L and R to be determined. So far, this setting is similar to the QZ flow except that the definition of the two matrices L and R needs to be further specified. The conventional condition that L and R be skew-symmetric so that $Q(t)$ and $Z(t)$ are orthogonal is certainly assumed in all cases, but we are more interested in specifying conditions on L and R so as to maintain the Hamiltonian structure. Besides, we are further interested in using L and R to establish limiting behaviour of $\mathscr{L}(t)$ that might be of some practical usages. We outline some general ideas below.

We first consider the sHH pencils. Suppose that $\mathscr{L}(0) = \mathcal{W}_0\lambda - \mathcal{H}_0$ is of the sHH structure to begin with. Write

$$\mathscr{L}(t) = \mathcal{W}(t)\lambda - \mathcal{H}(t).$$

In order that $\mathscr{L}(t)$ maintains the sHH structure for all t, it is necessary that $\mathcal{W}R - L\mathcal{W}$ and $\mathcal{H}R - L\mathcal{H}$ remain skew-Hamiltonian and Hamiltonian, respectively. A straightforward algebraic manipulation shows that a sufficient condition for this to happen is that

$$L = JR^\top J. \qquad (8.67)$$

Consequently, $Q(t)$ and $Z(t)$ can be interchanged via the relationship

$$Z(t) = JQ^\top(t)J, \qquad (8.68)$$

$$Q(t) = JZ^\top(t)J. \qquad (8.69)$$

Only one coordinate transformation of either (8.65) or (8.66) is needed for the isospectral flow of an sHH pencil.

For any given $2n \times 2n$ matrix X, define a new operator $\mathcal{P}_4$ by

$$\mathcal{P}_4(X) := \begin{bmatrix} \Pi_0(X_{11}) & -X_{21}^\top \\ X_{21} & -\Pi_0(X_{22}^\top) \end{bmatrix}. \tag{8.70}$$

Observe that $\mathcal{P}_4(X)$ is almost identical to $\Pi_0(X)$ except for a 'twist' at the $(2,2)$ block. Take the definitions

$$R := \mathcal{P}_4(\mathcal{W}^{-1}\mathcal{H}), \tag{8.71}$$

$$L := \mathcal{P}_4(\mathcal{H}\mathcal{W}^{-1}). \tag{8.72}$$

It is easy to see that the relationship

$$\mathcal{H}\mathcal{W}^{-1} = J(\mathcal{W}^{-1}\mathcal{H})^\top J \tag{8.73}$$

holds for every sHH pencil. A direct substitution then shows that the sufficient condition (8.67) is satisfied. In this way, we find that the dynamical system

$$\frac{d\mathscr{L}}{dt} = \mathscr{L}\mathcal{P}_4(\mathcal{W}^{-1}\mathcal{H}) - \mathcal{P}_4(\mathcal{H}\mathcal{W}^{-1})\mathscr{L}, \quad \mathscr{L}(0) = B_0\lambda - A_0, \tag{8.74}$$

defines an sHH flow which can be expressed as

$$\mathscr{L}(t) = J Z^\top(t) J (\mathcal{W}_0\lambda - \mathcal{H}_0) Z(t). \tag{8.75}$$

Had $\mathcal{P}_4$ been taken as Π_0, we would have precisely the standard QZ flow described earlier and the convergence behaviour of the QZ flow is well understood. With the little flip at the $(2,2)$ block in $\mathcal{P}_4$, we maintain the sHH structure and we can almost expect that a similar convergence behaviour will occur. We conceive the following conjecture from our numerical observation. Its assertion is in agreement with the sHH Schur form characterized in Benner *et al.* (2002). If the convergence can be proved, then we have a very simple way to realize the canonical form.

Conjecture 8.7. Suppose $\mathscr{L}(0)$ is an sHH pencil to begin with. Then the flow (8.75) with R defined by (8.71) maintains the sHH structure and converges to the canonical form

$$\widetilde{\mathscr{L}} = \begin{bmatrix} \widetilde{\mathcal{W}}_{11} & \widetilde{\mathcal{W}}_{12} \\ 0 & \widetilde{\mathcal{W}}_{11}^\top \end{bmatrix} \lambda - \begin{bmatrix} \widetilde{\mathcal{H}}_{11} & \widetilde{\mathcal{H}}_{12} \\ 0 & -\widetilde{\mathcal{H}}_{11}^\top \end{bmatrix},$$

where $\widetilde{\mathcal{W}}_{11}$ and $\widetilde{\mathcal{H}}_{11}$ are upper quasi-triangular, $\widetilde{\mathcal{W}}_{12}$ is skew-symmetric, and $\widetilde{\mathcal{H}}_{12}$ is symmetric, respectively.

We next consider the Hamiltonian pencils. It is easy to verify that $B\lambda - A$ is Hamiltonian if and only if $Q(B\lambda - A)Z$ is Hamiltonian for arbitrary nonsingular Q and symplectic Z. In order to maintain the Hamiltonian pencil, the R matrix in (8.66) must be Hamiltonian, but there is no restriction on L in (8.65). The only concern is somehow to ensure nice convergence.

For Hamiltonian pencils, both $B^{-1}A$ and $A^{-1}B$ are Hamiltonian matrices, but AB^{-1} and BA^{-1} are not. Based on our past experience, we thus propose to take $R = \mathcal{P}_1(B^{-1}A)$, which is a compromise of $\Pi_0(B^{-1}A)$ with the restriction (8.48) and makes Z orthogonal symplectic. There are no restrictions on L, so we use the QZ flow as a guide. In all, we propose the differential equation

$$\frac{\mathrm{d}\mathcal{L}}{\mathrm{d}t} = \mathcal{L}\mathcal{P}_1(B^{-1}A) - \Pi_0(AB^{-1})\mathcal{L}, \qquad (8.76)$$

which differs from the QZ flow defined in (8.9) at the $\mathcal{P}_1$ operator but keeps the pencil flow $\mathcal{L}(t)$ Hamiltonian for all t.

The limiting behaviour of (8.76) is somewhat more complicated to describe. For convenience, let Ξ denote the unit perdiagonal matrix whose entries are all zero but 1's along the north-east to south-west diagonal. We introduce the notion that a matrix X is *upper-left quasi-triangular* if the product $X\Xi$ is upper(-right) quasi-triangular in the usual sense. Again, the following conjecture is observed in our numerical experiments, but we have no proof for the moment.

Conjecture 8.8. Suppose the pencil $B_0\lambda - A_0$ is Hamiltonian. Then the flow defined by (8.76) remains a Hamiltonian pencil. Furthermore, we have the following.

(1) Suppose that $B_0\lambda - A_0$ has no purely imaginary eigenvalues. Then $\mathcal{L}(t)$ converges to the canonical form

$$\widehat{\mathcal{L}} = \begin{bmatrix} \widehat{B}_{11} & \widehat{B}_{12} \\ 0 & \widehat{B}_{22} \end{bmatrix} \lambda - \begin{bmatrix} \widehat{A}_{11} & \widehat{A}_{12} \\ 0 & \widehat{A}_{22} \end{bmatrix},$$

where $\widehat{A}_{11}$ and $\widehat{B}_{11}$ are upper quasi-triangular matrices with 1×1 or 2×2 blocks at the same corresponding locations, and $\widehat{A}_{22}$ and $\widehat{B}_{22}$ are upper-left quasi-triangular matrices with 1×1 or 2×2 blocks at the same corresponding locations.

(2) If $B_0\lambda - A_0$ has one pair of purely imaginary eigenvalues. Then $\mathcal{L}(t)$ converges to the same canonical form as above, with the exception of a non-zero entry at the $(n+1, n)$ position which is periodic in t.

Finally, we mention the following theorem concerning a general $2n \times 2n$ pencil (Benner, Mehrmann and Xu 1998).

Theorem 8.9. Given an arbitrary real $2n \times 2n$ pencil $B_0\lambda - A_0$, there exist an orthogonal matrix Q_3 and orthogonal symplectic matrices Q_1 and Q_2 such that

$$Q_3^\top B_0 Q_1 = \begin{bmatrix} \widetilde{B}_{11} & \widetilde{B}_{12} \\ 0 & \widetilde{B}_{22}^\top \end{bmatrix}, \quad Q_3^\top A_0 Q_2 = \begin{bmatrix} \widetilde{A}_{11} & \widetilde{A}_{12} \\ 0 & \widetilde{A}_{22}^\top \end{bmatrix}, \qquad (8.77)$$

where $\widetilde{B}_{ij}, \widetilde{A}_{ij} \in \mathbb{R}^{n \times n}$,, $\widetilde{B}_{11}, \widetilde{A}_{11}, \widetilde{B}_{22}$ are upper triangular and A_{22} is upper quasi-triangular.

Note that what is involved in Theorem 8.9 is a non-equivalence transformation, so generally it is not useful for eigenvalue preservation. However, in the case when $B_0 \lambda - A_0$ is Hamiltonian, then $Q_1^\top (B_0^{-1} A_0) Q_2$ is precisely the URV form for the Hamiltonian matrix $B_0^{-1} A_0$. The reference to Q_3 is completely annihilated. The above result therefore has been exploited as an effective way of eigenvalue computation for Hamiltonian pencils (Benner *et al.* 1998).

We are curious as to whether the canonical form described in (8.77) can be realized continuously. Defining $\mathcal{H}(t) = B^{-1}(t) A(t)$, we have already learned that the URV flow $\mathcal{H}(t)$ is governed by (8.61). In particular, we know that $Q_1(t)$ and $Q_2(t)$ should be governed by

$$\frac{dQ_1}{dt} = Q_1 \mathcal{P}_1((\mathcal{H} J \mathcal{H}^\top J)^{1/2}), \tag{8.78}$$

$$\frac{dQ_2}{dt} = Q_2 \mathcal{P}_2((\mathcal{H}^\top J \mathcal{H} J)^{1/2}), \tag{8.79}$$

respectively. It is not immediately clear how the dynamics for $Q_3(t)$ should be defined.

Consider the product

$$Z(t) := A(t) B^{-1}(t) = Q_3^\top \underbrace{A_0 Q_2 Q_1^\top B_0^{-1}}_{\mathscr{Z}} Q_3 = \begin{bmatrix} Z_{11} & Z_{12} \\ Z_{21} & Z_{22} \end{bmatrix}.$$

Note that $\mathscr{Z}(t)$ is not necessarily isospectral in t. However, the canonical form (8.77) motivates us to hope that, as t tends to infinity, the matrix $\mathscr{Z}(t)$ would ultimately exhibit the property that $Z_{11} = \widetilde{A}_{11} \widetilde{B}_{11}^{-1}$ is upper triangular, $Z_{21} = 0$ and $Z_{22} = \widetilde{A}_{22}^\top \widetilde{B}_{22}^{-\top}$ should be lower quasi-triangular. We suspect, therefore, that $Q_3(t)$ should be governed by the dynamical system

$$\frac{dQ_3}{dt} = Q_3 \mathcal{P}_4(Z), \tag{8.80}$$

where the operator $\mathcal{P}_4$ was defined earlier in (8.70). Assembling all together, we conjecture that the canonical form (8.77) can be realized via the dynamical system

$$\frac{dA}{dt} = A \mathcal{P}_2((A^\top B^{-\top} J B^{-1} A J)^{1/2}) - \mathcal{P}_4(AB^{-1})A, \quad A(0) = A_0, \tag{8.81}$$

$$\frac{dB}{dt} = B \mathcal{P}_1((B^{-1} A J A^\top B^{-\top} J)^{1/2}) - \mathcal{P}_4(AB^{-1})A, \quad B(0) = B_0. \tag{8.82}$$

If this conjecture is true, it would nicely express the complicated iterative algorithm described in Benner *et al.* (1998) in a concise form.

Table 8.2. Hierarchy of structure-preserving dynamical systems.

Initial structure	Dynamical system	Limiting behaviour	Operator
X_0 = staircase	$\dot{X} = [X, \Pi_0(X)]$	Ashlock *et al.* (1997)	$\Pi_0(X) = X^- - (X^-)^\top$
$B_0\lambda - A_0$ = staircase	$\dot{\mathscr{L}} = \mathscr{L}\Pi_0(Y^{-\top}X) - \Pi_0(XY^{-1})\mathscr{L}$	Conjecture 8.2	
B_0 = staircase	$\dot{B} = B\Pi_0(B^\top B) - \Pi_0(BB^\top)B$	Conjecture 8.3	D, N_R, N_L = controls
$B_0\lambda - A_0$ = Lancaster	$\dot{K} = \frac{1}{2}(CDK - KDC) + N_L^\top K + KN_R$ $\dot{C} = (MDK - KDM) + N_L^\top C + CN_R$ $\dot{M} = \frac{1}{2}(MDC - CDM) + N_L^\top M + MN_R$		
$\mathcal{H}_0$ = Hamiltonian	$\dot{\mathcal{H}} = [\mathcal{H}, \mathcal{P}_0(\mathcal{H})]$	Chu and Norris (1988)	$\mathcal{P}_0(X) = \begin{bmatrix} 0 & -X_{21}^\top \\ X_{21} & 0 \end{bmatrix}$
W_0 = skew-Hamiltonian	$\dot{W} = [W, \mathcal{P}_0(W^{1/2})]$		
$\mathcal{H}_0$ = Hamiltonian	$\dot{\mathcal{H}} = [\mathcal{H}, \mathcal{P}_1(\mathcal{H})]$	Conjecture 8.5	$\mathcal{P}_1(X) = \begin{bmatrix} \Pi_0(X_{11}^\top) & -X_{21} \\ X_{21} & \Pi_0(X_{11}) \end{bmatrix}$
W_0 = skew-Hamiltonian	$\dot{W} = [W, \mathcal{P}_1(W^{1/2})]$		
X_0 = general	$\dot{X} = X\mathcal{P}_3(X^\top X) - \mathcal{P}_3(XX^\top)X$	Conjecture 8.6	$\mathcal{P}_3$ = generalized $\mathcal{P}_0$
$\mathcal{H}_0$ = Hamiltonian	$\dot{X} = X\mathcal{P}_2((X^\top JXJ)^{1/2}) - \mathcal{P}_1((XJX^\top J)^{1/2})X$	URV flow	$\mathcal{P}_2(X) := \begin{bmatrix} -\Pi_0(X_{11}^\top) & X_{12} \\ -X_{12} & -\Pi_0(X_{11}^\top) \end{bmatrix}$
$W_0\lambda - \mathcal{H}_0$ = sHH	$\dot{\mathscr{L}} = \mathscr{L}\mathcal{P}_4(W^{-1}\mathcal{H}) - \mathcal{P}_4(HW^{-1})\mathscr{L}$	Conjecture 8.6	$\mathcal{P}_4(X) := \begin{bmatrix} \Pi_0(X_{11}) & -X_{21}^\top \\ X_{21} & -\Pi_0(X_{22}^\top) \end{bmatrix}$
$B_0\lambda - A_0$ = Hamiltonian	$\dot{\mathscr{L}} = \mathscr{L}\mathcal{P}_1(B^{-1}A) - \Pi_0(AB^{-1})\mathscr{L}$	Conjecture 8.8	
$B_0\lambda - A_0$ = general	$\dot{A} = A\mathcal{P}_2((A^\top B^{-\top}JB^{-1}AJ)^{1/2}) - \mathcal{P}_4(AB^{-1})A$ $\dot{B} = B\mathcal{P}_1((B^{-1}AJA^\top B^{-\top}J)^{1/2}) - \mathcal{P}_4(AB^{-1})A$	not tested	

It might be helpful to summarize the different dynamical systems discussed thus far in Table 8.2. Recall that the principal consideration in formulating these flows is to preserve the structure of the initial data. The special operators in the right-hand column of the table are designed for that purpose, all of which are variations of the operator Π_0. Only a few of these systems have their asymptotic behaviour understood in the literature. Those identified by a conjecture in the table have been extensively tested by numerical integrators, but no theory of asymptotic analysis is available for the present. If any of the conjectures is true, then the corresponding dynamical system often encapsulates a fairly complicated iterative process into a nice and simple mathematical expression. Be aware that we are not implying that the flows sampled at integer times will produce the same iterates as those generated by existing discrete methods; this coincidence might be too difficult to achieve for matrices with Hamiltonian structure. The only cases we know for sure about this coincidence are the QR, QZ and SVD flows. Nor are we inferring that these structure-preserving dynamical systems can easily be discretized with the resulting iterative schemes still preserving the original structure. We must stress that the link diagram in Figure 1.1 that we frequently refer to in this paper now has an added dimension of constraint, namely, structure preservation. Thus there is much room left for further investigation of these relationships.

8.5. Group structure

Needless to say, there are far too many other applications where it is desirable that a specific structure is maintained throughout a specified dynamical system. Like the canonical forms, the notion of 'structure' should be interpreted quite liberally. We have discussed only a few cases involving the spectrum, the singular values, the staircase, or the Hamiltonian structure from the linear algebra perspective. Obviously, it is never an overstatement that preserving volume, momentum, energy, symplecticity, or other kinds of physical quantities, is an extremely important task with significant consequences. The subject is simply so wide in scope that the author must humbly admit it is beyond his comprehension. We conclude this chapter by pointing out one more structure that has recently attracted tremendous interest.

The once-abstract notion of Lie theory is now a ubiquitous framework in many disciplines of sciences and engineering applications. In Section 7 we have also demonstrated how group actions often serve as the fundamental coordinate transformations leading to canonical forms. It should not come as a surprise, but rather a necessity, that many of the dynamical systems and numerical algorithms originally developed over Euclidean space need to be redeveloped over manifolds. By studying the underlying geometry, for example, critical algorithms such as the Newton and the conjugate gradient methods can be generalized to the Grassmann and the Stiefel manifolds in a natural way (Edelman *et al.* 1999).

We illustrate in this section how the Newton dynamics can take place on a Lie group (Owren and Welfert 2000). This notion typifies what we mean by a dynamical system that respects the group structure.

Let G be a Lie group and $\mathfrak{g}$ its corresponding Lie algebra. Keep in mind that elements in G can be abstract functionals or operators. Suppose we want to find 'zero(s)' of a given map,

$$f : G \to \mathfrak{g},$$

where the iterates are to stay on the manifold G. Given a current iterate $y_n \in G$, the Newton scheme can interpreted as solving the equation

$$\mathrm{d}f_{y_n}(u_n) + f(y_n) = 0, \tag{8.83}$$

for a tangent vector $u_n \in \mathfrak{g}$ and then updating to the next iterate via the exponential map

$$y_{n+1} = y_n \exp(u_n). \tag{8.84}$$

In the above, the differential

$$\mathrm{d}f_y : \mathcal{T}_y G \to \mathfrak{g}$$

can be interpreted as

$$\mathrm{d}f_y(u) = (\mathrm{d}/\mathrm{d}t)_{t=0} f(y \exp(tu)). \tag{8.85}$$

Alternatively, since all local charts of a Lie group can be obtained by translation, we can restrict ourselves to the local charts. In particular, it suffices to consider the 'local' representation of f at y_n,

$$\tilde{f} := f \circ L_{y_n} \circ \exp, \tag{8.86}$$

where $L_z(y) = zy$ with a fixed $z \in G$. This becomes a classical algebraic equation in Euclidean space. The Newton iteration involves the steps of solving the equation

$$\mathrm{d}\tilde{f}_{v_n}(u_n) + \tilde{f}(v_n) = 0, \tag{8.87}$$

for u_n, where v_n is the local parametrization, *i.e.*, the logarithm of y_n, updating in the linear space by $v_{n+1} = v_n + u_n$, and finally advancing to the new iterate on the manifold G by defining

$$y_{n+1} = y_n \exp(v_{n+1}). \tag{8.88}$$

Note that both formulations reduce to the standard method in the Euclidean case. It can be shown that under classical assumptions the proposed methods converge quadratically (Owren and Welfert 2000).

We think this framework can be repeatedly applied to generalize other types of algorithms originally designed for Euclidean space to Lie groups. How far this generalization should go, and how practical such extensions might be, are yet to be seen.

Acknowledgement

The author is deeply indebted to Professor Yoshimasa Nakamura and his research team at the Kyoto University for enthusiastically and gracefully relaying their knowledge and experiences on the Lotka–Volterra equation to him. The amazing connection among orthogonal polynomials, moment representation, τ functions, Hankel determinantal solution, the Toda lattice and the effective $vdLV$ algorithm would not have been understood by the author without their kind help.

REFERENCES

P.-A. Absil and K. Kurdyka (2006), 'On the stable equilibrium points of gradient systems', *Systems Control Lett.* **55**, 573–577.

P.-A. Absil, R. Mahony and B. Andrews (2005), 'Convergence of the iterates of descent methods for analytic cost functions', *SIAM J. Optim.* **16**, 531–547 (electronic).

N. I. Akhiezer (1965), *The Classical Moment Problem and Some Related Questions in Analysis*, Hafner, New York. Translated by N. Kemmer.

E. Allgower and K. Georg (1980), 'Simplicial and continuation methods for approximating fixed points and solutions to systems of equations', *SIAM Rev.* **22**, 28–85.

E. L. Allgower and K. Georg (2003), *Introduction to Numerical Continuation Methods*, Vol. 45 of *Classics in Applied Mathematics*, SIAM, Philadelphia, PA.

A. C. Antoulas (2005), *Approximation of Large-Scale Dynamical Systems*, Vol. 6 of *Advances in Design and Control*, SIAM, Philadelphia, PA.

A. I. Aptekarev, A. Branquinho and F. Marcellán (1997), 'Toda-type differential equations for the recurrence coefficients of orthogonal polynomials and Freud transformation', *J. Comput. Appl. Math.* **78**, 139–160.

P. Arbenz and G. H. Golub (1995), 'Matrix shapes invariant under the symmetric QR algorithm', *Numer. Linear Algebra Appl.* **2**, 87–93.

V. I. Arnold (1988), *Geometrical Methods in the Theory of Ordinary Differential Equations*, Vol. 250 of *Grundlehren der Mathematischen Wissenschaften (Fundamental Principles of Mathematical Sciences)*, second edn, Springer, New York. Translated from the Russian by Joseph Szücs (József M. Szűcs).

D. A. Ashlock, K. R. Driessel and I. R. Hentzel (1997), On matrix structures invariant under Toda-like isospectral flows, in *Proc. Fifth Conference of the International Linear Algebra Society, Atlanta 1995*, Vol. 254, pp. 29–48.

A. Baker (2002), *Matrix Groups: An Introduction to Lie Group Theory*, Springer Undergraduate Mathematics Series, Springer, London.

R. Barrett, M. Berry, T. F. Chan, J. Demmel, J. Donato, J. Dongarra, V. Eijkhout, R. Pozo, C. Romine and H. Van der Vorst (1994), *Templates for the Solution of Linear Systems: Building Blocks for Iterative Methods*, SIAM, Philadelphia, PA.

T. Beelen and P. Van Dooren (1990), Computational aspects of the Jordan canonical form, in *Reliable Numerical Computation*, Oxford University Press, New York, pp. 57–72.

P. Benner and D. Kressner (2006), 'Algorithm 854: Fortran 77 subroutines for computing the eigenvalues of Hamiltonian matrices II', *ACM Trans. Math. Software* **32**, 352–373.

P. Benner, R. Byers, V. Mehrmann and H. Xu (2002), 'Numerical computation of deflating subspaces of skew-Hamiltonian/Hamiltonian pencils', *SIAM J. Matrix Anal. Appl.* **24**, 165–190 (electronic).

P. Benner, D. Kressner and V. Mehrmann (2005), Skew-Hamiltonian and Hamiltonian eigenvalue problems: theory, algorithms and applications, in *Proc. Conference on Applied Mathematics and Scientific Computing*, Springer, Dordrecht, pp. 3–39.

P. Benner, V. Mehrmann and H. Xu (1998), 'A numerically stable, structure preserving method for computing the eigenvalues of real Hamiltonian or symplectic pencils', *Numer. Math.* **78**, 329–358.

A. Bhaya and E. Kaszkurewicz (2006), *Control Perspectives on Numerical Algorithms and Matrix Problems*, Vol. 10 of *Advances in Design and Control*, SIAM, Philadelphia, PA.

S. Blanes, F. Casas, J. A. Oteo and J. Ros (1998), 'Magnus and Fer expansions for matrix differential equations: The convergence problem', *J. Phys. A* **31**, 259–268.

A. M. Bloch and A. Iserles (2006), 'On an isospectral Lie-Poisson system and its Lie algebra', *Found. Comput. Math.* **6**, 121–144.

A. M. Bloch, R. W. Brockett and T. S. Ratiu (1992), 'Completely integrable gradient flows', *Comm. Math. Phys.* **147**, 57–74.

J. H. Bramble (1993), *Multigrid Methods*, Vol. 294 of *Pitman Research Notes in Mathematics Series*, Longman, Harlow.

W. L. Briggs (1987), *A Multigrid Tutorial*, SIAM, Philadelphia, PA.

R. W. Brockett (1991), 'Dynamical systems that sort lists, diagonalize matrices, and solve linear programming problems', *Linear Algebra Appl.* **146**, 79–91.

R. W. Brockett (1993), Differential geometry and the design of gradient algorithms, in *Differential Geometry: Partial Differential Equations on Manifolds, Los Angeles 1990*, Vol. 54 of *Proc. Sympos. Pure Math.*, AMS, Providence, RI, pp. 69–92.

A. Bunse-Gerstner, R. Byers, V. Mehrmann and N. K. Nichols (1991), 'Numerical computation of an analytic singular value decomposition of a matrix valued function', *Numer. Math.* **60**, 1–39.

M. P. Calvo, A. Iserles and A. Zanna (1997), 'Numerical solution of isospectral flows', *Math. Comp.* **66**, 1461–1486.

E. Celledoni and A. Iserles (2000), 'Approximating the exponential from a Lie algebra to a Lie group', *Math. Comp.* **69**, 1457–1480.

R. Chill (2003), 'On the Lojasiewicz–Simon gradient inequality', *J. Funct. Anal.* **201**, 572–601.

D. Chu, X. Liu and V. Mehrmann (2007), 'A numerical method for computing the Hamiltonian Schur form', *Numer. Math.* **105**, 375–412.

M. Chu (1986a), 'A continuous approximation to the generalized Schur decomposition', *Linear Algebra Appl.* **78**, 119–132.

M. T. Chu (1986b), 'A differential equation approach to the singular value decomposition of bidiagonal matrices', *Linear Algebra Appl.* **80**, 71–79.

M. T. Chu (1988), 'On the continuous realization of iterative processes', *SIAM Rev.* **30**, 375–387.

M. T. Chu (1991), 'A continuous Jacobi-like approach to the simultaneous reduction of real matrices', *Linear Algebra Appl.* **147**, 75–96.

M. T. Chu (1995), 'Constructing a Hermitian matrix from its diagonal entries and eigenvalues', *SIAM J. Matrix Anal. Appl.* **16**, 207–217.

M. T. Chu and N. Del Buono (2008a), 'Total decoupling of general quadratic pencils I: Theory', *J. Sound Vibration* **309**, 96–111.

M. T. Chu and N. Del Buono (2008b), 'Total decoupling of general quadratic pencils II: Structure preserving isospectral flows', *J. Sound Vibration* **309**, 112–128.

M. T. Chu and K. R. Driessel (1990), 'The projected gradient method for least squares matrix approximations with spectral constraints', *SIAM J. Numer. Anal.* **27**, 1050–1060.

M. T. Chu and R. E. Funderlic (2002), 'The centroid decomposition: Relationships between discrete variational decompositions and SVDs', *SIAM J. Matrix Anal. Appl.* **23**, 1025–1044 (electronic).

M. T. Chu and G. H. Golub (2002), Structured inverse eigenvalue problems, in *Acta Numerica*, Vol. 11, Cambridge University Press, pp. 1–71.

M. T. Chu and G. H. Golub (2005), *Inverse Eigenvalue Problems: Theory, Algorithms, and Applications*, Numerical Mathematics and Scientific Computation, Oxford University Press, New York.

M. T. Chu and Q. Guo (1998), 'A numerical method for the inverse stochastic spectrum problem', *SIAM J. Matrix Anal. Appl.* **19**, 1027–1039 (electronic).

M. T. Chu and L. K. Norris (1988), 'Isospectral flows and abstract matrix factorizations', *SIAM J. Numer. Anal.* **25**, 1383–1391.

M. T. Chu and N. T. Trendafilov (2001), 'The orthogonally constrained regression revisited', *J. Comput. Graph. Statist.* **10**, 746–771.

M. T. Chu and S.-F. Xu (2005), 'On computing minimal realizable spectral radii of non-negative matrices', *Numer. Linear Algebra Appl.* **12**, 77–86.

M. Chu, N. Del Buono, L. Lopez and T. Politi (2005), 'On the low-rank approximation of data on the unit sphere', *SIAM J. Matrix Anal. Appl.* **27**, 46–60 (electronic).

T. F. Cox and M. A. A. Cox (1994), *Multidimensional Scaling*, Vol. 59 of *Monographs on Statistics and Applied Probability*, Chapman & Hall, London.

M. L. Curtis (1984), *Matrix Groups*, second edn, Universitext, Springer, New York.

J. W. Daniel (1967), 'The conjugate gradient method for linear and nonlinear operator equations', *SIAM J. Numer. Anal.* **4**, 10–26.

E. Date, M. Kashiwara, M. Jimbo and T. Miwa (1983), Transformation groups for soliton equations, in *Nonlinear Integrable Systems: Classical Theory and Quantum Theory, Kyoto 1981*, World Scientific, Singapore, pp. 39–119.

P. Deift, J. Demmel, L. C. Li and C. Tomei (1991), 'The bidiagonal singular value decomposition and Hamiltonian mechanics', *SIAM J. Numer. Anal.* **28**, 1463–1516.

P. Deift, T. Nanda and C. Tomei (1983), 'Ordinary differential equations and the symmetric eigenvalue problem', *SIAM J. Numer. Anal.* **20**, 1–22.

N. Del Buono and L. Lopez (2002), 'Geometric integration on manifold of square oblique rotation matrices', *SIAM J. Matrix Anal. Appl.* **23**, 974–989 (electronic).

J. Della-Dora (1975), 'Numerical linear algorithms and group theory', *Linear Algebra Appl.* **10**, 267–283.

J. Demmel and W. Kahan (1990), 'Accurate singular values of bidiagonal matrices', *SIAM J. Sci. Statist. Comput.* **11**, 873–912.

J. Demmel, M. Gu, S. Eisenstat, I. Slapničar, K. Veselić and Z. Drmač (1999), 'Computing the singular value decomposition with high relative accuracy', *Linear Algebra Appl.* **299**, 21–80.

R. L. Devaney (1992), *A First Course in Chaotic Dynamical Systems: Theory and Experiment*, Addison-Wesley Studies in Nonlinearity, Addison-Wesley, Reading, MA.

L. Dieci, R. D. Russell and E. S. Van Vleck (1994), 'Unitary integrators and applications to continuous orthonormalization techniques', *SIAM J. Numer. Anal.* **31**, 261–281.

A. Edelman, T. A. Arias and S. T. Smith (1999), 'The geometry of algorithms with orthogonality constraints', *SIAM J. Matrix Anal. Appl.* **20**, 303–353 (electronic).

S. Elaydi (2005), *An Introduction to Difference Equations*, Undergraduate Texts in Mathematics, third edn, Springer, New York.

K. Engø (2003), 'Partitioned Runge–Kutta methods in Lie-group setting', *BIT* **43**, 21–39.

H. Faßbender, D. S. Mackey, N. Mackey and H. Xu (1999), 'Hamiltonian square roots of skew-Hamiltonian matrices', *Linear Algebra Appl.* **287**, 125–159.

L. Faybusovich (1991), 'Hamiltonian structure of dynamical systems which solve linear programming problems', *Phys. D* **53**, 217–232.

K. V. Fernando and B. N. Parlett (1994), 'Accurate singular values and differential qd algorithms', *Numer. Math.* **67**, 191–229.

J. G. F. Francis (1961/1962), 'The QR transformation: A unitary analogue to the LR transformation I', *Comput. J.* **4**, 265–271.

R. W. Freund and N. M. Nachtigal (1991), 'QMR: A quasi-minimal residual method for non-Hermitian linear systems', *Numer. Math.* **60**, 315–339.

O. Galor (2005), 'Discrete dynamical systems', *GE, Growth, Math Methods, Econ WPA*. Available at http://ideas.repec.org/p/wpa/wuwpge/0504001.html.

C. B. García and F. J. Gould (1980), 'Relations between several path following algorithms and local and global Newton methods', *SIAM Rev.* **22**, 263–274.

S. D. Garvey, M. I. Friswell and U. Prells (2002*a*), 'Co-ordinate transformations for second order systems I: General transformations', *J. Sound Vibration* **258**, 885–909.

S. D. Garvey, M. I. Friswell and U. Prells (2002*b*), 'Co-ordinate transformations for second order systems II: Elementary structure-preserving transformations', *J. Sound Vibration* **258**, 911–930.

C. W. Gear (1981), 'Numerical solution of ordinary differential equations: Is there anything left to do?', *SIAM Rev.* **23**, 10–24.

I. Gohberg, P. Lancaster and L. Rodman (1982), *Matrix Polynomials*, Academic Press, New York.

D. Goldberg (1991), 'What every computer scientist should know about floating-point arithmetic', *ACM Computing Surveys* **23**, 5–48.

G. Golub and W. Kahan (1965), 'Calculating the singular values and pseudo-inverse of a matrix', *J. Soc. Indust. Appl. Math. Ser. B, Numer. Anal.* **2**, 205–224.

G. H. Golub and C. F. Van Loan (1996), *Matrix Computations*, Johns Hopkins Studies in the Mathematical Sciences, third edn, Johns Hopkins University Press, Baltimore, MD.

G. H. Golub and J. H. Wilkinson (1976), 'Ill-conditioned eigensystems and the computation of the Jordan canonical form', *SIAM Rev.* **18**, 578–619.

A. Greenbaum (1997), *Iterative Methods for Solving Linear Systems*, Vol. 17 of *Frontiers in Applied Mathematics*, SIAM, Philadelphia, PA.

J. Guckenheimer (2002), Numerical analysis of dynamical systems, in *Handbook of Dynamical Systems*, Vol. 2, North-Holland, Amsterdam, pp. 345–390.

L. A. Hageman and D. M. Young (1981), *Applied Iterative Methods*, Academic Press, New York. Also, unabridged republication of the 1981 original: Dover, Mineola, NY (2004).

E. Hairer and G. Wanner (1996), *Solving Ordinary Differential Equations II: Stiff and Differential-Algebraic Problems*, Vol. 14 of *Springer Series in Computational Mathematics*, second edn, Springer, Berlin.

E. Hairer, C. Lubich and G. Wanner (2006), *Geometric Numerical Integration: Structure-Preserving Algorithms for Ordinary Differential Equations*, Vol. 31 of *Springer Series in Computational Mathematics*, second edn, Springer, Berlin.

E. Hairer, S. P. Nørsett and G. Wanner (1993), *Solving ordinary differential equations I: Nonstiff Problems*, Vol. 8 of *Springer Series in Computational Mathematics*, second edn, Springer, Berlin.

R. Hauser and J. Nedić (2007), 'On the relationship between the convergence rates of iterative and continuous processes', *SIAM J. Optim.* **18**, 52–64 (electronic).

U. Helmke and J. B. Moore (1994), *Optimization and Dynamical Systems*, Communications and Control Engineering Series, Springer, London.

U. Helmke and F. Wirth (2000), Controllability of the shifted inverse power iteration: the case of real shifts, in *International Conference on Differential Equations, Berlin 1999*, Vols 1, 2, World Scientific, River Edge, NJ, pp. 859–864.

U. Helmke and F. Wirth (2001), 'On controllability of the real shifted inverse power iteration', *Systems Control Lett.* **43**, 9–23.

U. Helmke, J. B. Moore and J. E. Perkins (1994), 'Dynamical systems that compute balanced realizations and the singular value decomposition', *SIAM J. Matrix Anal. Appl.* **15**, 733–754.

M. R. Hestenes and E. Stiefel (1952), 'Methods of conjugate gradients for solving linear systems', *J. Research Nat. Bur. Standards* **49**, 409–436.

N. J. Higham (1989), Matrix nearness problems and applications, in *Applications of Matrix Theory, Bradford 1988*, Vol. 22 of *Inst. Math. Appl. Conf. Ser., New Ser.*, Oxford University Press, New York, pp. 1–27.

R. Hirota, S. Tsujimoto and T. Imai (1993), Difference scheme of soliton equations, in *Future Directions of Nonlinear Dynamics in Physical and Biological Systems, Lyngby 1992*, Vol. 312 of *NATO Adv. Sci. Inst. Ser. B, Phys.*, Plenum, New York, pp. 7–15.

M. W. Hirsch and S. Smale (1979), 'On algorithms for solving $f(x) = 0$', *Comm. Pure Appl. Math.* **32**, 281–313.

R. A. Horn and C. R. Johnson (1990), *Matrix Analysis*, Cambridge University Press, Cambridge.

P. Horst (1965), *Factor Analysis of Data Matrices*, Holt, Rinehart and Winston, New York.

R. Howe (1983), 'Very basic Lie theory', *Amer. Math. Monthly* **90**, 600–623. Correction to 'Very basic Lie theory', **91** (1984), 247.

A. Iserles (2002), 'On the discretization of double-bracket flows', *Found. Comput. Math.* **2**, 305–329.

A. Iserles, H. Z. Munthe-Kaas, S. P. Nørsett and A. Zanna (2000), Lie-group methods, in *Acta Numerica*, Vol. 9, Cambridge University Press, pp. 215–365.

M. Iwasaki and Y. Nakamura (2002), 'On the convergence of a solution of the discrete Lotka–Volterra system', *Inverse Problems* **18**, 1569–1578.

M. Iwasaki and Y. Nakamura (2004), 'An application of the discrete Lotka–Volterra system with variable step-size to singular value computation', *Inverse Problems* **20**, 553–563.

M. Iwasaki and Y. Nakamura (2006), 'Accurate computation of singular values in terms of shifted integrable schemes', *Japan J. Indust. Appl. Math.* **23**, 239–259.

W. Kahan (1972), Conserving confluence curbs ill-condition. Technical report 6, Computer Science Department, University of California, Berkeley.

N. Karmarkar (1984), 'A new polynomial-time algorithm for linear programming', *Combinatorica* 4, 373–395.

C. T. Kelley and D. E. Keyes (1998), 'Convergence analysis of pseudo-transient continuation', *SIAM J. Numer. Anal.* **35**, 508–523 (electronic).

C. T. Kelley, L.-Z. Liao, L. Qi, M. T. Chu, J. P. Reese and C. Winton (2007), Projected pseudo-transient continuation. Preprint, North Carolina State University.

M. R. S. Kulenović and O. Merino (2002), *Discrete Dynamical Systems and Difference Equations with Mathematica*, Chapman & Hall/CRC, Boca Raton, FL.

P. D. Lax (1968), 'Integrals of nonlinear equations of evolution and solitary waves', *Comm. Pure Appl. Math.* **21**, 467–490.

W.-W. Lin and S.-F. Xu (2006), 'Convergence analysis of structure-preserving doubling algorithms for Riccati-type matrix equations', *SIAM J. Matrix Anal. Appl.* **28**, 26–39 (electronic).

W.-W. Lin, V. Mehrmann and H. Xu (1999), 'Canonical forms for Hamiltonian and symplectic matrices and pencils', *Linear Algebra Appl.* **302/303**, 469–533.

S. Lojasiewicz (1963), Une propriété topologique des sous-ensembles analytiques réels, in *Les Equations aux Dérivées Partielles, Paris 1962*, Éditions du Centre National de la Recherche Scientifique, Paris, pp. 87–89.

D. S. Mackey, N. Mackey and F. Tisseur (2003), 'Structured tools for structured matrices', *Electron. J. Linear Algebra* **10**, 106–145 (electronic).

C. Mehl (1999), 'Condensed forms for skew-Hamiltonian/Hamiltonian pencils', *SIAM J. Matrix Anal. Appl.* **21**, 454–476 (electronic).

V. Mehrmann and D. Watkins (2000), 'Structure-preserving methods for computing eigenpairs of large sparse skew-Hamiltonian/Hamiltonian pencils', *SIAM J. Sci. Comput.* **22**, 1905–1925 (electronic).

G. Meurant (2006), *The Lanczos and Conjugate Gradient Algorithms: From Theory to Finite Precision Computations*, Vol. 19 of *Software, Environments, and Tools*, SIAM, Philadelphia, PA.

A. Morgan (1987), *Solving Polynomial Systems Using Continuation for Engineering and Scientific Problems*, Prentice-Hall, Englewood Cliffs, NJ.

J. Moser (1975), 'Three integrable Hamiltonian systems connected with isospectral deformations', *Adv. Math.* **16**, 197–220.

W. A. Mulder and B. van Leer (1985), 'Experiments with implicit upwind methods for the Euler equations', *J. Comput. Phys.* **59**, 232–246.

H. Munthe-Kaas (1998), 'Runge–Kutta methods on Lie groups', *BIT* **38**, 92–111.

Y. Nakamura (2004), A new approach to numerical algorithms in terms of integrable systems, in *Proc. International Conference on Informatics Research for Development of Knowledge Society Infrastructure: ICKS 2004* (T. Ibaraki, T. Inui and K. K. Tanaka, eds), IEEE Computer Society Press, pp. 194–205.

Y. Nakamura (2006), *Functionality of Integrable Systems*, Kyoritsu Shuppan Co., Tokyo, Japan. (In Japanese.)

R. Orsi (2006), 'Numerical methods for solving inverse eigenvalue problems for nonnegative matrices', *SIAM J. Matrix Anal. Appl.* **28**, 190–212 (electronic).

J. M. Ortega and W. C. Rheinboldt (2000), *Iterative Solution of Nonlinear Equations in Several Variables*, Vol. 30 of *Classics in Applied Mathematics*, SIAM, Philadelphia, PA.

B. Owren and B. Welfert (2000), 'The Newton iteration on Lie groups', *BIT* **40**, 121–145.

C. Paige and C. Van Loan (1981), 'A Schur decomposition for Hamiltonian matrices', *Linear Algebra Appl.* **41**, 11–32.

B. N. Parlett (1974), 'The Rayleigh quotient iteration and some generalizations for nonnormal matrices', *Math. Comp.* **28**, 679–693.

B. N. Parlett and O. A. Marques (2000), An implementation of the dqds algorithm (positive case), in *Proc. International Workshop on Accurate Solution of Eigenvalue Problems, University Park 1998*, Vol. 309, pp. 217–259.

C. Pöppe (1989), 'General determinants and the τ function for the Kadomtsev–Petviashvili hierarchy', *Inverse Problems* **5**, 613–630. See also corrigenda: **5**, 1173.

F. A. Potra and S. J. Wright (2000), 'Interior-point methods', *J. Comput. Appl. Math.* **124**, 281–302.

A. Ruhe (1987), 'Closest normal matrix finally found!', *BIT* **27**, 585–598.

H. Rutishauser (1954), 'Der Quotienten-Differenzen-Algorithmus', *Z. Angew. Math. Physik* **5**, 233–251.

H. Rutishauser (1960), 'Über eine kubisch konvergente Variante der *LR*-Transformation', *Z. Angew. Math. Mech.* **40**, 49–54.

Y. Saad and M. H. Schultz (1986), 'GMRES: A generalized minimal residual algorithm for solving nonsymmetric linear systems', *SIAM J. Sci. Statist. Comput.* **7**, 856–869.

G. V. Savinov (1983), 'The conjugate gradient method for systems of nonlinear equations', *J. Math. Sci.* **23**, 2012–2017. Translated from *Zap. Naučn. Sem. Leningrad. Otdel. Mat. Inst. Steklov.* (LOMI), **70** (1977), 178–183.

H. Sedaghat (2003), *Nonlinear Difference Equations: Theory with Applications to Social Science Models*, Vol. 15 of *Mathematical Modelling: Theory and Applications*, Kluwer, Dordrecht.

R. Shaw (1982), *Linear Algebra and Group Representations I: Linear Algebra and Introduction to Group Representations*, Academic Press, London.

L. Simon (1983), 'Asymptotics for a class of nonlinear evolution equations, with applications to geometric problems', *Ann. of Math.* (2) **118**, 525–571.

S. Smale (1977), Convergent process of price adjustment and global Newton methods, in *Frontiers of Quantitative Economics: Invited papers, Econometric Soc., Third World Congress, Toronto 1975*, Vol. IIIA, North-Holland, Amsterdam, pp. 191–205.

V. I. Smirnov (1970), *Linear Algebra and Group Theory*, Dover, New York.

S. T. Smith (1991), 'Dynamical systems that perform the singular value decomposition', *Systems Control Lett.* **16**, 319–327.

G. W. Stewart (1993), 'On the early history of the singular value decomposition', *SIAM Rev.* **35**, 551–566.

A. M. Stuart and A. R. Humphries (1996), *Dynamical Systems and Numerical Analysis*, Vol. 2 of *Cambridge Monographs on Applied and Computational Mathematics*, Cambridge University Press, Cambridge.

W. W. Symes (1981/82), 'The *QR* algorithm and scattering for the finite nonperiodic Toda lattice', *Phys. D* **4**, 275–280.

G. Szegő (1975), *Orthogonal Polynomials*, fourth edn. Vol. XXIII of *AMS Colloquium Series*, AMS, Providence, RI.

M. Takata, M. Iwasaki, K. Kimura and Y. Nakamura (2005), An evaluation of singular value computation by the discrete Lotka–Volterra system, in *Proc. 2005 International Conference on Parallel and Distributed Processing Techniques and Applications: PDPTA2005*, Vol. II, pp. 410–416.

M. Takata, M. Iwasaki, K. Kimura and Y. Nakamura (2006), Performance of a new scheme for bidiagonal singular value decomposition of large scale, in *Proc. IASTED International Conference on Parallel and Distributed Computing and Networks: PDCN2006*, pp. 304–309.

T.-Y. Tam (2004), 'Gradient flows and double bracket equations', *Differential Geom. Appl.* **20**, 209–224.

F. Tisseur and K. Meerbergen (2001), 'The quadratic eigenvalue problem', *SIAM Rev.* **43**, 235–286 (electronic).

A. Toselli and O. Widlund (2005), *Domain Decomposition Methods: Algorithms and Theory*, Vol. 34 of *Springer Series in Computational Mathematics*, Springer, Berlin.

S. Tsujimoto (1995), Molecule solution of hungry Volterra equations, in *Soliton Theory and Its Applications* (J. Satsuma, ed.), University of Tokyo, Japan, pp. 53–56. (In Japanese.)

S. Tsujimoto, Y. Nakamura and M. Iwasaki (2001), 'The discrete Lotka–Volterra system computes singular values', *Inverse Problems* **17**, 53–58.

Y. Z. Tsypkin (1971), *Adaptation and Learning in Automatic Systems*, Vol. 73 of *Mathematics in Science and Engineering*, Academic Press, New York. Translated from the Russian by Z. J. Nikolic.

Y. Z. Tsypkin (1973), *Foundations of the Theory of Learning Systems*, Vol. 101 of *Mathematics in Science and Engineering*, Academic Press, New York/London. Translated from the Russian by Z. J. Nikolic.

H. A. van der Vorst (2003), *Iterative Krylov Methods for Large Linear Systems*, Vol. 13 of *Cambridge Monographs on Applied and Computational Mathematics*, Cambridge University Press, Cambridge.

C. F. Van Loan (1984), 'A symplectic method for approximating all the eigenvalues of a Hamiltonian matrix', *Linear Alg. Appl.* **61**, 233–253.

R. S. Varga (2000), *Matrix Iterative Analysis*, Vol. 27 of *Springer Series in Computational Mathematics*, expanded edn, Springer, Berlin.

D. S. Watkins (1982), 'Understanding the QR algorithm', *SIAM Rev.* **24**, 427–440.

D. S. Watkins and L. Elsner (1988), 'Self-similar flows', *Linear Algebra Appl.* **110**, 213–242.

J. Wimp (1984), *Computation with Recurrence Relations*, Applicable Mathematics Series, Pitman, Boston, MA.

K. Wright (1992), 'Differential equations for the analytic singular value decomposition of a matrix', *Numer. Math.* **63**, 283–295.

M. H. Wright (2005), 'The interior-point revolution in optimization: history, recent developments, and lasting consequences', *Bull. Amer. Math. Soc. (N.S.)* **42**, 39–56 (electronic).

S. J. Wright (1997), *Primal-Dual Interior-Point Methods*, SIAM, Philadelphia, PA.

H. Yabe and M. Takano (2004), 'Global convergence properties of nonlinear conjugate gradient methods with modified secant condition', *Comput. Optim. Appl.* **28**, 203–225.

Z. Zeng and T. Y. Li (2007), A numerical method for computing the Jordan canonical form. Preprint, Northeastern Illinois University.

H. Zha and Z. Zhang (1995), 'A note on constructing a symmetric matrix with specified diagonal entries and eigenvalues', *BIT* **35**, 448–451.

S. Zhang and Z. Deng (2005), 'Geometric integration methods for general nonlinear dynamic equation based on Magnus and Fer expansions', *Progr. Natur. Sci. (English edn)* **15**, 304–314.

Acta Numerica (2008), pp. 87–145
doi: 10.1017/S0962492906350015

© Cambridge University Press, 2008

Accurate and efficient expression evaluation and linear algebra

James Demmel*
Department of Mathematics and Computer Science Division,
University of California, Berkeley, CA 94720, USA

Ioana Dumitriu[†]
Department of Mathematics, University of Washington,
Seattle, WA 98195, USA

Olga Holtz[‡]
Department of Mathematics,
University of California, Berkeley, CA 94720, USA
and
Department of Mathematics, Technische Universität Berlin,
D-10623, Berlin, Germany

Plamen Koev[§]
Department of Mathematics, North Carolina State University,
Raleigh, NC 27695, USA

We survey and unify recent results on the existence of accurate algorithms for evaluating multivariate polynomials, and more generally for accurate numerical linear algebra with structured matrices. By 'accurate' we mean that the computed answer has relative error less than 1, *i.e.*, has some correct leading digits. We also address efficiency, by which we mean algorithms that run in polynomial time in the size of the input. Our results will depend strongly on the model of arithmetic: most of our results will use the so-called *traditional model* (TM), where the computed result of op(a, b), a binary operation like $a + b$, is given by op$(a, b) * (1 + \delta)$ where all we know is that $|\delta| \leq \varepsilon \ll 1$. Here ε is a constant also known as machine epsilon.

* Supported by NSF grants CCF-0444486, CNS 0325873, by DOE grant DE-FC02-06ER25786, and by the University of California, Berkeley, Richard Carl Dehmel Distinguished Professorship.
† Supported by the Miller Institute for Basic Research in Science.
‡ Supported by the Sofja Kovalevskaja programme of the Alexander von Humboldt Foundation.
§ Supported by NSF grants DMS-0314286, DMS-0411962 and DMS-0608306.

We will see a common reason for the following disparate problems to permit accurate and efficient algorithms using only the four basic arithmetic operations: finding the eigenvalues of a suitably discretized scalar elliptic PDE, finding eigenvalues of arbitrary products, inverses, or Schur complements of totally non-negative matrices (such as Cauchy and Vandermonde), and evaluating the Motzkin polynomial. Furthermore, in all these cases the high accuracy is 'deserved', *i.e.*, the answer is determined much more accurately by the data than the conventional condition number would suggest.

In contrast, we will see that evaluating even the simple polynomial $x+y+z$ accurately is impossible in the TM, using only the basic arithmetic operations. We give a set of necessary and sufficient conditions to decide whether a high-accuracy algorithm exists in the TM, and describe progress toward a decision procedure that will take any problem and provide either a high-accuracy algorithm or a proof that none exists.

When no accurate algorithm exists in the TM, it is natural to extend the set of available accurate operations by a library of additional operations, such as $x + y + z$, dot products, or indeed any enumerable set which could then be used to build further accurate algorithms. We show how our accurate algorithms and decision procedure for finding them extend to this case.

Finally, we address other models of arithmetic, and the relationship between (im)possibility in the TM and (in)efficient algorithms operating on numbers represented as bit strings.

CONTENTS

1. Introduction

A result of a computation will be called *accurate* if it has a small relative error, in particular less than 1 (*i.e.*, some leading digits must be correct). Now we can ask what the following problems have in common.

(1) Accurately evaluate the Motzkin polynomial

$$p(x, y, z) = z^3 + x^2 y^2 (x^2 + y^2 - 3z^2).$$

(2) Accurately compute the entries or eigenvalues of a matrix obtained by performing an arbitrary sequence of operations chosen from the set {multiplication, J-inversion, Schur complement, taking submatrices}, starting from a set of *totally non-negative* (TN) matrices such as the Hilbert matrix, TN generalized Vandermonde matrices, *etc.*

(3) Accurately find the eigenvalues of a suitably discretized scalar elliptic PDE.

We also ask how they all differ from the apparently much easier problem of evaluating $x + y + z$.

The answer will depend strongly on our model of arithmetic. For most of this paper we will use the *traditional model* (TM) of arithmetic, that the computed result of $op(a, b)$, a binary operation such as $a + b$, is given by $op(a, b) \cdot (1 + \delta)$, where all we know is that $|\delta| \leq \varepsilon \ll 1$. Here ε is a real constant also known as *machine precision*. We will refer to $\mathrm{rnd}(op(a, b)) \equiv op(a, b)(1 + \delta)$ as the *rounded result* of $op(a, b)$. We will distinguish between the cases where the other quantities (including δs) are all real, or all complex.

To see why some expressions may or may not be evaluable accurately in the TM, consider multiplying or dividing two numbers each known to relative error $\eta < 1$: then their rounded product or quotient is clearly correct with relative error $O(\max(\eta, \varepsilon))$. This also holds when adding two like-signed real numbers (or subtracting real numbers with opposite signs). In contrast, subtracting two like-signed real numbers $x - y$ can lead to *cancellation* of leading digits. If x and y themselves have non-zero relative error bounds, then depending on the extent of cancellation, $x - y$ may have an arbitrary relative error. On the other hand, if x and y are exact inputs, then $\mathrm{rnd}(x \pm y) = (x \pm y)(1 + \delta)$ is also known with small relative error. In other words, an easy sufficient (but not necessary!) condition in the TM for an algorithm to be accurate is 'no inaccurate cancellation' (NIC).

NIC. The algorithm only (1) multiplies, (2) divides, (3) adds (resp. subtracts) real numbers with like (resp. differing) signs, and otherwise only (4) adds or subtracts input data.

Sometimes we will also include the square root among our allowed operations in NIC.[1]

In the TM, with real numbers, the three problems listed above all have novel accurate algorithms that use only four basic arithmetic operations ($+$, $-$, $\times$ and $/$), comparison and branching, and satisfy NIC. Furthermore, the matrix algorithms are efficient, running in $O(n^3)$ time (we say more about

---

[1] However, square roots require more care in bounding the relative error. In floating-point arithmetic on most computers, computing $y = x^{1/2^{100}}$ by 100 square roots and then $z = y^{2^{100}}$ by 100 squarings yields $z = 1$ independently of $x > 0$.

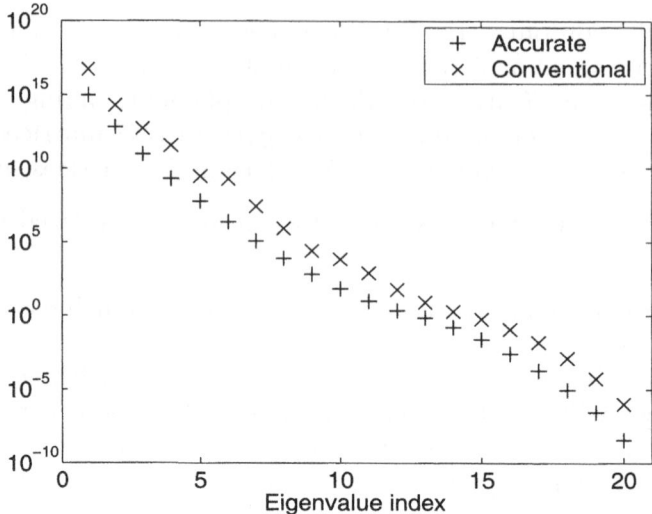

Figure 1.1. Eigenvalues of the 20th Schur complement of the 40-by-40 Vandermonde matrix $V_{ij} = i^{j-1}$, computed both using a conventional algorithm ($\times$) and and accurate algorithm ($+$).

efficiency below). These linear algebra algorithms depend on some recently discovered matrix factorizations and update formulas, and the algorithm for the Motzkin polynomial (surprisingly) fills a page with 8 cases. In contrast, with complex arithmetic, no accurate algorithms exist; nor is there an accurate algorithm using only these operations, in the real or complex case, for evaluating $x + y + z$ accurately.

For example, consider Figure 1.1, which shows the eigenvalues of a matrix obtained by taking the trailing 20-by-20 Schur complement of a 40-by-40 Vandermonde matrix. Both the eigenvalues computed by our algorithm (in standard double-precision floating-point arithmetic), and by a conventional algorithm are shown. Note that *every* eigenvalue computed by the conventional algorithm is wrong by orders of magnitude, whereas all ours are correct to nearly 14 digits, as confirmed by a very high-precision calculation.

Section 2 of this paper will survey a great many other examples of structured matrices where accurate and efficient linear algebra algorithms are possible using NIC as the main (but not only) tool; see Table 2.1 for a summary.

One may wonder whether this accuracy is 'overkill', because small uncertainties in the data might cause much larger uncertainties in the computed results. In this case, computing results to high accuracy would be more than the data deserves, and not worth any additional cost. Indeed, the usual condition numbers of the problems considered here are usually enormous. However, their *structured* condition numbers are often quite modest, justifying computing the answers to high accuracy. For example, while a Cauchy matrix $C_{ij} = 1/(x_i + y_j)$ such as the Hilbert matrix $(x_i = i = 1 + y_i)$ is considered badly conditioned since $\kappa(C) \equiv \|C\| \cdot \|C^{-1}\|$ can be very large, the entries of C^{-1} are actually much less sensitive functions of x_i and y_j than $\kappa(C)$ would indicate. Indeed, if the answer is given by a formula satisfying NIC, then the condition number can only be large when cancellation occurs when computing $x \pm y$ for uncertain input data x and y; each such expression adds the quantity $1/\text{rel\_gap}(x, y) \equiv (|x| + |y|)/|x \pm y|$ to the structured condition number. This is true of all the examples in Section 2, justifying their more accurate computation than would the usual condition number.

The profusion and diversity of these examples naturally raises the question as to what mathematical property they share that makes these algorithms possible. Section 3 of this paper addresses this, by describing progress towards a *decision procedure* for the more basic problem of deciding whether a given multivariate polynomial can be evaluated accurately using the basic rounded arithmetic operations, comparison, and branching. The answer will depend not just on the polynomial, but whether the data is real or complex, and on the domain of evaluation (a smaller domain may be easier than a larger one, if it eliminates difficult arguments). This decision procedure would yield simpler necessary and sufficient conditions (not identical in all cases) that tell us whether the algorithms in Section 2 (or others not yet discovered) must exist (we will use the fact that accurate determinants are necessary and often sufficient for accurate linear algebra). It will turn out that the results for real arithmetic are much more complicated than for complex arithmetic, where simple necessary and sufficient conditions may be stated (the answer is basically given by NIC above); this reflects the difference between algebraic geometry over the real and complex numbers.

One negative result of Section 3.3 will be the impossibility of evaluating $x + y + z$ using only the basic rounded arithmetic operations. This seems odd, since $x + y + z$ is so simple. But it is only simple if we use the fact that in practice (floating-point arithmetic), x, y and z are represented by finite bit strings that can be manipulated and analysed differently than by assuming only that $\text{rnd}(\text{op}(a, b)) = \text{op}(a, b)(1 + \delta)$ with $|\delta| \leq \varepsilon$. To go further we must extend our model of arithmetic. We do so in two ways.

Section 3.4 continues by adding so-called 'black-box' operations to the basic arithmetic operations. For example, one could assume that a subroutine for the accurate evaluation of $x + y + z$ (or of dot products, or of 3-by-3 determinants, *etc.*) also existed, and then ask the analogous question as to what other polynomials could be accurately evaluated, using this subroutine as a building block. This indeed models computational practice, where subroutine libraries of such black-box routines are provided in order to build accurate algorithms for other more complicated polynomials. In Section 3.4 we also describe how to extend our decision procedures when an arbitrary set of such black-box routines is available, and the question is whether another polynomial not already in the set can be evaluated accurately. A positive result will show that just the ability to compute 2-by-2 determinants accurately is enough to permit accurate and efficient linear algebra on the inverses of tridiagonal matrices. A negative result will be the impossibility of accurate linear algebra with Toeplitz matrices, given *any* set of block-box operations of bounded degree or with a bounded number of arguments.

Sections 3.3 and 3.4 go some way to describing the possibilities and limits of solving numerical problems accurately in practice. But 'in practice' means using finite representations with bits, *i.e.*, floating point, in which case accurate (even exact) polynomial evaluation is always possible, and the only question is cost. In Section 4, after a brief discussion of other arithmetic models, we will settle on one model we believe best captures the spirit of actual floating-point computation, but without limiting it to fixed word sizes: an arbitrary pair of integers (m, e) is used to represent the floating-point number $m \cdot 2^e$. In this model, we describe how the algorithms in Section 2 lead to efficient algorithms that run in time polynomial in the size of the inputs, the usual computer science notion of efficiency. In contrast, conventional algorithms, when simply run in high enough precision to get an accurate answer, do not run in polynomial time.

Finally, in Section 5 we consider the structured condition numbers for the problems we consider, which can be much smaller than the usual unstructured condition numbers and so justify accuracy computation. In prior work (Demmel 1987), the first author observed that for many problems the condition number of the condition number was approximately equal to the condition number of the original problem, and that this corresponded to the geometric property that the condition number was the reciprocal of the distance to the nearest ill-posed (or singular) problem. These observations apply here, with the following interesting consequence: for the examples considered here it is possible to compute the solution to a problem accurately if and only if it is possible to estimate its condition number accurately. An analogous phenomenon was observed in Demmel, Diament and Malajovich (2001).

2. Accurate and efficient algorithms for linear algebra

2.1. Introduction

The numerical linear algebra problems we will consider include computing the product of matrices, the Schur complement, the determinant or other minor, the inverse, the solution to a linear system or least-squares problem, and various matrix decompositions such as LDU (with or without pivoting) QR, SVD (singular value decomposition), and EVD (eigenvalue decomposition).

Conventional algorithms for these problems are at best only *backward stable*: when applied to a matrix A they compute the exact solution of a nearby problem $A + \delta A$, where $\|\delta A\| = \mathcal{O}(\varepsilon)\|A\|$, where $\| \cdot \|$ is some matrix norm and ε is machine epsilon. In consequence, the error in the computed solution depends on how sensitive the answer is to small changes in A, and is typically bounded in norm by $\frac{\|\delta A\|}{\|A\|}\kappa(A) = \mathcal{O}(\varepsilon)\kappa(A)$, where $\kappa(A)$ is a condition number (a scaled norm of the Jacobian of the solution map). Thus we have two ways to lose high relative accuracy: First, bounding the error only in norm may provide very weak bounds for tiny solution components; for example the error bound for the computed singular values guarantees an absolute error $|\sigma_{i,\text{true}} - \sigma_{i,\text{comp}}| = \mathcal{O}(\varepsilon) \max_i \sigma_{i,\text{true}}$, so that the large singular values have small relative errors, but not the small ones. Second, when $\kappa(A)$ is large, even large solution components may be inaccurate, as when inverting an ill-conditioned matrix.

However, these conventional algorithms ignore the *structure* of the matrix, which is critical to our approach. Rather than treating, say, a Cauchy matrix C as a collection of n^2 independent entries $C_{ij} = 1/(x_i + y_j)$, we treat it as a function of its $2n$ parameters x_i and y_j. Starting from these $2n$ parameters, we can find accurate expressions (because they satisfy NIC) for C's determinant $\det(C) = \prod_{i<j}(x_i - x_j)(y_i - y_j)/\prod_{i,j}(x_i + y_j)$ and other linear algebra problems. As mentioned in Section 1, expressions satisfying NIC also imply that their structured condition numbers can be arbitrarily smaller than their conventional condition numbers.

Now we outline our general approach to these problems. First we consider the problems whose solutions are rational functions of the parameters, such as computing a determinant or minor. Indeed, all these solutions can be expressed using minors or quotients of minors. For example, the entries of the inverse or LDU factorization are (quotients of) minors, the product AB can be extracted from

$$\begin{bmatrix} I & A & 0 \\ 0 & I & B \\ 0 & 0 & 1 \end{bmatrix}^{-1},$$

and the last column of

$$\begin{bmatrix} I & A & -b \\ A^T & 0 & 0 \\ 0 & 0 & 1 \end{bmatrix}^{-1}$$

contains the solution of the overdetermined least-squares problem $\min_x \|Ax - b\|_2$. Thus the ability to compute certain minors with high relative accuracy is sufficient to solve these linear algebra problems with high relative accuracy. Conversely, knowing a factorization such as LDU with high relative accuracy yields the determinant with similar accuracy (via the product $\pm \prod_i D_{ii}$). Thus we see that matrix structures that permit accurate computations of certain determinants are both necessary and sufficient for solution of these linear algebra problems with high relative accuracy. In this section we will identify a number of matrix structures that permit such accurate determinants to be calculated.

Second, we consider the EVD and SVD, which involve more general algebraic functions of the matrix entries. To compute these accurately, we need other tools, which we will summarize below in Section 2.2. Briefly, our approach will be to compute one of several other matrix decompositions using only rational operations (and possibly square roots), and then apply iterative schemes to these decompositions that have accuracy guarantees.

Efficient conventional algorithms (*i.e.*, using $\mathcal{O}(n^3)$ arithmetic operations) exist for each of the above problems and are available in free packages (*e.g.*, LAPACK (Anderson *et al.* 1999)) or embedded in commercial ones (*e.g.*, MATLAB (The MathWorks 1992)). So an extra challenge is to find not just accurate algorithms, but ones that also take $\mathcal{O}(n^3)$ operations.

Our results, using only NIC, are summarized in Table 2.1, which describes (in a $\mathcal{O}(\cdot)$ sense) the speed of the fastest-known accurate algorithm for each problem shown. There is one column for each linear algebra problem considered, and one row for each structured matrix class. The abbreviations not yet defined will be explained as we continue.

The rest of this section is organized as follows. Section 2.2 briefly presents accurate algorithms for the EVD and SVD. Section 2.3 walks through Table 2.1 row by row, again briefly explaining the results. Finally, Section 2.4 explains how much more is possible if we expand the class of formulas we may use beyond NIC in a certain disciplined way. This naturally raises the question of whether or not there is a systematic method to recognize such formulas, which is the final topic of this paper.

2.2. Tools for computing EVD and SVD accurately

2.2.1. Rank-revealing decompositions and SVD
The first accurate SVD algorithm depends on a *rank-revealing decomposition*, or RRD (Demmel *et al.* 1999), of matrix A, a factorization $A = XDY$ where D is non-singular and diagonal, and X and Y^T have full column rank and are 'well conditioned'. Note that A may be rectangular or singular. The most obvious example of an RRD is the SVD, where X and Y are as well conditioned as possible. Other examples where X and Y are (nearly always) well conditioned come from Gaussian elimination with complete pivoting $A = LDU$, or from QR with complete pivoting $A = QDR$;

more sophisticated pivoting techniques with better condition bounds on the unit triangular factors are available (Chan 1987, Chandrasekaran and Ipsen 1994, Gu and Eisenstat 1996, Hong and Pan 1992, Hwang, Lin and Yang 1992, Miranian and Gu 2003, Stewart 1993). An RRD $A = XDY$ has two attractive properties, as follows.

(a) Given the RRD, it is possible to compute the SVD to high relative accuracy in the following sense (Demmel *et al.* 1999, Section 3, Demmel and Koev 2001, Algorithm 2).

- The relative error in each singular value σ_i is bounded by

$$\mathcal{O}(\varepsilon \max(\kappa(X), \kappa(Y))),$$

  where $\kappa(X) = \|X\| \cdot \|X\|^{-1}$.

- The relative error in the ith computed (left or right) singular vector is bounded by

$$\mathcal{O}(\varepsilon \max(\kappa(X), \kappa(Y))) / \min_{j \neq i} \text{rel\_gap}(\sigma_i, \sigma_j).$$

  In other words, the condition number can only be large if the singular value agrees with another one to many leading digits, no matter how small they are in absolute value.

(b) These error bounds do not change if the RRD is known only approximately (either because of uncertainty in A or round-off in computing the RRD), as long as (Demmel *et al.* 1999, Theorem 2.1, Eisenstat and Ipsen 1995, Li 1999):

- we can compute $\hat{X}$ where $\|X - \hat{X}\| = \mathcal{O}(\varepsilon)\|X\|$,
- we can compute a diagonal $\hat{D}$ where $|D_{ii} - \hat{D}_{ii}| = \mathcal{O}(\varepsilon)|D_{ii}|$,
- we can compute $\hat{Y}$ where $\|Y - \hat{Y}\| = \mathcal{O}(\varepsilon)\|Y\|$.

In other words, we only need the factors X and Y with high absolute accuracy, not relative accuracy, a fact that will significantly expand the scope of applicability.

Among the various algorithms cited above for computing the SVD, we sketch one (Demmel *et al.* 1999, Algorithm 3.2), along with an explanation of its accuracy.

(1) Compute the SVD of XD using one-sided Jacobi, yielding $XD = \bar{U}\bar{\Sigma}\bar{V}^T$. Thus $A = \bar{U}\bar{\Sigma}\bar{V}^T Y$.

(2) Multiply $W = \bar{\Sigma}(\bar{V}^T Y)$, respecting parentheses. Thus $A = \bar{U}W$.

(3) Compute the SVD of W using one-sided Jacobi, yielding $W = \bar{\bar{U}}\Sigma V^T$. Thus $A = \bar{U}\bar{\bar{U}}\Sigma V^T$.

(4) Multiply $U = \bar{U}\bar{\bar{U}}$, yielding the SVD $A = U\Sigma V^T$.

Briefly, the reason this works is that in steps (1) and (3), which potentially combine numbers over very wide ranges of magnitude, one-sided Jacobi respects this scaling by, in step (1) for example, creating backward errors in column i of XD that are proportional to D_{ii} (Demmel and Veselić 1992, Drmač 1998, Mathias 1996). Furthermore, each step costs $\mathcal{O}(n^3)$ arithmetic operations.

2.2.2. Bidiagonal SVD

The second accurate SVD algorithm depends on a *bidiagonal reduction* (BR) of matrix A, a factorization $A = UBV^T$ where B is bidiagonal (non-zero on the main and first super-diagonal) and U and V are unitary. This is an intermediate factorization in the standard SVD algorithm. If the entries of B are determined to high relative accuracy, so is B's SVD in the same sense as the RRD determines the SVD as described above (but without any factor like $\max(\kappa(X), \kappa(Y))$ in the error bounds). Furthermore, accurate $\mathcal{O}(n^3)$ algorithms are available (Demmel and Kahan 1990, Parlett 1995).

2.2.3. Accurate EVD

Now we discuss the EVD. Clearly, if A is symmetric positive definite, and a symmetric RRD $A = XDX^T$ is available, then the SVD and EVD are identical. If A is symmetric indefinite but an accurate SVD is attainable, then the only remaining task is assigning correct signs to the singular values, which may be done using the algorithms of Dopico, Molera and Moro (2003). Algorithms for computing symmetric RRDs of certain symmetric structured matrices are presented in Koev and Dopico (2007) and Peláez and Moro (2006).

We also know of two accurate *non-symmetric* eigenvalue algorithms, for totally non-negative (TN) and for certain sign-regular matrices, which we call TN^J (see Section 2.3.6 for definitions).

In the TN case, the trick is to implicitly perform an accurate similarity transformation to a *symmetric* tridiagonal positive definite matrix which is available to us in factored form. The TN eigenvalue problem is thus reduced to the bidiagonal SVD problem.

The sign-regular TN^J matrices are similar to symmetric anti-bidiagonal matrices (Holtz 2005) (*i.e.*, the only non-zero entries are on the antidiagonal and one sub-antidiagonal). This similarity can be performed accurately by transforming implicitly an appropriate bidiagonal decomposition of the TN^J matrix. Finally, the eigenvalues of the anti-bidiagonal matrix are its singular values with appropriate signs known from theory.

2.3. Designing accurate algorithms for different structured classes

In this section we look at the particular approaches in designing accurate algorithms for different matrix classes in order to fill the rows of Table 2.1,

Table 2.1. Existing algorithms for accurate computations with various classes of structured matrices. Entries like n^2 are meant in a big-O sense; see Section 2.1 for details. 'No' means no accurate algorithms exist without using arbitrary precision arithmetic; see Section 3.5 for details.

Type of matrix	$\det A$	A^{-1}	Any minor	Gauss. elim.			RRD	QR	NE	$Az = b$	SVD	EVD	Reference
				NP	PP	CP							
Acyclic	n	n^2	n	n^2	n^2	n^2	n^2				n^3		Demmel et al. (1999)
DSTU	n^3	n^5	n^3	n^3	n^3	n^3	n^3				n^3		Demmel et al. (1999), Peláez and Moro (2006)
TSC	n	n^3	n	n^4	n^4	n^4	n^4				n^4		Demmel et al. (1999), Peláez and Moro (2006)
Diagonally dominant	n^3		No	n^3	n^3	n^3	n^3				n^3		Ye (2008a)
M-matrices	n^3	n^3	No	n^3	n^3	n^3	n^3				n^3		Alfa, Xue and Ye (2002), Demmel and Koev (2004b), O'Cinneide (1996), Peña (2004)
Cauchy (non-TN)	n^2	n^2	n^2	n^2	n^2	n^3	n^3		n^2		n^3		Boros et al. (1999), Demmel (1999)
Vandermonde (non-TN)	n^2		No				n^3		n^2		n^3		Björck and Pereyra (1970), Higham (1990), Demmel (1999), Demmel and Koev (2006)
Displacement rank one	n^2						n^3				n^3		Demmel (1999)
Totally non-negative	n	n^3	n^3	n^3	n^4	n^4	n^3	n^3	0	n^2	n^3	n^3	Koev (2005, 2007)
TN^J	n	n^3	n^3	n^3	n^4	n^4	n^3	n^3	0	n^2	n^3	n^3	Koev and Dopico (2007)
Toeplitz	No		No	No	No	No	No	No	No		No	No	Demmel et al. (2006)

explaining only a few in detail. Each row refers to a matrix class, and each column to a linear algebra problem. A table entry n^α means that an accurate linear algebra algorithm costing $\mathcal{O}(n^\alpha)$ arithmetic operations for the given problem and class exists. A 'No' entry means that no accurate algorithm using traditional arithmetic exists, and indeed no accurate algorithm exists without using arbitrary precision arithmetic, in a sense to be made precise in Section 3.5.

We begin by explaining some of the terser column headings. 'Any minor' means that an arbitrary minor of the matrix may be computed accurately, not just the determinant. 'Gauss. elim. NP' means *Gaussian elimination with no pivoting* (GENP), and similarly 'PP' and 'CP' refer to *partial pivoting* (GEPP) and *complete pivoting* (GECP), respectively. 'RRD' is a *rank-revealing decomposition* as described above (frequently, but not always, the same as GECP). 'NE' is *Neville elimination* (Gasca and Peña 1992), a variation on GENP where L and U are represented as products of bidiagonal matrices (corresponding to elimination where a multiple of row i is added to row $i+1$ to create one zero entry). $Az = b$ refers to solving $Az = b$ accurately given conditions on b (alternating signs in its components).

2.3.1. Acyclic matrices

A matrix A is called *acyclic* if its graph is: namely, the bipartite graph with one node for each row and one node for each column and an edge (i, j) if A_{ij} is non-zero. Acyclic matrices include bidiagonal matrices (see Section 2.2.2), and broken arrow matrices (which are non-zero only on the diagonal and one row or one column), among exponentially many other possibilities (Demmel and Gragg 1993).

Acyclic matrices are precisely the class of matrix sparsity patterns with the property that the Laplace expansion of each minor can have at most one non-zero term (Demmel and Gragg 1993). Thus every non-zero minor can be computed accurately as the product of n matrix entries. Any acyclic matrix is also a DSTU matrix (see the following section), and so the algorithms for DSTU matrices may be used.

2.3.2. DSTU (diagonal scaled totally unimodular) matrices

A matrix A is called *totally unimodular* (TU) if all its minors are $0, 1$, or -1. A matrix is *diagonally scaled totally unimodular* (DSTU) if it is of the form $A = D_1 Z D_2$, where D_1 and D_2 are diagonal and Z is totally unimodular.

Accurate LDU and SVD algorithms for DSTU matrices were presented in Demmel (1999) and are based on the following observation.

(1) The Schur complement of a DSTU matrix is DSTU.

(2) If, at any step in the inner loop of Gaussian elimination, the subtraction

$$a'_{ij} = a_{ij} - \frac{a_{ik}a_{kj}}{a_{kk}} \tag{2.1}$$

has two non-zero operands, then the result a'_{ij} must be exactly 0.

In other words, to make Gaussian elimination accurate, a one-line addition is required to test whether both a_{ij} and $\frac{a_{ik}a_{kj}}{a_{kk}}$ are non-zero, and to set $a'_{ij} = 0$ if they are. Then the modified Gaussian elimination satisfies NIC, yielding an accurate LDU decomposition. LDU with complete pivoting yields an accurate RRD (with $\kappa(L)$ and $\kappa(U)$ both bounded by $\mathcal{O}(n^2)$: Demmel *et al.* (1999, Theorem 10.2)), and an accurate RRD yields an accurate SVD as discussed in Section 2.2.1.

If a DSTU matrix is symmetric, Peláez and Moro (2006) derived accurate algorithms that preserve and exploit the symmetry in their matrices. They also presented such *symmetric* algorithms for TSC matrices discussed next.

DSTU matrices arise naturally in the formulation of eigenvalue problems for Sturm–Liouville equations (Demmel and Koev 2001), and more general scalar elliptic PDE with suitable finite element discretizations (Demmel *et al.* 1999). We discuss this further below in Section 2.4.

2.3.3. TSC (total signed compound) matrices

Let $\mathcal{S}$ be the set of all matrices with a given sparsity and sign pattern. $\mathcal{S}$ is called *sign non-singular* (SNS) if it contains only square matrices, and the Laplace expansion of the determinant of each $G \in \mathcal{S}$ is the sum of monomials of like-sign, with at least one non-zero monomial. $\mathcal{S}$ is called *total signed compound* (TSC) if every square submatrix of any $G \in \mathcal{S}$ is either SNS, or structurally singular (*i.e.*, no non-zero monomials appear in its determinant expansion). Acyclic matrices are obviously a special case of TSC matrices, with at most one monomial appearing in each minor.

According to Demmel *et al.* (1999, Lemma 7.2) any minor of a TSC matrix may be computed accurately using not more than $4n-1$ arithmetic operations (and not counting various graph traversal operations). With this computing the LDU decomposition of a TSC matrix is easy. If at any step of Gaussian elimination the subtraction in (2.1) is one of same-signed quantities, then a'_{ij} is recomputed as a quotient of minors, each of which is computed accurately as above. The total cost could go up to $\mathcal{O}(n^4)$, but this is still efficient, according to our convention.

2.3.4. Diagonally dominant and M-matrices

A matrix A is called (*row*) *diagonally dominant* if the sums $s_i = a_{ii} - \sum_{j \neq i} |a_{ij}|$ are non-negative for all rows i. If in addition its off-diagonal entries a_{ij} are non-positive (so that $s_i = \sum_j a_{ij}$), then it is called a (*row*) *diagonally dominant M-matrix.* It turns out that these off-diagonal matrix entries and the s_i, not the diagonal entries a_{ii}, are the right parameters for doing accurate linear algebra with this class of matrices. Intuitively, it is clear that the s_i are the natural parameters since the conditions $s_i \geq 0$ define the class.

We explain how to do accurate *LDU* decomposition with no pivoting or complete pivoting, in the case of a row diagonally dominant *M*-matrix. Briefly, the algorithm can be organized to satisfy NIC (see Demmel and Koev (2004*b*), O'Cinneide (1996) and Peña (2004) for details). For simplicity of notation, let the n^2 matrix parameters be $b_{ij} = -a_{ij}$ and s_i, so all are non-negative. The diagonal elements, a_{ii}, are readily available accurately as a sum of positive numbers:

$$a_{ii} = s_i + \sum_{i=1}^{n} b_{ij}. \tag{2.2}$$

The Schur complements computed using Gaussian elimination with complete or no pivoting inherit the diagonally dominant *M*-matrix structure. The parameters defining the Schur complement – the row sums (call them s_i') and off-diagonal elements (call them $a_{ij}' = -b_{ij}'$) – are rational functions with positive coefficients in the s_is and b_{ij}s:

$$s_i' = s_i + \frac{b_{i1}}{a_{11}}s_1, \qquad b_{ij}' = b_{ij} + \frac{b_{i1}}{a_{11}}b_{1j},$$

with a_{ii} given by (2.2). Since the above expressions satisfy NIC, the *LDU* decomposition computed using them will be accurate, as will the subsequent SVD.

Several improvements on this results have been made. Peña (2004) suggested an alternative diagonal pivoting strategy which guarantees L and U to be well conditioned (as opposed to 'well conditioned in practice' which is what Gaussian elimination with complete pivoting delivers). Ye (2008*a*, 2008*b*) generalized this approach to symmetric diagonally dominant matrices (removing the restriction on the signs of off-diagonal elements). It turns out that in the process of Gaussian elimination with complete pivoting, updating the s_i and the diagonal entries still satisfies NIC. However, there can be (arbitrary) cancellation in the off-diagonal entries. Nonetheless, Ye shows that the errors in the off-diagonal entries can be bounded in absolute value so as to be able to guarantee that L and U are computed with small norm-wise errors, which is all that is required for an RRD to, in turn, provide an accurate SVD.

2.3.5. Matrices with displacement rank one
Matrices A that satisfy the Sylvester equation

$$DA - AT = B,$$

where $B = uv^T$ is unit rank, are said to have *displacement rank one*. In the easiest case, when D and T are diagonal ($D = \mathrm{diag}(d_1, d_2, \ldots, d_n)$, $T = \mathrm{diag}(t_1, t_2, \ldots, t_n)$), A is a (quasi-Cauchy) matrix $a_{ij} = \frac{u_i v_j}{d_i - t_j}$ (Kailath and Olshevsky 1995, 1997).

The quasi-Cauchy structure is preserved in the process of Gaussian elimination with complete pivoting (Demmel 1999, Demmel *et al.* 1999). The explicit formula for a determinant (or a minor) of a (quasi-)Cauchy matrix satisfies NIC as mentioned before. In fact, Gaussian elimination can still be made accurate at a cost of $\mathcal{O}(n^3)$ just by changing the inner loop from (2.1) to

$$a'_{ij} = a_{ij} \cdot \frac{(d_i - d_k)(t_k - t_j)}{(d_k - t_j)(d_i - t_k)}.$$

This is the starting point in computing the SVD of many displacement rank-one matrices. The Vandermonde matrix $V = \left[x_i^{j-1}\right]_{i,j=1}^n$ has a displacement rank one, where $D = \mathrm{diag}(x_1, x_2, \ldots, x_n)$ and T is the lower shift matrix $t_{i,i-1} = 1, i = 1, 2, \ldots, n - 1, t_{1n} = 1$.

Then $DA - AT = (x_1^n - 1, x_2^n - 1, \ldots, x_n^n - 1)^T(0, 0, \ldots, 0, 1) \equiv B$. The matrix T is circulant (and a root of unity) and is diagonalized $T = Q\Lambda Q^*$ by the (unitary) matrix of the DFT $Q_{ij} = \alpha^{(i-1)(j-1)}$, where α is a primitive nth root of unity, with eigenvalues $\Lambda_{ii} = \alpha^{(i-1)(n-1)}$.

Thus $DA - AQ\Lambda Q^* = B$, and so $D(AQ) - (AQ)\Lambda = BQ$, i.e., AQ is a quasi-Cauchy matrix (since BQ still has rank one). Now from an accurate SVD of $AQ = U\Sigma V^*$ we automatically obtain an accurate SVD of $A = U\Sigma(QV)^*$. But note that we need both the constant matrices Q and Λ for this to work, which goes beyond NIC.

The same idea generalizes to other displacement rank-one matrices. For example, if $DA - AQ = B$ and D and T are unitarily diagonalizable, $D = QD_1Q^*$ and $T = SD_2S^*$, then

$$D_1(Q_1^*AQ_2) - (Q_1^*AQ_2)D_2 = (Q_1^*u)(v^TQ_2)$$

and $Q_1^*AQ_2$ is a quasi-Cauchy matrix. If the decompositions $D = QD_1Q^*$ and $T = SD_2S^*$, and the products Q_1^*u and v^TQ_2 can be formed accurately, then from an accurate SVD of the quasi-Cauchy matrix $Q_1^*AQ_2 = U\Sigma V^*$ we obtain an accurate SVD of A: $A = (Q_1U)\Sigma(Q_2V)^*$. This approach works, *e.g.*, for polynomial Vandermonde matrices involving orthogonal polynomials (Demmel and Koev 2006) – see also Demmel *et al.* (1999), Demmel (1999), Higham (1988) and Kailath and Olshevsky (1997) – but again requires one to know certain constants accurately, thus going beyond NIC.

2.3.6. *Totally non-negative and TN^J sign-regular matrices*

The matrices all of whose minors are non-negative are called *totally non-negative* (TN). Despite this seemingly severe restriction on the minors, TN matrices arise frequently in practice: a Vandermonde matrix with positive and increasing nodes, the Pascal matrix, and the Hilbert matrix are all examples of TN matrices. The first reference in the literature (that we are aware of) for accurate matrix computations dates back to 1963 for a Vandermonde matrix with positive and increasing nodes in an example of Kahan and Farkas (1963a, 1963c, 1963b). This phenomenon was rediscovered in the celebrated paper by Björck and Pereyra (1970) and later carefully analysed and generalized (Boros *et al.* 1999, Higham 1987, 1990, Marco and Martínez 2007, Demmel and Koev 2005, Martínez and Peña 1998, 1998, 2003). All these methods are based on explicit decompositions of the corresponding matrices where all entries of the decompositions may be computed with expressions satisfying NIC.

These ideas generalize to *any* TN matrix (Koev 2005, 2007) and are based on a structure theorem for TN matrices (Fallat 2001, Gasca and Peña 1992, 1996): any non-singular TN matrix can be decomposed as a product of non-negative bidiagonal factors

$$A = L^{(1)} L^{(2)} \cdots L^{(n-1)} D U^{(n-1)} \cdots U^{(1)}. \tag{2.3}$$

As mentioned before, this variation on Gaussian elimination, called Neville elimination, arises by eliminating all off-diagonal matrix entries by adding a multiple of row (resp. column) i to row (resp. column) $i + 1$ to zero out one entry, and eliminating entries diagonal by diagonal, from the outermost (with row, resp. column, multipliers stored in $L^{(1)}$, resp. $U^{(1)}$) to innermost (with row, resp. column, multipliers stored in $L^{(n-1)}$, resp. $U^{(n-1)}$). There are exactly n^2 independent non-negative parameters in the above decomposition. They parametrize the space of *all* TN matrices.

It turns out that it is possible to perform essentially all linear algebra on TN matrices by using only TN-preserving transformations. In other words, given the parametrization of A in (2.3), it is possible to accurately compute the parametrization of a submatrix (unsigned) inverse, Schur complement, converse, or product of two such matrices, all in $\mathcal{O}(n^3)$ time and satisfying NIC (Koev 2007). In other words, the ability to do accurate linear algebra is 'closed' under all these operations. Furthermore, based on NIC, it is possible to accurately reduce such a parametrized matrix to bidiagonal form, enabling an accurate SVD, and to accurately reduce it to tridiagonal form $T = BB^T$ by a similarity, reducing the non-symmetric eigenvalue problem to an accurate SVD (Koev 2005). Thus, virtually all linear algebra with TN matrices can be performed accurately.

The only remaining question is about the starting point of this approach – the accurate bidiagonal decompositions of the original matrix. The entries of the bidiagonal decomposition are products of quotients of initial minors (*i.e.*, contiguous minors that include the first row or column). Thus, for virtually all well-known TN matrices – Pascal, Vandermonde, Cauchy (as well as their products, Schur complements, *etc.*) – there are accurate formulas for their computation (Boros *et al.* 1999, Koev 2005, Martínez and Peña 1998, 2003).

A matrix is *sign-regular* (Gantmacher and Krein 2002) if all minors of the same order have the same sign (but not necessarily all positive as is the case with TN matrices). A row- or a column-reversed TN matrix is sign-regular, and the class of such matrices is denoted TN^J. Most linear algebra problems for TN^J matrices follow trivially from the corresponding TN algorithms, except for the eigenvalue algorithm (Koev 2007), which requires a TN^J-preserving transformation into a symmetric anti-bidiagonal matrix.

We believe that the eigenvalue algorithms for TN and TN^J are the first examples of accurate eigenvalue algorithms for non-symmetric matrices.

2.4. Going beyond NIC (no inaccurate cancellation)

We have cited several examples where we can do more general classes of accurate structured matrix computations by using more general building blocks than permitted by insisting on no inaccurate cancellation (NIC).

An accurate SVD of a Vandermonde matrix required knowing roots of unity accurately (or, more precisely, being able to perform the operation $x - \alpha$ accurately, where α is a root of unity). More general displacement rank-one problems required similar accurate operations for constants α drawn from eigenvalues from a fixed sequence of matrices, as well as the knowledge of the orthogonal eigenvectors of these matrices.

Most interestingly, by allowing ourselves to accurately compute a given set of polynomials, but all of bounded numbers of terms and degrees, we can extend our DSTU approach from being able to accurately find eigenvalues of only rather simply discretized differential equations, to being able to accurately compute all the eigenvalues of the scalar elliptic partial differential equation $\nabla \cdot (\theta \nabla u) + \lambda \rho u = 0$ on a domain Ω with zero Dirichlet boundary conditions, where $\theta(x)$ and $\rho(x)$ are scalar functions discretized on a general triangulated mesh in a standard way (isoperimetric finite elements on a triangulated mesh). In this case it is the smallest eigenvalues that are of physical interest, and they are accurately determined by the coefficients of the PDE. This result depends on a novel matrix factorization of the discretized differential operator in Boman, Hendrickson and Vavasis (2004).

It is examples such as these that encourage us to systematically ask what expressions we can accurately evaluate, including by allowing ourselves additional 'black boxes' as building blocks. This is the topic of the next section.

3. Accurate algorithms for polynomial evaluation

In this section we give a partial answer to the question 'When can a multivariate (real or complex) polynomial be evaluated accurately?' These results (except for Section 3.5.3) have been published, with completely rigorous proofs, in Demmel *et al.* (2006); we provide here intuitions and proof sketches.

To summarize the content of this section, we give (sometimes tight) necessary and sufficient conditions for accurate multivariate polynomial evaluation over given domains. These conditions depend strongly on the type of arithmetic chosen, specifically on the type of 'basic' operations allowed, as well as on the domain that the inputs are taken from (and also on whether the inputs belong to $\mathbb{R}^n$ or to $\mathbb{C}^n$).

Intuitively, accurate evaluation of small quantities is a more complicated issue than accurate evaluation of large quantities; thus the 'interesting' domains, as we will see, lie arbitrarily close to or intersect the *variety* of the polynomial (the set of points where the polynomial is 0). Evaluation on domains that are not of this type (but are otherwise sufficiently well behaved) is easy (see Section 3.2). Therefore, the variety plays a *necessary role*.

Example 3.1. To illustrate the role of the variety, we use the following example. Consider the 2-parameter family of polynomials

$$M_{jk}(x) = j \cdot x_3^6 + x_1^2 \cdot x_2^2 \cdot (j \cdot x_1^2 + j \cdot x_2^2 - k \cdot x_3^2),$$

where j and k are positive integers, and the domain of evaluation is $\mathbb{R}^3$. Assume that we allow only addition, subtraction and multiplication of two arguments as basic arithmetic operations, together with comparisons and branching.

When $k/j < 3$, $M_{jk}(x)$ is *positive definite*, *i.e.*, zero only at the origin and positive elsewhere. This will mean that $M_{jk}(x)$ is easy to evaluate accurately using a simple method discussed in Section 3.2.

When $k/j > 3$, then we will show that $M_{jk}(x)$ cannot be evaluated accurately by *any* algorithm using only addition, subtraction and multiplication of two arguments. This will follow from a simple necessary condition on the real variety $V_{\mathbb{R}}(M_{jk})$, the set of real x where $M_{jk}(x) = 0$: see Theorem 3.10.

When $k/j = 3$, *i.e.*, on the boundary between the above two cases, $M_{jk}(x)$ is a multiple of the Motzkin polynomial (Reznick 2000). The real variety $V_{\mathbb{R}}(M_{jk}) = \{x : |x_1| = |x_2| = |x_3|\}$ of this polynomial satisfies the necessary condition of Theorem 3.10, and, to our knowledge, the simplest accurate algorithm to evaluate it has 8 cases depending on the relative values of $|x_i \pm x_j|$. For example, on the branch defined by the inequalities $x_1 - x_3| \le |x_1 + x_3| \wedge |x_2 - x_3| \le x_2 + x_3|$, the algorithm evaluates p using the non-obvious formula

$$
\begin{aligned}
p(x_1, x_2, x_3) = {} & x_3^4 \cdot [4((x_1 - x_3)^2 + (x_2 - x_3)^2 + (x_1 - x_3)(x_2 - x_3))] \\
& + x_3^3 \cdot [2(2(x_1 - x_3)^3 + 5(x_2 - x_3)(x_1 - x_3)^2 \\
& \quad + 5(x_2 - x_3)^2(x_1 - x_3) + 2(x_2 - x_3)^3)]
\end{aligned}
$$

$$+ x_3^2 \cdot [(x_1 - x_3)^4 + 8(x_2 - x_3)(x_1 - x_3)^3 + (x_2 - x_3)^4$$
$$+ 9(x_2 - x_3)^2(x_1 - x_3)^2 + 8(x_2 - x_3)^3(x_1 - x_3)]$$
$$+ x_3 \cdot [2(x_2 - x_3)(x_1 - x_3)((x_1 - x_3)^3 + (x_2 - x_3)^3$$
$$+ 2(x_2 - x_3)(x_1 - x_3)^2 + 2(x_2 - x_3)^2(x_1 - x_3)]$$
$$+ (x_2 - x_3)^2(x_1 - x_3)^2((x_1 - x_3)^2 + (x_2 - x_3)^2).$$

In contrast to the real case, when the domain is $\mathbb{C}^3$, Theorem 3.10 will show that $M_{jk}(x)$ cannot be accurately evaluated using only addition, subtraction and multiplication.

The necessary conditions we obtain for accurate evaluability depend only on the variety of $p(x)$, but the variety alone is not always enough.

Example 3.2.   Consider the irreducible, homogeneous, degree $2d$, real polynomial

$$p(x) = (x_1^{2d} + x_2^{2d}) + (x_1^2 + x_2^2)(q(x_3, \ldots, x_n))^2,$$

where $q(\cdot)$ is homogeneous of degree $d-1$. The variety $V(p) = \{x_1 = x_2 = 0\}$ satisfies the necessary condition for accurate evaluability, but near $V(p)$ the polynomial $p(x)$ is 'dominated' by $(x_1^2 + x_2^2)(q(x_3, \ldots, x_n))^2$, so accurate evaluability of $p(x)$ depends on the accurate evaluability of $q(\cdot)$.

We may now apply the same principle to $q(\cdot)$, *etc.*, thus creating a decision tree of polynomials. Rather than a characterizing theorem, one might expect therefore that, in many cases, the answer can only be given by a recursive decision procedure, expanding $p(x)$ near the components of its variety and so on. We discuss this more in Section 3.3.

The rest of Section 3 is structured as follows. In Section 3.1, we formalize the type of algorithms we are interested in. Section 3.2 makes rigorous the intuition that accurate evaluation 'far from the variety' is possible. Section 3.3 considers the traditional model of arithmetic, on 'well-behaved' domains similar to the ones chosen for the algorithms of Section 2. This model has three basic operations, $+, -, \times$, and allows for exact negation. While not sufficient for the accurate evaluation *everywhere* of even simple polynomial expressions such as $x + y + z$, the traditional model is simple enough to allow us to give a characterization of accurately evaluable *complex* polynomials, as well as (generally distinct) necessary and sufficient conditions for accurate evaluability of real polynomials (sometimes these conditions are identical, and offer a complete characterization). In addition, for the real case, we show current progress toward constructing a decision procedure for accurate evaluability of real polynomials. Section 3.4 expands the practical scope of our analysis, since concluding that a computation is 'impossible' is not the end of the story; instead, this prompts the question of which additional computational building blocks would be needed to make it possible.

For example, current computers often have a 'fused multiply-add' instruction $x + y \cdot z$ that computes the answer with one rounding error, and there are software libraries that provide collections of accurately implemented polynomials needed for certain applications, *e.g.*, computational geometry (Shewchuk 1997). Given any such a collection of what we will call 'blackbox' operations (about which we assume only a small relative error), we will ask how much larger a set of polynomials can be evaluated accurately.

Finally, Section 3.5 discusses the implications of these results. Firstly, they shed some light on the existence of accurate algorithms for linear algebra operations like the ones described in Section 2: each such algorithm satisfies NIC (see Section 1, and thus also satisfies the necessary condition for accurate evaluability presented in Theorem 3.10). The apparently unrelated classes of structured matrices for which efficient and accurate linear algebra algorithms exist share a common underlying algebraic structure. Also, there may be other structured matrix classes sharing this property and for which accurate algorithms could be built. Secondly, our results show that some expressions or classes of problems *cannot* be accurately evaluated, even with an arbitrary set of bounded-degree black-box operations at our disposal. The practical implication of this is that, for certain types of problems, the use of arbitrarily high precision is necessary (see Section 4). Lastly, but perhaps most importantly, our results lay down a path toward the ultimate goal: a decision procedure (or 'compiler') which, given as inputs a polynomial p, a domain $\mathcal{D}$, and (perhaps) a set of black-box operations, either produces an accurate algorithm for the evaluation of p on $\mathcal{D}$ (including how to choose the machine precision ϵ for the desired relative error η: see Section 3.1), or exhibits a 'minimal' set of black-box operations that are still needed.

3.1. Formal statement and models of algorithms

We formalize here both the problem and the models of algorithms we will use. We introduce the notation $p_{\text{comp}}(x, \delta)$ for the output of the algorithm, and $\delta = (\delta_1, \delta_2, \dots, \delta_k)$ for the vector of rounding errors.

For example, consider the algorithm that computes $p(x) = x_1 + x_2 + x_3$ by performing two additions: it first adds x_1 to x_2, and then adds the result to x_3. If the first and second additions introduce the relative errors δ_1, respectively δ_2, we obtain that, for this algorithm,

$$
\begin{aligned}
p_{\text{comp}}(x, \delta) &= \big((x_1 + x_2)(1 + \delta_1) + x_3\big)(1 + \delta_2) \\
&= (x_1 + x_2 + x_3)(1 + \delta_2) + (x_1 + x_3)\delta_1(1 + \delta_2). \quad (3.1)
\end{aligned}
$$

We give below a formal description of the algorithms we consider. For more in-depth discussion of these assumptions and comparisons with other models of computations, see Section 4.

Definition 3.3.   All algorithms considered in this section will satisfy the following constraints.

(1) The inputs x are given exactly, rather than approximately.

(2) The algorithm always computes the output $p_{\text{comp}}(x, \delta)$ in finitely many steps and, moreover, computes the exact value of $p(x)$ when all rounding errors $\delta = 0$. This constraint excludes iterative algorithms which might produce an approximate value of $p(x)$ even when $\delta = 0$. Some of the reasons for this choice can be found in Section 2.2.

(3) The basic arithmetic operations beyond the traditional addition, subtraction and multiplication, if any, must be given explicitly. We refer to the case when additional polynomial operations are included as *extended arithmetic*. Constants are available to our algorithms only in the extended model and are also given explicitly.

(4) We consider algorithms both with and without comparisons and branching, since this choice may change the set of polynomials that we can accurately evaluate. In the branching case, note that $p_{\text{comp}}(x, \delta)$ will actually be piecewise polynomial.

(5) If the computed value of an operation depends only on the values of its operands, *i.e.*, if the same operands x and y of $\text{op}(x, y)$ always yield the same δ in $\text{rnd}(\text{op}(x, y)) = \text{op}(x, y) \cdot (1 + \delta)$, then we call our model *deterministic*; else it is *non-deterministic*. One can show that comparisons and branching let a non-deterministic machine simulate a deterministic one, and subsequently restrict our investigation to the easier non-deterministic model.

Finally, we must formalize what type of domains we consider. Although, in principle, any semi-algebraic set $\mathcal{D}$ could be examined, for simplicity we consider open domains $\mathcal{D}$, especially $\mathcal{D} = \mathbb{R}^n$ or $\mathcal{D} = \mathbb{C}^n$. We can now give the formal definition of accuracy.

Definition 3.4.   We say that $p_{\text{comp}}(x, \delta)$ is an *accurate algorithm* for the evaluation of $p(x)$ for $x \in \mathcal{D}$ if

$$\forall\, 0 < \eta < 1 \quad \ldots \text{ for any } \eta = \text{desired relative error}$$
$$\exists\, 0 < \epsilon < 1 \quad \ldots \text{ there is an } \epsilon = \text{machine precision}$$
$$\forall\, x \in \mathcal{D} \quad \ldots \text{ so that for all } x \text{ in the domain}$$
$$\forall\, |\delta_i| \leq \epsilon \quad \ldots \text{ and for all rounding errors bounded by } \epsilon$$
$$|p_{\text{comp}}(x, \delta) - p(x)| \leq \eta \cdot |p(x)| \ldots \text{ the relative error is at most } \eta.$$

Note that the algorithm proposed above, which produces the p_{comp} given in (3.1) for the evaluation of $x_1 + x_2 + x_3$, is not an accurate algorithm (consider the case when $x_1 + x_2 = -x_3$). This is not accidental (see Theorem 3.10).

Given an algorithm producing a polynomial p_{comp}, the problem of deciding whether it is accurate *is* a Tarski-decidable problem (Renegar 1992, Tarski 1951). What is unclear is whether the *existence* of an accurate algorithm for a given polynomial and domain is a Tarski-decidable problem, since we see no way to express 'there exists an algorithm' in the required format.

3.2. The bounded-from-below case (empty variety)

We consider the simpler case where the polynomial $p(x)$ to be evaluated is bounded (in absolute value) above and below, in an appropriate manner, on the domain $\mathcal{D}$ (this is what we referred to previously as 'far from the variety', *i.e.*, the set where the polynomial is 0). If the domain $\mathcal{D}$ is compact, we give here, with proof, the following theorem. (We let $\bar{\mathcal{D}}$ denote the closure of $\mathcal{D}$.)

Theorem 3.5. Let $p_{\mathrm{comp}}(x, \delta)$ be *any* algorithm computing $p(x)$ satisfying $p_{\mathrm{comp}}(x, 0) = p(x)$, *i.e.*, it computes the right value in the absence of rounding error. Let $p_{\min} := \inf_{x \in \bar{\mathcal{D}}} |p(x)|$. Suppose $\bar{\mathcal{D}}$ is compact and $p_{\min} > 0$. Then $p_{\mathrm{comp}}(x, \delta)$ is an accurate algorithm for $p(x)$ on $\mathcal{D}$.

Proof. Since the relative error on $\mathcal{D}$ is

$$|p_{\mathrm{comp}}(x, \delta) - p(x)| / |p(x)| \leq |p_{\mathrm{comp}}(x, \delta) - p(x)| / p_{\min},$$

it suffices to show that the right-hand side numerator approaches 0 uniformly as $\delta \to 0$. This follows by writing the value of $p_{\mathrm{comp}}(x, \delta)$ along any branch of the algorithm as

$$p_{\mathrm{comp}}(x, \delta) = p(x) + \sum_{\alpha > 0} p_\alpha(x) \delta^\alpha,$$

where $\alpha > 0$ is a multi-index with at least one component exceeding 0. By compactness of $\bar{\mathcal{D}}$, all p_α are bounded on $\bar{\mathcal{D}}$, and thus there exists some constant $C > 0$ such that

$$\left| \sum_{\alpha > 0} p_\alpha(x) \delta^\alpha \right| \leq C \sum_{\alpha > 0} |\delta|^\alpha.$$

The right-hand side goes to 0 uniformly as the upper bound ϵ on each $|\delta_i|$ goes to zero. $\qquad\square$

What about domains that are not compact, *e.g.*, not bounded? The proof above points to some of the issues that may occur: ratios $p_\alpha(x)/p(x)$ could become unbounded, even though $p_{\min} > 0$. Another way to see that requiring $p_{\min} > 0$ is not enough is to consider the polynomial

$$p(x) = 1 + (x_1 + x_2 + x_3)^2.$$

To evaluate this polynomial accurately, intuitively, one needs to evaluate $(x_1 + x_2 + x_3)^2$ accurately, once it is sufficiently large. If one uses only addition, subtraction, and multiplication, this is not possible. (These considerations will be made explicit in Section 3.3.3.)

There are, however, cases in which unboundedness is not an impediment. Consider the case of a homogeneous polynomial $p(x)$, to be evaluated on a homogeneous domain $\mathcal{D}$ (*i.e.*, a domain with the property that $x \in \mathcal{D}$ implies $\gamma x \in \mathcal{D}$, for any scalar γ). Due to the homogeneity of p, we can then restrict our analysis to $\mathcal{D} \cap S^{n-1}$ (the unit ball in $\mathbb{R}^n$), or $\mathcal{D} \cap S^{2n-1}$ (the unit ball in $\mathbb{C}^n$). On such domains we can use a compactness argument, as we did before.

Theorem 3.6. Let $p(x)$ be a homogeneous polynomial, let $\mathcal{D}$ be a homogeneous domain, and let S denote the unit ball in $\mathbb{R}^n$ (or $\mathbb{C}^n$). Let

$$p_{\min,\text{homo}} := \inf_{x \in \bar{\mathcal{D}} \cap S} |p(x)|.$$

Then $p(x)$ can be evaluated accurately if $p_{\min,\text{homo}} > 0$.

A simple, Horner-like scheme that provides an accurate $p_{\text{comp}}(x, \delta)$ in this case is given in Demmel *et al.* (2006), along with a proof.

3.3. Traditional arithmetic

In this section we consider the basic or traditional arithmetic over the real or complex fields, with the three basic operations $\{+, -, \times\}$, to which we add negation. The model of arithmetic is governed by the laws in Section 3.1, and has also been described in Section 2. We remind the reader that this arithmetic model *does not allow* the use of constants.

Section 3.3.1 describes the necessary condition for accurate evaluability over both real and complex domains. Section 3.3.2, respectively Section 3.3.3, deals with sufficient conditions for accurate evaluability over $\mathbb{C}^n$, respectively $\mathbb{R}^n$. We show that the necessary and sufficient conditions for accurate evaluation coincide in the complex case, in Section 3.3.2. Section 3.3.3 also describes progress toward understanding how to construct a decision procedure in the real case.

Throughout this section, we will make use of the following definition of allowability.

Definition 3.7. Let p be a polynomial over $\mathbb{R}^n$ or $\mathbb{C}^n$, with variety $V(p) := \{x : p(x) = 0\}$. We call $V(p)$ *allowable* if it can be represented as a union of intersections of hyperplanes of the form

$$Z_i = \{x \ : \ x_i = 0\}, \tag{3.2}$$
$$S_{ij} = \{x \ : \ x_i + x_j = 0\}, \tag{3.3}$$
$$D_{ij} = \{x \ : \ x_i - x_j = 0\}. \tag{3.4}$$

If $V(p)$ is not allowable, we call it *unallowable*.

The word 'allowable' in the definition above is used because, as we will see, polynomials with 'unallowable' varieties do not allow for the existence of accurate evaluation algorithms.

For a polynomial p, having an allowable variety $V(p)$ is obviously a Tarski-decidable property (following Tarski (1951)), since the number of unions of intersections of hyperplanes (3.2)–(3.4) is finite.

3.3.1. Necessity: real and complex

All the statements, proofs, and proof sketches in this section work equally well for both the real and the complex case, and thus we will treat them together.

Throughout this section we will denote the variable space by $\mathcal{S} \in \{\mathbb{R}^n, \mathbb{C}^n\}$.

To state and explain the main result of this section, we need to introduce some additional notions and notation.

Definition 3.8. Given a polynomial p over $\mathcal{S}$ with unallowable variety $V(p)$, consider all sets W that are finite intersections of allowable hyperplanes defined by (3.2), (3.3), (3.4), and subtract from $V(p)$ all those W for which $W \subset V(p)$. We call the remaining subset of the variety *points in general position* and denote it by $G(p)$.

If $V(p)$ is not allowable, then from Definition 3.8 it follows that $G(p) \neq \emptyset$.

Definition 3.9. Given $x \in \mathcal{S}$, define the set $\mathrm{Allow}(x)$ as the intersection of all allowable planes going through x,

$$\mathrm{Allow}(x) := \left(\cap_{x \in Z_i} Z_i\right) \cap \left(\cap_{x \in S_{ij}} S_{ij}\right) \cap \left(\cap_{x \in D_{ij}} D_{ij}\right),$$

with the understanding that

$$\mathrm{Allow}(x) := \mathcal{S} \quad \text{whenever} \quad x \notin Z_i, \ S_{ij}, \ D_{ij} \quad \text{for all} \ \ i, j.$$

Note that $\mathrm{Allow}(x)$ is a linear subspace of S.

In general, we are interested in the sets $\mathrm{Allow}(x)$ primarily when $x \in G(p)$. For each such x, $\mathrm{Allow}(x) \not\subset V(p)$, which follows directly from the definition of $G(p)$.

We can now state the main result of this section, which is a necessity condition for the evaluability of polynomials over domains. In the following, we denote by $\overline{\mathrm{Int}(\mathcal{D})}$ the closure of the interior of the domain $\mathcal{D}$.

Theorem 3.10.   Let p be a polynomial over a domain $\mathcal{D} \in \mathcal{S}$, such that $\mathcal{D} = \overline{\mathrm{Int}(\mathcal{D})}$. Let $G(p)$ be the set of points in general position on the variety $V(p)$. If $\mathrm{Int}(\mathcal{D}) \cap G(p) \neq \emptyset$, then p is not accurately evaluable on $\mathcal{D}$.

With a little more work one can see that 'failures' are not rare. More precisely, in the same circumstances as above, any algorithm attempting to compute p accurately on $\mathcal{D}$ will fail to do so consistently on a set of positive measure.

Corollary 3.11.   Let p and $\mathcal{D}$ as before, $x \in \mathrm{Int}(\mathcal{D}) \cap G(p)$, $\epsilon > 0$, $1 > \eta > 0$, and $p_{\mathrm{comp}}(\cdot, \delta)$ be the result of an algorithm attempting to compute p on $\mathcal{D}$ with error vector δ. Then there exists a set Δ_x *arbitrarily close* to x and a set Δ_δ of positive measure in $H_\epsilon := \{\delta \; : \; |\delta_i| \leq \epsilon\}$ such that $|p_{\mathrm{comp}} - p|/|p| > \eta$ when computed at any point $y \in \Delta_x$ using any vector of relative errors $\delta \in \Delta_\delta$.

For the benefit of the reader we give here a sketch of the proof of Theorem 3.10 in an informal style. Details and rigorous statements can be found in Demmel *et al.* (2006).

Proof of Theorem 3.10. The essential idea is to consider under what kind of circumstances can an algorithm in which every non-trivial operation introduces errors actually produce a perfect 0. Note that, by definition, for an algorithm to be accurate, it must compute $p(x)$ *exactly* when $x \in V(p)$, and it cannot output 0 for any $x \notin V(p)$.

For starters, think of the algorithm as a *directed acyclic graph* (DAG) with input, computational, branching, and output nodes – as in Aho, Hopcroft and Ullman (1975). Every computational node has two inputs (which may both come from a single other computational node). All computational nodes are labelled by $(\mathrm{op}(\cdot), \delta_i)$ with $\mathrm{op}(\cdot)$ representing the operation that takes place at that node. It means that at each node, the algorithm takes in two inputs, executes the operation, and multiplies the result by $(1 + \delta_i)$. Finally, for every branch of the algorithm, there is a single destination node, with one input and no output, whose input value is the result of the algorithm.

For simplicity, in this sketch we only consider non-branching algorithms.

Assume that $x \in G(p)$ is fixed, and let us examine the algorithm as a function of the error variables δ. Some computational nodes in this DAG might do 'trivial' work (work that, given the input x, outputs 0 for all choices of variables δ). For example, such a node might receive input from a single computational node, subtract it from itself, and thus output 0. Note that multiplication nodes cannot produce a 0 unless they receive a 0 as an input.

For all non-trivial computation nodes, the output result is a polynomial of δ (and thus it will only vanish on a set of δs of measure 0).

As such, for any $x \in G(p)$, there will be a positive measure set Δ of δs for which non-trivial nodes will not output 0. Let us now choose some δ in this set and then look at the computational output node. Since we assume that the algorithm is accurate, the output node must be 0, and therefore the output node must be of 'trivial' type. Let us track back zeros in the computation, marking the nodes where such zeros appear and propagate from. In other words, backward-reconstruct paths of zeros that lead to the output of the computation.

Zeros propagate forward by multiplication, or by the addition/subtraction of identical quantities; but how do the *first* zeros on such paths (from the perspective of the computation) get created? A quick analysis shows that there are only three possibilities: either they are sources (zero as an input), or come from nodes corresponding to the trivial operation of subtracting an input from itself $(q(\delta) - q(\delta)$, since the node that computed this input must have been non-trivial), or they correspond to the addition or subtraction of two equal source inputs $(x_i = x_j$ or $x_i = -x_j)$.

We illustrate these possibilities in Figure 3.1. The white nodes are 'trivial' nodes, labelled with the operation executed there and the error variable; for clarity, we dropped the indices on the variables δ_i, and chose not to represent certain parts of the graph. The grey nodes are non-trivial nodes. Arrows

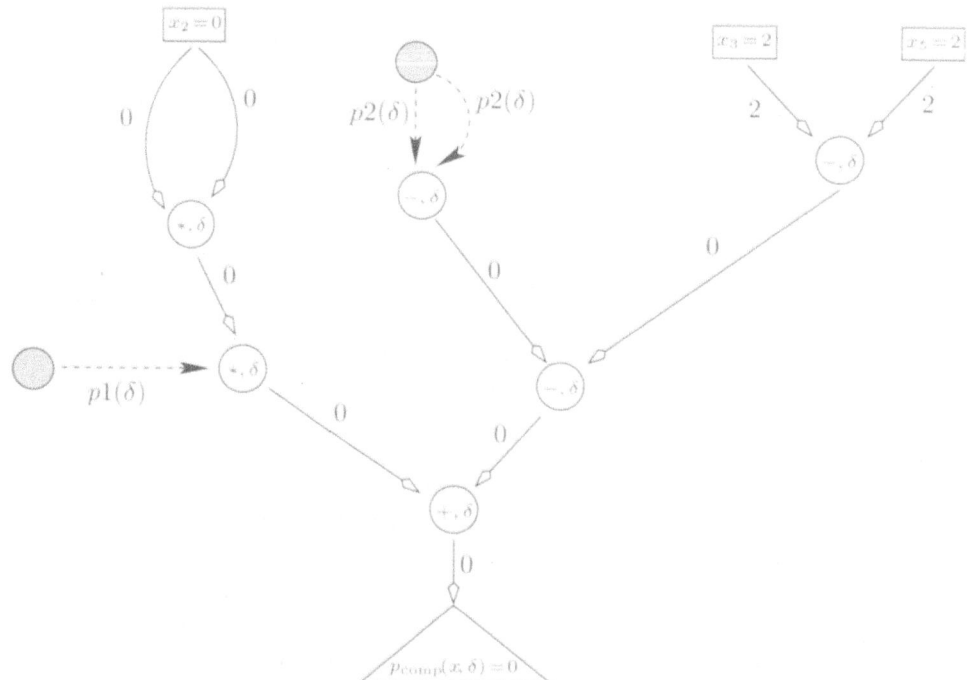

Figure 3.1. The three ways to produce zeros.

are labelled with the value they carry. Rectangles represent source nodes, and the triangle is the final output node.

The key observation is that *all of these zeros would be preserved if we replaced x with any $y \in \mathrm{Allow}(x)$*. In other words, if the algorithm outputs $p_{\mathrm{comp}}(x, \delta) = 0$, for some $\delta \in \Delta$, then it will also output $p_{\mathrm{comp}}(y, \delta) = 0$, for *all* $\delta \in \Delta$, and *all* $y \in \mathrm{Allow}(x)$.

For example, assume that the polynomial in Figure 3.1 is

$$p(x) = (x_1 + x_4 + x_6)^2 + x_2^4 + (x_3 - x_5)^2,$$

with unallowable variety

$$V(p) = \{x_1 + x_4 + x_6 = 0\} \cap \{x_2 = 0\} \cap \{x_3 = x_5\},$$

and that we want to compute p at $x = (1, 0, 2, 3, 2, -4) \in G(p)$. Then the result of the computation would be correct: $p_{\mathrm{comp}}(x, \delta) = 0$. However, this algorithm would also output $p_{\mathrm{comp}}(y, \delta) = 0$ for the point $y = (1, 0, 2, 3, 2, 4)$, which is in $\mathrm{Allow}(x) = \{x_2 = 0\} \cap \{x_3 = x_5\}$, but not in $V(p)$, since $p(y) = 16$.

Since $x \in G(p)$, $\mathrm{Allow}(x) \not\subseteq V(p)$, and thus the algorithm obtains 0 on points not in the variety, hence it fails.   □

3.3.2. *Sufficiency: the complex case*

Suppose we now restrict input values to be complex numbers and use the same algorithm types and the notion of accurate evaluability from the previous sections. By Theorem 3.10, for a polynomial p of n complex variables to be accurately evaluable over $\mathbb{C}^n$, it is necessary that its variety $V(p) := \{z \in \mathbb{C}^n : p(z) = 0\}$ be allowable.

We give and explain here a result that shows that this condition is also sufficient. This characterization is possible in the complex polynomial case because complex varieties are (pun intended) much simpler than real ones. In particular, Theorem 3.13 has no correspondent for real varieties, and therefore we cannot prove anything close to Theorem 3.12 for the real polynomial case.

Theorem 3.12. Let $p : \mathbb{C}^n \to \mathbb{C}$ be a polynomial with integer coefficients and zero constant term. Then p is accurately evaluable on $\mathcal{D} = \mathbb{C}^n$ if and only if the variety $V(p)$ is allowable.

To prove this, we first investigate allowable complex varieties. We start by recalling a basic fact about complex polynomial varieties (Theorem 3.13), which can, for example, be deduced from Theorem 3.7.4 in Taylor (2004, p. 53). Let V denote any complex variety. To say that $\dim_{\mathbb{C}}(V) = k$ means that, for each $z \in V$ and each $\delta > 0$, there exists $w \in V \cap B(z, \delta)$ such that w has a V-neighbourhood that is homeomorphic to a real $2k$-dimensional ball.

Theorem 3.13. Let p be a non-constant polynomial over $\mathbb{C}^n$. Then
$$\dim_{\mathbb{C}}(V(p)) = n - 1.$$

Corollary 3.14. Let $p : \mathbb{C}^n \to \mathbb{C}$ be a non-constant polynomial whose variety $V(p)$ is allowable. Then $V(p)$ is a union of allowable hyperplanes.

Proof. Since $V(p)$ is allowable, let $V(p) = \cup_j S_j$ be the (minimal) way to write $V(p)$ as an irredundant union of irredundant intersections of hyperplanes. Assume that, for some j_0, S_{j_0} is not a hyperplane but an (irredundant) intersection of hyperplanes. Let $z \in S_{j_0} \setminus \cup_{j \neq j_0} S_j$. Then, for some $\delta > 0$, $B(z,\delta) \cap V(p) \subset S_{j_0}$. Since $\dim_{\mathbb{C}}(S_{j_0}) < n - 1$, no point in $B(z,\delta) \cap V(p)$ has a $V(p)$-neighbourhood that is homeomorphic to a real $2(n-1)$-dimensional ball, which is a contradiction. $\square$

Corollary 3.15. If $p : \mathbb{C}^n \to \mathbb{C}$ is a polynomial whose variety $V(p)$ is allowable, then it is a product $p = c\prod_j p_j$, where each p_j is a power of x_i, $(x_i - x_j)$, or $(x_i + x_j)$.

Proof. By Corollary 3.14, the variety $V(p)$ is an irredundant union of allowable hyperplanes.

Choose a hyperplane H in that union. If $H = Z_{j_0}$ for some J_0, expand p into a Taylor series in x_{j_0}. If $H = D_{i_0 j_0}$ (or $H = S_{i_0 j_0}$) for some i_0, j_0, expand p into a Taylor series in $(x_{i_0} - x_{j_0})$ (or $(x_{i_0} + x_{j_0})$). In this case, the zeroth coefficient of p in the expansion must be the zero polynomial in x_j, $j \neq j_0$ (or $j \notin \{i_0, j_0\}$). Hence there is a k such that $p(x) = x_{j_0}^k \tilde{p}(x)$ in the first case, or $p(x) = (x_{i_0} \pm x_{j_0})^k \tilde{p}(x)$ in the second (third) one. In any case, we choose k maximal, so that $V(\tilde{p})$ does not include H.

It is easy to see that the variety $V(\tilde{p})$ must include $V(p) \setminus H$ (the union of all the other hyperplanes), whose dimension is $n - 1$. Moreover, $V(\tilde{p})$ (by Theorem 3.13) has dimension $n - 1$ and, by the maximality of k, does not include H.

If $V(\tilde{p}) \cap H := H'$ were non-empty, it would follow that $\dim(H') \leq n - 2$ (since it is included in the hyperplane H, and strictly smaller than H). This would contradict Theorem 3.13, which states that $\dim(V(\tilde{p})) = n - 1$. Therefore it must be that $V(\tilde{p}) \cap H = \emptyset$, and thus $V(\tilde{p})$ must equal $V(p) \setminus H$, the union of a smaller number of allowable hyperplanes.

Proceed inductively by factoring $\tilde{p}$ in the same fashion. $\square$

The crucial point in the proof above is that the $V(\tilde{p}) \cap H$ *must be* $\emptyset$, due to Theorem 3.13. The same argument would fail in the real case: to illustrate this, consider the polynomial $p(x_1, x_2, x_3) = x_1^4 + x_1^2(x_2 + x_3)^2$. The variety $V(p) = \{x_1 = 0\}$ has dimension 2, but, after factoring out x_1^2, the variety of the remaining polynomial, $\tilde{p} = x_1^2 + (x_2 + x_3)^2$, is given by $\{x_1 = 0\} \cup \{x_2 + x_3 = 0\}$, which has dimension 1. We can now prove Theorem 3.12.

Proof of Theorem 3.12. By Corollary 3.15, $p = c\prod_j p_j$, with each p_j a power of x_k or $(x_k \pm x_l)$. It also follows that c must be an integer since all coefficients of p are integers. Since each of the factors is accurately evaluable, and we can get any integer constant c in front of p by repeated addition (followed, if need be, by negation), which are again accurate operations, the algorithm that forms their product and then adds/negates to obtain c evaluates p accurately. $\qquad\square$

Theorem 3.12 implies that only homogeneous polynomials are accurately evaluable over $\mathbb{C}^n$.

3.3.3. *Sufficiency: toward a decision procedure for the real case*

In this section we relate the accurate evaluability of a polynomial to the accurate evaluability of its 'dominant terms', and explore a possible avenue toward a decision procedure to establish the former via a recursive/inductive procedure based on the latter.

We consider only homogeneous polynomials, for reasons outlined in Section 3.2, and we also consider separately the branching and non-branching cases. Most of the section is devoted to non-branching algorithms, but we do need branching for our statements at the end; we keep the reader informed of all changes in the assumptions.

To accurately compute a homogeneous polynomial of degree d using a non-branching algorithm, one needs to use a homogeneous algorithm, described by the following definition and lemma, to be used later in Section 3.3.5.

Definition 3.16. We call an algorithm $p_{\mathrm{comp}}(x, \delta)$ with error set δ for computing $p(x)$ *homogeneous of degree d* if:

(1) the final output is of degree d in x,

(2) no output of a computational node exceeds degree d in x,

(3) the output of every computational node is homogeneous in x.

Lemma 3.17. If $p(x)$ is a homogeneous polynomial of degree d and if a non-branching algorithm evaluates $p(x)$ accurately by computing $p_{\mathrm{comp}}(x,\delta)$, the algorithm must itself be homogeneous of degree d.

The proof combines the relative errors $|p_{\mathrm{comp}}(x, \delta) - p(x)|/|p(x)|$, treated as in the proof of Theorem 3.5, and an analysis of the algorithm as a DAG, as in Section 3.3.1.

Owing to the complexity of the issues, the rest of this section is subdivided into four parts.

- Section 3.3.4 makes rigorous the notion of dominance and explains how to find the dominant terms by using various simple linear changes of variables.

- In Section 3.3.5, we explain how to 'prune' an algorithm to manufacture an algorithm that evaluates one of its dominant terms, and we establish that accurate evaluation of the dominant terms identified in Section 3.3.4 is *necessary* for the accurate evaluation of the polynomial.

- Section 3.3.6 establishes that accurate evaluation of a special set of dominant terms, together with the slices of space where they dominate, is sufficient for accurate evaluation of the polynomial.

- Finally, Section 3.3.7 discusses obstacles to a complete inductive procedure.

3.3.4. Dominance

We now describe what we mean by 'dominant terms' of the polynomial. Given an allowable variety $V(P)$, we fix an irreducible component of $V(p)$. Any such component is described by linear allowable constraints. We note (see Demmel *et al.* (2006)) that any given component of $V(p)$ can be put into the form $x_1 = x_2 = \cdots = x_k = 0$, using what we call a standard change of variables: standard changes of variables are linear transformations of the variables, which are intuitively simple, but whose exact combinatorial definition is long and we choose to leave it out.

After a standard change of variables, we look at the component $x_1 = x_2 = \cdots = x_k = 0$. We can assume that the polynomial $p(x)$ can be written (almost following MATLAB notation) as

$$p(x) = \sum_{\lambda \in \Lambda} c_\lambda x_{[1:k]}^\lambda q_\lambda(x_{[k+1:n]}),$$

where we write $x_{[1:k]} := (x_1, \ldots, x_k), x_{[k+1:n]} := (x_{k+1}, \ldots, x_n)$. Also, we let Λ be the set of all multi-indices $\lambda := (\lambda_1, \ldots, \lambda_k)$ appearing above.

To determine all dominant terms associated with the component $x_1 = x_2 = \cdots = x_k = 0$, consider the Newton polytope P of the polynomial p with respect to the variables x_1 through x_k only, *i.e.*, the convex hull of the exponent vectors $\lambda \in \Lambda$ (see, *e.g.*, Miller and Sturmfels (2005, p. 71)). Next, consider the normal fan $N(P)$ of P (see Ziegler (1995, pp. 192–193)) consisting of the cones of all row vectors η whose dot products with $x \in P$ are maximal for x on a fixed face of P. That means that, for every non-empty face F of P, we take

$N_F :=$

$$\left\{ \eta = (n_1, \ldots, n_k) \in (\mathbb{R}^k) : F \subseteq \left\{ x \in P : \eta x \left(:= \sum_{j=1}^k n_j x_j \right) = \max_{y \in P} \eta y \right\} \right\}$$

and

$$N(P) := \{N_F : F \text{ is a face of } P\}.$$

Finally, consider the intersection of the negative of the normal fan $-N(P)$ and the non-negative quadrant $\mathbb{R}_+^k$. This splits the first quadrant $\mathbb{R}_+^k$ into several regions S_{Λ_j} according to which subsets Λ_j of exponents λ 'dominate' close to the considered component of the variety $V(p)$, in the following sense.

Definition 3.18.   Let Λ_j be a subset of Λ that determines a face of the Newton polytope P of p such that the negative of its normal cone $-N(P)$ intersects $(\mathbb{R}^k)_+$ non-trivially (not only at the origin). Define $S_{\Lambda_j} \in (\mathbb{R}^k)_+$ to be the set of all non-negative row vectors η such that

$$\eta\lambda_1 = \eta\lambda_2 < \eta\lambda, \quad \forall\lambda_1, \lambda_2 \in \Lambda_j, \quad \text{and} \quad \lambda \in \Lambda \setminus \Lambda_j.$$

Note that if x_1 through x_k are small, then the exponential change of variables $x_j \mapsto -\log|x_j|$ gives rise to a correspondence between the non-negative part of $-N(P)$ and the space of original variables $x_{[1:k]}$. We map the sets S_{Λ_j} back into a neighbourhood of 0 in $\mathbb{R}^k$ by lifting.

Definition 3.19.   Let $F_{\Lambda_j} \subseteq [-1,1]^k$ be the set of all points $x_{[1:k]} \in \mathbb{R}^k$ such that

$$\eta := (-\log|x_1|, \ldots, -\log|x_k|) \in S_{\Lambda_j}.$$

For any j, the closure of F_{Λ_j} contains the origin in $\mathbb{R}^k$. Given a point $x_{[1:k]} \in F_{\Lambda_j}$, and given $\eta = (n_1, n_2, \ldots, n_k) \in S_{\Lambda_j}$, for any $t \in (0,1)$, the vector $(x_1 t^{n_1}, \ldots, x_k t^{n_k})$ is in F_{Λ_j}. Indeed, if $(-\log|x_1|, \ldots, -\log|x_k|) \in S_{\Lambda_j}$, then so is $(-\log|x_1|, \ldots, -\log|x_k|) - \log|t|\eta$, since all equalities and inequalities that define S_{Λ_j} will be preserved, the latter because $\log|t| < 0$.

Example 3.20.   Consider the following polynomial:

$$p(x_1, x_2, x_3) = x_2^8 x_3^{12} + x_1^2 x_2^2 x_3^{14} + x_1^8 x_3^{12} + x_1^6 x_2^{14} + x_1^{10} x_2^6 x_3^4.$$

This polynomial is positive and easy to evaluate accurately; the reason we have chosen it is to illustrate the Newton polytope, its normal fan, and the sets F_{Λ_j} and S_{Λ_j} defined above.

For this example,

$$V(p) = \{x_1 = x_2 = 0\} \cup \{x_1 = x_3 = 0\} \cup \{x_2 = x_3 = 0\}.$$

We examine the behaviour of the polynomial near the $x_1 = x_2 = 0$ component of the variety (*i.e.*, we consider x_3 to be large). Note that only the first three monomial terms, $x_2^8 x_3^{12}$, $x_1^2 x_2^2 x_3^{14}$, and $x_1^8 x_3^{12}$ will play an important role, since if $x_1, x_2 \ll 1$, $x_1^6 x_2^{14} \ll x_2^8 x_3^{12}$, respectively, $x_1^{10} x_2^6 x_3^4 \ll x_1^8 x_3^{12}$.

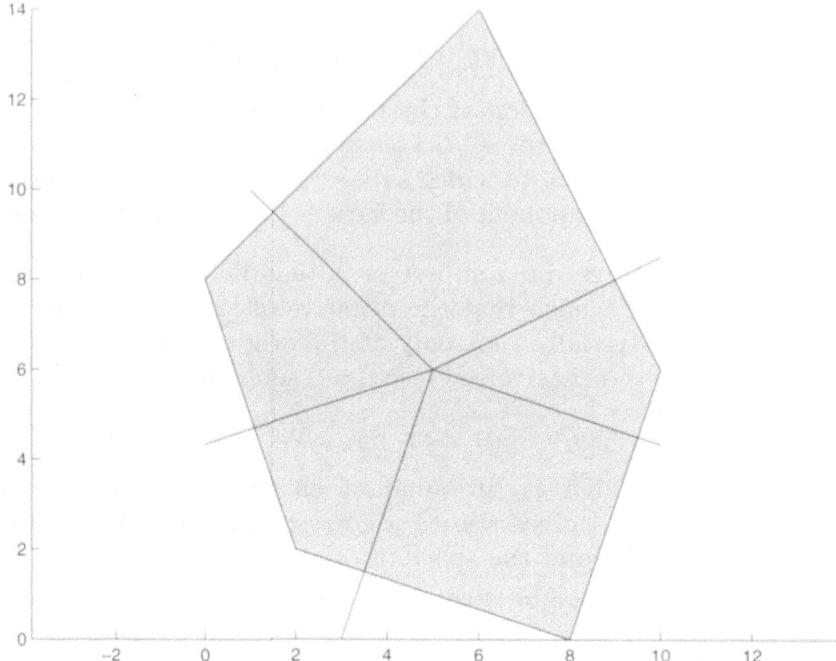

Figure 3.2. The Newton polytope P and its normal fan $N(P)$.

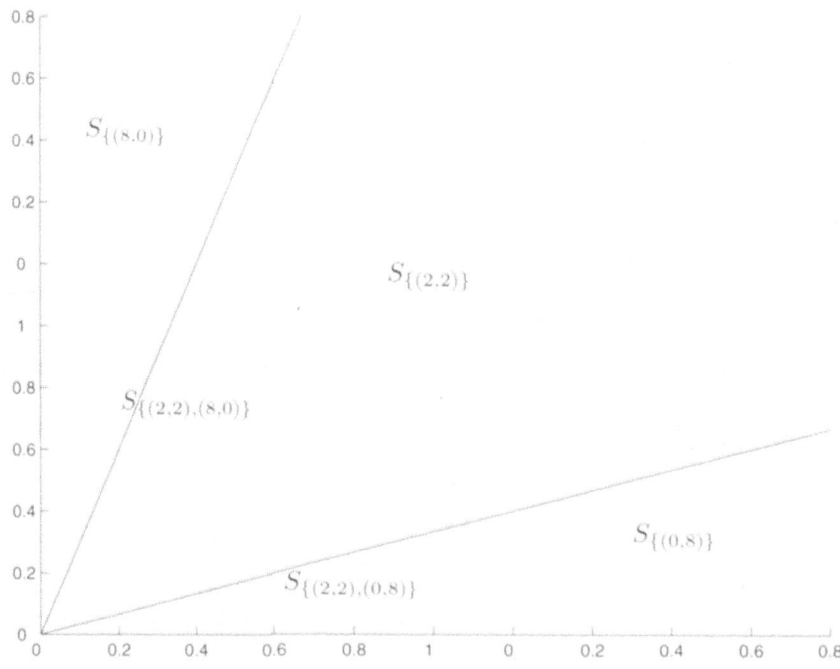

Figure 3.3. The intersection $-N(P) \cap \mathbb{R}_+^k$ and the regions S_{Λ_j}.

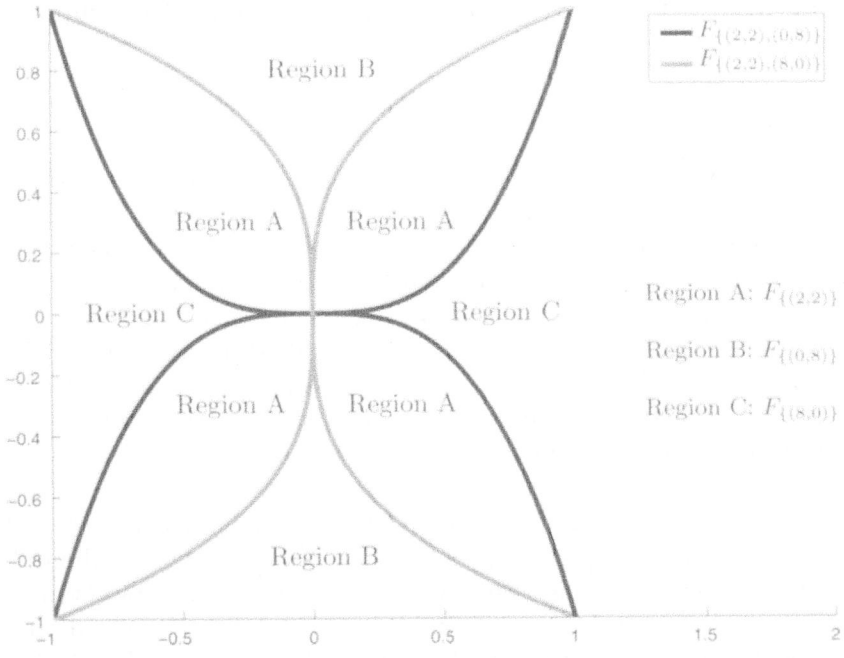

Figure 3.4. The regions F_{Λ_j}.

Figures 3.2, 3.3 and 3.4 show the Newton polytope P of p with respect to the variables x_1, x_2, its normal fan $N(P)$, the intersection $-N(P) \cap R_+^2$, the regions S_{Λ_j}, and the regions F_{Λ_j}.

Definition 3.21.   We define the *dominant term* of $p(x)$ corresponding to the component $x_1 = \cdots = x_k = 0$ and the region F_{Λ_j} by

$$p_{\mathrm{dom}_j}(x) := \sum_{\lambda \in \Lambda_j} c_\lambda x_{[1:k]}^\lambda q_\lambda(x_{[k+1:n]}).$$

The following observations about dominant terms are immediate.

Lemma 3.22.   Let $\eta = (n_1, \ldots, n_k) \in S_{\Lambda_j}$ and let $d_j := \sum_{\lambda_i \in \Lambda_j} \lambda_i n_i$. Let x^0 be fixed and let

$$x(t) := (x_1(t), \ldots, x_n(t)), \qquad x_j(t) := \begin{cases} t^{n_j} x_j^0, & j = 1, \ldots, k, \\ x_j^0, & j = k+1, \ldots, n. \end{cases}$$

Then $p_{\mathrm{dom}_j}(x(t))$ has degree d_j in t and is the lowest-degree term of $p(x(t))$ in t, that is,

$$p(x(t)) = p_{\mathrm{dom}_j}(x(t)) + o(t^{d_j}) \quad \text{as } t \to 0, \qquad \deg_t p_{\mathrm{dom}_j}(x(t)) = d_j.$$

Corollary 3.23. Under the assumptions of Lemma 3.22, suppose that $p_{\mathrm{dom}_j}(x^0) \neq 0$. Then

$$\lim_{t \to 0} \frac{p_{\mathrm{dom}_j}(x(t))}{p(x(t))} = 1.$$

The next question is whether the term p_{dom_j} indeed dominates the remaining terms of p in the region F_{Λ_j}, in the sense that $p_{\mathrm{dom}_j}(x)/p(x)$ is close to 1 sufficiently close to $x_1 = \cdots = x_k = 0$. Indeed, we show that each dominant term p_{dom_j}, such that the convex hull of Λ_j is a facet of the Newton polytope of p and whose variety $V(p_{\mathrm{dom}_j})$ does not have a component strictly larger than the set $x_1 = \cdots = x_k = 0$, dominates the remaining terms in p, not only in F_{Λ_j}, but in a certain *slice* $\tilde{F}_{\Lambda_j}$ around F_{Λ_j}. These dominant terms, corresponding to larger sets Λ_j, are the useful ones, since they pick up terms relevant not only in the region F_{Λ_j} but also in its neighbourhood.

In Example 3.20 above, the useful dominant terms correspond to the regions $F_{\{(2,2),(8,0)\}}$ and $F_{\{(2,2),(0,8)\}}$ (the only relevant edges of the polygon). This points to the fact that we should be ultimately interested only in dominant terms corresponding to the facets, *i.e.*, the highest-dimensional faces, of the Newton polytope of p. Note that the convex hull of Λ_j is a facet of the Newton polytope N if and only if the set S_{Λ_j} is a one-dimensional ray.

The next lemma will be instrumental for our results in Section 3.3.6. It shows that each dominant term p_{dom_j} such that the convex hull of Λ_j is a facet of the Newton polytope of p and whose variety $V(p_{\mathrm{dom}_j})$ does not have a component strictly larger than the set $x_1 = \cdots = x_k = 0$ indeed dominates the remaining terms in p in a certain 'slice' $\tilde{F}_{\Lambda_j}$ around F_{Λ_j}.

Lemma 3.24. Let p_{dom_j} be the dominant term of a homogeneous polynomial p corresponding to the component $x_1 = \cdots = x_k = 0$ of the variety $V(p)$ and to the set Λ_j whose convex hull is a facet of the Newton polytope N.

Let $\tilde{S}_{\Lambda_j}$ be any closed pointed cone in $(\mathbb{R}^k)_+$ with vertex at 0 that does not intersect other one-dimensional rays S_{Λ_l}, $l \neq j$, and contains $S_{\Lambda_j} \setminus \{0\}$ in its interior. Let $\tilde{F}_{\Lambda_j}$ be the closure of the set

$$\{x_{[1:k]} \in [-1,1]^k : (-\log|x_1|, \ldots, -\log|x_k|) \in \tilde{S}_{\Lambda_j}\}. \tag{3.5}$$

Suppose the variety $V(p_{\mathrm{dom}_j})$ of p_{dom_j} is allowable and intersects $\tilde{F}_{\Lambda_j}$ only at 0. Let $\|\cdot\|$ be any norm. Then, for any $\delta = \delta(j) > 0$, there exists $\varepsilon = \varepsilon(j) > 0$ such that

$$\left| \frac{p_{\mathrm{dom}_j}(x_{[1:k]}, x_{[k+1:n]})}{p(x_{[1:k]}, x_{[k+1:n]})} - 1 \right| < \delta \quad \text{whenever} \quad \frac{\|x_{[1:k]}\|}{\|x_{[k+1:n]}\|} \leq \varepsilon \text{ and } x_{[1:k]} \in \tilde{F}_{\Lambda_j}.$$

$$\tag{3.6}$$

For a proof of Lemma 3.24, the reader is referred to Demmel *et al.* (2006).

The above discussion of dominance was based on the transformation of a given irreducible component of the variety to the form $x_1 = \cdots = x_k = 0$. We must reiterate that the identification of dominant terms becomes possible only after a suitable change of variables C is used to put a given irreducible component into the standard form $x_1 = \cdots = x_k = 0$ and then the sets Λ_j are determined. Note, however, that the polynomial p_{dom_j} is given in terms of the original variables, *i.e.*, as a sum of monomials in the original variables x_q and sums/differences $x_q \pm x_r$. We therefore use the more precise notation $p_{\mathrm{dom}_j,C}$ in the rest of this section.

Definition 3.25.   Without loss of generality, we can assume that any standard change of variables has the form

$$x = (x_{[1:k_1]}, x_{[k_1+1:k_2]}, \ldots, x_{[k_{l-1}+1:k_l]})$$

$$\mapsto \widetilde{x} = (\widetilde{x}_{[1:k_1]}, \widetilde{x}_{[k_1+1:k_2]}, \ldots, \widetilde{x}_{[k_{l-1}+1:k_l]}),$$

$$\text{where } \widetilde{x}_{k_m+1} := x_{k_m+1}, \quad \widetilde{x}_{k_m+2} := x_{k_m+2} - \sigma_{k_m+2}x_{k_m+1}, \ldots,$$

$$\widetilde{x}_{k_m+1} := x_{k_m+1} - \sigma_{k_m+1}x_{k_m+1}, \quad k_0 := 0, \quad \sigma_r = \pm 1 \quad \text{for all pertinent } r.$$

Note also that we can think of the vectors $\eta \in S_{\Lambda_j}$ as being indexed by integers 1 through k_l, *i.e.*, $\eta = (n_1, \ldots, n_{k_l})$. Moreover, to define pruning in the next subsection we will assume that

$$n_{k_m+1} \leq n_r \quad \text{for all } r = k_m + 2, \ldots, k_{m+1} \quad \text{and for all } m = 0, \ldots, l - 1. \tag{3.7}$$

3.3.5. Pruning

We show here how to convert an accurate algorithm that evaluates a polynomial p into an accurate algorithm that evaluates a selected dominant term $p_{\mathrm{dom}_j,C}$. This will imply that being able to evaluate dominant terms accurately is a necessary condition for being able to evaluate the original polynomial accurately.

This process, which we will refer to as *pruning*, will consist of deleting some vertices and edges and redirecting certain other edges in the DAG that represents the algorithm. We explain the pruning process informally and through an example; for the rigorous definition, see Demmel *et al.* (2006).

Starting at the sources, we process each node provided that both of its inputs have been processed already (acyclicity insures that this can be done). Then, at any node u which performs an addition or subtraction of two inputs from nodes v and w of different degrees, we delete the node and the in-edge from the input of smaller degree (say v) and redirect the out-edge from u to w (the node with the larger degree output). Then we go backward and delete all nodes and/or edges on that sub-DAG, up to the source nodes. We denote the output of the pruned algorithm by $p_{\mathrm{dom}_j,C,\mathrm{comp}}(x,\delta)$.

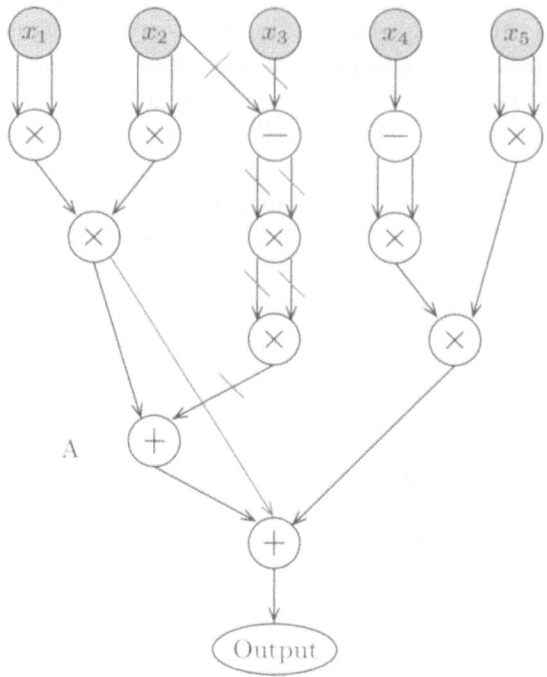

Figure 3.5. Pruning an algorithm for
$p(x) = x_1^2 x_2^2 + (x_2 - x_3)^4 + (x_3 - x_4)^2 x_5^2$.

Example 3.26. Figure 3.5 shows an example of pruning an algorithm that evaluates the polynomial

$$x_1^2 x_2^2 + (x_2 - x_3)^4 + (x_3 - x_4)^2 x_5^2$$

using the substitution

$$(tx_1, x_2, tx_3 + x_2, tx_4 + x_2, x_5)$$

near the component

$$x_1 = 0, \quad x_2 = x_3 = x_4.$$

The result of pruning is an algorithm that evaluates the dominant term

$$x_1^2 x_2^2 + (x_3 - x_4)^2 x_5^2.$$

The node A has two sub-DAGs leading to it; the right one (going back to the sources x_2 and x_3) is pruned due to the fact that it computes $(x_2 - x_3)^4$, a quantity of order $O(t^4)$, whereas the other produces $x_1^2 x_2^2$, a quantity of order $O(t^2)$.

The output of the original algorithm is given by

$$p_{\text{comp}}(x, \delta) = \big[(x_1^2(1 + \delta_1)x_2^2(1 + \delta_2)(1 + \delta_3)$$
$$+ (x_2 - x_3)^4(1 + \delta_4)^4(1 + \delta_5)^2(1 + \delta_6)\big](1 + \delta_7)$$
$$+ \big[(x_3 - x_4)^2(1 + \delta_8)^2(1 + \delta_9)x_5^2(1 + \delta_{10})(1 + \delta_{11})\big](1 + \delta_{12}).$$

The output of the pruned algorithm is

$$p_{\text{dom}_j,C,\text{comp}}(x, \delta) = \big[x_1^2 x_2^2(1 + \delta_1)(1 + \delta_2)(1 + \delta_3))(1 + \delta_7) + (x_3 - x_4)^2 x_5^2$$
$$\times (1 + \delta_8)^2(1 + \delta_9)(1 + \delta_{10})(1 + \delta_{11})\big](1 + \delta_{12}).$$

We formalize the main result regarding the pruning process below.

Theorem 3.27. Suppose a non-branching algorithm evaluates a polynomial p accurately on $\mathbb{R}^n$ by computing $p_{\text{comp}}(x, \delta)$. Suppose C is a standard change of variables (as in Definition 3.25) associated with an irreducible component of $V(p)$. Let $p_{\text{dom}_j,C}$ be one of the corresponding dominant terms of p and let S_{Λ_j} satisfy (3.7). Then the pruned algorithm with output $p_{\text{dom}_j,C,\text{comp}}(x, \delta)$ evaluates $p_{\text{dom}_j,C}$ accurately on $\mathbb{R}^n$. In other words, being able to compute all such $p_{\text{dom}_j,C}$ for all components of the variety $V(p)$ and all standard changes of variables C accurately is a necessary condition for computing p accurately.

3.3.6. Sufficiency of evaluating dominant terms

Our next goal is to prove a converse to Theorem 3.27; however, strictly speaking, the results that follow do not provide a true converse, since branching is needed to construct an algorithm that evaluates a polynomial p accurately from algorithms that evaluate its dominant terms accurately. Recall that Theorem 3.27 involves non-branching algorithms.

We make two assumptions: that our polynomial p is homogeneous and irreducible. The latter assumption effectively reduces the problem to that of accurate evaluation of a non-negative polynomial, due to the following lemma.

Lemma 3.28. If a polynomial p is irreducible and has an allowable variety $V(p)$, then it is either a constant multiple of a linear form that defines an allowable hyperplane, or it does not change its sign in $\mathbb{R}^n$.

Hence, we can restrict ourselves to the case of a homogeneous, irreducible, non-negative polynomial over the entire $\mathbb{R}^n$. For this case, we have the following theorem.

Theorem 3.29. Let p be a homogeneous non-negative polynomial whose variety $V(p)$ is allowable. Suppose that all dominant terms $p_{\text{dom}_j,C}$ for all components of the variety $V(p)$, all standard changes of variables C and all subsets Λ_j satisfying (3.7) are accurately evaluable. Then there exists a branching algorithm that evaluates p accurately over $\mathbb{R}^n$.

Proof of Theorem 3.29. We first show how to evaluate p accurately in a neighbourhood of each irreducible component of its variety $V(p)$. We next evaluate p accurately off these neighbourhoods of $V(p)$. The final algorithm will involve branching depending on which region the input belongs to, and the subsequent execution of the corresponding subroutine.

Consider a particular irreducible component V_0 of the variety $V(p)$; using a standard change of variables C, we map V_0 to a set of the form $\tilde{x}_1 = \cdots = \tilde{x}_k = 0$. We create an ϵ-neighbourhood of V_0 where we can evaluate p accurately; this neighbourhood is built up from semi-algebraic ϵ-neighbourhoods. More precisely, for each V_0, we can find a collection (S_j) of semi-algebraic sets, all determined by polynomial inequalities with integer coefficients, and the corresponding numbers ϵ_j, so that the polynomial p can be evaluated with desired accuracy η in each ϵ_j-neighbourhood of V_0 within the piece S_j. Moreover, testing whether a particular point x is within ϵ_j of V_0 within S_j can be done by branching based on polynomial inequalities with integer coefficients.

The final algorithm will be organized as follows. Given an input x, determine by branching whether x is in S_j and within the corresponding ϵ_j of a component V_0. If that is the case, evaluate $p(x)$ using the algorithm that is accurate in S_j in that neighbourhood of V_0. For x not in any of the neighbourhoods, evaluate p by Horner's rule. Since the polynomial p is strictly positive off the neighbourhoods of the components of its variety, the reasoning of Section 3.2 applies, showing that the Horner's rule algorithm is accurate. If x is on the boundary of a set S_j, any applicable algorithm will do, since the inequalities we use are not strict. Thus the resulting algorithm for evaluating p will have the desired accuracy η.   □

3.3.7. *Obstacles to a complete inductive procedure*

The results of the previous sections suggest the existence of an inductive procedure that could be used to determine whether or not a given polynomial is accurately evaluable by reducing the problem for the original polynomial p to the same problem for its dominant terms, then their dominant terms, and so forth, going all the way to 'base' cases: monomials or other polynomials that are easy to analyse. In order to work, the dominant terms would have to be simpler, or smaller, by some measure, than the original polynomial; this would require finding an induction variable that gets reduced at each step.

The most obvious two choices are the number of variables or the degree of the polynomial under consideration; unfortunately, there are cases when both fail to decrease. Furthermore, the dominant term may even coincide with the polynomial itself. For example, if

$$p(x) = A(x_{[3:n]})x_1^2 + B(x_{[3:n]})x_1 x_2 + C(x_{[3:n]})x_2^2,$$

where A, B, C are non-negative polynomials in x_3 through x_n, then the only useful dominant term of p in the neighbourhood of the set $x_1 = x_2 = 0$ is the polynomial p itself. For this case, analysing the dominant term yields no progress whatsoever.

Another possibility is induction on domains or slices of space, but we do not yet envision how to make this idea precise, since we do not know exactly when a given polynomial is accurately evaluable on a given domain.

Further work to establish a full decision procedure is therefore highly desirable.

3.4. Extended arithmetic

In this section, we consider adding 'black-box' real or complex polynomial operations to the basic, traditional model. We describe this type of operation below.

Definition 3.30. We call a black-box operation any type of operation that takes a number of inputs (real or complex) $x_1, \ldots, x_k$ and produces an output q such that q is a polynomial in $x_1, \ldots, x_k$.

Example 3.31. $q(x_1, x_2, x_3) = x_1 + x_2 x_3$.

Note that $+, -$, and $\cdot$ are all black-box operations on two inputs.

Consider a fixed set of multivariate polynomials $\{q_j : j \in J\}$ with real or complex inputs (perhaps infinite). In the extended arithmetic model, the operations allowed are the black-box operations $q_1, \ldots, q_k$, and negation. With the exception of negation, which is exact, all the others yield $\mathrm{rnd}(\mathrm{op}(a_1, \ldots, a_l)) = \mathrm{op}(a_1, \ldots, a_l)(1+\delta)$, with $|\delta| < \epsilon$ (ϵ being the machine precision). We consider the same arithmetic models as in Section 3.1, with this extended class of operations.

3.4.1. Necessity: real and complex

In order to analyse the way in which the necessity condition for having an allowable variety (Theorem 3.10) changes under these extended assumptions, we need to introduce a new, more general definition of allowability.

Essentially, a black box for computing p can be used for computing other polynomials, namely all the polynomials obtainable from p via permuting, repeating, negating, and zeroing some subset of the variables. Therefore each black box accounts for a potentially larger set of polynomials that can be evaluated with *a single* rounding error, using that black box, and we must consider all of them in our analysis. Note that in the traditional case (when we had addition, subtraction, and multiplication of two numbers as our black boxes) our set of three operations was closed under the aforementioned changes.

The definition below formalizes the set of polynomials obtainable from a given one, through this process of negation, repetition, permutation, and zeroing of variables.

Recall that we denote by $\mathcal{S}$ the space of variables (which may be either $\mathbb{R}^n$ or $\mathbb{C}^n$). From now on we will denote the set $\{1, \ldots, n\}$ by $\mathcal{K}$, and the set of pairs $(i, j) \in \mathcal{K} \times \mathcal{K}$ such that $i < j$ by $\mathcal{K}^2_<$.

Definition 3.32. Let $p(x_1, \ldots, x_n)$ be a multivariate polynomial over $\mathcal{S}$ with variety $V(p)$. Let $\mathcal{K}_Z \subseteq \mathcal{K}$, and let $\mathcal{K}_D, \mathcal{K}_S \subseteq \mathcal{K}^2_<$. Modify p as follows: impose conditions of the type Z_i for each $i \in \mathcal{K}_Z$, and of type D_{ij}, respectively S_{ij}, on all pairs of variables in $\mathcal{K}_D$, respectively $\mathcal{K}_S$. Rewrite p subject to those conditions (e.g., set $X_i = 0$ for all $i \in \mathcal{K}_Z$), and denote it by $\tilde{p}$, and denote by $\mathcal{K}_R$ the set of remaining independent variables (use the convention which eliminates the second variable in each pair in $\mathcal{K}_D$ or $\mathcal{K}_S$).

Choose a set $T \subseteq \mathcal{K}_R$, and let

$$V_{T, \mathcal{K}_Z, \mathcal{K}_D, \mathcal{K}_S}(p) = \cap_\alpha V(q_\alpha),$$

where the polynomials q_α are the coefficients of the expansion of $\tilde{p}$ in the variables x_T:

$$\tilde{p}(x_1, \ldots, x_k) = \sum_\alpha q_\alpha x_T^\alpha,$$

with q_α being polynomials in $x_{\mathcal{K}_R \setminus T}$ only.

Finally, let $\mathcal{K}_N$ be a subset of $\mathcal{K}_R \setminus T$. We negate each variable in $\mathcal{K}_N$, and let $V_{T, \mathcal{K}_Z, \mathcal{K}_D, \mathcal{K}_S, \mathcal{K}_N}(p)$ be the variety obtained from $V_{T, \mathcal{K}_Z, \mathcal{K}_D, \mathcal{K}_S}(p)$, with each variable in $\mathcal{K}_N$ negated.

For simplicity, we denote a set $(T, \mathcal{K}_Z, \mathcal{K}_D, \mathcal{K}_S, \mathcal{K}_N)$ by $\mathcal{I}$.

We illustrate this process by the following example.

Example 3.33. Let $p(x, y, z) = x + y \cdot z$ (the fused multiply-add). We record below some of the possibilities for the subvarieties $V_\mathcal{I}(p)$; the sets $\mathcal{I} = (T, \mathcal{K}_Z, \mathcal{K}_D, \mathcal{K}_S, \mathcal{K}_N)$ are implicit:

$$\begin{aligned}
V(p(x, 0, z)) &= \{x = 0\}, \\
V(p(x, x, x)) &= \{x = 0\} \cup \{x = -1\}, \\
V(p(0, y, z)) &= \{y = 0\} \cup \{z = 0\}, \\
V(p(x, y, -x)) &= \{x = 0\} \cup \{y = 1\}, \\
V(p(x, y, y)) &= \{x + y^2 = 0\}, \\
V(p(x, y, -z)) &= \{x - yz = 0\}.
\end{aligned}$$

We include the 'traditional' operations in the arithmetic by the definitions $q_{-2}(x_1, x_2) = x_1 x_2$, $q_{-1}(x_1, x_2) = x_1 + x_2$, and $q_0(x_1, x_2) = x_1 - x_2$, and note that the sets

$$Z_i = \{x \ : \ x_i = 0\}, \tag{3.8}$$

$$S_{ij} = \{x \ : \ x_i + x_j = 0\}, \tag{3.9}$$

$$D_{ij} = \{x \ : \ x_i - x_j = 0\} \tag{3.10}$$

describe all non-trivial sets of type $V_{\mathcal{I}}$, for q_{-2}, q_{-1}, and q_0.

We will assume from now on that the black-box operations q_j with $j \in J$ (J may be infinite, and $\{-2, -1, 0\} \subset J$) are given and fixed.

Definition 3.34.   We call any set $V_{\mathcal{I}}(q_j)$ with $\mathcal{I} = (T, \mathcal{K}_Z, \mathcal{K}_D, \mathcal{K}_S, \mathcal{K}_N)$ as defined above and q_j a black-box operation *basic q-allowable*.

We call any set R *irreducible q-allowable* if it is an irreducible component of a (finite) intersection of basic q-allowable sets, *i.e.*, when R is irreducible and

$$R \subseteq \cap_l Q_l,$$

where each Q_l is a basic q-allowable set.

We call any set Q *q-allowable* if it is a (finite) union of irreducible q-allowable sets, *i.e.*,

$$Q = \cup_j R_j,$$

where each R_j is an irreducible q-allowable set.

Any set R which is not q-allowable we call *q-unallowable*.

Note that the above definition of q-allowability is closed under taking union, intersection, and irreducible components. This parallels the definition of allowability for the classical arithmetic case: in the classical case, every allowable set was already irreducible (being an intersection of hyperplanes).

Definition 3.35.   Given a polynomial p with q-unallowable variety $V(p)$, consider all sets W that are q-allowable (as in Definition 3.34), and subtract from $V(p)$ those W for which $W \subset V(p)$. We call the remaining subset of the variety *points in general position* and denote it by $\mathcal{G}(p)$.

Since $V(p)$ is q-unallowable, $\mathcal{G}(p)$ is non-empty.

Definition 3.36.   Given $x \in \mathcal{S}$, define the set q-Allow(x) as the intersection of all basic q-allowable sets going through x:

$$q\text{-Allow}(x) := \cap_{j \in J} \left(\cap_{\mathcal{I} \ : \ x \in V_{\mathcal{I}}(q_j)} V_{\mathcal{I}}(q_j) \right),$$

for all possible choices of $\mathcal{I}$. The intersection in parentheses is $\mathcal{S}$ whenever $x \notin V_{\mathcal{I}}(q_j)$ for all $\mathcal{I}$.

Note that when $x \in \mathcal{G}(p)$, q-Allow$(x) \not\subseteq \mathcal{G}(p)$.

We can now state our necessity condition.

Theorem 3.37. Given the black-box operations $\{q_j : j \in J\}$, and the model of arithmetic described above, let p be a polynomial defined over a domain $\mathcal{D} \subset \mathcal{S}$. Let $\mathcal{G}(p)$ be the set of points in general position on the variety $V(p)$. If there exists $x \in \mathcal{D} \cap \mathcal{G}(p)$ such that $q\text{-Allow}(x) \cap \text{Int}(\mathcal{D}) \neq \emptyset$, then p is not accurately evaluable on $\mathcal{D}$.

Proof of Theorem 3.37. The proof mimics the proof of Theorem 3.10; once again, we trace back zeros to what we now call q-allowable conditions, and make use of the DAG structure of the algorithm. In the non-branching case, we obtain that if the algorithm is run on an input $x \in G(p)$, then either $p_{\text{comp}}(x, \delta) \neq 0$ for almost all δ, or $p_{\text{comp}}(y, \delta) = 0$ for all $y \in \text{Allow}(x) \setminus V(p)$ and for all δ. The proof for the branching case is again a refinement of the proof for the non-branching one. $\square$

Note that if we consider only algorithms without branching, Theorem 3.37 remains true in the tighter case when we drop the irreducibility constraint from the definition of allowability.

We can also show that, arbitrarily close to any point $x \in \mathcal{G}(p)$, we can find sets S of positive measure such that the relative accuracy of the algorithm when run with inputs in S is either 1 or ∞; a result identical to Corollary 3.11 can also be proved for the extended arithmetic case.

3.4.2. Sufficiency: the complex case

In this section we obtain a sufficiency condition for the accurate evaluability of a complex polynomial, given a black-box arithmetic with operations $\{q_j \mid j \in J\}$ (J may be an infinite set).

Throughout this section, we assume our black-box operations include q^c, which consists of multiplication by a complex constant: $q^c(x) = c \cdot x$. Note that this operation is natural, and can be performed accurately given only a suitably accurate approximation of c.

We believe that the sufficiency condition we obtain here is not a necessary one, in general, but it does subsume the sufficiency condition we found for the basic complex case with classical arithmetic $\{+, -, \cdot\}$.

Theorem 3.38. (General case)[2] Given a polynomial $p : \mathbb{C}^n \to \mathbb{C}$, with $V(p)$ a finite union of irreducible varieties $V_{\mathcal{I}}(q_j)$, for $j \in J$, and $\mathcal{I}$ as above, then p is accurately evaluable.

Theorem 3.39. (Affine case) If all black-box operations q_j, $j \in J$ are affine, then a polynomial $p : \mathbb{C}^n \to \mathbb{C}$ is accurately evaluable if and only if $V(p)$ is a union of varieties $V_{\mathcal{I}}(q_j)$, for $j \in J$ and $\mathcal{I}$ as in Definition 3.32.

---

[2] This condition was stated in a slightly weaker form in Demmel *et al.* (2006).

The proofs follow easily from Lemma 3.40.

Lemma 3.40. If all varieties $V_{\mathcal{I}}(q_j))$ in the union defined by $V(p)$ are irreducible (in particular, if they are affine), then p is a product $p = c \prod_j p_j$, where each p_j is a power of q_j or a polynomial obtained from q_j by repeating, negating, or zeroing some of the variables; c is a complex constant. The argument is identical to the one we gave for the proof of Corollary 3.15, and it hinges on the irreducibility of the varieties $V_{\mathcal{I}}(q_j))$ in the union.

Note that Theorem 3.39 is a more general necessary and sufficient condition than Theorem 3.12, which only considered having q_{-2}, q_{-1}, and q_0 as operations, and restricted the polynomials to have integer coefficients (thus eliminating the need for q^c).

3.5. Numerical linear algebra consequences

Here we examine the results of Section 2, in light of Section 3. We take another look at Table 2.1, explaining the strong 'No' entries there. Those entries mean that no accurate algorithms exist even given an arbitrary set of black-box operations of bounded degree or with a bounded number of arguments. In other words, arbitrary precision arithmetic is needed for their accurate solution. This is the case for Toeplitz matrices because, as discussed earlier, we cannot evaluate their determinants accurately, and determinants are necessary to get the indicated entries accurately. Fully off-diagonal submatrices of diagonally dominant matrices are completely unstructured matrices, and so with irreducible determinants of unbounded degree. The same is true of M-matrices, except that the submatrix entries are non-positive. Minors of submatrices of non-TN Vandermonde have factors that are general Schur functions of arbitrary arguments, which can be irreducible of unbounded degree. We suspect that many other entries should also be 'No'.

3.5.1. Validation of our results

If we examine the matrix classes in Table 2.1, we see that their determinants are rational functions whose sets of zeros and of poles are allowable in traditional arithmetic. By considering numerators and denominators of these rational functions separately we see that both can be computed accurately (and then, provided that the denominator is not 0, their ratio can be computed accurately). Incorporating division more formally into our model to identify necessary and sufficient conditions for accurate evaluability of rational functions is the subject of ongoing work.

3.5.2. Negative results: accurate evaluation is impossible

Here we examine two classes of matrices for which some or all linear algebra operations are impossible given any set of black boxes with a bounded number of arguments: Toeplitz and various classes of Vandermonde that we define later.

We prove our results by reducing the problem of doing accurate linear algebra to that of accurately evaluating the determinant and certain minors (recall that the latter is a necessary condition for the former). What these results say roughly is that, if one wants to construct an accurate algorithm for finding the inverse that works for Toeplitz or Vandermonde matrices as a class, one needs to use arbitrary precision (more on this in Section 4).

We start by examining a more general problem. If the determinants $p_n(x) = \det M^{n \times n}(x)$ of a class of n-by-n structured matrices M do not satisfy the necessity conditions described in Theorem 3.37 for *any* enumerable set of black-box operations (perhaps with other properties, like bounded degree), then we can conclude that accurate algorithms of the sort described in the above citations are impossible.

In particular, to satisfy these necessity conditions would require that the varieties $V(p_n)$ be allowable (or q-allowable). For example, if V is a Vandermonde matrix, then $\det(V) = \prod_{i<j}(x_i - x_j)$ satisfies this condition, using only subtraction and multiplication.

The following theorem states a negative condition (which guarantees impossibility of existence for algorithm using *any* enumerable set of black-box operations of bounded degree).

Theorem 3.41. Let $M(x)$ be an n-by-n structured complex matrix with determinant $p_n(x)$ as described above. Suppose that for any n, $p_n(x)$ has an irreducible factor $\hat{p}_n(x)$ whose degree tends to infinity as n tends to infinity. Then, for any enumerable set of black-box arithmetic operations of bounded degree, for sufficiently large n it is impossible to accurately evaluate $p_n(x)$ over the complex numbers.

Proof. Let $q_1, \ldots, q_m$ be any finite set of black-box operations. To obtain a contradiction, suppose the complex variety $V(p_n)$ satisfies the necessary conditions of Theorem 3.37, *i.e.*, that $V(p_n)$ is allowable. This means that $V(p_n)$, which includes the hypersurface $V(\hat{p}_n)$ as an irreducible component, can be written as the union of irreducible q-allowable sets (by Definition 3.34). This means that $V(\hat{p}_n)$ must itself be equal to an irreducible q-allowable set (a hypersurface), since representations as unions of irreducible sets are unique. The irreducible q-allowable sets of codimension 1 are defined by single irreducible polynomials, which are in turn derived by the process of setting variables equal to one another, to one another's negation, or zero (as described in Definitions 3.32 and 3.34), and so have bounded degree. This contradicts the unboundedness of the degree of $V(\hat{p}_n)$.   □

In the next theorems we apply this result to the set of Toeplitz matrices. We use the following notation. Let T be an n-by-n Toeplitz matrix, with x_j

on the jth diagonal, so x_0 is on the main diagonal, x_{n-1} is in the top right corner, and x_{1-n} is in the bottom left corner. We give the following result without proof; for a proof, see Demmel *et al.* (2006).

Theorem 3.42. The determinant of a Toeplitz matrix T is irreducible over any field.

Therefore, for complex Toeplitz matrices, we have the following corollary.

Corollary 3.43. The determinants of the set of complex Toeplitz matrices cannot be evaluated accurately using any enumerable set of bounded-degree black-box operations.

In the real case, irreducibility of p_n is not enough to conclude that p_n cannot be evaluated accurately, because $V_{\mathbb{R}}(p_n)$ may still be allowable (and even vanish). So we consider another necessary condition for allowability. Since all black boxes have a finite number of arguments, their associated codimension-1 irreducible components must have the property that whether $x \in V_{\mathcal{I}}(q_j)$ depends on only a finite number of components of x. Thus, to prove that the hypersurface $V_{\mathbb{R}}(p_n)$ is not allowable, it suffices to find at least one regular point x^* in $V_{\mathbb{R}}(p_n)$ such that the tangent hyperplane at x^* is not parallel to sufficiently many coordinate directions, *i.e.*, membership in $V_{\mathbb{R}}(p_n)$ depends on more variables than any $V_{\mathcal{I}}(q_j)$. This is easy to do for real Toeplitz matrices.

Theorem 3.44. Let V be the variety of the determinant of real singular Toeplitz matrices. Then V has codimension 1, and at almost all regular points, its tangent hyperplane is parallel to no coordinate directions.

Corollary 3.45. The determinants of the set of real Toeplitz matrices cannot be evaluated accurately using any enumerable set of bounded-degree black-box operations.

Proofs of these results can be found in Demmel *et al.* (2006). Corollaries 3.43 and 3.45 imply that accurate linear algebra (in the sense of Section 2) is impossible on the class of Toeplitz matrices (either real or complex) in bounded precision.

We consider now the class of polynomial Vandermonde matrices V, where $V_{ij} = P_{j-1}(x_i)$ is a polynomial function of x_i, with $1 \le i, j \le n$. This class includes the standard Vandermonde (where $P_{j-1}(x_i) = x_i^{j-1}$) and many others.

Consider a generalized Vandermonde matrix where $P_{j-1}(x_i) = x_i^{j-1+\lambda_{n-i}}$ with $0 \le \lambda_1 \le \lambda_2 \le \cdots \le \lambda_n$. The tuple $\lambda = (\lambda_1, \lambda_2, \ldots, \lambda_n)$ is called a *partition*. Any square submatrix of such a generalized Vandermonde matrix is also a generalized Vandermonde matrix. A generalized Vandermonde matrix is known to have determinant of the form $s_\lambda(x) \prod_{i<j}(x_i - x_j)$, where $s_\lambda(x)$ is a polynomial of degree $|\lambda| = \sum_i \lambda_i$, and called a Schur function (Macdonald 1998). In infinitely many variables (not our situation) the Schur function is irreducible (Farahat 1958), but in finitely many variables, the Schur function is sometimes irreducible and sometimes not (but there are irreducible Schur functions of arbitrarily high degree); see Stanley (1999, Exercise 7.30).

We can thus derive the following theorem and corollary.

Theorem 3.46. By Theorem 3.41, no enumerable set of black-box operations of bounded degree can compute all Schur functions accurately when the x_i are complex.

Corollary 3.47. No enumerable set of black-box operations of bounded degree or of bounded number of arguments exists that will accurately evaluate all minors of complex generalized Vandermonde matrices in the generic case.

If we restrict the domain $\mathcal{D}$ to be non-negative real numbers, then the situation changes: the non-negativity of the coefficients of the Schur functions shows that they are positive in $\mathcal{D}$, and indeed the generalized Vandermonde matrix is totally positive (Karlin 1968).

Combined with the homogeneity of the Schur function, Theorem 3.6 implies that the Schur function, and so determinants (and minors) of totally positive generalized Vandermonde matrices can be evaluated accurately in classical arithmetic (and the algorithms mentioned in Section 2 are more efficient than the algorithm used in proving Theorem 3.6).

Now consider a polynomial Vandermonde matrix V_P defined by a family $\{P_k(x)\}_{k\in\mathbb{N}}$ of polynomials such that $\deg(P_k) = k$, and $V_P(i,j) = P_{j-1}(x_i)$. Note that these are included in the class of generalized Vandermonde matrices, and that the difference lies in the fact that for polynomial Vandermonde, the sequence of degrees is increasing and without gaps.

Note that any V_P can be written as $V_P = VC$, with V being a regular Vandermonde matrix, and C being an upper triangular matrix of coefficients of the polynomials P_k, i.e.,

$$P_{j-1}(x) = \sum_{i=1}^{j} C(i,j)x^{i-1}, \quad \forall 1 \le j \le n.$$

Let $c_{i-1} := \tilde{D}(i,i)$, for all $1 \le i \le n$, denote the highest-order coefficients of the polynomials $P_0(x), \ldots, P_{n-1}(x)$.

The following two results are proved informally in Demmel *et al.* (2006, Section 5).

Theorem 3.48. The set of principal minors of polynomial Vandermonde matrices includes polynomials which have irreducible factors of arbitrarily large degree.

Corollary 3.49. By Theorem 3.41, the set of polynomial Vandermonde matrices contains matrices whose inverses cannot be evaluated accurately even with the addition of any enumerable set of bounded-degree black boxes.

We can also say something about the LDU factorizations of polynomial Vandermonde matrices. With the matrix C being the upper triangular matrix of coefficients of the polynomials P_k, we can write $C = \tilde{D}\tilde{C}$, with $\tilde{D}$ being the diagonal matrix of highest-order coefficients, i.e., $\tilde{D}(i,i) = C(i,i)$ for all $1 \leq i \leq n$. We will assume that the matrices C and $\tilde{D}$ are given to us exactly.

If we let $V_P = L_P D_P U_P$ and $V = LDU$, it follows that

$$L_P = L,$$
$$D_P = D\tilde{D},$$
$$U_P = \tilde{D}^{-1}UC.$$

Since we cannot compute L accurately in the general Vandermonde case, it follows that we cannot compute L_P accurately in the polynomial Vandermonde case. Likewise, neither the SVD nor the symmetric eigenvalue decomposition (EVD) are computable accurately, but if the polynomials are certain orthogonal polynomials, then the accurate SVD is possible (Demmel and Koev 2006), and an accurate symmetric EVD may also be possible (Dopico *et al.* 2003).

3.5.3. Positive results: using extended arithmetic

Table 2.1 gathers together structured matrix classes for which it has been established whether – and which – accurate linear algebra algorithms exist. For some matrix classes, it was deduced that accurate class-algorithms do not exist, from the fact that a necessary condition (having an accurately evaluable determinant) was violated.

In this section, we explain how we can use the sufficiency condition for complex matrices, developed in Section 3.4.2.

Consider complex polynomial Cauchy matrices, defined (in their simplest form) as follows. Let p and q be complex polynomials of one variable. Now, using MATLAB notation, let

$$x_i := p(\widehat{x_i}), \quad \forall 1 \leq i \leq m,$$
$$y_j := q(\widehat{y_j}), \quad \forall 1 \leq j \leq m.$$

Definition 3.50. We call the matrix $C = (C_{ij})$ with $C_{ij} = \frac{1}{x_i+y_j}$ where x_i and y_j are, as above, a polynomial Cauchy matrix.

Definition 3.51. Let

$$Q^-(\widehat{x}_i, \widehat{y}_j) = p(\widehat{x}_i) - q(\widehat{y}_j),$$
$$Q^+(\widehat{x}_i, \widehat{y}_j) = p(\widehat{x}_i) + q(\widehat{y}_j),$$

be complex polynomials over $\mathbb{C}^2$.

Recall that the determinant of the Cauchy matrix C is

$$\det C = \frac{\prod_{i,j}(x_i - x_j)(y_i - y_j)}{\prod_{i,j}(x_i + y_j)}. \tag{3.11}$$

Although our models of arithmetic do not incorporate division, computers do perform division by a non-zero number as an accurate operation. Therefore, given accurate division and black-box algorithms for computing the polynomials Q^- and Q^+, one immediately has a simple and accurate algorithm to evaluate *any* minor for the matrix C, and therefore any linear algebra operations can be easily performed on C (this algorithm is guaranteed by Theorem 3.38).

In fact, we can obtain a much more general result.

Theorem 3.52. Let Φ be a formula satisfying NIC and depending on variables $x_1, \ldots, x_n$. Let p be a polynomial (resp. let $\{p_i\}_1^n$ be a set of polynomials), and let $x_i = p(A(i, 1 : m))$ (resp. $p_i(A(i, 1 : m))$) for some matrix of parameters A.

We can accurately evaluate Φ on the new set of inputs depending on the parameters of A, provided that we build three (resp. $m^2 + 2m$) black boxes, computing

$$\begin{cases} p, \\ Q^+(y_1, \ldots, y_n, z_1, \ldots, z_n) = p(y_1, \ldots, y_n) + p(z_1, \ldots, p_n), \\ Q^-(y_1, \ldots, y_n, z_1, \ldots, z_n) = p(y_1, \ldots, y_n) - p(z_1, \ldots, p_n), \end{cases}$$

respectively, for all $1 \le i \le j \le m$,

$$\begin{cases} p_i, \\ Q_{ij}^+(y_1, \ldots, y_n, z_1, \ldots, z_n) = p_i(y_1, \ldots, y_n) + p_j(z_1, \ldots, p_n), \\ Q_{ij}^-(y_1, \ldots, y_n, z_1, \ldots, z_n) = p_i(y_1, \ldots, y_n) - p_j(z_1, \ldots, p_n). \end{cases}$$

Another class of matrices which admit accurate linear algebra algorithms in extended arithmetic are the Green's matrices, which arise from discrete representations of Sturm–Liouville equations. These matrices are inverses of irreducible tridiagonal matrices.

Generic Green's matrices have a simple four-vector representation (see, for example, Ikebe (1979) and Nabben (1999)), as

$$
F_{i,j} = \begin{cases} a_i b_j, & \text{if } i \geq j, \\ c_i d_j, & \text{if } i < j, \end{cases}
$$

for $a = (a_1, \ldots a_n)$, $b = (b_1, \ldots, b_n)$, $c = (c_1, \ldots, c_n)$, $d = (d_1, \ldots, d_n)$, and $1 \leq i, j \leq n$.

The case when $a = c$ and $b = d$, i.e., the symmetric case, has been particularly well studied (see Gantmacher and Krein (2002) and Karlin (1968)), and we describe it in a bit more detail.

We use the notation $X\begin{pmatrix} i_1 & i_2 & \cdots & i_p \\ j_1 & j_2 & \cdots & j_p \end{pmatrix}$ for the minor of the matrix X corresponding to rows $i_1, \ldots, i_p$ and columns $j_1, \ldots, j_p$, and $\left| \begin{smallmatrix} x & y \\ z & t \end{smallmatrix} \right|$ for the determinant $(xt - yz)$.

All minors of symmetric Green's matrices have the simple representation (following Karlin (1968))

$$
G\begin{pmatrix} i_1 & i_2 & \cdots & i_p \\ j_1 & j_2 & \cdots & j_p \end{pmatrix} = a_{k_1} \begin{vmatrix} a_{k_2} & a_{l_1} \\ b_{k_2} & b_{l_1} \end{vmatrix} \begin{vmatrix} a_{k_3} & a_{l_2} \\ b_{k_3} & b_{l_2} \end{vmatrix} \cdots \begin{vmatrix} a_{k_p} & a_{l_{p-1}} \\ b_{k_p} & b_{l_{p-1}} \end{vmatrix} b_{l_p},
$$

where $k_m = \min(i_m, j_m)$ and $l_m = \max(i_m, j_m)$.

Similarly, all minors of generic Green's matrices can be shown (by a simple inductive argument) to be either 0 or products of linear and quadratic factors. Here, by 'linear factor' we mean a factor of the type a_i, b_j, c_k, or d_l, and by 'quadratic factor' we mean a factor of the type $xt - yz$, with x, y, t, z being entries of a, b, c, d.

We can then conclude that, given a black box computing $p(x, y, z, t) := xt - yz$ accurately, by Theorem 3.38 one can compute all minors of generic Green's matrices. Therefore, as was observed in Demmel and Koev (2001), one can evaluate all the minors of generic Green's matrices, and consequently perform linear algebra accurately.

Green's matrices belong to the class of *hierarchically semi-separable* (HSS) matrices. There are many definitions of the latter, one of them being that HSS matrices of order $k \in \mathbb{N}$ are matrices for which any off-diagonal submatrix has rank no bigger than k. Other examples are tridiagonal matrices, banded matrices, inverses of banded matrices, *etc.* The HSS matrices are extremely useful as preconditioners, and arise in many applications. Since determinants of tridiagonal matrices with independent indeterminates as entries are irreducible, and tridiagonals are special cases of HSS matrices, some (and perhaps all) HSS matrices do have irreducible determinants.

Still, we believe that further investigation of the large class of HSS matrices may yield other examples of subclasses for which simple black-box operations could be constructed in order to accurately compute minors, and therefore, be able to perform linear algebra accurately.

4. Other models of arithmetic

Though the arithmetic models in this paper use real (or complex) numbers and rounding errors, our goal is to draw conclusions about practical finite precision computation, *i.e.*, with numbers represented as finite bit strings (*e.g.*, floating-point numbers). In such a bit model, all rational functions of the arguments can be computed accurately, even exactly, because the arguments are rational; the only question is cost. In this section we draw conclusions about cost from our analysis.

We would like to quantify our intuition that, for example, it is much cheaper to accurately compute the determinant of an n-by-n Vandermonde matrix with the familiar formula than with Gaussian elimination with sufficiently high-precision arithmetic. We do not mean the difference between $O(n^2)$ and $O(n^3)$ arithmetic operations, but the difference in cost between low-precision and high-precision arithmetic. To quantify this cost, we need to pick a number representation.

We will assume that 'failure' is not allowed, *i.e.*, neither overflow nor underflow is permitted, so that intermediate (and final) results can grow or shrink in magnitude as needed to complete the computation.

We claim that the natural representation to use is the pair of integers (e, m) to represent $m \cdot 2^e$, *i.e.*, binary floating point. Pros and cons of various number models are discussed in Demmel *et al.* (2006), but we restrict ourselves here to explaining why we choose floating as opposed to fixed point, which is also widely used for analysis (in fixed point, $m \cdot 2^e$ would be represented using up to e explicit zeros before or after the bits representing m).

One can of course represent the same set of (binary) rational numbers in both fixed and floating point, but floating point is much more compressed: it takes about $\log_2 |e| + \log_2 |m|$ bits to represent (e, m), but about $|e| + \log_2 |m|$ bits to represent $m \cdot 2^e$ in fixed point, which is exponentially larger.

First, as a result of this possibly exponentially greater use of space by fixed point, it is possible for a sequence of n fixed-point arithmetic operations to take time exponential in n (repeated squaring doubles the length of result at each step, even if only a fixed number of the most significant bits are kept). In contrast, n floating-point arithmetic operations, with fixed relative error, take time that grows at worst like $O(n^2)$ (attained by repeated squaring again, which adds one bit to e at each squaring). In particular, any of the expressions in earlier sections of this paper can be evaluated in polynomial time in the size of the expression, and the size of their floating-point arguments.

Second, this exponentially greater use of space in fixed point means that algorithms can appear 'artificially' cheaper, because they are only polynomial in the input size $|e| + \log_2 |m|$, whereas they would not be polynomial as a function of the input size measured as $\log_2 |e| + \log_2 |m|$. (This is analogous to asking whether an algorithm with integer inputs runs in polynomial time or not, depending on whether the inputs are represented in unary or binary.) For example, it is possible to accurately compute the determinant of a general matrix with fixed-point entries in polynomial time in the size of the input (Clarkson 1992), but we know of no such polynomial-time algorithm with floating-point entries. Running a conventional determinant algorithm (*e.g.*, Gaussian elimination with pivoting) in high enough precision would require roughly $\log_2 \kappa(A) = \log_2(\|A\| \cdot \|A^{-1}\|)$ bits of precision, which can grow like $|e|$ rather than $\log_2 |e|$; *e.g.*, consider

$$A = \begin{bmatrix} y - x & y \\ y & y + x \end{bmatrix}$$

for $y \gg x$, where $\det(A) = -x^2$.

Indeed, the obvious 'witness' to identify a singular matrix, a null vector, can have exponentially more non-zero bits than the matrix, as the following example shows. Consider the $(2n+1)$-by-$(2n+1)$ tridiagonal matrix T with 1s on the subdiagonal, -1s on the superdiagonal, and

$$\operatorname{diag}(T) = [x_1, x_2, \ldots, x_{n-1}, x_n, 0, -x_n, -x_{n-1}, \ldots, -x_2, -x_1].$$

It is easy to confirm that T is singular, with right null vector

$$v = [1, p_1, p_2, \ldots, p_{2n}],$$

where $p_i = \det(T(1:i, 1:i))$ is a leading principal minor. If we let $x_i = 2^{e_i}$ with $e_1 = 0$, $e_2 = 1$, and $e_i \geq e_{i-1} + e_{i-2}$, then one can confirm for $i \leq n$ that p_i is an integer with f_i non-zero bits, where $f_1 = 1$, $f_2 = 2$, and $f_i = f_{i-1} + f_{i-2}$ is the Fibonacci sequence. Since f_i grows exponentially, the null vector v has exponentially many bits as a function of n, whereas the size of T is at most $O(n \log e_n)$, which can be as small as $O(n^2)$.

Another way to see the difference between fixed and floating point is to consider the simple expression $\prod_{i=1}^n (1 + x_i)$. If the x_i are supplied in fixed point, the entire expression can be computed exactly in polynomial time. However, in floating point, though the leading bits and trailing bits are easy, computing some of the bits is as hard as computing the permanent, a problem widely believed to have exponential complexity in n (Valiant 1979).

Here is the reduction to the permanent.[3] Let A be an n-by-n matrix whose entries are 0s and 1s. The permanent is the same as the determinant, except that all terms in the Laplace expansion are added, instead of some being added and some subtracted. Let r_i and c_j be independent indeterminates,

---

[3] We acknowledge Benjamin Diament for having discovered the result relating floating-point complexity to the permanent.

and consider the multivariate polynomial

$$p(r_1, \ldots, r_n, c_1, \ldots, c_n) = \prod_{A_{ij} \neq 0} (1 + r_i c_j). \qquad (4.1)$$

Then the coefficient k of $\prod_{i=1}^{n} r_i c_i$ in the expansion of p can be seen to be the permanent. Next we replace r_i and c_j by sufficiently widely spaced powers of 2, so that every coefficient of every term in the expansion of p appears in non-overlapping bits of p evaluated at these powers of 2. Since no coefficient can exceed 2^{n^2}, and since the sequence of exponents $(f_n, \ldots, f_1, e_n, \ldots, e_1)$ in any term $\prod_{i=1}^{n} r_i^{e_i} c_i^{f_i}$ of p can be thought of as the unique expansion of a number in base $n + 1$, one can see that choosing $r_i = 2^{n^2(n+1)^{i-1}}$ and $c_j = 2^{n^2(n+1)^{n+j-1}}$ suffices. The biggest-possible product $r_i c_j$ is

$$r_n c_n = 2^{n^2((n+1)^{n-1} + (n+1)^{2n-1})} \leq 2^{2n^2(n+1)^{2n}},$$

where the exponent takes at most $\log_2(2n^2(n+1)^{2n}) = O(n \log n)$ bits to represent, so all the arguments $r_i c_j$ in the product in (4.1) take $O(n^3 \log n)$ bits to represent.

Now we consider 'black-box arithmetic', whose purpose is to model the use of subroutine libraries with selected high-accuracy operations. We claim that any multivariate polynomial ('black box') with t terms of maximum degree d, can be evaluated accurately in polynomial time as a function of d, t and the size of the input floating-point numbers. The algorithm is simply to evaluate each term exactly, and then sum them in decreasing order of exponents, using a register of about $\log_2 t$ bits more than needed to store the longer term exactly (Demmel and Koev 2004a, Demmel and Hida 2003). In particular, any enumerable collection of black boxes that are all bounded in degree d and number of terms t can all be thought of as running in time polynomial in the size of their floating-point arguments, just like the basic operations of addition, subtraction and multiplication. If the number of terms t is proportional to the number of inputs (e.g., dot products of vectors of length t), then the cost is still polynomial in the input size.

In summary, in a natural floating-point model of arithmetic, the algorithms we have discussed run in polynomial time in the size of the inputs, whereas simply running a conventional algorithm in sufficiently high precision arithmetic to get the answer accurately can take exponentially longer. We know of no guaranteed polynomial-time alternatives to our algorithms.

5. Structured condition numbers

In this section we begin by recalling some attractive properties of structured condition numbers for problems that we can solve accurately, and discuss possible generalizations. If our problem is evaluating the function

$p(x_1, \ldots, x_n)$, then the structured condition number κ_{struct} is simply the derivative of the relative change in p with respect to relative changes in its arguments:

$$\kappa_{\text{struct}} = \frac{\left\| (x_1 \frac{\partial p}{\partial x_1}, \ldots, x_n \frac{\partial p}{\partial x_n}) \right\|}{|p|}, \tag{5.1}$$

where any vector norm may be used in the numerator.

The simplest case, as before, is for problems described by Theorem 5.12 and Corollary 5.15, which say that in the complex case, a necessary and sufficient condition for accurate evaluation of complex $p(x)$ using only traditional arithmetic ($\pm$ and $\times$) is that $V(p)$ be allowable, in which case $p(x)$ factors completely into factors of the forms x_i^α, and $(x_i \pm x_j)^\beta$, where α and β are fixed integers. This covers many of the linear algebra examples in Section 2. Given such a simple expression it is easy to evaluate the structured condition number: each factor x_i^α adds α to $(x_i \frac{\partial p}{\partial x_i})/p$, and each factor $(x_i \pm x_j)^\beta$ adds

$$|\beta x_i/(x_i \pm x_j)| \le |\beta|/\text{rel\_gap}(x_i, \mp x_j).$$

Slightly more generally, for expressions satisfying NIC, *e.g.*, including real expressions that only add like-signed values, analogous conclusions can be drawn. This is because factors that only add like-signed values can only make bounded contributions to the condition number.

Given a structured condition number for a decomposition such as LDU with complete pivoting (an RRD), this essentially becomes a structured condition number for the SVD (Demmel *et al.* 1999, Theorem 2.1).

Now we consider the set of *ill-posed problems*, *i.e.*, the ones whose structured condition numbers are infinite. Examining (5.1), we see that $p = 0$ is a necessary condition, *i.e.*, the ill-posed problems are a subset of $V(p)$. (If $p(x)$ were rational, we would include the poles as well.) For every term $|\beta|/\text{rel\_gap}(x_i, \mp x_j)$ in the structured condition number, the corresponding ill-posed set is defined by $x_i = \mp x_j$. All of $V(p)$ is not necessarily ill-posed, since, for example, small relative changes in x only cause small relative changes in $p(x) = x^\alpha$.

It is natural to ask if there is a relationship between the *distance to the nearest ill-posed problem*, *i.e.*, the smallest relative change to the x_i that make the problem ill-posed, and its structured condition number (Demmel 1987). It is easy to see that for any term $|\beta|/\text{rel\_gap}(x_i, \mp x_j)$ in the structured condition number, the smallest relative changes to x_i and $\mp x_j$ that make it infinite are close to $\text{rel\_gap}(x_i, \mp x_j)$ when it is small. In other words, the structured condition number is close to the reciprocal of the distance to the nearest ill-posed problem, measured by the smallest relative change to the arguments x_i. This helps explain geometrically why the structured condition number can be so much smaller than the unstructured one:

it takes, for example, a much larger perturbation to make $x_i = i - 1/2$ and $x_j = j - 1/2$ equal than the smallest singular value of the Hilbert matrix $H_{ij} = 1/(x_i + x_j)$.

This reciprocal-condition-number property, that the reciprocal of the condition number is approximately the distance to the nearest ill-posed problem, is common in numerical analysis (Demmel 1987, Rump 1999a, 2003a). The following simple asymptotic argument shows why.

If the structured condition number (5.1) is very large, then some component

$$\left| x_i \frac{\partial p}{\partial x_i} / p \right| \gg 1,$$

that is,

$$\left| p / \frac{\partial p}{\partial x_i} \right| \ll |x_i|,$$

or in other words one step of Newton's method

$$x_i^{\text{new}} = x_i - p / \frac{\partial p}{\partial x_i}$$

to find a root of $p = 0$ will take a very small step. Therefore it is plausible that this step $p / \frac{\partial p}{\partial x_i}$ is very close to the smallest (absolute) distance to the variety in the x_i direction (or the multiplicity of the root times $p / \frac{\partial p}{\partial x_i}$ is very close to the distance) and dividing by $|x_i|$ yields the relative distance.

Now let us go beyond expressions evaluable accurately just using NIC. Consider the case of a real positive polynomial or empty variety, as discussed in Section 3.2. The analysis in Theorem 3.5 (resp. Theorem 3.6) shows that the relative condition number will grow like $1/p_{\min}$ (resp. $1/p_{\min,\text{homo}}$), the reciprocal of the smallest value $p(x)$ can take on the appropriate domain. So the relative condition number can be arbitrarily large, but in the absence of a variety intersecting the domain it remains bounded.

Based on these examples and analysis, we conjecture that for traditional arithmetic, the following two statements hold.

(1) The reciprocal of the structured condition number is an approximation of the relative distance from x to the nearest ill-posed problem, perhaps asymptotically.

(2) This relative distance is approximately given by $\text{rel\_gap}(x_i, \mp x_j)$ for some i and j.

This reciprocal-condition-number property is quite robust as the arguments above suggest, and does not necessarily depend on accurate evaluability. For example, if $p(x) = (x_1 + x_2 + x_3)^\alpha$ then its structured condition number is $\alpha \|x\| / |x_1 + x_2 + x_3|$, and $|x_1 + x_2 + x_3| / \|x\|_1$ is indeed the relative distance. However, the reciprocal-condition-number property is not universal but depends on the structure we impose (Rump 1998, 1999b, 2003b). Just as this reciprocal-condition-number property is equivalent to the statement that computing the condition number is as sensitive a problem as solving the original problem, we conjecture that the structured condition number κ_{struct} can only be computed accurately if the original problem p can be, at least in the interesting case when κ_{struct} is large. This seems reasonable since $p(x)$ ends up in the denominator of κ_{struct}, so we need to evaluate p accurately near its zeros (or poles). But the numerators $\partial p/\partial x_i$ could be anything, and perhaps even have zeros on unallowable varieties, so to be more precise we conjecture that p can be evaluated accurately in some open neighbourhood of its zeros (or poles) if and only if κ_{struct} can be.

6. Conclusions

In this paper, we have made the case for accurate evaluation of polynomial expressions and accurate linear algebra; we have shown that such evaluation is desirable (Section 1), significant (Section 4) and often realizable efficiently (Section 2). We have listed, in Section 2, many types of structured matrices that have been analysed from an accuracy perspective in the numerical linear algebra literature, while in Section 3 we identified the common algebraic structure that made them analysable in the first place.

There are limits to how much we can hope to extend the class of structured matrices for which linear algebra can be performed accurately; the 'negative examples' of Section 3.5 show that, for some classes of matrices, accuracy cannot be achieved in finite precision, and both Sections 2 and 3 mention problems that are impossible to solve in 'traditional' arithmetic. The former should be seen as 'hard' barriers, but the latter should be seen as a challenge, both from theoretical and computational perspectives. The theory should aim to provide answers to the question of how to extend one's arithmetic by adding 'black-box' operations, in order to make these structured problems solvable (as we do for the examples of Section 2.3); the computation should design software implementing such 'black boxes'.

In summary, accurate evaluation is an important area of scientific computing, which has been advanced by the recent results presented here. Plenty of work remains in adding to both the theoretical framework (which apparently requires familiarity with 'pure' mathematical fields such as algebraic geometry, topology, and analysis) and to the practical one (software implementation).

REFERENCES

A. V. Aho, J. E. Hopcroft and J. D. Ullman (1975), *The Design and Analysis of Computer Algorithms*, second printing, Addison-Wesley Series in Computer Science and Information Processing.

A. S. Alfa, J. Xue and Q. Ye (2002), 'Accurate computation of the smallest eigenvalue of a diagonally dominant M-matrix', *Math. Comp.* **71**, 217–236 (electronic).

E. Anderson, Z. Bai, C. Bischof, J. Demmel, J. Dongarra, J. Du Croz, A. Greenbaum, S. Hammarling, A. McKenney, S. Blackford and D. Sorensen (1999), *LAPACK Users' Guide*, third edn, SIAM, Philadelphia.

Å. Björck and V. Pereyra (1970), 'Solution of Vandermonde systems of equations', *Math. Comp.* **24**, 893–903.

E. Boman, B. Hendrickson and S. Vavasis (2004), 'Solving elliptic finite element systems in near-linear time with support preconditioners', arXiv.org:cs/0407022.

T. Boros, T. Kailath and V. Olshevsky (1999), 'A fast Björck–Pereyra-type algorithm for parallel solution of Cauchy linear equations', *Linear Algebra Appl.* **302/303**, 265–293.

T. Chan (1987), 'Rank revealing QR factorizations', *Linear Algebra Appl.* **88/89**, 67–82.

S. Chandrasekaran and I. Ipsen (1994), 'On rank-revealing QR factorizations', *SIAM J. Matrix Anal. Appl.*

K. Clarkson (1992), Safe and effective determinant evaluation, in *33rd Annual Symposium on Foundations of Computer Science*, pp. 387–395.

J. Demmel (1987), 'On condition numbers and the distance to the nearest ill-posed problem', *Numer. Math.* **51**, 251–289.

J. Demmel (1999), 'Accurate singular value decompositions of structured matrices', *SIAM J. Matrix Anal. Appl.* **21**, 562–580 (electronic).

J. Demmel and W. Gragg (1993), 'On computing accurate singular values and eigenvalues of acyclic matrices', *Linear Algebra Appl.* **185**, 203–218.

J. Demmel and Y. Hida (2003), 'Accurate and efficient floating point summation', *SIAM J. Sci. Comput.* **25**, 1214–1248.

J. Demmel and W. Kahan (1990), 'Accurate singular values of bidiagonal matrices', *SIAM J. Sci. Statist. Comput.* **11**, 873–912.

J. Demmel and P. Koev (2001), Necessary and sufficient conditions for accurate and efficient rational function evaluation and factorizations of rational matrices. In *Structured Matrices in Mathematics, Computer Science, and Engineering II* (Boulder, CO, 1999), Vol. 281 of *Contemporary Mathematics*, AMS, Providence, RI, pp. 117–143.

J. Demmel and P. Koev (2004a), Accurate and efficient algorithms for floating point computation. In *Applied Mathematics Entering the 21st Century*, SIAM, Philadelphia, PA, pp. 73–88.

J. Demmel and P. Koev (2004b), 'Accurate SVDs of weakly diagonally dominant M-matrices', *Numer. Math.* **98**, 99–104.

J. Demmel and P. Koev (2005), 'The accurate and efficient solution of a totally positive generalized Vandermonde linear system', *SIAM J. Matrix Anal. Appl.* **27**, 142–152 (electronic).

J. Demmel and P. Koev (2006), 'Accurate SVDs of polynomial Vandermonde matrices involving orthonormal polynomials', *Linear Algebra Appl.* **417**, 382–396.

J. Demmel and K. Veselić (1992), 'Jacobi's method is more accurate than QR', *SIAM J. Matrix Anal. Appl.* **13**, 1204–1246.

J. Demmel, B. Diament and G. Malajovich (2001), 'On the complexity of computing error bounds', in *Found. Comput. Math.* **1**, 101–125.

J. Demmel, I. Dumitriu and O. Holtz (2006), Toward accurate polynomial evaluation in rounded arithmetic. In *Foundations of Computational Mathematics* (Santander 2005), Vol. 331 of *London Mathematical Society Lecture Notes*, Cambridge University Press, Cambridge, pp. 36–105.

J. Demmel, M. Gu, S. Eisenstat, I. Slapničar, K. Veselić and Z. Drmač (1999), 'Computing the singular value decomposition with high relative accuracy', *Linear Algebra Appl.* **299**, 21–80.

F. M. Dopico, J. M. Molera and J. Moro (2003), 'An orthogonal high relative accuracy algorithm for the symmetric eigenproblem', *SIAM J. Matrix Anal. Appl.* **25**, 301–351 (electronic).

Z. Drmač (1998), 'Accurate computation of the product induced singular value decomposition with applications', *SIAM J. Numer. Anal.* **35**, 1969–1994.

S. Eisenstat and I. Ipsen (1995), 'Relative perturbation techniques for singular value problems', *SIAM J. Numer. Anal.*

S. M. Fallat (2001), 'Bidiagonal factorizations of totally nonnegative matrices', *Amer. Math. Monthly* **108**, 697–712.

H. K. Farahat (1958), 'On Schur functions', *Proc. London Math. Soc.* (3) **8**, 621–630.

F. P. Gantmacher and M. G. Krein (2002), *Oscillation Matrices and Kernels and Small Vibrations of Mechanical Systems*, revised edn, AMS Chelsea Publishing, Providence, RI. Translation based on the 1941 Russian original.

M. Gasca and J. M. Peña (1992), 'Total positivity and Neville elimination', *Linear Algebra Appl.* **165**, 25–44.

M. Gasca and J. M. Peña (1996), On factorizations of totally positive matrices, in *Total Positivity and its Applications*, Kluwer, Dordrecht, pp. 109–130.

M. Gu and S. Eisenstat (1996), 'An efficient algorithm for computing a strong rank-revealing QR factorization', *SIAM J. Sci. Comput.* **17**, 848–869.

N. J. Higham (1987), 'Error analysis of the Björck–Pereyra algorithms for solving Vandermonde systems', *Numer. Math.* **50**, 613–632.

N. J. Higham (1988), 'Fast solution of Vandermonde-like systems involving orthogonal polynomials', *IMA J. Numer. Anal.* **8**, 473–486.

N. J. Higham (1990), 'Stability analysis of algorithms for solving confluent Vandermonde-like systems', *SIAM J. Matrix Anal. Appl.* **11**, 23–41.

O. Holtz (2005), 'The inverse eigenvalue problem for symmetric anti-bidiagonal matrices', *Linear Algebra Appl.* **408**, 268–274.

P. Hong and C. T. Pan (1992), 'Rank-revealing QR factorizations and the singular value decomposition', *Math. Comp.* **58**, 213–232.

T.-M. Hwang, W.-W. Lin and E. K. Yang (1992), 'Rank revealing LU factorization', *Linear Algebra Appl.* **175**, 115–141.

Y. Ikebe (1979), 'On inverses of Hessenberg matrices', *Linear Algebra Appl.* **24**, 93–97.

W. Kahan and I. Farkas (1963a), 'Algorithm 167: Calculation of confluent divided differences', *Commun. ACM* **6**, 164–165.

W. Kahan and I. Farkas (1963b), 'Algorithm 168: Newton interpolation with backward divided differences', *Commun. ACM* **6**, 165.

W. Kahan and I. Farkas (1963c), 'Algorithm 169: Newton interpolation with forward divided differences', *Commun. ACM* **6**, 165.

T. Kailath and V. Olshevsky (1995), 'Displacement structure approach to Chebyshev–Vandermonde and related matrices', *Integral Equations Operator Theory* **22**, 65–92.

T. Kailath and V. Olshevsky (1997), 'Displacement-structure approach to polynomial Vandermonde and related matrices', *Linear Algebra Appl.* **261**, 49–90.

S. Karlin (1968), *Total Positivity*, Vol. I, Stanford University Press, Stanford, CA.

P. Koev (2005), 'Accurate eigenvalues and SVDs of totally nonnegative matrices', *SIAM J. Matrix Anal. Appl.* **27**, 1–23 (electronic).

P. Koev (2007), 'Accurate computations with totally nonnegative matrices', *SIAM J. Matrix Anal. Appl.* **29**, 731–751.

P. Koev and F. Dopico (2007), 'Accurate eigenvalues of certain sign regular matrices', *Linear Algebra Appl.* **424**, 435–447.

R.-C. Li (1999), 'Relative perturbation theory II: Eigenspace and singular subspace variations', *SIAM J. Matrix Anal. Appl.* **20**, 471–492 (electronic).

I. G. Macdonald (1998), *Symmetric Functions and Orthogonal Polynomials*, Vol. 12 of *University Lecture Series*, AMS, Providence, RI.

A. Marco and J.-J. Martínez (2007), 'A fast and accurate algorithm for solving Bernstein–Vandermonde linear systems', *Linear Algebra Appl.* **422**, 616–628.

J. J. Martínez and J. M. Peña (1998), 'Fast algorithms of Björck–Pereyra type for solving Cauchy–Vandermonde linear systems', *Appl. Numer. Math.* **26**, 343–352.

J. J. Martínez and J. M. Peña (1998), 'Factorizations of Cauchy–Vandermonde matrices', *Linear Algebra Appl.* **284**, 229–237.

J. J. Martínez and J. M. Peña (2003), Factorizations of Cauchy–Vandermonde matrices with one multiple pole. In *Recent Research on Pure and Applied Algebra*, Nova Scientific, Hauppauge, NY, pp. 85–95.

The MathWorks (1992), *MATLAB Reference Guide*, The MathWorks, Natick, MA.

R. Mathias (1996), 'Accurate eigensystem computations by Jacobi methods', *SIAM J. Matrix Anal. Appl.* **16**, 977–1003.

E. Miller and B. Sturmfels (2005), *Combinatorial Commutative Algebra*, Vol. 227 of *Graduate Texts in Mathematics*, Springer, New York.

L. Miranian and M. Gu (2003), 'Strong rank revealing LU factorizations', *Linear Algebra Appl.* **367**, 1–16.

R. Nabben (1999), 'Decay rates of the inverse of nonsymmetric tridiagonal and band matrices', *SIAM J. Matrix Anal. Appl.* **20**, 820–837.

C. O'Cinneide (1996), 'Relative-error for the LU decomposition via the GTH algorithm', *Numer. Math.* **73**, 507–519.

B. Parlett (1995), The new qd algorithms, in *Acta Numerica*, Vol. 4, Cambridge University Press, pp. 459–491.

M. J. Peláez and J. Moro (2006), 'Accurate factorization and eigenvalue algorithms for symmetric DSTU and TSC matrices', *SIAM J. Matrix Anal. Appl.* **28**, 1173–1198 (electronic).

J. M. Peña (2004), 'LDU decompositions with L and U well conditioned', *Electron. Trans. Numer. Anal.* **18**, 198–208 (electronic).

J. Renegar (1992), 'On the computational complexity and geometry of the first-order theory of the reals I: Introduction. Preliminaries. The geometry of semialgebraic sets. The decision problem for the existential theory of the reals', *J. Symbolic Comput.* **13**, 255–299.

B. Reznick (2000), *Some Concrete Aspects of Hilbert's 17th Problem*, Vol. 253 of *Contemporary Mathematics*, AMS.

S. Rump (1998), 'Structured perturbations and symmetric matrices', *Linear Algebra Appl.* **278**, 121–132.

S. Rump (1999*a*), 'Ill-conditioned matrices are componentwise near to singularity', *SIAM Review* **41**, 102–112.

S. Rump (1999*b*), 'Ill-conditionedness need not be componentwise near to ill-posedness for least squares problems', *BIT* **39**, 143–151.

S. Rump (2003*a*), 'Structured perturbations I: Normwise distances', *SIAM J. Matrix Anal. Appl.* **25**, 1–30.

S. Rump (2003*b*), 'Structured perturbations II: Componentwise distances', *SIAM J. Matrix Anal. Appl.* **25**, 31–56.

J. R. Shewchuk (1997), 'Adaptive precision floating-point arithmetic and fast robust geometric predicates', *Discrete Comput. Geom.* **18**, 305–363.

R. P. Stanley (1999), *Enumerative Combinatorics 2*, Vol. 62 of *Cambridge Studies in Advanced Mathematics*, Cambridge University Press.

G. W. Stewart (1993), 'Updating a rank-revealing ULV decomposition', *SIAM J. Matrix Anal. Appl.* **14**, 494–499.

A. Tarski (1951), *A Decision Method for Elementary Algebra and Geometry*, University of California Press, Berkeley.

J. Taylor (2004), *Several Complex Variables with Connections to Algebraic Geometry and Lie Groups*, AMS Series on Graduate Studies in Mathematics, AMS.

L. G. Valiant (1979), 'The complexity of computing the permanent', *Theoret. Comput. Sci.* **8**, 189–201.

Q. Ye (2008*a*), 'Computing singular values of diagonally dominant matrices to high relative accuracy', *Math. Comp.*, to appear.

Q. Ye (2008*b*), 'Relative perturbation bounds for eigenvalues of symmetric positive definite diagonally dominant matrices', *SIAM J. Matrix Anal. Appl.*, to appear.

G. M. Ziegler (1995), *Lectures on Polytopes*, Vol. 152 of *Graduate Texts in Mathematics*, Springer, New York.

Acta Numerica (2008), pp. 147–190
doi: 10.1017/S0962492906360011

Asymptotic and numerical homogenization

B. Engquist and P. E. Souganidis
Department of Mathematics,
The University of Texas at Austin,
Austin, TX 78712, USA
E-mail: {engquist}{souganid}@math.utexas.edu

Homogenization is an important mathematical framework for developing effective models of differential equations with oscillations. We include in the presentation techniques for deriving effective equations, a brief discussion on analysis of related limit processes and numerical methods that are based on homogenization principles. We concentrate on first- and second-order partial differential equations and present results concerning both periodic and random media for linear as well as nonlinear problems. In the numerical sections, we comment on computations of multi-scale problems in general and then focus on projection-based numerical homogenization and the heterogeneous multi-scale method.

CONTENTS

1. Introduction

In this paper we present a very broad overview of the general subject of homogenization. The term homogenization is used in many areas for mechanisms that make a mixture or a process the same (homogeneous) throughout a domain. In mathematics the term is most commonly used in connection with the study of problems (differential equations) with rapidly variable coefficients. There are many variations and extensions, some purely theoretical and others of important practical value. Overall, the idea is to consider a complex multi-scale system, analyse its limit properties and approximate it by a simpler homogenized one.

Consider a general differential equation

$$F_\varepsilon(u^\varepsilon, x) = 0$$

depending on a small parameter ε with solution $u^\varepsilon : \mathbb{R}^d \to \mathbb{R}$. The natural mathematical questions are whether the family $(u^\varepsilon)_{\varepsilon>0}$ converges, as $\varepsilon \to 0$, in some topology to a limit u_0, and if so, whether that limit satisfies a separate 'homogenized' equation

$$\bar{F}(u_0, x) = 0.$$

To give a flavour of the problem and to introduce some of the basic concepts, we consider first the very simple example of a two-point boundary value problem:

$$\begin{cases} -(a^\varepsilon(x)u_x^\varepsilon)_x = f & \text{in } (0,1), \\ u^\varepsilon(0) = u^\varepsilon(1) = 0, \end{cases} \tag{1.1}$$

with

$$a^\varepsilon(x) = a(x/\varepsilon) > 0 \quad \text{and} \quad 1\text{-periodic in } y.$$

The coefficient a_ε is highly oscillatory when ε is small, because of the periodicity of a, and, as $\varepsilon \to 0$, converges weakly to its arithmetic mean.

It is straightforward to check, after integrating the differential equation in (1.1), dividing by a^ε and integrating again, that the solution u^ε of (1.1) is given by

$$u^\varepsilon(x) = -\int_0^x \left(a^\varepsilon(\xi)^{-1} \int_0^\xi f(\eta)\,d\eta + C^\varepsilon \right) d\xi,$$

with the constant of integration

$$C^\varepsilon = -\int_0^1 \left(a^\varepsilon(\xi)^{-1} \int_0^\xi f(\eta)\,d\eta \right) d\xi \Big/ \int_0^1 a^\varepsilon(\xi)^{-1}\,d\xi$$

determined by the boundary condition at $x = 1$.

Given this explicit formula of the solution, we now let $\varepsilon \to 0$ to find that

$$u^\varepsilon(x) \to u_0(x) = -a_H \int_0^x \left(\int_0^\xi f(\eta)\, d\eta + C \right) d\xi, \tag{1.2}$$

and

$$C^\varepsilon \to C = -\int_0^1 \left(\int_0^\xi f(\eta)\, d\eta \right) d\xi, \tag{1.3}$$

where

$$a_H = \left(\int_0^1 a(y)^{-1}\, dy \right)^{-1}.$$

It follows that u_0 satisfies the homogenized two-boundary value problem where a^ε is replaced by a_H, which is typically different from the arithmetic mean of a^ε,

$$\begin{cases} -a_H u_{0xx} = f & \text{in } (0,1), \\ u_0(0) = u_0(1) = 0. \end{cases} \tag{1.4}$$

The coefficient a_H can be seen as a homogenized or effective material coefficient, if a^ε represents an original material property, such as, for example, the conductivity of a composite material.

Problem (1.4) is of course much easier to analyse than the original (1.1). It is also much better suited for computations since any discretization does not need to resolve the small ε-scale of (1.1).

The problem can easily become more complicated. Consider, for example, the non-divergence-form two-point boundary value problem

$$\begin{cases} -u_{xx}^\varepsilon + \dfrac{1}{\varepsilon} b^\varepsilon(x) u_x^\varepsilon = f & \text{in } (0,1), \\ u^\varepsilon(0) = u^\varepsilon(1) = 0, \end{cases} \tag{1.5}$$

where

$$b^\varepsilon(x) = b(x/\varepsilon) \quad \text{with } b \text{ 1-periodic and} \quad \int_0^1 b(y)\, dy = 0. \tag{1.6}$$

As before, b^ε is highly oscillatory when ε is small, and converges weakly, as $\varepsilon \to 0$, to its arithmetic mean.

It is again an exercise, albeit slightly more elaborate, to check that the solution u^ε is given by

$$u^\varepsilon(x) = \int_0^x e^{B(y/\varepsilon)}\, dy\, C_\varepsilon - \int_0^x e^{B(y/\varepsilon)} \int_0^y e^{-B(z/\varepsilon)} f(z)\, dz, \tag{1.7}$$

with the constant of integration

$$C^{\varepsilon} = \left(\int_0^1 e^{B(\xi/\varepsilon)} \, dy \right)^{-1} \int_0^1 e^{B(x/\varepsilon)} \int_0^{\xi} e^{-B(y/\varepsilon)} f(y) \, dy \, d\xi$$

determined by the boundary condition at $x = 1$, and

$$B(y) = \int_0^y b(\xi) \, d\xi.$$

It is possible to check directly that, as $\varepsilon \to 0$, $u^{\varepsilon} \to u_0$, where u_0 solves the averaged boundary value problem

$$\begin{cases} -a_H u_{0xx} = f & \text{in } (0,1), \\ u_0(0) = u_0(1) = 0, \end{cases} \tag{1.8}$$

with the homogenized coefficient a_H given by

$$a_H = \left[\int_0^1 e^{-B(y)} \, dy \int_0^1 e^{B(y)} \, dy \right]^{-1}. \tag{1.9}$$

Second-order equations in non-divergence form are closely related, via Itô's calculus, to solutions of stochastic differential equations (SDE for short). If $(W_t)_{t \in \mathbb{R}}$ is a standard Brownian motion, the SDE corresponding to (1.5) is

$$dX_t^{\varepsilon} = -\frac{1}{\varepsilon} b^{\varepsilon}(X_t^{\varepsilon}) \, dt + \sqrt{2} \, dW_t. \tag{1.10}$$

The asymptotics, as $\varepsilon \to 0$, of the solution of (1.10) are closely related to the behaviour, as $\varepsilon \to 0$, of the u^{ε} of (1.5). As a matter of fact there is a probabilistic proof which yields that, as $\varepsilon \to 0$, $X_t^{\varepsilon} \to X_t^0$, in the appropriate sense, where

$$dX_t^0 = \sqrt{2 a_H} \, dW_t \quad \text{with } X_0^0 = x, \tag{1.11}$$

with a_H given by (1.9).

In general it is not possible to derive closed-form expressions for the limiting solution, and other techniques must be used. It is also not always the case that the homogenized equations have the same form as the original solution, as can be seen in the homogenization of scalar conservation laws.

The general theory deals with more sophisticated homogenization problems in higher dimensions and more complicated linear and nonlinear equations in periodic, almost periodic and, more generally, random environments. In this paper we discuss only linear and fully nonlinear first- and second-order equations in periodic and random settings.

There are several approaches to the study of homogenization depending on the exact form of the problem at the level of the ε-scale, and the nature of the environment (periodic, random, etc.). In this paper we discuss a number of them, but not all. The first method, which is used most often, especially at the formal level, is based on the construction of asymptotic expansions. This approach provides the homogenized equation but it is occasionally difficult to make rigorous due to lack of regularity. The second method, known as the energy method, is applied to problems which admit a variational formulation. A third method, known as two-scale convergence, is often used very effectively in variational problems and combines elements of the first two. A fourth method is based on probabilistic arguments and is used, of course, when the problem has a probabilistic interpretation. When dealing with nonlinear problems it is often necessary to adapt the general methodologies to accommodate the specific nonlinear structure. Some of the approaches used here include Γ-convergence, an effective tool for problems with variational formulation, and the perturbed-test function method typically used for (nonlinear) non-variational problems.

Even if some ideas in homogenization date further back, the early developments of the 1970s are fundamental. This is very well presented in the text from 1978 by Bensoussan, Lions and Papanicolaou (1978), which develops a systematic framework for asymptotic analysis. Influential contemporary contributions are the multi-scale analysis of Keller (1977), the homogenization techniques of Babuška (1976), the G-convergence theory of De Giorgi and Spagnolo (1973) and the analysis of Murat and Tartar (1997).

Lions, Papanicolaou and Varadhan (1983) were the first to consider the homogenization of first- and second-order fully nonlinear equations in the periodic setting. An influential contribution to the subject were the papers by Evans (1989, 1992). The literature for random homogenization is more limited. Papanicolaou and Varadhan (1979, 1981) and Kozlov (1985) were the first to consider the homogenization of uniformly elliptic linear divergence-form and non-divergence-form elliptic operators. The first nonlinear results in the variational setting were obtained by Dal Maso and Modica (1986) and in the non-variational setting by Souganidis (1999), Lions and Souganidis (2003), Kosygina, Rezakhanlou and Varadhan (2006) and Caffarelli, Souganidis and Wang (2005).

After the early period the development of homogenization has accelerated. In one direction there are many new results on nonlinear problems and stochastic equations, some of which are mentioned above. There is now also a wide literature on applications in solid and fluid mechanics. In another direction, theoretical tools developed in the context of homogenization, such as compensated compactness Γ-, G- and H-convergence, have been of great value in other areas of partial differential equations. Some of these later results will be discussed below but it is outside the scope of this paper to give an extensive review of the literature. We refer, for example, to the recent texts by Jikov, Kozlov and Oleinik (1991), Allaire et al. (1993), Cioranescu and Donato (2000), Marchenko and Khruslov (2006) and Pavliotis and Stewart (2007).

The purpose of this paper is to give easily accessible examples of the basic theory and recent progress. An important aspect is also the coupling of asymptotic and numerical methods. The first part (Sections 2 to 7) describes the general problem and is devoted to the analytic aspects of the theory, while the last three sections are devoted to numerical homogenization. In Section 2 we discuss the homogenization theory for linear divergence-form second-order elliptic partial differential equations (PDEs) in the periodic setting. Linear second-order elliptic PDEs in non-divergence form are discussed in Section 3. In Section 4 we consider nonlinear first- and second-order PDEs, in either divergence or non-divergence form, always in periodic media. In Section 5 we provide a general overview of the theory in the stationary ergodic environments, while in Section 6 we present results about rates of convergence. A few applications are given in Section 7 to illustrate the variety of homogenization results. In the second part (Sections 8 to 10) we start with a general discussion of numerical homogenization in Section 8, followed by two special techniques in Sections 9 and 10.

We conclude by remarking that homogenization is a very broad topic which cannot in all fairness be described in a few pages. The reader should keep in mind that here we only attempt to provide a short introduction and hope to stimulate further study of the subject. The same applies to the references. There are literally thousands of papers devoted to the different parts of the theory. Here we only refer to the ones which are relevant to the particular problems we discuss.

A final remark is that throughout the paper we will denote by C positive constants which are independent of ε.

2. Periodic homogenization for linear divergence-form second-order elliptic PDEs

We consider here the divergence-form elliptic boundary value problem

$$\begin{cases} -\mathrm{div}\left(A\left(\dfrac{x}{\varepsilon}\right) Du^{\varepsilon} \right) = f & \text{in } U, \\ u^{\varepsilon} = 0 & \text{on } \partial U, \end{cases} \tag{2.1}$$

which, using the summation convention, is rewritten as

$$\begin{cases} -\left(a_{ij}\left(\dfrac{x}{\varepsilon}\right) u^{\varepsilon}_{x_j} \right)_{x_i} = f & \text{in } U, \\ u^{\varepsilon} = 0 & \text{on } \partial U. \end{cases} \tag{2.2}$$

The matrix function $A = (a_{ij})_{1\leq i,j\leq d}$ is assumed to be

$$\text{symmetric, continuous and } Y\text{-periodic}, \tag{2.3}$$

where Y is the unit cube, and

$$\begin{cases} \text{uniformly elliptic, } i.e., \text{ there exist } \lambda, \Lambda > 0 \text{ such that,} \\ \quad \text{for all } \xi, y \in \mathbb{R}^N, \quad \Lambda|\xi|^2 \geqq a_{ij}(y)\xi_i\xi_j \geqq \lambda|\xi|^2, \end{cases} \tag{2.4}$$

where $|\cdot|$ denotes the Euclidean norm and we have used the usual summation convention.

A function $u^\varepsilon \in H_0^1(U)$, the Sobolev space of functions vanishing on the boundary ∂U which, together with their derivatives, are square-integrable in U, is defined to be a solution of (2.1) if, for all $v \in H_0^1(U)$,

$$a_\varepsilon(u^\varepsilon, v) = (f, v), \tag{2.5}$$

where, for $f \in L^2(U)$ and $u, v \in H_0^1(U)$,

$$(f, v) = \int_U fv\,\mathrm{d}x \quad \text{and} \quad a_\varepsilon(u, v) = \int_U a_{ij}\left(\frac{x}{\varepsilon}\right) u_{x_j} v_{x_i}\,\mathrm{d}x.$$

The problem is well defined in $H_0^1(U)$. For $f \in L^2(U)$, it admits a unique solution $u^\varepsilon \in H_0^1(U)$ satisfying the estimate

$$\|u^\varepsilon\|_{H_0^1(U)} \leqq C\|f\|_{L^2(U)}. \tag{2.6}$$

The two-point boundary value problem (1.1) is a special case of (2.1). Its variational formulation is

$$\int_0^1 a\left(\frac{x}{\varepsilon}\right) u_x^\varepsilon v_x\,\mathrm{d}x = \int_0^1 fv\,\mathrm{d}x \quad \text{for all } v \in H_0^1(0,1). \tag{2.7}$$

To explain the difficulties arising in the study of the behaviour, as $\varepsilon \to 0$, of (2.1), we proceed with (1.1). Taking $v = u^\varepsilon$ in (2.7), we find

$$\|u^\varepsilon\|_{H_0^1(0,1)} \leqq C.$$

Therefore we may extract subsequences, still denoted by u^ε, such that, as $\varepsilon \to 0$,

$$u^\varepsilon \to u \quad \text{weakly in } H_0^1(0,1).$$

Recall also that, as $\varepsilon \to 0$, we have weak $*$ convergence

$$a\left(\frac{\cdot}{\varepsilon}\right) \to \langle a \rangle = \int_0^1 a(y)\,\mathrm{d}y \quad \text{in } L^\infty(0,1).$$

It is therefore natural to expect that, in the limit $\varepsilon \to 0$, we have

$$-(\langle a \rangle u_x)_x = f \quad \text{in } (0,1),$$

an equation which is *not* satisfied, as we already know from the discussion in the Introduction.

Next we present an argument that is not based on having an exact formula and yields the correct answer. To this end, let

$$\xi^\varepsilon = a^\varepsilon u_x^\varepsilon.$$

Since the a^εs are bounded, it follows that the family $(\xi^\varepsilon)_{\varepsilon>0}$ is bounded in $L^2(0,1)$, and, since it satisfies

$$-\xi_x^\varepsilon = f,$$

is actually bounded in $H^1(0,1)$. Therefore, along subsequences $\varepsilon \to 0$,

$$\xi^\varepsilon \to \xi \text{ in } L^2(0,1) \quad \text{and} \quad \xi_x^\varepsilon \to \xi_x \text{ weakly in } L^2(0,1).$$

In view of the above,

$$\frac{1}{a^\varepsilon}\xi^\varepsilon \rightharpoonup \left\langle \frac{1}{a} \right\rangle \xi \text{ weakly in } L^2(0,1) \quad \text{and} \quad -\xi_x = f.$$

Since

$$\frac{1}{a^\varepsilon}\xi^\varepsilon = u_x^\varepsilon,$$

we conclude that

$$u_x = \left\langle \frac{1}{a} \right\rangle \xi \quad \text{and} \quad -\left(\left\langle \frac{1}{a} \right\rangle^{-1} u_x \right)_x = f.$$

We now present the method of the asymptotic expansion, which is very convenient and is a common technique for obtaining homogenized equations. The justification may need other general tools.

To study the asymptotic behaviour of the solutions of (2.1), we form a two-scale expansion in ε and then match terms of the same order in ε.

We begin with the *ansatz* that

$$u^\varepsilon(x) = u_0(x, x/\varepsilon) + \varepsilon u_1(x, x/\varepsilon) + \varepsilon^2 u_2(x, x/\varepsilon) + \cdots, \qquad (2.8)$$

with $u_j(x, y)$ $(j = 0, 1, 2, \ldots)$ Y-periodic in y.

If we regard x and y as separate variables, applied to a function $\phi(x, x/\varepsilon)$, the differentiation operator D_x becomes $D_x + \frac{1}{\varepsilon}D_y$. Then, if

$$A^\varepsilon v = -\operatorname{div}\left(A\left(\frac{x}{\varepsilon}\right) Dv \right),$$

we may write

$$A^\varepsilon = \varepsilon^{-2}A_1 + \varepsilon^{-1}A_2 + \varepsilon^0 A_3 + \cdots,$$

with

$$\begin{cases} A_1 = -\operatorname{div}_y(A(y)D_y), \\ A_2 = -\operatorname{div}_y(A(y)D_x) - \operatorname{div}_x(A(y)D_y), \\ A_3 = -\operatorname{div}_x(A(y)D_x). \end{cases} \qquad (2.9)$$

Using the assumed expansion, the equation yields

$$\begin{cases} \text{(i)} & A_1 u_0 = 0, \\ \text{(ii)} & A_1 u_1 + A_2 u_0 = 0, \\ \text{(iii)} & A_1 u_2 + A_2 u_1 + A_3 u_0 = f. \end{cases} \tag{2.10}$$

The homogenized operator is then constructed from (2.10).

Before we continue with the analysis, we recall the classical Fredholm alternative argument, which states that

$$\begin{cases} B\phi = F \text{ in } Y, \\ \phi \ Y\text{-periodic,} \end{cases} \tag{2.11}$$

where B is a general second-order operator in either divergence or non-divergence form, has a solution if and only if

$$\int_Y F(y)m(y)\,\mathrm{d}y = 0,$$

where m is an invariant measure, $i.e.$, the unique solution of the adjoint equation

$$\begin{cases} B^* m = 0 \text{ in } Y, \\ m \ Y\text{-periodic,} \end{cases} \tag{2.12}$$

with

$$m > 0 \quad \text{and} \quad \int_Y m(y)\,\mathrm{d}y = 1. \tag{2.13}$$

We return now to the asymptotic analysis of (2.1).

Since the only periodic solutions of (2.10(i)) are constants in y, we have

$$u_0(x,y) = u(x),$$

and therefore (2.10(ii)) reduces to

$$\begin{cases} -\mathrm{div}_y(A(y)D_y u_1) = \mathrm{div}_y(A(y)D_x u_0) \text{ in } Y, \\ u_1 \ Y\text{-periodic in } y, \end{cases} \tag{2.14}$$

which, in view of the previous discussion, admits a unique solution by Fredholm's alternative, up to addition of a constant. Indeed, here for (2.14) we can take $m \equiv 1$. It is then immediate that

$$\int_Y \mathrm{div}_y(A(y)D_x u_0)m\,\mathrm{d}y = 0.$$

The special form of the right-hand side of (2.14) allows us to represent u_1 as

$$u_1(x,y) = w \cdot D_x u_0 + \tilde{u}(x), \tag{2.15}$$

where each component of $w = (w_1, \ldots, w_d)$ is the unique (up to additive constants) solution of the cell problem

$$\begin{cases} -\mathrm{div}_y(A(y)D_y w_i) = \mathrm{div}_y(Ae_i) & \text{in } U, \\ w_i \ Y\text{-periodic in } y. \end{cases} \tag{2.16}$$

Applying the Fredholm alternative once again to (2.10(iii)), recalling that the average of $f - A_2 u_1 - A_3 u_0$ over Y must be zero, and replacing u_1 by (2.15) leads to the homogenized equation

$$\begin{cases} -\mathrm{div}(\bar{A}Du_0) = f & \text{in } U, \\ u_0 = 0 & \text{on } \partial U, \end{cases} \tag{2.17}$$

with the effective matrix $\bar{A} = (\bar{a}_{ij})_{1 \leq i,j \leq d}$ given by

$$\bar{a}_{ij} = \int_Y a_{ij}(y)(D_y w_i + e_i)(D_y w_j + e_j)\,\mathrm{d}y. \tag{2.18}$$

Notice that the matrix $\bar{A}$ is uniformly elliptic. Moreover, since it is constant, (2.17) can also be written in the non-divergence form

$$\begin{cases} -\,\mathrm{tr}\,\bar{A}D^2 u_0 = f & \text{in } U, \\ u_0 = 0 & \text{on } \partial U. \end{cases}$$

Although the method of the asymptotic expansion is simple, it is often difficult to implement, since it is based on establishing the expansion (2.8). It is, however, very useful to guess the homogenized problem. A second step is then required to prove the actual convergence of the u^εs to u_0. The latter can be established by several methods, such as maximum principle, Γ- or G-convergence, *etc.* An effective method, known as the energy method, was introduced in Tartar (1977). It is based on the choice of appropriate test functions in the variational formulation (2.5) of (2.1) which says that, for all $v \in H_0^1(U)$,

$$\int_U \left(A\left(\frac{x}{\varepsilon}\right)Du^\varepsilon(x) \cdot Dv(x) \right) \mathrm{d}x = \int_U f(x)v(x)\,\mathrm{d}x. \tag{2.19}$$

As already discussed in the one-dimensional case, it is not possible to pass to the limit $\varepsilon \to 0$ in the left-hand side of (2.19) for every v, since the families $(A(\frac{\cdot}{\varepsilon}))_{\varepsilon>0}$ and $(Du^\varepsilon)_{\varepsilon>0}$ are only weakly convergent. The idea of the energy method is to use an appropriate family v^ε of test functions to pass in the limit $\varepsilon \to 0$ taking advantage of the 'compensated compactness' which takes place. This family of test functions is given by

$$v^\varepsilon(x) = v(x) + \varepsilon \tilde{w}\left(\frac{x}{\varepsilon}\right) \cdot Dv(x), \tag{2.20}$$

with v a smooth H_0^1 function and $\tilde{w} = (w_1, \ldots, w_d)$ the solution of the adjoint cell problem

$$\begin{cases} -\mathrm{div}(A^*(y)(D_y \tilde{w}_i + e_i)) = 0 \ \text{ in } Y, \\ \tilde{w}_i \ Y\text{-periodic in } y, \end{cases}$$

which, once again, exists in view of Fredholm's alternative. Inserting (2.20) in (2.19) allows us to eliminate the 'bad' terms and prove the convergence.

We next present an argument justifying the formal expansion (2.8) under the additional assumption that the coefficients, f and, hence, the solution u^ε are smooth. Under these assumptions it is possible to show that, as $\varepsilon \to 0$, $u^\varepsilon \to u$ uniformly in $\bar{U}$.

To this end, take u_1 as in (2.15) with $\tilde{u}_1 \equiv 0$ and set

$$z_\varepsilon = u^\varepsilon - (u + \varepsilon u_1 + \varepsilon^2 u_2).$$

It follows that

$$A^\varepsilon z_\varepsilon = -\varepsilon \tilde{z}_\varepsilon,$$

with

$$\tilde{z}_\varepsilon = A_2 w + A_3 u_1 + \varepsilon A_3 u_2.$$

If f is smooth, then u_0, u_1 and u_2 are smooth, and hence

$$|\tilde{z}_\varepsilon| \leqq C \ \text{ in } U.$$

On the boundary ∂U we have

$$z_\varepsilon = -(\varepsilon u_1 + \varepsilon^2 u_2),$$

and therefore

$$|z_\varepsilon| \leqq C\varepsilon \ \text{ on } \partial U.$$

It follows from the maximum principle that

$$|z_\varepsilon| \leqq C\varepsilon \ \text{ in } \bar{U}$$

and, finally,

$$|u^\varepsilon - u| \leqq C\varepsilon \ \text{ in } \bar{U}.$$

The general result is as follows.

Theorem 2.1. Assume (2.3) and (2.4) and let u^ε and u_0 be, respectively, the solutions of (2.1) and (2.17). Then, as $\varepsilon \to 0$, $u^\varepsilon \to u_0$ weakly in H_0^1. In addition, $u^\varepsilon - u_0 - \varepsilon u_1(\frac{\cdot}{\varepsilon}) \to 0$ strongly in H_0^1, where u_1 is the solution of (2.14).

The proof of the weak convergence follows along the lines discussed earlier. The strong convergence without the correction εu_1 is not true. We refer to Bensoussan *et al.* (1978) for an extensive discussion and the detailed proof.

Next we describe briefly another method, known as two-scale convergence and developed in Nguetseng (1989) and Allaire (1992), which, in some sense, blends the asymptotic expansion and energy methods and yields the homogenization in one step.

To describe the method it is necessary to introduce some notation. We write $C_p^\infty(Y)$ for the functions in $C^\infty(\mathbb{R}^d)$ that are Y-periodic, while $C_p(Y)$ is the space of continuous Y-periodic functions. Finally, $C_c^\infty(U; C_p^\infty(Y))$ is the space of infinitely smooth and compactly supported functions with values in $C_p^\infty(Y)$.

We say that a family $(u^\varepsilon)_{\varepsilon>0}$ in $L^2(U)$ is two-scale convergent to $u_0 \in L^2(U \times Y)$ if, for all $\psi \in C_c^\infty(U; C_p^\infty(Y))$,

$$\lim_{\varepsilon \to 0} \int_U u^\varepsilon(x)\psi\left(x, \frac{x}{\varepsilon}\right) \mathrm{d}x = \int_U \int_Y u_0(x,y)\psi(x,y)\,\mathrm{d}x\,\mathrm{d}y. \qquad (2.21)$$

We next summarize the key results on two-scale convergence. For each bounded family $(u^\varepsilon)_{\varepsilon>0}$ in $L^2(U)$ there exists a subsequence, still denoted by u^ε, and $u_0 \in L^2(U \times Y)$ such that, as $\varepsilon \to 0$, u^ε is two-scale convergent to u_0. Moreover, as $\varepsilon \to 0$,

$$u^\varepsilon \to u = \int_Y u_0(\cdot, y)\,\mathrm{d}y \quad \text{weakly in } L^2(U),$$

and, if u_0 is smooth, and, as $\varepsilon \to 0$,

$$\|u^\varepsilon\|_{L^2(U)} \to \|u_0\|_{L^2(U \times Y)},$$

then,

$$u^\varepsilon(\cdot) - u_0\left(\cdot, \frac{\cdot}{\varepsilon}\right) \to 0 \quad \text{strongly in } L^2(U).$$

Finally, if $(u^\varepsilon)_{\varepsilon>0}$ is bounded in $H^1(U)$, then there exist $u \in H^1(U)$ and $u_1 \in L^2(U; H_p^1(Y))$ such that, up to subsequences, u^ε and Du^ε are two-scale convergent, as $\varepsilon \to 0$, to u and $D_x u + D_y u_1$ respectively.

We now sketch how two-scale convergence can be used to study the asymptotic behaviour, as $\varepsilon \to 0$, of (2.1). All the arguments below work up to subsequences.

In view of (2.6) and our earlier general results on two-scale convergence, we know that there exist $u_0 \in H_0^1(U)$ and $u_1 \in L^2(U; H_p^1(Y))$ such that u^ε and Du^ε are two-scale convergent to u and $D_x u + D_y u_1$ respectively. The expectation is that u_ε should behave as $u + \varepsilon u_1(\cdot, \frac{\cdot}{\varepsilon})$.

Using test functions such as $\phi + \varepsilon\phi_1(\cdot, \frac{\cdot}{\varepsilon})$ in (2.19), with ϕ and ϕ_1 smooth and Y-periodic, yields

$$\int_U A\left(\frac{x}{\varepsilon}\right) Du^\varepsilon \cdot \left[D\phi(x) + D_y\phi_1\left(x, \frac{x}{\varepsilon}\right) + \varepsilon D_x\phi_1\left(x, \frac{x}{\varepsilon}\right)\right] \mathrm{d}x$$

$$= \int_U f(x)\left[\phi(x) + \varepsilon\phi_1\left(x, \frac{x}{\varepsilon}\right)\right] \mathrm{d}x.$$

Passing to the limit, as $\varepsilon \to 0$, in the definition of two-scale convergence for Du_ε gives

$$
\int_U \int_Y A(y)[Du(x) + D_y u_1(x, y)] \cdot [D\phi(x) + D_y \phi_1(x, y)] \, dx \, dy
$$
$$
= \int_U f(x)\phi(x) \, dx. \tag{2.22}
$$

A straightforward application of the Lax–Milgram theorem implies that there exists a unique solution (u, u_1) of (2.22) in $H_0^1(U) \times L^2(U; H_p^1(Y))$. Hence, the whole family u^ε and Du^ε are two-scale convergent to u and $D_x u + D_y u_1$, respectively. Moreover, it follows by integration that (2.22) is a variational formulation for the system

$$
\begin{cases}
-\operatorname{div}_y(A(y)(D_x u + D_y u_1)) = 0 \ \text{ in } U \times Y, \\
-\operatorname{div}_x \left[\int_Y A(y)(D_x u + D_y u_1) \, dy \right] = f \ \text{ in } U, \\
u = 0 \ \text{ on } \partial U, \\
u_1(x, \cdot) \ \ Y\text{-periodic in } y,
\end{cases} \tag{2.23}
$$

which is equivalent to the usual homogenized and cell problem equations derived earlier.

To obtain the strong convergence result in Theorem 2.1 using the two-scale method, it suffices to observe that $u_1(x, \frac{x}{\varepsilon}) = Dw(\frac{x}{\varepsilon}) \cdot D_x u(x)$ is in $L^2(U)$ and can be used as a test function for two-scale convergence.

3. Periodic homogenization for linear second-order elliptic PDEs in non-divergence form

We next consider uniformly elliptic non-divergence-form equations and, in particular, the problem

$$
\begin{cases}
A^\varepsilon u^\varepsilon = -a_{ij}\left(\dfrac{x}{\varepsilon}\right) u^\varepsilon_{x_i x_j} + \dfrac{1}{\varepsilon} b_j\left(\dfrac{x}{\varepsilon}\right) u^\varepsilon_{x_j} = f \ \text{ in } U, \\
u^\varepsilon = 0 \ \text{ on } \partial U,
\end{cases} \tag{3.1}
$$

assuming in addition to (2.3) and (2.4) that

$$
b \text{ is } Y\text{-periodic and bounded.} \tag{3.2}
$$

Notice that, if $b_j = -(a_{ij})_{y_j}$, then (3.1) is the problem considered in the previous section.

The behaviour, as $\varepsilon \to 0$, of (3.1) can be studied using probabilistic methodology applied to homogenization (Bensoussan *et al.* 1978).

Since it requires considerable terminology, we choose here to present its PDE analogue. We again use the method of asymptotic expansions. We set

$$A^\varepsilon = \varepsilon^{-2} A_1 + \varepsilon^{-1} A_2 + \varepsilon^0 A_3, \tag{3.3}$$

where

$$\begin{cases} A_1 = -a_{ij}(y)\partial^2_{y_i y_j} + b_i(y)\partial_{y_i}, \\ A_2 = -a_{ij}(y)\partial^2_{y_i y_j} - a_{ij}(y)\partial^2_{x_i y_j} + b_i \partial_{x_i}, \\ A_3 = -a_{ij}(y)\partial^2_{x_i x_j}, \end{cases} \tag{3.4}$$

and use the *ansatz*

$$u^\varepsilon(x) = u_0\left(x, \frac{x}{\varepsilon}\right) + \varepsilon u_1\left(x, \frac{x}{\varepsilon}\right) + \varepsilon^2 u_2\left(x, \frac{x}{\varepsilon}\right) + \cdots, \tag{3.5}$$

with u_0, u_1, u_2 periodic in $y = x/\varepsilon$, which leads to the equations

$$\begin{cases} \text{(i)} & A_1 u_0 = 0, \\ \text{(ii)} & A_1 u_1 + A_2 u_0 = 0, \\ \text{(iii)} & A_1 u_2 + A_2 u_1 + A_3 u_0 = f. \end{cases} \tag{3.6}$$

As in Section 2, it is immediate that, since u_0 is assumed to be periodic in y and A_1 is uniformly elliptic, we must have

$$u_0(x, y) = u_0(x). \tag{3.7}$$

Then (3.6(ii)) becomes

$$A_1 u_1 + b(y) \cdot D u_0 = 0. \tag{3.8}$$

We consider again the adjoint problem

$$\begin{cases} A_1^* m = -(a_{ij}(y)m)_{y_i y_j} - \operatorname{div}_y(bm) = 0 \ \text{ in } Y, \\ m \ \ Y\text{-periodic}, \end{cases} \tag{3.9}$$

which has a unique solution m such that

$$m > 0 \quad \text{and} \quad \int_Y m(y)\, dy = 1. \tag{3.10}$$

Fredholm's alternative once again says that (3.8) has a periodic solution provided the b_is satisfy, for each $i = 1, \dots, d$, the compatibility (centering) condition

$$\int b_i(y)m(y)\, dy = 0. \tag{3.11}$$

Assuming (3.11) and using the form of A_1, we may assume that u_1 has the form

$$u_1(x, y) = -\chi^i(y)u_{0x_i} + \tilde{u}_1(x), \tag{3.12}$$

where, for each $i = 1, \ldots, d$, χ^i satisfies

$$A_1 \chi^i + b_i(y) = 0 \quad \text{in } Y \quad \text{and} \quad \chi^i \text{ is } Y\text{-periodic.} \tag{3.13}$$

Then (3.6(iii)) can be rewritten as

$$A_1 u_2 = f + u_{0x_i x_j} \left(a_{ij} - a_{kj} \chi^i_{y_k} - a_{ki} \chi^j_{y_k} + \frac{1}{2}(b_i \chi^j + b_j \chi^i) \right). \tag{3.14}$$

We recall that it is possible to find a solution as long as the right-hand side of (3.14) is orthogonal to the invariant measure m. This leads to the homogenized equation

$$-\bar{a}_{ij} u_{x_i x_j} = f \quad \text{in } U, \tag{3.15}$$

with

$$\bar{a}_{ij} = \int_Y m(y) \left[a_{ij} - a_{kj} \chi^i_{y_k} - a_{kj} \chi^j_{y_k} + \frac{1}{2}(b_i \chi^j + b_j \chi^i) \right] dy.$$

Direct computations lead to the identity

$$\bar{a}_{ij} = \int m \left[a_{ij} + a_{k\ell} \chi^j_{y_k} \chi^i_{y_\ell} - a_{kj} \chi^i_{y_k} - a_{ij} \chi^j_{y_k} \right] dy. \tag{3.16}$$

Using the uniform ellipticity of the a_{ij}s it is now possible to show that the matrix $\bar{A} = (\bar{a}_{ij})_{1 \leq i,j \leq d}$ is also uniformly elliptic.

It is an instructive exercise, which we leave to the interested reader, to check that for the simple one-dimensional problem considered in the Introduction, (3.16) is the same as (1.9).

Theorem 3.1. Assume (2.3), (2.4), (3.2) and (3.11). Let u^ε and u_0 be the solutions of (3.1) and (3.15), respectively. If $f \in C^3(U)$, then

$$|u^\varepsilon - u_0| \leq C\varepsilon \quad \text{on } \bar{U}. \tag{3.17}$$

We briefly sketch the proof and refer to Bensoussan *et al.* (1978) for the details. To this end, we return to the issue of the solvability of (3.14). We take $\tilde{u}_1 = 0$ and write

$$u_2(x, y) = u_{0x_i x_j} \chi^{ij}(y), \tag{3.18}$$

where the periodic functions χ^{ij} solve, for $i, j = 1, \ldots, d$,

$$A_1 \chi^{ij} = a_{ij} - a_{kj} \chi^i_{y_k} - a_{ki} \chi^j_{y_k} + \frac{1}{2}(b_i \chi^j + b_j \chi^i) - \bar{a}_{ij}. \tag{3.19}$$

With this choice it is easy to check that

$$A^\varepsilon \tilde{u}^\varepsilon = f + \varepsilon A_2 u_2 + \varepsilon A_3 u_1 + \varepsilon^2 A_3 u_2 = f + \varepsilon g_\varepsilon,$$

with

$$\begin{aligned} g_\varepsilon = &-a_{ij}(\chi^{k\ell}_{y_i} u_{0x_j x_k x_\ell} + \chi^{k\ell}_{y_j} u_{0x_i x_k x_\ell}) \\ &+ b_i \chi^{k\ell} u_{0x_i x_k x_\ell} + a_{ij} \chi^\ell u_{0x_i x_j x_\ell} - \varepsilon a_{ij} \chi^{k\ell} u_{0x_i x_j x_k x_\ell}. \end{aligned}$$

If f is sufficiently smooth, then the solution u of

$$\begin{cases} -\bar{a}_{ij}u_{0x_ix_j} = f & \text{in } U, \\ u_0 = 0 & \text{on } \partial U, \end{cases}$$

is also smooth, and thus

$$|g_\varepsilon| \leqq C \quad \text{on } \bar{U}.$$

Therefore, if we set

$$z_\varepsilon = u^\varepsilon - \tilde{u}^\varepsilon,$$

we see that

$$A^\varepsilon z_\varepsilon = \varepsilon g_\varepsilon \quad \text{in } U \quad \text{with} \quad z_\varepsilon = -\varepsilon u_1 - \varepsilon^2 u_2 \quad \text{on } \partial U,$$

and hence

$$|z_\varepsilon| \leqq C\varepsilon \quad \text{on } \bar{U}.$$

We conclude the discussion by remarking once again that the key step in the analysis was solving the cell problem. The necessary solvability solution then leads to the homogenized operator.

4. Nonlinear periodic homogenization

In this section we consider the homogenization of nonlinear equations both in the variational and non-variational settings. As before, we concentrate on second-order problems. The approach in the variational setting is based on the general method of Γ-convergence (see Dal Maso (1993) for the general theory). The theory for non-divergence-form equations is based on viscosity solutions and the so-called perturbed test function method (see Lions et al. (1983) for the first result, and then Evans (1989)).

We begin with the variational setting. The goal is to study the behaviour, as $\varepsilon \to 0$, of functionals of the form

$$F_\varepsilon(u) = \int_U f\left(\frac{x}{\varepsilon}, Du(x)\right) \mathrm{d}x, \tag{4.1}$$

defined on, for example, $W^{1,p}(U)$ and their associated calculus of variations problems

$$\min\left\{\int_U f\left(\frac{x}{\varepsilon}, Du\right) \mathrm{d}x : u \in W_0^{1,p}(U)\right\}. \tag{4.2}$$

Functionals and minimization problems like (4.1) and (4.2) model phenomena in continuum mechanics in the presence of micro-structures with ε the scale of the medium. The integrand f is, typically, the energy density. For example, F_ε may be the stored energy of an elastic material and u a deformation, or u may be the difference of potential in a composed material occupying U.

The main physical question is whether the medium modelled by F_ε behaves, in the limit $\varepsilon \to 0$, as a homogeneous one. If this is the case, then there must exist an energy density f_0, independent of x and U, such that the minima of (4.2) converge to the minima of

$$\min\left\{ \int_U f_0(Du(x))\,\mathrm{d}x : u \in W_0^{1,p}(U) \right\}. \tag{4.3}$$

The convergence of the minimum values and, perhaps, the minimizing functions of (4.2) are obtained as a consequence of the convergence, as $\varepsilon \to 0$, of (4.1) to the homogenized functional

$$F_0(u) = \int_U f_0(Du(x))\,\mathrm{d}x$$

in the sense of the Γ-convergence introduced by De Giorgi and Franzoni (1975). The book by Dal Maso (1993) is an excellent reference for the whole theory of Γ-convergence.

Next we briefly describe some of the basic facts. We begin with the definition of Γ-convergence on a metric space. Let $F_\varepsilon : X \to [0, \infty]$ be defined on a metric space (X, d). The family $(F_\varepsilon)_{\varepsilon > 0}$ Γ-converges to F, as $\varepsilon \to 0$, if:

(i) for every $x \in X$ and every family $(x_\varepsilon)_{\varepsilon > 0}$ such that $x_\varepsilon \to x$, as $\varepsilon \to 0$, we have $F(x) \leq \liminf_\varepsilon F_\varepsilon(x_\varepsilon)$, and

(ii) for every $x \in X$ there exists a family $(x_\varepsilon)_{\varepsilon > 0}$ such that $x_\varepsilon \to x$, as $\varepsilon \to 0$, and $F(x) = \lim_\varepsilon F_\varepsilon(x_\varepsilon)$.

The fundamental result of Γ-convergence says that if $(F_\varepsilon)_{\varepsilon > 0}$ is d-equicoercive and Γ-converges on X to F, then

$$\min\{F(x) : x \in X\} = \liminf_\varepsilon \{F_\varepsilon(x) : x \in X\},$$

and, moreover, if x_ε is a minizer of (4.2) and $x_\varepsilon \to x$, as $\varepsilon \to 0$, then x is a minimizer of (4.3).

We now concentrate on the Γ-convergence of functionals of the form (4.1). To this end, let U be a bounded subset of $\mathbb{R}^d$ and $p > 1$. As far as the integrand $f : U \times \mathbb{R}^d \to \mathbb{R}$ is concerned, it is assumed that

$$\begin{cases} \text{(i)} \quad \text{for every } \xi \in \mathbb{R}^d, \ f(\cdot, \xi) \text{ is measurable and } Y\text{-periodic,} \\[6pt] \text{(ii)} \quad \text{for almost every } y \in \mathbb{R}^d, \ f(y, \cdot) \text{ is convex in } \mathbb{R}^d, \\[6pt] \text{(iii)} \quad \text{for some } c > 0, \text{ almost every } y \in \mathbb{R}^d \text{ and every } \xi \in \mathbb{R}^d, \\[6pt] \qquad 0 \leq f(y, \xi) \leq c(|\xi|^p + 1) \quad \text{and} \quad f(y, 2\xi) \leq cf(y, \xi) + c. \end{cases} \tag{4.4}$$

For $\varepsilon > 0$ consider the functional $F_\varepsilon : L^p(U) \to [0, \infty]$ defined by

$$F_\varepsilon(u) = \begin{cases} \int_U f\left(\dfrac{x}{\varepsilon}, Du\right) dx & \text{if } u \in W^{1,p}(U), \\ +\infty & \text{otherwise,} \end{cases} \qquad (4.5)$$

and let $f_0 : \mathbb{R}^d \to [0, +\infty)$ be defined by

$$f_0(\xi) = \inf_{v \in W^{1,p}_{\text{per}}(Y)} \int_Y f(y, \xi + Dv(y)) \, dy, \qquad (4.6)$$

where $W^{1,p}_{\text{per}}(Y)$ denotes the set of all Y-periodic functions in $W^{1,p}_{\text{loc}}(\mathbb{R}^d)$.
Finally, consider $F_0 : L^p(\mathbb{R}^d) \to \mathbb{R}$ given by

$$F_0(u) = \begin{cases} \int_U f_0(Du) \, dx & \text{if } u \in W^{1,p}(U), \\ +\infty & \text{otherwise.} \end{cases} \qquad (4.7)$$

Theorem 4.1. Assume that f satisfies (4.4) and let F_0 be defined by (4.7) for f_0 given by (4.6). Then, for every sequence $\varepsilon_n \to 0$, the sequence $(F_{\varepsilon_n})_{\varepsilon_n > 0}$ Γ-converges to F_0.

We now turn our attention to non-variational homogenization problems. Before we enter into details we point out the main difference between linear and nonlinear theories. In the methods described in Sections 2 and 3, the key step was the solvability of the auxiliary cell problems obtained after expanding. This solvability was accomplished using Fredholm's alternative and the invariant measures, which are positive solutions of the adjoint operator. Cell problems also arise in nonlinear problems. In such settings, however, there is no notion of adjoint operator and Fredholm's alternative. It is therefore necessary to solve the cell problem directly.

For definiteness we consider the boundary value problem

$$\begin{cases} F\left(D^2 u^\varepsilon, \dfrac{x}{\varepsilon}, x\right) = 0 & \text{in } U, \\ u^\varepsilon = 0 & \text{on } \partial U. \end{cases} \qquad (4.8)$$

We assume that

$$F \in C(S^d \times \mathbb{R}^d \times U) \quad \text{is } Y\text{-periodic,} \qquad (4.9)$$

and

$$\begin{cases} \text{uniformly elliptic, } i.e., \text{ there exist } \lambda, \Lambda > 0 \text{ such that,} \\ \text{for all } (y, x) \in \mathbb{R}^d \times U \text{ and all } X, Y \in S^d \text{ with } Y \geq 0, \\ \lambda \|Y\| \leq F(X, y, x) - F(X + Y, y, x) \leq \Lambda \|Y\|, \end{cases} \qquad (4.10)$$

where S^d denotes the space of $d \times d$ symmetric matrices.

Solutions of (4.8) are understood in the Crandall–Lions viscosity sense. For completeness we recall the definition below; for details we refer to Crandall, Ishii and Lions (1992).

Definition 4.2. An upper semi-continuous (respectively, lower semi-continuous) function u is a viscosity subsolution (respectively, supersolution) of $G(D^2u, Du, Du, u, x) = 0$ in U if, for all smooth ϕ and all local maximum (respectively, minimum) points x of $u - \phi$, we have $G(D^2\phi, D\phi, u(x), x) \leqq 0$ (respectively, $G(D^2\phi, D\phi, u(x), x) \geqq 0$). A continuous function is a solution, if it is both a subsolution and supersolution.

It turns out that viscosity solutions are the correct class of weak solutions for fully nonlinear (degenerate) elliptic second-order equations.

We return now to the homogenization of (4.8) and assume, for simplicity, that F is independent of x. We follow the asymptotic expansion approach and consider the *ansatz*

$$u^\varepsilon(x) = u_0\left(x, \frac{x}{\varepsilon}\right) + \varepsilon u_1\left(x, \frac{x}{\varepsilon}\right) + \varepsilon^2 u_2\left(x, \frac{x}{\varepsilon}\right) + \cdots,$$

with $u_0, u_1, u_2, \ldots$, periodic in y. Substituting in (4.8) we find

$$F\left(\frac{1}{\varepsilon^2}D_y^2 u_0 + \frac{1}{\varepsilon}(D_y^2 u_1 + D_{x,y}^2 u_0) + D_x^2 u_0 + D_{x,y}^2 u_1 + D_y^2 u_2 + \cdots, \frac{x}{\varepsilon}\right) = 0.$$

A heuristic argument based on the ellipticity of F yields that both u_0 and u_1 must be independent of y and, hence, the expansion reduces to

$$F\left(D_y^2 u_2 + D_x^2 u_0, \frac{x}{\varepsilon}\right) = 0.$$

The goal is then to find u_2 so that $F(D_y^2 u_2 + D_x^2 u_0, \frac{x}{\varepsilon})$ is a constant $\bar{F}(D_x^2 u_0)$ independent of $y = x/\varepsilon$. This leads to the cell problem

$$\begin{cases} \text{for each } P \in S^d \text{ there exists a unique constant } \bar{F}(P) \\ \text{such that there exists a periodic solution } v \text{ of} \qquad\qquad (4.11) \\ \quad F(P + D_y^2 v, y) = \bar{F}(P) \text{ in } \mathbb{R}^d. \end{cases}$$

We recall the previous discussion on the fundamental difference between linear and nonlinear problems. In the linear setting the cell problem is solved using Fredholm's alternative, with the constant $\bar{F}(P)$ arising as the necessary compatibility condition for solvability. In the nonlinear problem, however, Fredholm's alternative has no meaning and it becomes necessary to solve (4.11) directly. The cell problem can be thought of as a nonlinear eigenvalue problem with $\bar{F}(P)$ the eigenvalue and the solution v, which is usually called the corrector, as the eigenfunction.

The cell problem is solved using the approximate problem

$$\alpha v_\alpha + F(D^2 v_\alpha + P, y) = 0 \quad \text{in } \mathbb{R}^d,$$

obtaining appropriate bounds and passing to the limit $\alpha \to 0$. The constant $\bar{F}(P)$ is obtained as the limit of $-\alpha v_\alpha(0)$. This step requires non-trivial arguments from the theory of viscosity solutions which are beyond the scope of this review. We refer to Lions *et al.* (1983), Evans (1989), *etc.*, for a discussion.

It is worth remarking that this approximate problem is not artificial. Indeed, if

$$v^\alpha(y) = \alpha v_\alpha \left(\frac{y}{\alpha^{1/2}} \right),$$

then

$$v^\alpha + F\left(D^2 v^\alpha + p, \frac{y}{\alpha^{1/2}} \right) = 0 \quad \text{in } \mathbb{R}^d.$$

If there is homogenization, then, as $\alpha \to 0$, $v^\alpha \to \bar{v}$ locally uniformly in $\mathbb{R}^d$, with

$$\bar{v} + \bar{F}(D^2 \bar{v} + P) = 0 \quad \text{in } \mathbb{R}^d,$$

in which case, since $\bar{v}$ is bounded, we must have $\bar{v} = -\bar{F}(P)$.

To state the result we consider the boundary value problem

$$\begin{cases} \bar{F}(D^2 u_0) = 0 & \text{in } U, \\ u_0 = 0 & \text{on } \partial U. \end{cases} \tag{4.12}$$

The following was proved in Evans (1989).

Theorem 4.3. Let u^ε and u_0 be, respectively, the solutions of (4.8) and (4.12), with $\bar{F}$ given by (4.11). Then, as $\varepsilon \to 0$, $u^\varepsilon \to u_0$ uniformly in $\bar{U}$.

The proof is relatively simple, therefore we briefly sketch its main steps.

It is not difficult to show, under some technical assumptions on F, that there exists $C > 0$ such that

$$|u^\varepsilon| \leqq C \quad \text{on } \bar{U}.$$

We define next the so-called 'relaxed' half-limit, a very important tool in the theory of viscosity solutions, introduced by Barles and Perthame (1988). They are

$$u^*(x) = \limsup_{y \to x, \, \varepsilon \to 0} u^\varepsilon(y) \quad \text{and} \quad u_*(x) = \liminf_{y \to x, \, \varepsilon \to 0} u^\varepsilon(y).$$

It is of course immediate that $u_* \leqq u^*$ in U. Next we show that u^* and u_* are, respectively, sub- and supersolution of (4.12). The comparison property of viscosity solutions then yields that we must have $u^* \leqq u_*$, and hence $u_0 = u_* = u^*$, a fact which yields that, as $\varepsilon \to 0$, $u^\varepsilon \to u_0$ in U.

We employ the so-called 'perturbed test function' method (Evans 1989) to show that u^* is a subsolution. The argument for u_* is similar. We would like to point out to the reader the similarity of the argument below to one used for linear equations. The main difference is that, due to the lack of higher regularity of solutions to fully nonlinear equations, it is not possible to justify the asymptotics directly. Instead this is done at the level of the test functions.

Let ϕ be a test function and suppose that $x_0 \in U$ is a strict local maximum of $u^* - \phi$ in $B(x_0, \delta)$ for some small $\delta > 0$. Moreover, we assume that $u^*(x_0) = \phi(x_0)$. We argue by contradiction and assume that

$$\bar{F}(D^2\phi(x_0)) = \sigma > 0.$$

Let ψ be the solution of the cell problem

$$F(D_y^2\psi + D^2\phi(x_0), y) = \bar{F}(D^2\phi(x_0)).$$

It follows[1] that the function

$$v^\varepsilon(x) = \phi(x) + \varepsilon^2 \psi\left(\frac{x}{\varepsilon}\right)$$

is a supersolution of

$$F\left(D^2v^\varepsilon, \frac{x}{\varepsilon}\right) \geqq 0 \quad \text{in } B(x_0, \delta).$$

The comparison of viscosity solutions then yields that

$$u^\varepsilon(x_0) - v^\varepsilon(x_0) \leqq \max_{B(x_0,\delta)}(u^\varepsilon - v^\varepsilon) \leqq \max_{\partial B(x_0,\delta)}(u^\varepsilon - v^\varepsilon).$$

Since $v^\varepsilon \to \phi$, as $\varepsilon \to 0$, we find

$$u^*(x_0) - \phi(x_0) \leqq \max_{\partial B(x_0,\delta)}(u^* - \phi),$$

which is a contradiction, since $u^*(x_0) = \phi(x_0)$ and $\max_{\partial B(x_0,\delta)}(u^* - \phi) < 0$.

At this point we remark that, if the corrector v is not periodic, which will be the case in general, it is necessary for the above argument to work as well as for the uniqueness of $\bar{F}(P)$ to obtain correctors which are strictly sub-quadratic at infinity. This is not always possible: see Lions and Souganidis (2003).

We also remark that the approach described above can be used to study the homogenization of several variants of (4.8), including completely degenerate (Hamilton–Jacobi) equations.

---

[1] This argument, which is non-rigorous since ψ is not necessarily smooth, can be justified using arguments from the theory of viscosity solutions.

5. The random setting

In this section we outline several recent developments to the theory of fully nonlinear first- and second-order partial differential equations in stationary ergodic settings. The general problem is the almost sure (a.s. for short) behaviour, as $\varepsilon \to 0$, of the solution of u^ε of equations of the general form

$$
\begin{cases}
F\left(D^2 u^\varepsilon, \varepsilon D^2 u^\varepsilon, Du^\varepsilon, u^\varepsilon, x, \dfrac{x}{\varepsilon}, \omega\right) = 0 \ \ \text{in } U, \\
u^\varepsilon = g \ \ \text{on } \partial U,
\end{cases}
\tag{5.1}
$$

where F and g satisfy all the necessary assumptions for (5.1) to have, for each $\varepsilon > 0$ and $\omega \in \Omega$, the underlying probability space, a unique viscosity solution $u^\varepsilon(\cdot, \omega) \in C(\bar{U})$. The key assumptions for the homogenization are that

$$
F \text{ is stationary with respect to } (y, \omega), \tag{5.2}
$$

and

$$
\begin{cases}
\text{the underlying measure-preserving transformation} \\
\quad \tau_y : \Omega \to \Omega \ \ \text{is ergodic.}
\end{cases}
\tag{5.3}
$$

Recall that a random field ξ is stationary if, for any finite set of points $x_1, \ldots, x_k \in \mathbb{R}^d$ and any $h \in \mathbb{R}^d$, the distribution of the random vector

$$
\xi(x_1 + h), \ldots, \xi(x_k + h)
$$

does not depend on h. If $\xi : \mathbb{R}^d \times \Omega \to \mathbb{R}$ is a stationary random field, where (Ω, μ) is the underlying probability space, then it can be represented in the form

$$
\xi(x, \omega) = \tilde{\xi}(\tau_x \omega),
$$

for some fixed random variable $\tilde{\xi} : \Omega \to \mathbb{R}$ and a measure-preserving transformation $\tau_x : \Omega \to \Omega$ with $x \in \mathbb{R}^d$.

A measure-preserving transformation $(\tau_x)_{x \in \mathbb{R}^d}$ is ergodic if all translation-invariant subsets of Ω have probability either zero or one. We shall also say that a stochastic process is stationary ergodic if it is stationary and the measure-preserving group is ergodic.

Stationary ergodic media are rather general. The classical periodic and almost periodic settings can be thought of as special cases. But the general setting includes other configurations, such as random chessboards with tiles of arbitrary random size.

The goal is to show that there exists an effective first- or second-order (depending only on the particular form of F in (5.1)) nonlinearity $\bar{F}$ such that, if $u_0 \in C(\bar{U})$ solves

$$\begin{cases} \bar{F}(D^2 u_0, Du_0, u_0, x) = 0 & \text{in } U, \\ u_0 = g & \text{on } \partial U, \end{cases} \tag{5.4}$$

then, as $\varepsilon \to 0$ and a.s. in ω, $u^\varepsilon(\cdot, \omega) \to u_0$.

We recall that most of the results in periodic homogenization are based on the fact that it is possible to solve the associated macroscopic (cell) problem. In the almost periodic and random settings the macroscopic problem is set in the whole space. The inherited lack of compactness then creates several difficulties. In the almost periodic case these difficulties can be overcome using the fact that such a setting is essentially compact. Exact solutions of the macroscopic cell problems can be replaced by approximate ones (see Ishii (1999), Lions and Souganidis (2005a)). The situation is, however, quite different in the random setting. In general, it is not possible to solve either exactly or approximately the macroscopic problem. To overcome this very serious difficulty, it is necessary to develop and follow a different strategy which makes use of the ergodic theorem and its nonlinear version (the subadditive ergodic theorem).

Papanicolaou and Varadhan (1979, 1981), Kozlov (1985), Zhikov (1993) and Yurinskii (1980, 1982) (see also Jikov et al. (1991)) were the first to consider the problem of homogenizing linear, uniformly elliptic/parabolic operators. The first nonlinear result in the variational setting was obtained by Dal Maso and Modica (1986). The homogenization of fully nonlinear, convex, first-order (Hamilton–Jacobi) equations was considered in Souganidis (1999) and Rezankhanlou and Tarver (2000). Lions and Souganidis (2005a) studied the homogenization of, possibly degenerate, viscous Hamilton–Jacobi equations and Kosygina et al. (2006) considered the same problem but in the uniformly elliptic/parabolic setting. Lions and Souganidis (2003, 2005a) showed that the associated macroscopic problems do not have solutions. The homogenization of fully nonlinear, uniformly elliptic equations was studied by Caffarelli et al. (2005). Finally, Caffarelli and Souganidis (2008) recently obtained uniform error estimates for the homogenization of uniformly elliptic problems, which are described in Section 6.

The first problem we consider here is the viscous Hamilton–Jacobi equation

$$-\varepsilon \operatorname{tr} A\left(\frac{y}{\varepsilon}, \omega\right) D^2 u^\varepsilon + H\left(Du^\varepsilon, \frac{x}{\varepsilon}, \omega\right) + u^\varepsilon = 0 \quad \text{in } \mathbb{R}^d, \tag{5.5}$$

where

$$\begin{cases} A(\cdot, \omega) \in C(S^d) \text{ and } H(\cdot, \cdot, \omega) \in C(\mathbb{R}^d \times \mathbb{R}^d) \\ \text{are stationary ergodic} \end{cases} \tag{5.6}$$

and

$$A \text{ is degenerate elliptic and } H \text{ is coercive and convex.} \qquad (5.7)$$

Since A is assumed to be degenerate elliptic, the next theorem yields as a special case the homogenization of Hamilton–Jacobi equations of the form

$$H\left(Du^{\varepsilon}, \frac{x}{\varepsilon}, \omega\right) + u^{\varepsilon} = 0 \quad \text{in } \mathbb{R}^{d}. \qquad (5.8)$$

The main result is as follows.

Theorem 5.1. There exists $\bar{H} \in C(\mathbb{R}^{d})$ coercive and convex such that, if u_0 is the solution of $u_0 + \bar{H}(Du_0) = 0$ in $\mathbb{R}^{d}$, then, as $\varepsilon \to 0$ and a.s., $u^{\varepsilon}(\cdot, \omega) \to u_0$ in $C(\mathbb{R}^{d})$. Moreover,

$$\bar{H}(p) = \inf_{\phi \in \mathcal{S}} \sup_{y} \left[-\varepsilon \operatorname{tr} A(y, \omega) D^2\phi + H(D\phi + p, y, \omega) \right],$$

where $\mathcal{S}$ is the set of continuous random fields ψ that are a.s. strictly sublinear at infinity.

The proof of the theorem in Lions and Souganidis (2005 a) is based strongly on the (stochastic) control interpretation of (5.6), which is available due to the assumptions that H is convex and A independent of Du, and the use of the (sub-additive) ergodic theorem. For $A = I$ another proof and different formula for $\bar{H}$ were obtained in Kosygina *et al.* (2006). For general A, even completely degenerate, this formula is generalized in Lions and Souganidis (2005 b). It should be noted that convexity plays absolutely no role in the periodic/almost periodic settings. What happens without convexity in the random setting is an open problem. Recently Lions and Souganidis (2008) obtained a simple proof of Theorem 5.2 which, although it relies on the convexity of H, does not use the control interpretation at all.

Consider next the boundary value problem

$$\begin{cases} F\left(D^2u^{\varepsilon}, Du^{\varepsilon}, x, \dfrac{x}{\varepsilon}, \omega\right) = 0 \quad \text{in } U, \\[2mm] u^{\varepsilon} = g \quad \text{on } \partial U, \end{cases} \qquad (5.9)$$

where

$$F \text{ is stationary ergodic and uniformly elliptic.} \qquad (5.10)$$

The main result obtained in Caffarelli *et al.* (2005) is as follows.

Theorem 5.2. There exists a uniformly elliptic $\bar{F} \in C(S^{d})$ such that, if $u_0 \in C(\bar{U})$ is the solution of $\bar{F}(D^2u_0, x) = 0$ in U and $u_0 = g$ on ∂U, then, as $\varepsilon \to 0$, and a.s. in ω, $u^{\varepsilon}(\cdot, \omega) \to u_0$ in $C(\bar{U})$.

Moreover, for each $(P, x) \in S^d \times U$,

$$F(P, x) = \lim_{R \to \infty} \inf_{\psi} \sup_{|y| \leq R} F(D^2\psi, x, y, \omega),$$

with the infimum taken over functions which are strictly subquadratic at infinity.

The proof of Theorem 5.2 is completely different from that of Theorem 5.1. It is based on identifying, for each level $\mu \in \mathbb{R}$, all the matrices $P \in C(S^D)$ such that $\bar{F}(P) \leq \mu$ or $\bar{F}(P) \geq \mu$. This is accomplished using the obstacle problem with quadratic obstacles and studying the ergodic properties of the contact set.

6. Rates of convergence

We describe here a number of results concerning rates of convergence for the homogenization of first- and second-order equations.

We begin with periodic Hamilton–Jacobi equations and, in particular, the problem

$$u^\varepsilon + H\left(Du^\varepsilon, \frac{x}{\varepsilon}, x\right) = 0 \ \text{ in } \mathbb{R}^d. \tag{6.1}$$

The following result was proved by Capuzzo-Dolcetta and Ishii (2001).

Theorem 6.1. Assume that H is convex, coercive and periodic and let $u^\varepsilon, u_0 \in C^{0,1}(\mathbb{R}^N)$ be the solutions of (6.1) and the homogenized equation, respectively. There exists a constant $C > 0$, depending only on H, such that

$$|u^\varepsilon - u_0| \leqq C \varepsilon^{1/3}.$$

The rate of convergence for the periodic homogenization of

$$\begin{cases} F\left(D^2 u^\varepsilon, \frac{x}{\varepsilon}\right) = f \ \text{ in } U, \\ u^\varepsilon = g \ \text{ on } \partial U, \end{cases} \tag{6.2}$$

was obtained recently by Caffarelli and Souganidis (2008).

Theorem 6.2. Assume that F is uniformly elliptic and periodic and let $u^\varepsilon \in C(\bar{U})$ and $u_0 \in C^{0,1}(\bar{U})$ be, respectively, the solutions of the oscillating and homogenized equations. There exist $\alpha \in (0, 1)$ depending on the ellipticity constants and the dimension, and $C > 0$ depending, in addition, on the Lipschitz constant of u_0, such that

$$|u^\varepsilon - u_0| \leqq C \varepsilon^\alpha \ \text{ in } \bar{U}.$$

The proof of Theorem 6.1 is based on classical arguments from the theory of viscosity solutions, while Theorem 6.2 is based on a general methodology, developed in Caffarelli and Souganidis (2007), which is based on regularity results concerning viscosity solutions of uniformly elliptic PDEs and the notion of δ-viscosity solutions.

We turn now to the random setting. Here it is necessary to introduce a *rate of mixing* somewhere in the assumptions. To this end, we assume that the nonlinearity F is strongly mixing with algebraic rate, *i.e.*, we assume that

$$
\begin{cases}
\text{there exists } \phi : [0, \infty) \to [0, \infty) \text{ such that } \phi(r)r^{-\alpha} \to 0 \\
\text{for some } \alpha > 0 \text{ and for all } P \in S^d \text{ and } x, y \in \mathbb{R}^d \\
|E(F(P, x, \cdot)F(P, y, \cdot)) - (EF(P, x, \cdot))(EF(P, y, \cdot))| \\
\quad \leqq \phi(|x - y|)(EF^2(P, x, \cdot))^{1/2}(EF^2(P, y, \cdot))^{1/2},
\end{cases}
\tag{6.3}
$$

where E denotes the expectation in the probability space.

The following result on the linear equation

$$
\begin{cases}
-a_{ij}\left(\dfrac{x}{\varepsilon}, \omega\right)u^{\varepsilon}_{x_i x_j} = f \quad \text{in } U, \\
u^{\varepsilon} = g \quad \text{on } \partial U,
\end{cases}
\tag{6.4}
$$

was proved in Yurinskii (1982).

Theorem 6.3. Assume that the matrix $A = (a_{ij})_{1 \leqq i,j \leqq d}$ is uniformly elliptic, stationary and strongly mixing. Let u^{ε}, u_0 be the solutions of (6.4) and the homogenized equation respectively. There exist $\alpha \in (0, 1)$, depending only on the ellipticity constant, and $C > 0$ depending on the ellipticity constants and the Lipschitz constant of u_0, such that

$$
|u^{\varepsilon} - u_0| \leqq C \varepsilon^{\alpha} \quad \text{on } \bar{U}.
$$

Finally, concerning the fully nonlinear equation

$$
\begin{cases}
F\left(D^2 u^{\varepsilon}, \dfrac{x}{\varepsilon}, \omega\right) = f \quad \text{in } U, \\
u^{\varepsilon} = g \quad \text{on } \partial U,
\end{cases}
\tag{6.5}
$$

the following was proved in Caffarelli and Souganidis (2008).

Theorem 6.4. Assume that F is uniformly elliptic, stationary and strongly mixing with algebraic rate. Let $u^{\varepsilon} \in C(\bar{U})$ and $u_0 \in C^{0,1}(\bar{U})$ be the solutions of (6.5) and the homogenized equation, respectively. There exist positive constants C and c depending on the ellipticity, the Lipschitz constant of F, the dimension, the mixing rate and F, but not ε, such that, for all $\varepsilon \in (0,1)$, there exist $A_\varepsilon \subset \Omega$ such that

$$\mu(A_\varepsilon) \leqq C\,\varepsilon^{c|\ln\varepsilon|^{-1/2}} \quad \text{and} \quad |u^\varepsilon - u_0| \leqq C\,\varepsilon^{c|\ln\varepsilon|^{-1/2}} \quad \text{in } \Omega \setminus A_\varepsilon.$$

7. Applications

We briefly discuss here a few applications of the homogenization results presented earlier.

The first application concerns motion of interfaces in random media. A typical problem is the evolution of the boundary Γ_t of an open subset Ω_t of $\mathbb{R}^N$ with velocity V in the direction of the normal vector n given by

$$V = -\varepsilon\delta \operatorname{tr} Dn + v\left(n, \frac{x}{\varepsilon}\right).$$

Using the level set formulation for generalized front propagation, the problem reduces to the study of the a.s. behaviour, as $\varepsilon \to 0$, of the solution $u^\varepsilon(\cdot,\omega)$ of the corresponding level set PDE:

$$\begin{cases} u_t^\varepsilon - \varepsilon\delta \operatorname{tr}(I - \widehat{Du^\varepsilon} \otimes \widehat{Du^\varepsilon})D^2 u^\varepsilon + v\left(\widehat{Du^\varepsilon}, \dfrac{x}{\varepsilon}, \omega\right)|Du^\varepsilon| = 0 \\[2mm] \qquad \text{in } \mathbb{R}^N \times (0,\infty), \\[2mm] u^\varepsilon = u_0 \quad \text{on } \mathbb{R}^N \times \{0\}, \end{cases} \qquad (7.1)$$

where $\Omega_0 = \{u_0 > 0\}$ and, for $p \in \mathbb{R}^N \setminus \{0\}$, $\hat{p} = |p|^{-1}p$.

There is no known result in the random case for (7.1) if $\delta \neq 0$. The reason is that, in view of the dependence on the gradient, the results of Lions and Souganidis (2005b) do not apply here. The behaviour of (7.1) for $\delta > 0$ in the periodic/almost periodic setting was analysed in Lions and Souganidis (2005b). More recently, Cardaliaguet, Lions and Souganidis (2008) looked at special cases of (7.1) and studied in detail what happens when the assumptions of Lions and Souganidis (2005b) are not satisfied. Finally, some related results were obtained by Dirr, Karali and Yip (2007).

The only known result in the random case obtained in Souganidis (1999) and Lions and Souganidis (2003) is as follows.

Theorem 7.1. Let $\delta = 0$. Assume that the map $p \mapsto v(\hat{p}, y)|p|$ is convex and $|v| \geq v_0 > 0$. There exists a convex $\bar{H} \in C(\mathbb{R}^N)$ such that, if $u_0 \in UC(\mathbb{R}^N \times [0,\infty))$ is the unique solution of $u_0 + \bar{H}(Du_0) = 0$ in $\mathbb{R}^N \times [0,\infty)$, then, as $\varepsilon \to 0$ and a.s. in ω, $u^\varepsilon(\cdot,\omega) \to u_0$ in $C(\mathbb{R}^N \times [0,\infty))$. In particular, as $\varepsilon \to 0$, and a.s. in ω, $\Gamma_t^\varepsilon \to \Gamma_t^0$ in the Hausdorff metric, where Γ_t^0 is moving with normal velocity $V = -\bar{H}(n)$.

An interesting question is what happens when the velocity v vanishes and changes sign. Mathematically this is a very challenging problem, since the positivity of v is critical, at the technical level, for obtaining the necessary estimates.

The typical problem is the equation

$$\begin{cases} u_t^\varepsilon + v\left(\dfrac{x}{\varepsilon}\right)|Du^\varepsilon| = 0 \ \ \text{in } \mathbb{R}^d \times (0, \infty), \\ u^\varepsilon = g \ \ \text{on } \mathbb{R}^d, \end{cases} \tag{7.2}$$

which is the level set PDE giving the evolution of $\Omega_0 = \{x \in \mathbb{R}^d : u_0(x) > 0\}$ with normal velocity

$$V = v\left(\frac{x}{\varepsilon}\right).$$

As far as the velocity v is concerned we assume that

$$v \text{ is } Y\text{-periodic and changes sign.} \tag{7.3}$$

The following was proved in Cardaliaguet *et al.* (2008). We refer to Craciun and Bhattachayra (2003) for related numerical results.

Theorem 7.2. Assume (7.3). Then, for every $R, T > 0$, as $\varepsilon \to 0$, we have $u^\varepsilon \to u_0$, and

$$u^\varepsilon \to u_0 = \theta g + (1 - \theta)u_1,$$

in the weak $*$ convergence sense in $L^\infty(B_R \times (\theta, T))$, where $\theta = |\{v < 0\} \cap Y|$ and u_1 is the solution of the homogenized equation with velocity v^+.

The heuristics behind the result is that the front cannot penetrate the $\{v < 0\}$ region while it keeps propagating in all places. As a result it wraps around $\{v = 0\}$ and eventually keeps going 'leaving a piece behind'. This latter phenomenon is characterized by the weak convergence.

The next example is about large deviations of diffusion processes in random environments. This is a very general topic, which cannot be discussed in any generality here. Instead we present a special case. Giving any references is beyond the scope of this paper.

Consider the diffusion process $(X_t^\varepsilon)_{t \geq 0}$ which evolves according to the SDE

$$\begin{cases} \mathrm{d}X_t^\varepsilon = b\left(\dfrac{X_t^\varepsilon}{\varepsilon}, \omega\right) + \sqrt{2\varepsilon}\Sigma\left(\dfrac{X_t^\varepsilon}{\varepsilon}, \omega\right)\mathrm{d}B_t \quad\quad (t > 0), \\ X_0^\varepsilon = x, \end{cases}$$

where $(B_t)_{t \geq 0}$ is a standard M-dimensional Brownian motion on a different probability space (Ω_0, F_0, P_0), b is a Lipschitz-continuous stationary ergodic vector field and Σ is a Lipschitz-continuous and stationary ergodic $d \times M$ matrix.

The medium is modelled by the stationary ergodic potential $V : \mathbb{R}^d \times \Omega \to [0, \infty)$ and the behaviour of the diffusion is governed by the weighted probability

$$Q^\varepsilon_{t,\omega}(d\omega_0) = S^{-1}_{t,\omega} \exp \left\{ -\varepsilon^{-1} \int_0^t V \left(\frac{X^\varepsilon_s}{\varepsilon}, \omega \right) \right\} P_0(d\omega_0),$$

where $S_{t,\omega}$ is a normalizing factor.

To formulate a typical large-deviation result it is necessary to consider, for $v_0 \in BUC(\mathbb{R}^N)$, the initial value problem

$$\begin{cases} v^\varepsilon_t - \varepsilon \, \text{tr}(\Sigma\Sigma^T)\left(\frac{x}{\varepsilon}, \omega \right) D^2 v^\varepsilon + H(Du^\varepsilon, \frac{x}{\varepsilon}, \omega) = 0 \;\; \text{in } \mathbb{R}^N \times (0, \infty), \\ v^\varepsilon = v_0 \;\; \text{on } \mathbb{R}^N \times \{0\}, \end{cases}$$

where, for $(p, y, \omega) \in \mathbb{R}^N \times \mathbb{R}^N \times \Omega$,

$$H(p, y, \omega) = \text{tr}((\Sigma\Sigma^T)(y, \omega)p \otimes p) - b(y, \omega) \cdot p - V(y, \omega). \tag{7.4}$$

Theorem 5.1 applied to this equation yields the existence of $\bar{H} : \mathbb{R}^N \to \mathbb{R}$ such that, if $\bar{v} \in BUC(\mathbb{R}^N \times [0, \infty))$ solves $\bar{v}_t + \bar{H}(D\bar{v}) = 0$ in $\mathbb{R}^N \times [0, \infty)$ with $\bar{v} = v_0$ on $\mathbb{R}^N$, then, as $\varepsilon \to 0$ and a.s. in ω, $v^\varepsilon \to \bar{v}$ in $C(\mathbb{R}^N)$.

As a consequence the following large-deviation principle holds.

Theorem 7.3. Let $\bar{L}$ be the convex dual of the effective Hamiltonian $\bar{H}$ corresponding to (7.4). For any Borel subset A of $\mathbb{R}^N$ with interior A^0 and a.s. in ω,

$$-t \inf_{A^0} \bar{L}(t^{-1}(x - y)) \leq \varliminf_{\varepsilon \to 0} \varepsilon \log Q_{t,\omega}(X^\varepsilon_t \in A)$$

$$\leq \varlimsup_{\varepsilon \to 0} \varepsilon \log Q_{t,\omega}(X^\varepsilon_t \in A) \leq -t \inf_{y \in A} \left[\bar{L}(t^{-1}(x - y)). \right]$$

Another application is related to combustion and the propagation of fronts arising as asymptotic limits of reaction–diffusion equations in a random environment. The particular problem of interest is the a.s. asymptotics, as $\varepsilon \to 0$, of the solution u^ε of the KPP-type equation

$$\begin{cases} u^\varepsilon_t - L^\varepsilon u^\varepsilon = \varepsilon^{-1} f \left(u^\varepsilon, \frac{x}{\varepsilon}, \omega \right) \;\; \text{in } \mathbb{R}^N \times (0, T), \\ u^\varepsilon = g \;\; \text{on } \mathbb{R}^N \times \{0\}. \end{cases}$$

Here L^ε is a general second-order uniformly elliptic operator

$$L^\varepsilon v = -\varepsilon \, \text{tr} \, A \left(\frac{x}{\varepsilon}, \omega \right) D^2 v + b \left(\frac{x}{\varepsilon}, \omega \right) \cdot Dv,$$

with $A(\cdot, \omega) \in C^{0,1}(S^d)$ and, $b(\cdot, \omega) \in C^{0,1}(\mathbb{R}^d)$ stationary ergodic and $f(\cdot, \cdot, \omega) \in C^{0,1}(\mathbb{R} \times \mathbb{R}^d)$.

The nonlinearity f is assumed to be of KPP-type, $i.e.$,

$$\begin{cases} f(0,y,\omega) = f(1,y,\omega) = 0, \ \ f(u,y,\omega) > 0 \ \ \text{if } u \in (0,1), \ \ \text{and} \\ f(u,y,\omega) \leqq c(y,\omega)u \ \ \text{with } c(y,\omega) = f_u(0,y,\omega). \end{cases}$$

Theorem 7.4. There exists $\bar{H}$ such that, as $\varepsilon \to 0$ and a.s. in ω, locally uniformly in $\mathbb{R}^N \times (0,\infty)$ and exponentially fast,

$$u^\varepsilon(x,t,\omega) \to \begin{cases} 1 & \text{in int}\{\bar{v} = 0\}, \\ 0 & \text{in } \{\bar{v} > 0\}, \end{cases}$$

where $\bar{v}$ is the unique viscosity solution of the variational inequality

$$\begin{cases} \min[\bar{v}_t + \bar{H}(D\bar{v},x), \bar{v}] = 0 \ \ \text{in } \mathbb{R}^N \times (0,\infty), \\ \bar{v} = \begin{cases} 0 & \text{in } \{g > 0\}, \\ +\infty & \text{in } \{g < 0\}. \end{cases} \end{cases}$$

We conclude with an example concerning the homogenization of a linear transport equation. Although we did not discuss such equations, we include this discussion here as an example of the fact that the homogenized equation may develop futures not existing at the level ε of the oscillations. In particular we present a simple hyperbolic example where the homogenized equation contains an integral term describing a memory effect.

Consider the 2D-linear advection equation

$$u_t^\varepsilon(x,y,t) + a^\varepsilon(y)u_x^\varepsilon(x,y,t) = 0,$$

with initial conditions $u(x,y,0) = g(x,y)$ and a^ε bounded. This simple model helps in explaining certain fingering effects of flow in layered subsurface reservoirs (Hou 2003)

It is easy to write the solution explicitly,

$$u^\varepsilon(x,y,t) = g(x - a^\varepsilon(y),t,y),$$

but not as simple to derive the homogenized solution for the weak limit u^0 of u^ε.

Using the Laplace transform, Tartar (1989) showed that the weak limit satisfies the initial value problem

$$u_t^0 + a(y)u_x^0 = \int_0^t \int_{-\infty}^\infty \frac{\partial^2}{\partial x^2} u^0(x - \lambda(t-s),y,s)\,\mathrm{d}\mu_y(\lambda)\,\mathrm{d}s,$$
$$u^0(x,y,0) = g(x,y),$$

where $a(y)$ is the weak limit of $a^\varepsilon(y)$ and μ_y is the Young measure of $a^\varepsilon(y)$, as $\varepsilon \to 0$.

8. Numerical homogenization

Very often there is no known closed form of the homogenized or effective equations but we may still be in a situation where such an equation exists. The option is then to use numerical techniques to generate the homogenized equation or the effect of using a homogenized equation. The goal would typically be that of numerically solving a differential equation, which involves a wide range of scales. If the range of scales is very large the computational cost of direct numerical solution is prohibitive and some approximation is needed.

Let us briefly comment on the computational complexity. Consider a differential equation with the size of the domain of the independent variables of order one. Let the problem be of the type discussed in this paper, where material fluctuations with wave length $O(\varepsilon)$ produce a solution that also has oscillations with wave length $O(\varepsilon)$.

We also know from the asymptotic analysis of the earlier sections that the detailed interaction of oscillations is essential for the structure of the homogenized equation. This means that a direct numerical method must accurately represent the oscillations. From the Shannon sampling theorem (Shannon 1949), we then see that the number of unknowns must at least be of order $O(\varepsilon^{-d})$, where d is the dimension of the space of independent variables.

There are basically four classes of numerical multi-scale methods:

(1) the classical numerical multi-scale methods that aim to solve the full problem efficiently,

(2) generation of the effective equation by numerical solutions of cell problems,

(3) techniques that start from the original full problem and generate a reduced model,

(4) methods that on the finest scale only sample the original problem in order to reduce the overall computational complexity.

In class (1) we have, for example, the well-established methods of multi-grid and the fast multi-pole method. The discretized full problem with all scales is solved efficiently such that the computational cost is essentially proportional to the number of unknowns. There is a weak relation to homogenization in multi-grid methods in that the coarse-grid operator should approximate the homogenized differential equation (Engquist and Luo 1997). For the fast multi-pole method the far-field interaction is described by low-dimensional operators for which averaging plays a role. These methods are, however, not as closely related to homogenization as the methods we will focus on, and they will not be further discussed here.

For class (2) we assume it is known that there exists a homogenized equation based on the solution of cell problems. If there is no explicit formula for the cell problem solution it is natural to approximate that solution numerically. If the material properties are periodic this technique obviously applies. For more general settings it is still practically useful, as in the work of Durlofsky on homogenization or upscaling in oil reservoir modelling. See, for example, the discussion in Durlofsky (1998). The computational domain is divided into subdomains and a local cell problem is solved in each subdomain. The cell problems are solved by fine discretizations that resolve any oscillation and the solution is then used to define a global effective equation, which can be solved numerically on a coarser grid.

In class (3) the computational cost is at least as large as in (1), but the goal is now to generate a simplified problem that can then efficiently be solved many times. This simplified problem can have significance in itself or it can be used, for example, in control applications. The so-called model reduction techniques belong to this class.

A typical model reduction problem starts by a system of ordinary differential equations describing the state $x(t)$ and output $y(t)$ for a given input or control signal $u(t)$:

$$\frac{\mathrm{d}x(t)}{\mathrm{d}t} = Ax(t) + Bu(t), \qquad x \in \mathbb{R}^n, \ u \in \mathbb{R}^m$$

$$y(t) = Cx(t) + Du(t), \qquad y \in \mathbb{R}^p, \ n \gg m, p.$$

The $n \times n$ matrix A is assumed to be very large and the goal is to replace this system with a reduced one. The reduced system should have a much lower dimension n but approximately the same relation between u and y for a relevant set of u-values. The large size of A typically originates from a discretization of a differential operator in space that may describe a multi-scale problem of the type discussed in this paper. For this type and related problems there is extensive literature: see, for example, Obinata and Anderson (2001).

The multi-scale finite element method (MSFEM) also belongs to this class. It was originated by Babuška (see, for example, Babuška, Caloz and Osborn (1994)), and further refined by Hou (see, for example, Hou (2003)). In this approach a numerical cell problem is not used to determine an effective equation but to generate new finite element basis functions. These basis functions cover a domain of several oscillations. Hou and others have successfully applied MSFEM to oil reservoir modelling problems (Hou 2003).

Projection-based numerical homogenization, which is briefly discussed in the next section, is another technique that belongs to this class.

In class (4) the computational cost for direct solution of the full problem is too large and resolution of the finest scale can only be done in selected domains. The heterogeneous multi-scale method, or HMM (Engquist *et al.* 2007), is a framework for this class, and HMM is presented in Section 10. Another similar strategy is the equation-free technique by Kevrekidis *et al.* (2003). Both these methods rely on refined numerical approximation over sampled domains.

It should be pointed out that there is a strong similarity between the analytical homogenization described in the earlier sections and some of the numerical methods below. For example, projection-based numerical homogenization uses averaging, and adds a high-frequency correction term. The solution of cell problems is central to the heterogeneous multi-scale method and to some of the methods briefly mentioned above.

9. Projection-based numerical homogenization

In projection-based homogenization the fully discretized problem is projected onto a lower-dimensional space. The number of unknowns will be lower and the relevant operator condensed. Early work along these lines using wavelets were presented in Beylkin and Brewster (1995), Dorobantu and Engquist (1998) and Gilbert (1998), but we will follow the development as given in Andersson, Engquist, Ledfelt and Runborg (1999) and Engquist and Runborg (2001, 2001). Consider first the simple two-point boundary value problem that was introduced in Section 1,

$$\begin{cases} -(a^{\varepsilon}(x)u_x^{\varepsilon})_x = f & \text{in } (0,1), \\ u^{\varepsilon}(0) = u^{\varepsilon}(1) = 0, \end{cases} \tag{9.1}$$

$$a^{\varepsilon}(x) = a(x/\varepsilon) > 0, \qquad a(y) \text{ is 1-periodic.}$$

Let us approximate this problem by centred divided differences,

$$L_h^{\varepsilon}u_h^{\varepsilon} = \frac{1}{h}\left(a^{\varepsilon}(x_{j+1/2})\frac{u_{j+1} - u_j}{h} - a^{\varepsilon}(x_{j-1/2})\frac{u_j - u_{j-1}}{h}\right) = f_h, \tag{9.2}$$

$$\text{for } j = 1, 2, \ldots, J-1,$$

where $u_0 = u_J = 0$ and

$$x_j = jh, \quad j = 0, 1, \ldots, J \quad \text{and} \quad Jh = 1.$$

Here

$$u_{\varepsilon,h} = (u_1, u_2, \ldots, u_{J-1})^T \quad \text{and} \quad f_h = \left(f(x_1), f(x_2), \ldots, f(x_{J-1})\right)^T,$$

and the linear operator $L_{\varepsilon,h}$ can be viewed as a finite difference operator or as a $(J-1) \times (J-1)$ matrix.

Our goal, as in general model reduction, is to transform (9.2) into a lower-dimensional problem that is easier to solve. In the example above the best way would be to use the homogenized equation (9.1) and discretize that.

The stepsize h could then be chosen without consideration of the small wave length ε in the oscillations.

In general we do not have access to an explicit homogenized form of the equation and a fully numerical process is needed. We first outline a projection process for homogenization of the original ∞-dimensional problem that easily can be adjusted to finite-dimensional applications, for example of the form (9.2).

Consider an equation $Lu = f$ where L is a linear operator, f a right-hand side and u a solution that contains fine scales. Let P be a projection operator onto a subspace where the fine scales in the original solution do not exist. Our objective is to find the (projection-generated) *homogenized* operator $\bar{L}$ such that $\bar{L}Pu = f$ for all f such that $Pf = f$. (When $Pf \neq f$ we also need to find the homogenized right-hand side $\bar{f}$.) We confine ourselves to the case of Hilbert spaces.

Let X be a Hilbert space of functions, typically a Sobolev space. Let $X_0 \subset X$ be a closed subspace representing the coarse part of the functions, and denote by P_x the orthogonal (and symmetric) projection operator in X onto X_0. Let the spaces X_0 and $X_0^\perp$ inherit the inner product and norm of X, so that $\|u\|_X = \|u\|_{X_0}$ and $(u, v)_X = (u, v)_{X_0}$ when $u, v \in X_0$, and similarly for $X_0^\perp$. In addition, set $Q_x = I_x - P_x$, where I_x is the identity operator in X, and introduce the unitary operator $\mathcal{W}_x$ on X defined by

$$\mathcal{W}_x : X \mapsto X_0^\perp \times X_0, \qquad \mathcal{W}_x u = \begin{pmatrix} Q_x u \\ P_x u \end{pmatrix}. \tag{9.3}$$

In the same way, define the corresponding operators P_y, Q_y and $\mathcal{W}_y$ for another Hilbert space Y with subspace Y_0. Let $\mathcal{L}(X, Y)$ be the set of bounded linear maps from X to Y. For an operator $L \in \mathcal{L}(X, Y)$, we have

$$\mathcal{W}_y L \mathcal{W}_x^* \begin{pmatrix} u \\ v \end{pmatrix} = \mathcal{W}_y L (P_x + Q_x)(u + v) = \begin{pmatrix} Q_y L (P_x + Q_x)(u + v) \\ P_y L (P_x + Q_x)(u + v) \end{pmatrix}$$
$$= \begin{pmatrix} Q_y L (P_x v + Q_x u) \\ P_y L (P_x v + Q_x u) \end{pmatrix} \equiv \begin{pmatrix} A & B \\ C & D \end{pmatrix} \begin{pmatrix} u \\ v \end{pmatrix}, \tag{9.4}$$

where

$$A = Q_y L Q_x \in \mathcal{L}(X_0^\perp, Y_0^\perp), \quad B = Q_y L P_x \in \mathcal{L}(X_0, Y_0^\perp),$$
$$C = P_y L Q_x \in L(X_0^\perp, Y_0), \quad D = P_y L P_x \in \mathcal{L}(X_0, Y_0). \tag{9.5}$$

When A is invertible the following definition can be stated.

Definition 9.1. Suppose $L \in \mathcal{L}(X, Y)$ and $f \in Y$. When A in (9.4), (9.5) is invertible (one-to-one and onto), we define the *homogenized* operator $\bar{L} : X_0 \mapsto Y_0$ as the Schur complement with respect to the decomposition in (9.4),

$$\bar{L} = D - CA^{-1}B, \tag{9.6}$$

and the homogenized right-hand side as

$$\bar{f} = P_y f - CA^{-1}Q_y f. \tag{9.7}$$

We will write $\bar{L}_{X,X_0}$ and $\bar{f}_{X,X_0}$ when there is a need to display explicitly between which spaces the homogenization step is made. From Definition 9.1 we immediately have the following result.

Lemma 9.2. Suppose $Lu = f$, where $L \in \mathcal{L}(X,Y)$, $u \in X$ and $f \in Y$. If A^{-1} exists,

$$\bar{L}P_x u = \bar{f}. \tag{9.8}$$

Proof. Since $L_u = f$, we get

$$\mathcal{W}_y L \mathcal{W}_x^* \mathcal{W}_x u = \mathcal{W}_y f \quad \Rightarrow \quad \begin{pmatrix} A & B \\ C & D \end{pmatrix} \begin{pmatrix} Q_x u \\ P_x u \end{pmatrix} = \begin{pmatrix} Q_y f \\ P_y f \end{pmatrix}. \tag{9.9}$$

Moreover, since A is invertible, this system can be reduced with Gaussian elimination. It yields (9.8). $\qquad\square$

The homogenized operator expressed in terms of projections takes the form

$$\bar{L} = D - CA^{-1}B = PLP - PLQ(QLQ)^{-1}QLP.$$

In the elliptic case, there is a striking similarity between the Schur complement in Definition 9.1 and the classical homogenized operator in (2.18), repeated here for convenience:

$$PLP - PLQ(QLQ)^{-1}QLP, \tag{9.10}$$

$$\nabla \left(\int_{I_d} G(y)\,\mathrm{d}y \right) \nabla - \nabla \left(\int_{I_d} G(y) \frac{\mathrm{d}\chi(y)}{\mathrm{d}y}\,\mathrm{d}y \right) \nabla. \tag{9.11}$$

Both are written as the average of the original operator minus a correction term, which is computed in a similar way for both operators. For the analytic case, a local elliptic cell problem is solved to get $G\partial_y \chi$, while in the projection case, a positive operator $A = QLQ$ defined on a subspace is inverted to obtain $LQA^{-1}B$. The average over the terms is obtained by integration in the analytical case, and by applying P in the projection case.

In the practical and computational setting L and u are finite-dimensional as in the model problem (9.2). A natural way to define the projections discussed above is to use a wavelet basis. Discrete and coarse scales are well defined and the localization properties of wavelets are also practical (Beylkin and Brewster 1995, Engquist and Runborg 2001). In the definition of $\bar{L}$ above, B, C and D are sparse matrices approximating differential operators. The matrix A^{-1} is dense but well approximated by a sparse matrix (Beylkin, Coifman and Rokhlin 1991). This means that $\bar{L}$ is an approximation of a sparse matrix and can be seen as a numerical homogenization.

10. The heterogeneous multi-scale method

The heterogeneous multi-scale method (HMM) is a framework for developing and analysing numerical techniques that couple different models with
different ranges of scales in the same simulation. It was first introduced by
E and Engquist (2003a). See also the shorter and less technical presentation in E and Engquist (2003b) and the more extensive survey in Engquist
et al. (2007).

Let us turn to the structure of HMM. The general setting is as follows.
We are given a microscopic system whose state variable is denoted by u,
together with a micro-model, which can be abstractly written as

$$f(u, d) = 0, \qquad (10.1)$$

where d is the data given by auxiliary conditions, such as initial and boundary conditions for the problem. We are not interested in the microscopic
details of u, but rather the macroscopic state of the system, which we denote
by U. It satisfies some abstract macroscopic equation:

$$F(U, D) = 0, \qquad (10.2)$$

where D stands for the macroscopic data that are necessary for the model
to be complete. We could here view (10.1) as the original problem with an
ε-scale and view (10.2) as the effective or homogenized equation. Note that
F does not need to be known.

Let Q denote the compression operator that maps u to U, and let R be
any operator that reconstructs u from U, that is,

$$Qu = U, \qquad RU = u. \qquad (10.3)$$

Thus Q and R should satisfy $QR = I$, where I is the identity operator. Q is
called a compression operator instead of a projection operator since it can
be more general than projection, *e.g.*, it can be a general coarse-graining
operator, as in bio-molecular modelling. The terminology of reconstruction operator is adopted from Godunov schemes for nonlinear conservation
laws (LeVeque 1990) and gas-kinetic schemes (Xu and Prendergast 1994).
Compression and reconstruction operators are similar to the projection and
prolongation operators used in multi-grid methods, or the restriction and
lifting operators in Kevrekidis *et al.* (2003).

Examples of Q and R were in given in E and Engquist (2003a). The goal
of HMM is to compute U using the abstract form of F and the micro-scale
model. It consists of two main components.

1. Selection of a macroscopic solver
Even though the macroscopic model is not available completely or is invalid
on part of the computational domain, any available knowledge of the form
of F is used to select a suitable macroscopic solver.

2. Estimating the missing macro-scale data D using the micro-scale model
This is typically done in two steps.

(a) *Constrained micro-scale simulation.* At each point where some macro-scale data are needed, perform a series of constrained microscopic simulations. The micro-scale solution needs to be constrained so that it is consistent with the local macroscopic state, *i.e.*, $d = d(U)$. In practice, this is often the most important technical step.

(b) *Data processing.* Use the micro-scale data generated from the microscopic simulations to extract the needed macro-scale data.

Data estimation can be performed either 'on the fly' or in a pre-processing step. The latter is often advantageous if the needed data depend on very few variables. Before we turn to concrete examples, we should emphasize that HMM is not a specific method: it is a framework for designing methods. For any particular problem, there is usually a considerable amount of work, such as designing the constrained microscopic solvers, that is necessary to turn HMM into a specific numerical method. In the remainder of this section, we will discuss examples of how HMM can be used for some relatively simple problems.

Let us first discuss the elliptic problem of Section 3. Consider the 2D case,

$$\begin{cases} -\left(a_{ij}\left(\dfrac{x}{\varepsilon}\right)u^\varepsilon_{x_j}\right)_{x_i} = f \text{ in } U, \\ u^\varepsilon = 0 \text{ on } \partial U. \end{cases} \tag{10.4}$$

Here ε is a small parameter that signifies explicitly the multi-scale nature of the coefficients. It is the ratio between the scale of the coefficient and the scale of the computational domain D. Here we present an approach based on the finite volume method. This is a simplified version of the methods presented in Abdulle and E (2003). Similar ideas can also be found in Durlofsky (1991).

As the macro-scale solver, we choose a finite volume method on a macro-scale grid, and we will let $\Delta x, \Delta y$ be the grid size. The grid points are at the centre of the cells, and the fluxes are defined at the boundaries of the cells. The macro-scale scheme is simply that on each cell, the total fluxes are balanced by the total source or sink terms,

$$-J_{i-\frac{1}{2}j} + J_{i+\frac{1}{2}j} - J_{i,j-\frac{1}{2}} + J_{i,j+\frac{1}{2}} = \int_{K_{i,j}} f(x)\,\mathrm{d}x. \tag{10.5}$$

Here K denotes the (i,j)th cell.

The data that need to be estimated are the fluxes. This is done as follows. At each point where the fluxes are needed, we solve the original micro-scale model (10.4) on a square domain of size δ, with the following boundary condition: $u^\varepsilon(x) - U(x)$ is periodic, where $U(x)$ is a linear function constructed from the macro-state at the two neighbouring cells; *e.g.*, for computing $J_{i+1,j}$, we have

$$
\begin{aligned}
U(x, y) = {} & \frac{1}{2}(U_{j,k} + U_{j+1,k}) + \frac{U_{j+1,k} - U_{j,k}}{\Delta x}(x - x_{j+\frac{1}{2}}) \\
& + \frac{U_{j+1,k+1} - U_{j,k+1} - (U_{j+1,k-1} - U_{j,k-1})}{4\Delta y}(y - y_k).
\end{aligned}
\tag{10.6}
$$

We then use

$$
J_{j+\frac{1}{2},k} = \frac{1}{\delta^2}\int_{I_\delta} j_1^\varepsilon \, dx, \quad J_{j,k+\frac{1}{2}} = \frac{1}{\delta^2}\int_{I_\delta} j_2^\varepsilon \, dx, \qquad U_{j+1,k} - U_{j,k}, \tag{10.7}
$$

where $(j_1^\varepsilon(x), j_2^\varepsilon(x)) = (a_{1k}u_{x_k}^\varepsilon, a_{2k}u_{x_k}^\varepsilon)$, to compute an approximation to the needed flux. The periodic boundary condition for the micro-scale problem is not the only choice: other boundary conditions might be used. A more thorough discussion is found in Yue and E (2008). To reduce the influence of the boundary conditions, a weight function can be inserted in (10.7).

To implement this idea, note that the Js are linear functions of $\{U_{i,j}\}$. Therefore to compute the fluxes, we first solve the local problems with U replaced by the nodal basis functions: $\Phi_{\ell,m}$ is the nodal basis function (vector) associated with the (ℓ, m)th cell if $\Phi_{\ell,m}$ is zero everywhere except at the (ℓ, m)th cell where it is 1. For each such basis function, there are only a few local problems that need to be solved, since the basis function vanishes on most cells. Since U can be written as a linear combination of these nodal basis functions, the fluxes corresponding to U can also be written as a linear combination of the fluxes correspond to these nodal basis functions. In this way, (10.5) is turned into a system of linear equations for U.

Now how do we choose δ? Clearly the smaller the δ, the less costly the algorithm. If the original problem (10.4) has scale separation, *i.e.*, the micro-scale length ε is much smaller than $O(1)$, then we can choose δ such that $\varepsilon \ll \delta \ll 1$. This results in savings of cost for HMM, compared with solving the original micro-scale problem on the whole domain D.

The correctors in the homogenization theory describe the low-frequency contribution from interaction of highly oscillatory functions. That interaction is represented by the micro-scale problems in the δ-domains. There is a natural correspondence to the analytic cell problems.

We chose to use a finite volume method for our standard elliptic equation as the first example because it gives a good background for the following examples: a finite element method for the elliptic equation and a finite volume method for nonlinear conservation laws.

In the finite element approximation of (10.4), the macro-scale solver can be chosen simply as the standard C^0 piecewise linear finite element method over a macroscopic triangulation $\mathcal{T}_H$ of mesh size H. We will denote by X_H the macroscopic finite element space which could be the standard piecewise linear finite elements over $\mathcal{T}_H$.

The data that need to be estimated from the micro-scale model are contained in the stiffness matrix on $\mathcal{T}_H : A = (A_{ij})$, where

$$A_{ij} = \int_\Omega \nabla\Phi_i(\mathbf{x})\bar{A}(\mathbf{x})\nabla\Phi_j(\mathbf{x})\,\mathrm{d}\mathbf{x}. \qquad (10.8)$$

Here $\bar{A}(\mathbf{x})$ is the homogenized conductivity tensor and $\{\Phi_i(\mathbf{x})\}$ are the basis functions for X_H. Had we known $\bar{A}(\mathbf{x})$, we could have evaluated A_{ij} simply by numerical quadrature: if $f_{ij}(\mathbf{x}) = \nabla\Phi_i(\mathbf{x})\bar{A}(\mathbf{x})\nabla\Phi_j(\mathbf{x})$, then

$$A_{ij} = \int_\Omega f_{ij}(\mathbf{x})\,\mathrm{d}\mathbf{x} \simeq \sum_{T\in\mathcal{T}_H} |T| \sum_{\mathbf{x}_k\in T} \omega_k f_{ij}(\mathbf{x}_k), \qquad (10.9)$$

where $\{\mathbf{x}_k\}$ and $\{\omega_k\}$ are the quadrature points and weights respectively, while $|T|$ is the area of the element T.

In the absence of explicit knowledge of $\bar{A}(\mathbf{x})$, our problem reduces to the approximation of the values of $\{\bar{A}(\mathbf{x}_k)\}$. This will be done by solving the original micro-scale model locally around each quadrature point $\{\mathbf{x}_k\}$.

Let $I_\delta(\mathbf{x}_k) \ni \mathbf{x}_k$ be a square of size δ. Consider

$$-\left(a_{ij}\left(\frac{x}{\varepsilon}\right)\phi^\varepsilon_{x_j}\right)_{x_i} = 0, \qquad \mathbf{x} \in I_\delta(\mathbf{x}_k). \qquad (10.10)$$

The main objective is to probe efficiently the micro-scale behaviour under the constraint that the average gradient of the solution ϕ^ε is fixed to be a given constant vector. Having solutions to this local problem, we can define the effective conductivity tensor at $\mathbf{x}_k$ by the relation

$$\left\langle A\left(\frac{x}{\varepsilon}\right)\nabla\phi^\varepsilon\right\rangle_{I_\delta} = \bar{A}(\mathbf{x}_k)\langle\nabla\phi^\varepsilon\rangle_{I_\delta}, \qquad (10.11)$$

where $\langle v\rangle_{I_\delta} = (1/|I_\delta|)\int_{I_\delta} v(\mathbf{x})\,\mathrm{d}\mathbf{x}$. The basis of this procedure is the theory of Section 2. The homogenization theorems allow us to define the effective (or homogenized) conductivity tensor, by considering the infinite volume limit of the solutions of the micro-scale problem subject to the constraint that the average gradient remains fixed. The effective tensor is defined by an average relation of the type (10.11) in the infinite volume limit, i.e.,

$$L = \frac{\delta}{\varepsilon} \to \infty.$$

In the special case when the micro-structure is periodic, the infinite volume problem reduces to a periodic problem.

Let us now consider the coupling between gas dynamics on the macro-scale and molecular dynamics (MD) on the micro-scale. The macroscopic equations are the usual conservation laws of density, momentum and energy. In one dimension, it can be expressed in a generic form:

$$\partial_t \mathbf{u} + \partial_x \mathbf{f} = 0. \tag{10.12}$$

Here $\mathbf{f}$ is the flux. Traditional gas dynamics models assume that $\mathbf{f}$ is a known function of $\mathbf{u}$. Here we do not make that assumption. Instead we will extract $\mathbf{f}$ from an underlying atomistic model, namely, molecular dynamics (MD).

As the macro-scale solver, we select a finite volume method. One example is the Lax–Friedrichs scheme on a staggered grid:

$$\mathbf{u}_{j+1/2}^{n+1} = \frac{\mathbf{u}_j^n + \mathbf{u}_{j+1}^n}{2} - \frac{\Delta t}{\Delta x}(\mathbf{f}_{j+1}^n - \mathbf{f}_j^n). \tag{10.13}$$

The data that need to be estimated from MD are again the fluxes. This is done by performing a constrained MD simulation locally at the cell boundaries, which are the cell centres for the previous time step. The constraints are that the average density, momentum and energy of the MD system should agree with the local macro-state at the current time step n. This is realized by initializing the MD with such constraints and applying the periodic boundary condition afterwards. Using the Irving–Kirkwood formula, which relates the fluxes to the MD data, we can then extract the macroscale fluxes by time-averaging the MD data. Ensemble averaging may also be used.

In this way we can rely on the more fundamental equations of molecular dynamics rather than on a flux function based on an empirical equation of state. Note that the micro-scale solver here plays the role of a Riemann solver in a standard conservation law scheme.

Remark. Recall the connection between projection-based and analytic homogenization. In both methods the homogenized operators were derived from a direct averaging and a correction for the high-frequency interaction. The correction involved the solution of a cell problem in the analytic case and the solution of a system of linear equations for the wavelet-based methodology. Here with HMM there is no such decomposition into averaging and correction, but the notion of a cell problem is quite clear. It is in the localized micro-scale problem where the high-frequency interaction is resolved and then transmitted to the model for the coarse scale. The compression steps typically include averaging, but that averaging step is taken after the local or cell problem is solved.

Acknowledgements

The material in this review is to large degree based on the work of the authors with various collaborators, and it is our pleasure to acknowledge Luis Caffarelli, Weinan E, Xiantao Li, Pierre-Louis Lions, Weiqing Ren,

Olof Runborg and Eric Vanden-Eiden. The research of Bjorn Engquist was supported by NSF grant DMS-0714612 and that of P. E. Souganidis by NSF grant DMS-0555826.

REFERENCES

A. Abdulle and W. E (2003), 'Finite difference heterogeneous multi-scale method for homogenization problems', *J. Comput. Phys.* **191**, 18–39.

G. Allaire (1992), 'Homogenization and two-scale convergence', *SIAM J. Math. Anal.* **23**, 1482–1518.

G. Allaire, A. Braides, G. Buttazzo, A. Defranceschi and L. Gibiansky (1993), School on Homogenization. SISSA Ref. 140/73/M.

U. Andersson, B. Engquist, G. Ledfelt and O. Runborg (1999), 'A contribution to wavelet-based subgrid modeling', *Appl. Comput. Harmon. Anal.* **7**, 151–164.

I. Babuška (1976), Homogenization and its applications: Mathematical and computational problems. In *Numerical Solution of Partial Differential Equations III*, Academic Press, pp. 89–116.

I. Babuška, G. Caloz and E. Osborn (1994), 'Special finite element methods for a class of second order elliptic problems with rough coefficients', *SIAM J. Numer. Anal.* **31**, 945–981.

G. Barles and B. Perthame (1988), 'Exit time problems in optimal control and vanishing viscosity solutions of Hamilton–Jacobi equations', *SIAM J. Control Optim.* **26**, 1133–1148.

A. Bensoussan, J.-L. Lions and G. Papanicolaou (1978), *Asymptotic Analysis for Periodic Structures*, North-Holland.

G. Beylkin and M. Brewster (1995), 'A multiresolution strategy for numerical homogenization', *Appl. Comput. Harmon. Anal.* **2**, 327–349.

G. Beylkin, R. Coifman and V. Rokhlin (1991), 'Fast wavelet transforms and numerical algorithms I', *Comm. Pure Appl. Math.* **44**, 141–183.

L. A. Caffarelli and P. E. Souganidis (2007), Error estimates for the homogenization of uniformly elliptic pde in strongly mixing random media. Preprint.

L. A. Caffarelli and P. E. Souganidis (2008), 'A rate of convergence for monotone finite difference approximations to fully nonlinear uniformly elliptic PDE', *Comm. Pure Appl. Math.* **61**, 1–17.

L. A. Caffarelli, P. E. Souganidis and L. Wang (2005), 'Stochastic homogenization for fully nonlinear, second-order partial differential equations', *Comm. Pure Appl. Math.* **30**, 319–361.

I. Capuzzo-Dolcetta and H. Ishii (2001), 'On the rate of convergence in homogenization of Hamilton–Jacobi equations', *Indiana U. Math. J.* **50**, 110–129.

P. Cardaliaguet, P.-L. Lions and P. E. Souganidis (2008), 'A discussion about the homogenization of moving interfaces', *J. Mathématique Pure et Appliqué*, to appear.

D. Cioranescu and P. Donato (2000), *An Introduction to Homogenization*, Oxford University Press.

B. Craciun and K. Bhattachayra (2003), 'Homogenization of a Hamilton–Jacobi equation associated with the geometric motion of an interface', *Proc. Roy. Soc. Edinburgh* A **133**, 773–805.

M. G. Crandall, H. Ishii and P.-L. Lions (1992), 'User's guide to viscosity solutions of second order partial differential equations', *Bull. Amer. Math. Soc.* **27**, 1–67.

G. Dal Maso (1993), *An Introduction to Γ-Convergence*, Birkhäuser.

G. Dal Maso and L. Modica (1986), 'Nonlinear stochastic homogenization and ergodic theory', *J. Reine Angew. Math.* **368**, 28–42.

I. Daubechies (1991), *Ten Lectures on Wavelets*, SIAM.

E. De Giorgi and T. Franzoni (1975), 'Su un tipo di convergenza variazionale', *Atti. Acad. Naz. Lincei Rend. Cl. Sci. Mat.* **58**, 842–850.

E. De Giorgi and S. Spagnolo (1973), 'Sulla convergenza degli integrali dell'energia par operatori ellittici del secundo ordine', *Boll. Un. Mat. Ital.* **4**, 391–411.

N. Dirr, G. Karali and A. Yip (2007), Pulsating wave for mean curvature flow in homogeneous medium. Preprint.

M. Dorobantu and B. Engquist (1998), 'Wavelet-based numerical homogenization', *SIAM J. Numer. Anal.* **35**, 540–559.

L. J. Durlofsky (1991), 'Numerical calculation of equivalent grid block permeability tensors for heterogeneous porous media', *Water. Resour. Res.* **27**, 699–708.

L. J. Durlofsky (1998), 'Coarse scale models of two-phase flow in heterogeneous reservoirs: Volume averaged equations and their relationship to existing up-scaling techniques', *Comput. Geosci.* **2**, 73–92.

W. E and B. Engquist (2003*a*), 'The heterogeneous multi-scale method', *Commun. Math. Sci.* **1**, 87–133.

W. E and B. Engquist (2003*b*), 'Multi-scale modeling and computation', *Notices Amer. Math. Soc.* **50**, 1062–1070.

W. E, B. Engquist, X. Li, W. Ren and E. Vanden-Eijnden (2007), 'Heterogeneous multiscale methods', *Commun. Comput. Phys.* **2**, 367–450.

B. Engquist and E. Luo (1997), 'Convergence of a multigrid method for elliptic equations with highly oscillatory coefficients', *SIAM J. Numer. Anal.* **34**, 2254–2273.

B. Engquist and O. Runborg (2001), Wavelet-based numerical homogenization with applications. In *Proc. Conference on Multiscale and Multiresolution Methods: Theory and Applications*, Vol. 20 of *Lecture Notes in Computational Science and Engineering*, Springer, pp. 97–148.

B. Engquist and O. Runborg (2002), Projection generated homogenization. In *Proc. Conference on Multiscale Problems in Science and Technology*, Springer, pp. 129–150.

L. C. Evans (1989), 'The perturbed test function method for viscosity solutions of nonlinear PDE', *Proc. Roy. Soc. Edinburgh* A **111**, 359–375.

L. C. Evans (1992), 'Periodic homogenization of certain fully nonlinear partial differential equations', *Proc. Roy. Soc. Edinburgh* A **120**, 245–265.

A. C. Gilbert (1998), 'A comparison of multiresolution and classical one-dimensional homogenization schemes', *Appl. Comput. Harmon. Anal.* **5**, 1–35.

T. Y. Hou (2003), Numerical approximations to multiscale solutions in partial differential equations. In *Frontiers in Numerical Analysis* (J. F. Blowey, A. W. Craig and T. Shardlow, eds), Springer, pp. 241–302.

T. Y. Hou and X.-H. Wu (1997), 'A multiscale finite element method for elliptic problems in composite materials and porous media', *J. Comput. Phys.* **134**, 169–189.

T. Y. Hou, X.-H. Wu and Z. Cai (1999), 'Convergence of a multiscale finite element method for elliptic problems with rapidly oscillating coefficients', *Math. Comp.* **68**, 913–943.

T. J. R. Hughes, G. R. Feijo'o, L. Mazzei and J.-B. Quicy (1999), 'The variational multiscale method: A paradigm for computational mechanics', *Comput. Methods Appl. Mech. Engrg* **166**, 515–533.

H. Ishii (1999), Homogenization of the Cauchy problem for Hamilton–Jacobi equations. In *Stochastic Analysis, Control, Optimization and Applications*, Systems & Control: Foundations & Applications, Birkhäuser, Boston, pp. 305–324.

V. V. Jikov, S. M. Kozlov and O. A. Oleinik (1991), *Homogenization of Differential Operators and Integral Functions*, Springer.

J. B. Keller (1977), Effective behavior of heterogeneous media. In *Statistical Mechanics and Statistical Methods in Theory and Application*, Plenum, pp. 631–644.

I. G. Kevrekidis, C. W. Gear, J. M. Hyman, P. G. Kevrekidis, O. Runborg and C. Theodoropoulos (2003), 'Equation-free, coarse-ground multiscale computation: Enabling microscopic simulators to perform system-level analysis', *Commun. Math. Sci.* **1**, 715–762.

S. Knapek (1999), 'Matrix-dependent multigrid-homogenization for diffusion problems', *SIAM J. Sci. Statist. Comput.* **20**, 512–533.

E. Kosygina, F. Rezakhanlou and S. R. S. Varadhan (2006), 'Stochastic homogenization for Hamilton–Jacobi–Bellman equations', *Comm. Pure Appl. Math.* **59**, 1489–1521.

S. M. Kozlov (1985), 'The method of averaging and walk in inhomogeneous environments', *Russian Math. Surveys* **40**, 73–145.

R. LeVeque (1990), *Numerical Methods for Conservation Laws*, Birkhäuser.

P.-L. Lions and P. E. Souganidis (2003), 'Correctors for the homogenization of Hamilton–Jacobi equations in a stationary ergodic setting', *Comm. Pure Appl. Math.* **LVI**, 1501–1524.

P.-L. Lions and P. E. Souganidis (2005a), 'Homogenization of degenerate second-order PDE in periodic and almost periodic environments and applications', *Ann. Inst. H. Poincaré, Anal. Nonlineaire* **22**, 667–677.

P.-L. Lions and P. E. Souganidis (2005b), 'Homogenization for "viscous" Hamilton–Jacobi equations in stationary, ergodic media', *Comm. Partial Differential Equations* **30**, 335–376.

P.-L. Lions and P. E. Souganidis (2008), Homogenization of Hamilton–Jacobi and viscous Hamilton–Jacobi equations in stationary, ergodic environments revisited. Preprint.

P.-L. Lions, G. Papanicolaou and S. R. S. Varadhan (1983), Homogenization of Hamilton–Jacobi equations. Unpublished.

V. A. Marchenko and E. Y. Khruslov (2006), *Homogenization of Partial Differential Equations*, Vol. 46 of *Progress in Mathematical Physics*, Birkhäuser.

F. Murat and L. Tartar (1977), Calculus of variations and homogenization. In *Topics in the Mathematical Modelling of Composite Materials* (A. Cherkaev and R. V. Kohn, eds), Birkhäuser, Basel, pp. 139–173. Originally in French from 1985.

N. Neuss, W. Jäger and G. Wittum (2000), 'Homogenization and multigrid', *Computing* **66**, 1–21.

G. Nguetseng (1989), 'A general convergence result for a functional related to the theory of homogenization', *SIAM J. Math. Anal.* **20**, 608–623.

G. Obinata and D. O. Anderson (2001), *Model Reduction for Control System Design*, Springer.

G. Papanicolaou and S. R. S. Varadhan (1979), Boundary value problems with rapidly oscillating random coefficients. In *Proc. Colloq. on Random Fields: Rigorous Results in Statistical Mechanics and Quantum Field Theory* (J. Fritz, J. L. Lebaritz and D. Szasz, eds), Vol. 10 of *Colloquia Mathematica Societ. Janos Bolyai*, pp. 835–873.

G. Papanicolaou and S. R. S. Varadhan (1981), Diffusion with random coefficients. In *Essays in Statistics and Probability* (P. R. Krishnaiah, ed.), North-Holland.

G. A. Pavliotis and A. M. Stewart (2007), *Multiscale Methods: Averaging and Homogenization*, Springer.

F. Rezankhanlou and J. Tarver (2000), 'Homogenization for stochastic Hamilton–Jacobi equations', *Arch. Ration. Mech. Anal.* **151**, 277–309.

C. E. Shannon (1949), 'Communication in the presence of noise', *Proc. Inst. Radio Engineers* **37**, 10–21.

P. E. Souganidis (1999), 'Stochastic homogenization of Hamilton–Jacobi equations and some applications', *Asympt. Anal.* **20**, 1–11.

L. Tartar (1977), Cours Peccot au Collège de France. Unpublished.

L. Tartar (1989), Nonlocal effects induced by homogenization. In *PDE and Calculus of Variation*, Birkhäuser, pp. 925–938.

K. Xu and K. H. Prendergast (1994), 'Numerical Navier–Stokes solutions from gas kinetic theory', *J. Comput. Phys.* **114**, 9–17.

X. Y. Yue and W. E (2008), 'The local microscale problem in the multiscale modelling of strongly heterogeneous media: Effect of boundary conditions and cell size', *J. Comput. Phys.*, to appear.

V. V. Yurinskii (1980), 'On the homogenization of boundary value problems with random coefficients', *Sibir. Matem. Zh.* **21**, 209–223. English translation: *Siber. Math. J.* **21** (1981), 470–482.

V. V. Yurinskii (1982), 'On the homogenization of non-divergent second order equations with random coefficients', *Sibir. Matem. Zh.* **23**, 176–188. English translation: *Siber. Math. J.* **23** (1982), 276–287.

V. V. Zhikov (1993), 'Asymptotic problems related to a second-order parabolic equation in nondivergence form with randomly homogeneous coefficients' (Russian), *Differentsial'nye Uravneniya* **29**, 859–869. English translation: *Differential Equations* **29** (1993), 735–744.

Acta Numerica (2008), pp. 191–234
doi: 10.1017/S0962492906370018

© Cambridge University Press, 2008

Interior-point methods for optimization

Arkadi S. Nemirovski

School of Industrial and Systems Engineering,
Georgia Institute of Technology,
Atlanta, Georgia 30332, USA
E-mail: arkadi.nemirovski@isye.gatech.edu

Michael J. Todd

School of Operations Research and Information Engineering,
Cornell University, Ithaca, NY 14853, USA
E-mail: mjt7@cornell.edu

This article describes the current state of the art of interior-point methods
(IPMs) for convex, conic, and general nonlinear optimization. We discuss the
theory, outline the algorithms, and comment on the applicability of this class
of methods, which have revolutionized the field over the last twenty years.

CONTENTS

1. Introduction

During the last twenty years, there has been a revolution in the methods used to solve optimization problems. In the early 1980s, sequential
quadratic programming and augmented Lagrangian methods were favoured
for nonlinear problems, while the simplex method was basically unchal-
lenged for linear programming. Since then, modern interior-point methods
(IPMs) have infused virtually every area of continuous optimization, and
have forced great improvements in the earlier methods. The aim of this
article is to describe interior-point methods and their application to convex
programming, special conic programming problems (including linear and
semidefinite programming), and general possibly non-convex programming.

We have also tried to complement the earlier articles in this journal by Wright (1992), Lewis and Overton (1996), and Todd (2001).

Almost twenty-five years ago, Karmarkar (1984) proposed his projective method to solve linear programming problems: from a theoretical point of view, this was a polynomial-time algorithm, in contrast to Dantzig's simplex method. Moreover, with some refinements it proved a very worthy competitor in practical computation, and substantial improvements to both interior-point and simplex methods have led to the routine solution of problems (with hundreds of thousands of constraints and variables) that were considered untouchable previously. Most commercial software, for example CPlex (Bixby 2002) and XpressMP (Guéret, Prins and Sevaux 2002), includes interior-point as well as simplex options.

The majority of the early papers following Karmarkar's dealt exclusively with linear programming and its near-relatives, convex quadratic programming and the (monotone) linear complementarity problem. Gill, Murray, Saunders, Tomlin and Wright (1986) showed the strong connection to earlier barrier methods in nonlinear programming; Renegar (1988) and Gonzaga (1989) introduced path-following methods with an improved iteration complexity; and Megiddo (1989) suggested, and Monteiro and Adler (1989) and Kojima, Mizuno and Yoshise (1989) realized, primal–dual versions of these algorithms, which are the most successful in practice.

At the same time, Nesterov and Nemirovski were investigating the new methods from a more fundamental viewpoint: What are the basic properties that lead to polynomial-time complexity? It turned out that the key property is that the barrier function should be *self-concordant*. This seemed to provide a clear, complexity-based criterion to delineate the class of optimization problems that could be solved in a provably efficient way using the new methods. The culmination of this work was the book by Nesterov and Nemirovski (1994), whose complexity emphasis contrasted with the classic text on barrier methods by Fiacco and McCormick (1968).

Fiacco and McCormick describe the history of (exterior) penalty and barrier (sometimes called interior penalty) methods; other useful references are Nash (1998) and Forsgren, Gill and Wright (2002). Very briefly, Courant (1943) first proposed penalty methods, while Frisch (1955) suggested the logarithmic barrier method and Carroll (1961) the inverse barrier method (which inspired Fiacco and McCormick). While these methods were among the most successful for solving constrained nonlinear optimization problems in the 1960s, they lost favour in the late 1960s and 1970s when it became apparent that the subproblems that needed to be solved became increasingly ill-conditioned as the solution was approached.

The new research alleviated these fears to some extent, at least for certain problems. In addition, the ill-conditioning turned out to be relatively benign: see, *e.g.*, Wright (1992) and Forsgren *et al.* (2002). Moreover,

Nesterov and Nemirovski (1994) showed that, at least in principle, *any* convex optimization problem could be provided with a self-concordant barrier. This was purely an existence result, however, as the generated barrier could not be efficiently evaluated in general. (So we should qualify our earlier statement: the class of optimization problems to which the new methods can be efficiently applied consists of those with a *computationally tractable* self-concordant barrier.) To contrast with the general case, Nesterov and Nemirovski listed a considerable number of important problems where computationally tractable self-concordant barriers were available, and provided a calculus for constructing such functions for more complicated sets. A very significant special case was that of the positive semidefinite cone, leading to semidefinite programming. Independently, Alizadeh (1995) developed an efficient interior-point method for semidefinite programming, with the motivation of obtaining strong bounds for combinatorial optimization problems.

The theory of self-concordant barriers is limited to convex optimization. However, this limitation has become less burdensome as more and more scientific and engineering problems have been shown to be amenable to convex optimization formulations. Researchers in control theory have been much influenced by the ability to solve semidefinite programming problems (or linear matrix inequalities, in their terminology) arising in their field: see Boyd, El Ghaoui, Feron and Balakrishnan (1994). Moreover, a number of seemingly non-convex problems arising in engineering design can be reformulated as convex optimization problems: see Boyd and Vandenberghe (2004) and Ben-Tal and Nemirovski (2001).

Besides the books we have cited, other useful references include the lecture notes of Nemirovski (2004) and the books of Nesterov (2003) and Renegar (2001) for general convex programming; for mostly linear programming, the books of Roos, Terlaky and Vial (1997), Vanderbei (2007), Wright (1997) and Ye (1997); for semidefinite programming, the handbook of Wolkowicz, Saigal and Vandenberghe (2000); and for general nonlinear programming, the survey articles of Forsgren *et al.* (2002) and Gould, Orban and Toint (2005).

In Section 2, we discuss self-concordant barriers and their properties, and then describe interior-point methods for both general convex optimization problems and conic problems, as well as the calculus of self-concordant barriers. Section 3 treats conic optimization in detail, concentrating on symmetric or self-scaled cones, including the non-negative orthant (linear programming) and the positive semidefinite cone (semidefinite programming). We also briefly discuss some recent developments in hyperbolicity cones, global polynomial optimization, and copositive programming. Finally, Section 4 is concerned with the application of interior-point methods to general, possibly non-convex, nonlinear optimization. These methods are used in some of the most effective codes for such problems, such as IPOPT

(Wächter and Biegler 2006), KNITRO (Byrd, Nocedal and Waltz 2006), and LOQO (Vanderbei and Shanno 1999).

We have concentrated on the theory and application in structured convex programming of interior-point methods, since the polynomial-time complexity of these methods and its range of applicability have been a major focus of the research of the last twenty years. For further coverage of interior-point methods for general nonlinear programming we recommend the survey articles of Forsgren *et al.* (2002) and Gould, Orban and Toint (2005). Also, to convey the main ideas of the methods, we have given short shrift to important topics including attaining feasibility from infeasible initial points, dealing with infeasible problems, and superlinear convergence. The literature on interior-point methods is huge, and the area is still very active; the reader wishing to follow the latest research is advised to visit the Optimization Online website www.optimization-online.org/ and the Interior-Point Methods Online page at www-unix.mcs.anl.gov/otc/InteriorPoint/. A very useful source is Helmberg's semidefinite programming page www-user.tu-chemnitz.de/~helmberg/semidef.html. Software for optimization problems, including special-purpose algorithms for semidefinite and second-order cone programming, is available at the Network Enabled Optimization System (NEOS) homepage neos.mcs.anl.gov/neos/solvers/index.html.

2. The self-concordance-based approach to IPMs

Preliminaries

The first *path-following* interior-point polynomial-time methods for linear programming, analysed by Renegar (1988) and Gonzaga (1989), turned out to belong to the very well-known *interior penalty scheme* going back to Fiacco and McCormick (1968). Consider a convex program

$$\min\{c^T x : x \in X\}, \tag{2.1}$$

X being a closed convex domain (*i.e.*, a closed convex set with a non-empty interior) in $\mathbb{R}^n$; this is one of the universal forms of a convex program. In order to solve the problem with a path-following scheme, one equips X with an *interior penalty* or *barrier* function F – a smooth and strongly convex[1] function defined on int X such that $F(x_k) \to +\infty$ on every sequence of points $x_k \in$ int X converging to a point $\bar{x} \in \partial X$ – and considers the *barrier family* of functions

$$F_t(x) = tc^T x + F(x), \tag{2.2}$$

where $t > 0$ is the *penalty parameter*. Under mild assumptions (*e.g.*, when X is bounded), every function F_t attains its minimum on int X at a unique point $x_*(t)$, and the *central path* $\{x_*(t) : t \geq 0\}$ converges, as $t \to \infty$,

---

[1] Hessian positive definite everywhere.

to the optimal set of (2.1). The path-following scheme for solving (2.1) suggests 'tracing' this path as $t \to \infty$ according to the following conceptual algorithm:

> Given the current iterate $(t_k > 0, x_k \in \text{int}\, X)$ with x_k 'reasonably close' to $x_*(t_k)$, we
>
> (a) replace the current value t_k of the penalty parameter with a larger value t_{k+1}; and
>
> (b) run an algorithm for minimizing $F_{t_{k+1}}(\cdot)$, starting at x_k, until a point x_{k+1} close to $x_*(t_{k+1}) = \text{argmin}_{\text{int}\, X}\, F_{t_{k+1}}(\cdot)$ is found.

As a result, we get a new iterate (t_{k+1}, x_{k+1}) 'close to the path' and loop to step $k + 1$.

The main advantage of the scheme described above is that $x_*(t)$ is, essentially, the unconstrained minimizer of F_t, which allows the use in (b) of basically any method for smooth convex unconstrained minimization, *e.g.*, the Newton method. Note, however, that the classical theory of the path-following scheme did not suggest its polynomiality; rather, the standard theory of unconstrained minimization predicted slow-down of the process as the penalty parameter grows. In sharp contrast to this common wisdom, both Renegar and Gonzaga proved that, when applied to the *logarithmic barrier* $F(x) = -\sum_i \ln(b_i - a_i^T x)$ for a polyhedral set $X = \{x : a_i^T x \le b_i, 1 \le i \le m\}$, a Newton-method-based implementation of the path-following scheme can be made polynomial. These breakthrough results were obtained via an *ad hoc* analysis of the behaviour of the Newton method as applied to the logarithmic barrier (augmented by a linear term). In a short time Nesterov realized what intrinsic properties of the standard log-barrier are responsible for this polynomiality, and this crucial understanding led to the general *self-concordance-based* theory of polynomial-time interior-point methods developed in Nesterov and Nemirovski (1994); this theory explained the nature of existing interior-point methods (IPMs) for LP and allowed the extension of these methods to the entire field of convex programming. We now provide an overview of the basic results of this theory.[2]

2.1. Self-concordance

In retrospect, the notion of self-concordance can be extracted from analysis of the classical results on the local quadratic convergence of Newton's

---

[2] Up to minor refinements which can be found in Nemirovski (2004), all results quoted in the next subsection without explicit references are taken from Nesterov and Nemirovski (1994).

method as applied to a smooth convex function f with non-singular Hessian. These results state that a quantitative description of the domain of quadratic convergence depends on (a) the condition number of $\nabla^2 f$ evaluated at the minimizer x_*, and (b) the Lipschitz constant of $\nabla^2 f$. In hindsight, such a description seems unnatural, since it is 'frame-dependent': it heavily depends on an *ad hoc* choice of the Euclidean structure in $\mathbb{R}^n$; indeed, both the condition number of $\nabla^2 f(x_*)$ and the Lipschitz constant of $\nabla^2 f(\cdot)$ depend on this structure, which is in sharp contrast to the affine invariance of the Newton method itself. At the same time, a smooth strongly convex function f by itself defines at every point x a Euclidean structure $\langle u, v \rangle_{f,x} = D^2 f(x)[u, v]$. With respect to this structure, $\nabla^2 f(x)$ is as well-conditioned as it could be – it is just the unit matrix. The idea of Nesterov was to use this local Euclidean structure, intrinsically linked to the function f we intend to minimize, in order to quantify the Lipschitz constant of $\nabla^2 f$, with the ultimate goal of getting a 'frame-independent' description of the behaviour of the Newton method. The resulting notion of self-concordance is defined as follows.

Definition 2.1. Let $X \subset \mathbb{R}^n$ be a closed convex domain. A function $f : \operatorname{int} X \to \mathbb{R}$ is called *self-concordant* (SC) on X if

 (i) f is a three times continuously differentiable convex function with $f(x_k) \to \infty$ if $x_k \to \bar{x} \in \partial X$; and
 (ii) f satisfies the differential inequality

$$|D^3 f(x)[h, h, h]| \leq 2\big(D^2 f(x)[h, h]\big)^{3/2}, \quad \forall x \in \operatorname{int} X, \ h \in \mathbb{R}^n. \quad (2.3)$$

Given a real $\vartheta \geq 1$, F is called a ϑ-*self-concordant barrier* (ϑ-SCB) for X if F is self-concordant on X and, in addition,

$$|DF(x)[h]| \leq \vartheta^{1/2}\big(D^2 F(x)[h, h]\big)^{1/2}, \quad \forall x \in \operatorname{int} X, \ h \in \mathbb{R}^n. \quad (2.4)$$

(As above, we will use f for a general SC function and F for an SCB in what follows.) Note that the powers $3/2$ and $1/2$ in (2.3) and (2.4) are a must, since both sides of the inequalities should be of the same homogeneity degree with respect to h. In contrast to this, the two sides of (2.3) are of different homogeneity degrees with respect to f, meaning that if f satisfies a relation of the type (2.3) with some constant factor on the right-hand side, we can always make this factor equal to 2 by scaling f appropriately. The advantage of the specific factor 2 is that with this definition, the function $x \mapsto -\ln(x) : \mathbb{R}_{++} \to \mathbb{R}$ becomes a 1-SCB for $\mathbb{R}_+$ directly, without any scaling, and this function is the main building block of the theory we are presenting. Finally, we remark that (2.3) and (2.4) have a very transparent interpretation: they mean that $D^2 f$ and F are Lipschitz-continuous, with constants 2 and $\vartheta^{1/2}$, in the local Euclidean (semi)norm $\|h\|_{f,x} = \sqrt{\langle h, h \rangle_{f,x}} = \sqrt{h^T \nabla^2 f(x) h}$ defined by f or similarly by F.

It turns out that self-concordant functions possess nice local properties and are perfectly well suited to Newton minimization. We are about to present the most important of the related results. In what follows, f is an SC function on a closed convex domain X.

2.1.0. Bounds on third derivatives and the recession space of SC functions

For all $x \in \operatorname{int} X$ and all $h_1, h_2, h_3 \in \mathbb{R}^n$, we have

$$|D^3 f(x)[h_1, h_2, h_3]| \le 2\|h_1\|_{f,x}\|h_2\|_{f,x}\|h_3\|_{f,x}.$$

The *recession subspace* $E_f = \{h : D^2 f(x)[h, h] = 0\}$ of f is independent of $x \in \operatorname{int} X$, and $X = X + E_f$. In particular, if $\nabla^2 f(x)$ is positive definite at some point in $\operatorname{int} X$, then $\nabla^2 f(x)$ is positive definite for all $x \in \operatorname{int} X$ (in this case, f is called a *non-degenerate* SC function; this is always the case when X does not contain lines).

It is convenient to write $A \succ 0$ ($A \succeq 0$) to denote that the symmetric matrix A is positive definite (semidefinite), and $A \succeq B$ and $B \preceq A$ ($A \succ B$ and $B \prec A$) if $A - B \succeq 0$ ($A - B \succ 0$).

2.1.1. Dikin's ellipsoid and the local behaviour of f

For every $x \in \operatorname{int} X$, the *unit Dikin ellipsoid* of f $\{y : \|y - x\|_{f,x} \le 1\}$ is contained in X, and within this ellipsoid, f is nicely approximated by its second-order Taylor expansion:

$$r := \|h\|_{f,x} < 1 \Rightarrow$$

$$(1 - r)^2 \nabla^2 f(x) \preceq \nabla^2 f(x + h) \preceq \frac{1}{(1 - r)^2} \nabla^2 f(x), \qquad (2.5)$$

$$f(x) + \nabla f(x)^T h + \rho(-r) \le \quad f(x + h) \quad \le f(x) + \nabla f(x)^T h + \rho(r),$$

where $\rho(s) := -\ln(1 - s) - s = s^2/2 + s^3/3 + \cdots$. (Indeed, the lower bound in the last line holds true for all h such that $x + h \in \operatorname{int} X$.)

2.1.2. The Newton decrement and the damped Newton method

Let f be non-degenerate. Then $\|\cdot\|_{f,x}$ is a norm, and its conjugate norm is $\|\eta\|_{f,x}^* = \max\{h^T \eta : \|h\|_{f,x} \le 1\} = \sqrt{\eta^T [\nabla^2 f(x)]^{-1} \eta}$. The quantity

$$\lambda(x, f) := \|\nabla f(x)\|_{f,x}^* = \|[\nabla^2 f(x)]^{-1} \nabla f(x)\|_{f,x}$$

$$= \max_h \{Df(x)[h] : D^2 f(x)[h, h] \le 1\},$$

called the *Newton decrement* of f at x, is a finite continuous function of $x \in \operatorname{int} X$ which vanishes exactly at the (unique, if any) minimizer x_f of f on $\operatorname{int} X$; this function can be considered as the 'observable' measure of proximity of x to x_f. The Newton decrement possesses the following properties:

$$\lambda(x, f) < 1 \Rightarrow \begin{cases} \operatorname{argmin}_{\operatorname{int} X} f \neq \emptyset, & (a) \\ f(x) - \min_{\operatorname{int} X} f \leq \rho(\lambda(x, f)), & (b) \\ \|x_f - x\|_{f,x} \leq \frac{\lambda(x,f)}{1 - \lambda(x,f)}, & (c) \\ \|x_f - x\|_{f,x_f} \leq \frac{\lambda(x,f)}{1 - \lambda(x,f)}. & (d) \end{cases} \tag{2.6}$$

In particular, when it is at most $1/2$, the Newton decrement is, within an absolute constant factor, the same as $\|x - x_f\|_{f,x}$, $\|x - x_f\|_{f,x_f}$, and $\sqrt{f(x) - \min_{\operatorname{int} X} f}$.

The *damped Newton method* as applied to f is the iterative process

$$x_{k+1} = x_k - \frac{1}{1 + \lambda(x_k, f)} [\nabla^2 f(x_k)]^{-1} \nabla f(x_k) \tag{2.7}$$

starting at a point $x_0 \in \operatorname{int} X$. The damped Newton method is well defined: all its iterates belong to $\operatorname{int} X$. Besides this, setting $\lambda_j := \lambda(x_j, f)$, we have

$$\lambda_{k+1} \leq 2\lambda_k^2 \quad \text{and} \quad f(x_k) - f(x_{k+1}) \geq \rho(-\lambda_k) = \lambda_k - \ln(1 + \lambda_k). \tag{2.8}$$

As a consequence of (2.8) and (2.7), we get the following 'frame- and data-independent' description of the convergence properties of the damped Newton method as applied to an SC function f: the domain of quadratic convergence is $\{x : \lambda(x, f) \leq 1/4\}$; after this domain is reached, every step of the method nearly squares the Newton decrement, the $\|\cdot\|_{f,x_f}$-distance to the minimizer and the residual in terms of f. Before the domain is reached, every step of the method decreases the objective by at least $\Omega(1) = 1/4 - \ln(5/4)$. It follows that a non-degenerate SC function admits its minimum on the interior of its domain if and only if it is bounded below, and if and only if $\lambda(x, f) < 1$ for certain x. Whenever this is the case, for every $\epsilon \in (0, 0.1]$ the number of steps N of the damped Newton method which ensures that $f(x_k) \leq \min_{\operatorname{int} X} f + \epsilon$ does not exceed $O(1) [\ln \ln(1/\epsilon) + f(x_0) - \min_{\operatorname{int} X} f]$. (Here and below, $O(1)$ denotes a suitably chosen absolute constant.)

2.1.3. Self-concordance and Legendre transformations

Let f be non-degenerate. Then the domain $\{y : f_*(y) < \infty\}$ of the (modified) Legendre transformation

$$f_*(y) = \sup_{x \in \operatorname{int} X} \left[-y^T x - f(x)\right]$$

of f is an open convex set, f_* is self-concordant on the closure X_* of this set, and the mappings $x \mapsto -\nabla f(x)$ and $y \mapsto -\nabla f_*(y)$ are bijections of $\operatorname{int} X$ and $\operatorname{int} X_*$ that are inverse to each other. Besides this, X_* is a closed cone with a non-empty interior, specifically, the cone dual to the recession cone of X.

We next list specific properties of SCBs not shared by more general SC functions. In what follows, F is a non-degenerate ϑ-SCB for a closed convex domain X.

2.1.4. Non-degeneracy, semiboundedness, attaining minimum
F is non-degenerate if and only if X does not contain lines. We have

$$\forall(x \in \text{int}\, X, y \in X) : \nabla F(x)^T (y - x) \leq \vartheta \qquad (2.9)$$

(semiboundedness) and

$$\forall(x \in \text{int}\, X, y \in X \text{ with } \nabla F(x)^T(y-x) \geq 0) : \|y-x\|_{F,x} \leq \vartheta+2\sqrt{\vartheta}. \quad (2.10)$$

F attains its minimum on $\text{int}\, X$ if and only if X is bounded; otherwise $\lambda(x, F) \geq 1$ for all $x \in \text{int}\, X$.

2.1.5. Useful bounds
For $x \in \text{int}\, X$, let $\pi_x(y) = \inf\{t : t > 0, x + t^{-1}(y - x) \in X\}$ be the Minkowski function of X with respect to x. We have

$$\forall(x, y \in \text{int}\, X) : \begin{cases} F(y) \leq F(x) + \vartheta \ln\big(\frac{1}{1-\pi_x(y)}\big), \\ F(y) \geq F(x) + \nabla F(x)^T(y - x) + \ln\big(\frac{1}{1-\pi_x(y)}\big) - \pi_x(y). \end{cases}$$
$$(2.11)$$

2.1.6. Existence of the central path and its convergence to the optimal set
Consider problem (2.1) and assume that the domain X of the problem is equipped with a self-concordant barrier F, and the level sets of the objective $\{x \in X : c^T x \leq \alpha\}$ are bounded. In the situation in question, F is non-degenerate, $c^T x$ attains its minimum on X, the central path

$$x_*(t) := \operatorname*{argmin}_{x \in \text{int}\, X} F_t(x), \quad F_t(x) := tc^T x + F(x), \quad t > 0,$$

is well-defined, all functions F_t are self-concordant on X and

$$\epsilon(x_*(t)) := c^T x_*(t) - \min_{x \in X} c^T x \leq \frac{\vartheta}{t}, \quad t > 0. \qquad (2.12)$$

Moreover, if $\lambda(x, F_t) \leq \frac{1}{2}$ for some $t > 0$, then

$$\epsilon(x) \leq \frac{\vartheta + \sqrt{\vartheta}}{t}. \qquad (2.13)$$

Let us derive the claims in 2.1.6 from the preceding facts, mainly in order to explain why these facts are important. By 2.1.1, F is non-degenerate, since X does not contain lines. The fact that all F_t are SC is evident from the definition: self-concordance is clearly preserved when adding a linear or convex quadratic function to an SC one. Further, the level sets of the objective on X are bounded, so that the objective attains its minimum over X at some point x_* and, as is easily seen, is coercive on X: $c^T x \geq \alpha + \beta \|x\|$ for all $x \in X$ with appropriate constants $\beta > 0$ and α ($\| \cdot \|$, without subscripts, always denotes the Euclidean norm). Now fix a point $\bar{y}$ in int X; then $\pi_x(\bar{y}) \leq \frac{\|\bar{y}-x\|}{r+\|\bar{y}-x\|}$ for all $x \in$ int X, where $r > 0$ is such that a $\| \cdot \|$-ball of radius r centred at $\bar{y}$ belongs to X. Invoking the first line of (2.11) with $y = \bar{y}$, we conclude that $F(x) \geq F(\bar{y}) + \vartheta \ln(\frac{r}{r+\|x-\bar{y}\|})$ for all $x \in$ int X. Recalling that the objective is coercive, we conclude that $F_t(x) \to \infty$ as $x \in X$ and $\|x\| \to \infty$, so that the level sets of F_t are bounded. Since F_t is, along with F, an interior penalty for X, these sets are in fact compact subsets of int X, whence F_t attains its minimum on int X. Since F_t is convex and non-degenerate along with F, the minimizer is unique; thus, the central path is well-defined. To verify (2.12), note that $\nabla F(x_*(t)) = -tc$, whence $c^T(x_*(t) - y) = t^{-1}\nabla F(x_*(t))^T(y - x_*(t))$. By (2.9), the right-hand side in this equality is at most ϑ/t, provided $y \in X$, and (2.12) follows. Finally, when $\lambda(x, F_t) \leq 1/2$, then $\|x - x_*(t)\|_{F_t, x_*(t)} \leq 1$ by (2.6.d) as applied to F_t instead of F. Since $\| \cdot \|_{F_t, u} \equiv \| \cdot \|_{F, u}$, we get $\|x - x_*(t)\|_{F, x_*(t)} \leq 1$, whence

$$c^T(x - x_*(t)) \leq \|c\|_{F, x_*(t)}^* \|x - x_*(t)\|_{F, x_*(t)}$$

$$\leq \|c\|_{F, x_*(t)}^* = t^{-1}\|\nabla F(x_*(t))\|_{F, x_*(t)}^* \leq t^{-1}\sqrt{\vartheta},$$

which combines with (2.12) to imply (2.13).

2.2. A primal polynomial-time path-following method

As an immediate consequence of the above results, we arrive at the following important result.

Theorem 2.1. Consider problem (2.1) and assume that the level sets of the objective are bounded, and we are given a ϑ-SCB F for X; according to 2.1.6, c and F define a central path $x_*(\cdot)$. Suppose we also have at our disposal a starting pair $(t_0 > 0, x_0 \in$ int $X)$ which is close to the path in the sense that $\lambda(x_0, F_{t_0}) \leq 0.1$, and consider the following implementation (the basic path-following algorithm) of the path-following scheme:

$$(t_k, x_k) \mapsto \begin{cases} t_{k+1} = (1 + 0.1\vartheta^{-1/2})t_k, \\ x_{k+1} = x_k - \frac{1}{1+\lambda(x_k, F_{t_{k+1}})}[\nabla^2 F(x_k)]^{-1}\nabla F_{t_{k+1}}(x_k). \end{cases} \quad (2.14)$$

This recurrence is well-defined (*i.e.*, $x_k \in$ int X for all k), maintains closeness to the path (*i.e.*, $\lambda(x_k, F_{t_k}) \leq 0.1$ for all k) and ensures the efficiency estimate

$$\forall k : c^T x_k - \min_{x \in X} c^T x \leq \frac{\vartheta + \sqrt{\vartheta}}{t_k} \leq \frac{\vartheta + \sqrt{\vartheta}}{t_0} \exp\left\{-\frac{0.095}{\sqrt{\vartheta}}k\right\}. \quad (2.15)$$

In particular, for every $\epsilon > 0$, it takes at most

$$N(\epsilon) = O(1)\sqrt{\vartheta}\ln\left(\frac{\vartheta}{t_0\epsilon} + 2\right)$$

steps of the recurrence to get a strictly feasible (*i.e.*, in int X) solution to the problem with residual in terms of the objective at most ϵ.

Proof. In view of 2.1.2 and 2.1.6, all we need to prove by induction on k is that (2.14) maintains closeness to the path. Assume that $(t = t_k, x = x_k)$ is close to the path, and let us verify that the same is true for $(t_+ = t_{k+1}, x_+ = x_{k+1})$. Taking into account that $\|\cdot\|_{F_t,u} = \|\cdot\|_{F,u}$ and similarly for the conjugate norms, we have $0.1 \geq \lambda(x, F_t) = \|tc + \nabla F(x)\|_{F,x}^*$ so that

$$t\|c\|_{F,x}^* \leq 0.1 + \|\nabla F(x)\|_{F,x}^*$$
$$\leq 0.1 + \vartheta^{1/2},$$

using the fact that F is a ϑ-SCB. This implies that

$$\lambda(x, F_{t_+}) = \|[tc + \nabla F(x)] + (t_+ - t)c\|_{F,x}^*$$
$$\leq \|tc + \nabla F(x)\|_{F,x}^* + (t_+/t - 1)t\|c\|_{F,x}^*$$
$$\leq 0.1 + 0.1\vartheta^{-1/2}t\|c\|_{F,x}^* \leq 0.1[1 + \vartheta^{-1/2}[0.1 + \vartheta^{1/2}]] \leq 0.21.$$

Finally, we obtain

$$\lambda(x_+, F_{t_+}) \leq 2\lambda^2(x, F_{t_+}) \leq 0.1,$$

using (2.8). $\qquad\qquad\qquad\qquad\qquad\qquad\qquad\qquad\qquad\qquad\qquad\square$

Remarks. **A** The algorithm presented in Theorem 2.1 is, in a sense, incomplete: it does not explain how to approach the central path in order to start path-tracing. There are many ways to resolve this issue. Assume, *e.g.*, that X is bounded and we know in advance a point $y \in$ int X. When X is bounded, every linear form $g^T x$ generates a central path, and we can easily find such a path passing through y: with $g = -\nabla F(y)$, the corresponding path passes through y when $t = 1$. Now, as $t \to +0$, all paths converge to the minimizer x_F of F over X, and thus approach each other. At the same time, we can as easily trace the paths backwards as trace them forwards – with the parameter updating rule $t_{k+1} = (1 - 0.1\vartheta^{-1/2})t_k$, the recurrence in Theorem 2.1 still maintains closeness to the path, now along a sequence of values of the parameter t decreasing geometrically. Thus, we can trace the auxiliary path passing through y backwards until coming close to the path of interest, and then start tracing the latter path forwards. A simple analysis demonstrates that with simple on-line termination and switching rules, the resulting algorithm, for every $\epsilon > 0$, produces a strictly feasible

ϵ-solution to the problem at the price of no more than

$$O(1)\sqrt{\vartheta}\ln\left(\frac{\vartheta\mathcal{V}}{(1 - \pi_{x_F}(y))\epsilon} + 2\right)$$

Newton steps of both phases, where $\mathcal{V} = \max_{x \in X} c^T x - \min_{x \in X} c^T x$.

B The outlined path-following algorithm, using properly chosen SC barriers, yields the currently best polynomial-time complexity bounds for basically all 'well-structured' generic convex programs, such as those of linear, second-order cone, semidefinite, and geometric programming, to name just a few. At the same time, from a practical perspective a severe shortcoming of the algorithm is its worst-case-oriented nature: as presented, it will *always* perform according to its worst-case theoretical complexity bounds. There exist implementations of IPMs that are much more powerful in practice, using more aggressive parameter updating policies that are adjusted during the course of the algorithm. All known algorithms of this type are *primal–dual*: they work simultaneously on the problem and its dual, and nearly all of them, including all those implemented so far in professional software, work with conic problems, specifically, those of linear, second-order cone, and semidefinite programming (the only exceptions are the cone-free primal–dual methods proposed in Nemirovski and Tunçel (2005); these methods, however, have not yet been implemented). Our next goal is to describe the general theory of primal–dual interior-point methods for conic problems.

2.3. Interior-point methods for conic problems

Interior-point methods for conic problems are associated with specific ϑ-SC barriers for cones, those satisfying the so-called *logarithmic homogeneity* condition.

Definition 2.2. Let $\mathbf{K} \subset \mathbb{R}^n$ be a cone (from now on, all cones are closed and convex, have non-empty interiors, and contain no lines). A ϑ-self-concordant barrier F for $\mathbf{K}$ is called *logarithmically homogeneous* (an LHSCB), if

$$\forall(\tau > 0, x \in \text{int } \mathbf{K}) : F(\tau x) = F(x) - \vartheta \ln \tau. \qquad (2.16)$$

In fact, *every* self-concordant function on a cone $\mathbf{K}$ satisfying the identity (2.16) is automatically a ϑ-SCB for $\mathbf{K}$, since whenever a smooth function F satisfies the identity (2.16), we have

$$\forall(x \in \text{int } K) : \nabla F(x)^T x = -\vartheta, \quad \nabla^2 F(x)x = -\nabla F(x); \qquad (2.17)$$

it follows that when F is self-concordant, we have

$$\lambda(x, F) = \sqrt{\nabla F(x)^T [\nabla^2 F(x)]^{-1} \nabla F(x)} = \sqrt{-\nabla F(x)^T x} = \sqrt{\vartheta},$$

meaning that F is indeed a ϑ-SCB for $\mathbf{K}$. A nice and important fact is that the (modified) Legendre transformation $F_*(s)$ of a ϑ-LHSCB F for a cone $\mathbf{K}$ is a ϑ-LHSCB for the cone

$$\mathbf{K}_* := \{s \in \mathbb{R}^n : s^T x \geq 0, \; \forall x \in \mathbf{K}\} \tag{2.18}$$

dual to $\mathbf{K}$. The resulting symmetry of LHSCBs complements the symmetry between cones and their duals. Moreover, we have the following result.

Proposition 2.1. The mappings $x \mapsto -\nabla F(x)$ and $s \mapsto -\nabla F_*(s)$ are inverse bijections between int $\mathbf{K}$ and int $\mathbf{K}_*$, and these bijections are homogeneous of degree -1: $-\nabla F(\tau x) = -\tau^{-1} \nabla F(x)$, $x \in$ int $\mathbf{K}$, $\tau > 0$, and similarly for F_*. Finally, $\nabla^2 F$ and $\nabla^2 F_*$ are homogeneous of degree -2, with $\nabla^2 F_*(-\nabla F(x)) = [\nabla^2 F(x)]^{-1}$ and $\nabla^2 F(-\nabla F_*(s)) = [\nabla^2 F_*(s)]^{-1}$.

Now assume that we want to solve a primal–dual pair of conic problems

$$\min_x \{c^T x : Ax = b, x \in \mathbf{K}\} \quad (P),$$
$$\max_{y,s} \{b^T y : A^T y + s = c, s \in \mathbf{K}_*\} \quad (D), \tag{2.19}$$

where the rows of A are linearly independent and both problems have strictly feasible solutions (*i.e.*, feasible solutions with $x \in$ int $\mathbf{K}$ and $s \in$ int $\mathbf{K}_*$). Assume also that we have at our disposal a ϑ-LHSCB F for $\mathbf{K}$ along with its Legendre transformation F_*, which is a ϑ-LHSCB for $\mathbf{K}_*$. (P) can be treated as a problem of the form (2.1), with the affine set $L = \{x : Ax = b\}$ playing the role of the 'universe' $\mathbb{R}^n$ and $\mathbf{K} \cap L$ in the role of X. It is easily seen that the restriction of F to int $\mathbf{K} \cap L$ is a ϑ-SCB for the resulting problem (2.1) (see rule $\mathbf{D}$ in Section 2.4), and that this is a problem with bounded level sets. As a result, we can define the *primal central path* $\{x_*(t)\}$, which comprises strictly feasible solutions to (P) and converges, as $t \to \infty$, to the primal optimal set. Similarly, setting $Y = \{y : c - A^T y \in \mathbf{K}_*\}$, the dual problem can be written in the form of (2.1), namely, as $\min_{y \in Y} [-b]^T y$. The domain Y of this problem can also be equipped with a ϑ-SCB, namely, $F_*(c - A^T y)$, and again the problem has bounded level sets, so that we can define the associated central path $\{y_*(t)\}$. This induces the *dual central path* $\{s_*(t) := c - A^T y_*(t)\}$; the latter path comprises interior points of $\mathbf{K}_*$. We have arrived at the *primal–dual central path* $\{z_*(t) := (x_*(t), s_*(t))\}$ 'living' in the interior of $\mathbf{K} \times \mathbf{K}_*$. It is easily seen that *for every $t > 0$, the point $z_*(t)$ is uniquely defined by the following restrictions on its components x, s*:

$x \in X^o := $ int $\mathbf{K} \cap \{x : Ax = b\}$ [strict primal feasibility],

$s \in S^o := $ int $\mathbf{K}_* \cap \{s : \exists y$ such that $A^T y + s = c\}$ [strict dual feasibility],

$$\left.\begin{array}{l} s = -t^{-1} \nabla F(x) \\ x = -t^{-1} \nabla F_*(s) \end{array}\right\} \text{ [augmented complementary slackness].} \tag{2.20}$$

Note that, by Proposition 2.1, each of the complementary slackness equations implies the other, so that we could eliminate either one of them; we keep both to highlight the primal–dual symmetry.

Primal–dual path-following interior-point methods trace simultaneously the primal and dual central paths basically in the same fashion as the method described in Theorem 2.1. It turns out that tracing the paths together is much more advantageous than tracing only one of them. In our general setting these advantages permit, for example,

- adaptive long-step strategies for path-tracing (Nesterov 1997);
- an elegant way ('self-dual embedding'; see, *e.g.*, Ye, Todd and Mizuno (1994), Xu, Hung and Ye (1996), Andersen and Ye (1999), Luo, Sturm and Zhang (2000), de Klerk, Roos and Terlaky (1997) and Potra and Sheng (1998)) to initialize path-tracing even in the case when no strictly feasible solutions to (P) and (D) are available in advance; and
- building certificates of strict (*i.e.*, preserved by small perturbations of the data) primal or dual infeasibility (Nesterov, Todd and Ye 1999) when it holds, *etc.*

Primal–dual IPMs achieve their full power when the underlying cones are *self-scaled*, which is the case in linear, second-order cone, and semidefinite programming, considered in depth in Section 3. In the remaining part of this subsection, we overview, still in the general setting, another family of primal–dual IPMs, those based on *potential reduction*.

Potential-reduction interior-point methods

We now present two potential-reduction IPMs which are straightforward conic generalizations, developed by Nesterov and Nemirovski (1994), of algorithms originally proposed for LP.

Karmarkar's Algorithm. The first polynomial-time interior-point method for LP was discovered by Karmarkar (1984). The conic generalization of the algorithm is as follows. Assume that we want to solve a strictly feasible problem (P) in the following special setting: the primal feasible set $X := \{x \in \mathbf{K} : Ax = b\}$ is bounded, the optimal value is known to be 0, and we know a strictly feasible primal starting point $\bar{x}$ (using conic duality, every strictly primal–dual feasible conic problem can be transformed into this form). Lastly, $\mathbf{K}$ is equipped with a ϑ-LHSCB F. It is immediately seen that under our assumptions $b \neq 0$, so that we lose nothing when assuming that the first equality constraint reads $e^T x = 1$ for some vector e. Subtracting this equality with appropriate coefficients from the remaining equality constraints in (P), we can make all these constraints homogeneous, thus representing the problem in the form

$$\min_x \{c^T x : x \in L \cap \mathbf{K}, e^T x = 1\},$$

where L is a linear subspace in $\mathbb{R}^n$. Note that since X is bounded, we have $e^T x > 0$ for every $0 \neq x \in (L \cap \mathbf{K})$. If we exclude the trivial case $c^T \bar{x} = 0$ (here already $\bar{x}$ is an optimal solution), $c^T x$ is positive on the relative interior $X^o := X \cap \operatorname{int} \mathbf{K}$ of X, so that the projective transformation $x \mapsto p(x) := x/c^T x$ is well defined on X^o; this transformation maps X^o onto the relative interior $Z^o := Z \cap \operatorname{int} \mathbf{K}$ of the set $Z := \{z : z \in L \cap \mathbf{K}, c^T z = 1\}$, the inverse transformation being $z \mapsto z/e^T z$. The point is that Z *is unbounded*, since otherwise the linear form $e^T z$ would be bounded and positive on Z due to $e^T x > 0$ for $0 \neq x \in L \cap \mathbf{K}$, and so $c^T x$ would be bounded away from 0 on X^o, which is not the case. All we need is to generate a sequence $z_k \in Z^o$ such that $\|z_k\| \to \infty$ as $k \to \infty$; indeed, for such a sequence we clearly have $e^T z_k \to \infty$ and $c^T z_k = 1$, whence the points $x_k = z_k/e^T z_k$, which are feasible solutions to the problem of interest, satisfy $c^T x_k \to 0 = \min_{x \in X} c^T x$ as $k \to \infty$. This is how we 'run to ∞ along Z' using Karmarkar's algorithm. Let $G(z)$ be the restriction of F to Z^o. Treating Z as a subset of its affine hull $\operatorname{Aff}(Z)$, so that Z is a closed convex domain in a certain $\mathbb{R}^n$, we find that G is a ϑ-SCB for Z (see rule $\mathbf{D}$ in Section 2.4). Since Z, along with $\mathbf{K}$, does not contain lines, G is non-degenerate and therefore $\lambda(z, G) \geq 1$ for all $z \in Z^o$ by 2.1.4 (recall that Z is unbounded); applying 2.1.2, we conclude that *the step $z \mapsto z^+(z)$ of the damped Newton method as applied to G, z maps Z^o into Z^o and reduces G by at least the absolute constant* $\delta := 1 - \ln(2) > 0$. It follows that applying the damped Newton method to G, we push G to $-\infty$, and therefore indeed run to ∞ along Z. To get an explicit efficiency estimate, let us look at the *Karmarkar potential function* $\phi : X^o \to \mathbb{R}$ defined by $\phi(x) = \vartheta \ln(c^T x) + F(x)$; note that $\phi(x) = G(p(x))$ due to the ϑ-logarithmical homogeneity of F. It follows that *the basic Karmarkar step $x \mapsto x_+(x) = p^{-1}(z^+(p(x)))$ maps X^o into itself and reduces the potential by at least δ*. In Karmarkar's algorithm, one iterates this step (usually augmented by a line search aimed at getting a larger reduction in the potential than the guaranteed reduction δ) starting with $x_0 := \bar{x}$, thus generating a sequence $\{x_k\}_{k=0}^\infty$ of strictly feasible solutions to (P) such that $\phi(x_k) \leq \phi(\bar{x}) - k\delta = F(\bar{x}) + \vartheta \ln(c^T \bar{x}) - k\delta$. Recalling that X is bounded, so that F is bounded below on X^o by 2.1.4, we have also $F(x) \geq \widehat{F} := \min_{x \in X^o} F(x)$, whence $F(\bar{x}) + \vartheta \ln(c^T \bar{x}) - k\delta \geq \phi(x_k) \geq \widehat{F} + \vartheta \ln(c^T x_k)$. We arrive at the efficiency estimate

$$c^T x_k = c^T x_k - \min_{x \in X} c^T x \leq c^T \bar{x} \exp\left(\frac{F(\bar{x}) - \widehat{F} - k\delta}{\vartheta}\right),$$

meaning that, for every $\epsilon \in (0, 1)$, at most $\left\lfloor \frac{[F(\bar{x}) - \widehat{F}] + \vartheta \ln(1/\epsilon)}{\delta} \right\rfloor + 1$ steps of the method are needed to arrive at a strictly feasible solution x_k with $c^T x_k = c^T x_k - \min_X c^T x \leq \epsilon c^T \bar{x}$. The advantage of the Karmarkar algorithm, as compared to that in Theorem 2.1, is that our only interest now is driving the (on-line-observable) potential function $\phi(x)$ to $-\infty$ as rapidly as possible, while staying strictly feasible at all times; this can be done, *e.g.*, by augmenting the basic step with an appropriate line search, which usually leads to a much larger reduction in $\phi(\cdot)$ at each step than the reduction δ guaranteed by the theory. As a result, the practical performance of Karmarkar's algorithm is typically much better than predicted by the theoretical complexity estimate above. On the negative side, the latter estimate is worse than that for the basic path-following method from Theorem 2.1: now the complexity is proportional to ϑ rather than to $\vartheta^{1/2}$ and $\vartheta \geq 1$ may well be large. To circumvent this difficulty, we now present a *primal–dual* potential-reduction algorithm, extending to the general conic case the algorithm of Ye (1991) originally developed for LP.

Primal–dual potential-reduction algorithm. This algorithm is a 'genuine primal–dual one'; it works on a strictly feasible pair (2.19) of conic problems and associates with this pair the generalized Tanabe–Todd–Ye (Tanabe 1988, Todd and Ye 1990) *primal–dual potential function* $p : X^o \times S^o \to \mathbb{R}$ defined by

$$p(x,s) := (\vartheta + \sqrt{\vartheta}) \ln(s^T x) + F(x) + F_*(s) =: p_0(x,s) + \sqrt{\vartheta} \ln(s^T x). \quad (2.21)$$

It is easily seen that p_0 is bounded below on $X^o \times S^o$ and the set of minimizers of p_0 on $X^o \times S^o$ is exactly the primal–dual central path, where p_0 takes the value $p_* = \vartheta \ln(\vartheta) - \vartheta$. It follows that *for* $(x,s) \in X^o \times S^o$, *the duality gap* $s^T x$ *can be bounded in terms of* p:

$$(x,s) \in X^o \times S^o \Rightarrow s^T x \leq \exp\{\vartheta^{-1/2}[p(x,s) - p_*]\}. \quad (2.22)$$

(It can be readily checked – see Proposition 3.1 – that $c^T x - b^T y = s^T x \geq 0$ for any feasible x and (y,s), so $s^T x$ bounds the distance from optimality of both the primal and dual objective function values.)

Hence all we need in order to approach primal–dual optimality is a 'basic primal–dual step': an update $(x,s) \mapsto (x_+, s_+) : X^o \times S^o \to X^o \times S^o$ which 'substantially' reduces the potential p, at least by a positive absolute constant δ. Iterating this update (perhaps augmented by a line search aimed at further reduction in p) starting with a given initial point $(x_0, s_0) \in X^o \times S^o$, we get a sequence of strictly feasible primal solutions x_k and dual slacks $s_k \in \text{int } \mathbf{K}_*$ (which can be immediately extended to dual feasible solutions (y_k, s_k)), such that $p(x_k, s_k) \leq p(x_0, s_0) - k\delta$, which combines with (2.22) to yield the efficiency estimate

$$s_k^T x_k \leq \exp\{\vartheta^{-1/2}[p_0(x_0, s_0) - p_*]\} \exp\{-\delta \vartheta^{-1/2} k\} s_0^T x_0.$$

Now it takes only $O(1)\sqrt{\vartheta}$ steps to reduce the (upper bound on the) duality gap by an absolute constant factor, and we end up with complexity bounds almost identical to those in Theorem 2.1.

It remains to explain how to make a basic primal–dual step. This can be done as follows. With a fixed positive threshold $\bar{\lambda}$, given $(x, s) \in X^o \times S^o$, we linearize the logarithmic term in the potential in x and in s, thus getting the functions

$$\xi \mapsto p^x(\xi) = (\vartheta + \sqrt{\vartheta}) \frac{s^T \xi}{s^T x} + F(\xi) + \text{const}_x : \text{int } \mathbf{K} \to \mathbb{R},$$

$$\sigma \mapsto p^s(\sigma) = (\vartheta + \sqrt{\vartheta}) \frac{\sigma^T x}{s^T x} + F_*(\sigma) + \text{const}_s : \text{int } \mathbf{K}_* \to \mathbb{R},$$

which are non-degenerate self-concordant functions on $\mathbf{K}$ and $\mathbf{K}_*$, respectively. We compute the Newton direction $d_x = \text{argmin}_d \{ d^T \nabla p^x(x) + \frac{1}{2} d^T \nabla^2 p^x(x) d : x + d \in \text{Aff}(X) \}$ of $p^x|_X$ at $\xi = x$ along with the corresponding Newton decrement $\lambda := \lambda(x, p^x|_X) = \sqrt{-\nabla p^x(x)^T d_x}$. When $\lambda \geq \bar{\lambda}$, one can set $s_+ = s$ and take for x_+ the damped Newton iterate $x + (1 + \lambda)^{-1} d_x$ of x, the Newton method being applied to $p^x|_X$. When $\lambda < \bar{\lambda}$, one can set

$$x_+ = x \quad \text{and} \quad s_+ = \frac{s^T x}{\vartheta + \sqrt{\vartheta}} [-\nabla F(x) - \nabla^2 F(x) d_x].$$

It can be shown that with a properly chosen *absolute constant* $\bar{\lambda} > 0$, this update indeed ensures that $(x_+, s_+) \in X^o \times S^o$ and $p(x_+, s_+) \leq p(x, s) - \delta$, where $\delta > 0$ depends solely on $\bar{\lambda}$. Note that the same is true for the 'symmetric' updating obtained by similar construction with the primal and dual problems swapped, and one is welcome to use the better (the one with a larger reduction in the potential) of these two updates or their line-search augmentations.

2.4. The calculus of self-concordant barriers

The practical significance of the nice results we have described depends heavily on our ability to equip the problem we are interested in (a convex program (2.1), or a primal–dual pair of conic programs (2.19)) with self-concordant barrier(s). *In principle* this can always be done: every closed convex domain $X \subset \mathbb{R}^n$ admits an $O(1)n$-SCB; when the domain is a cone, this barrier can be chosen to be logarithmically homogeneous. Assuming without loss of generality that X does not contain lines, one can take as such a barrier the function

$$F(x) = O(1) \ln \text{mes}_n \{ y : y^T (z - x) \leq 1, \ \forall z \in X \},$$

where mes_n denotes n-dimensional (Jordan or Lebesgue) measure. This function has a transparent geometric interpretation: the set whose measure we are taking is the polar of $X - x$. When X is a cone (closed, convex, containing no lines and with a non-empty interior), the *universal barrier* given by the expression above is automatically logarithmically homogeneous.

From a practical perspective, the existence theorem just formulated is not of much interest – the universal barrier is usually pretty difficult to compute, and in the rare cases when this is possible, it may be non-optimal in terms of its self-concordance parameter. Fortunately, there exists a simple and fully algorithmic 'calculus' of self-concordant barriers which allows us to build systematically explicit efficiently computable SCBs for seemingly all generic convex programs associated with 'computationally tractable' domains. We start with the list of the most basic rules (essentially, the only ones needed in practice) of 'self-concordant calculus'.

A If F is a ϑ-SCB for X and $\alpha \geq 1$, then αF is an $(\alpha\vartheta)$-SCB for X.

B *Direct products.* Let F_i, $i = 1, \ldots, m$, be ϑ_i-SCBs for closed convex domains $X_i \subset \mathbb{R}^{n_i}$. The 'direct sum' $F(x^1, \ldots, x^m) = \sum_i F_i(x^i)$ of these barriers is a $(\sum_i \vartheta_i)$-SCB for the direct product $X = X_1 \times \cdots \times X_m$ of the sets.

C *Intersection.* Let F_i, $i = 1, \ldots, m$, be ϑ_i-SCBs for closed convex domains $X_i \subset \mathbb{R}^n$, and let the set $X = \bigcap_i X_i$ possess a non-empty interior. Then $F(x) = \sum_i F_i(x)$ is a $(\sum_i \vartheta_i)$-SCB for X.

D *Inverse affine image.* Let F be a ϑ-SCB for a closed convex domain $X \subset \mathbb{R}^n$, and $y \mapsto Ay+b : \mathbb{R}^k \to \mathbb{R}^n$ be an affine mapping whose image intersects int X. Then the function $G(y) = F(Ay + b)$ is a ϑ-SCB for the closed convex domain $Y = \{y : Ay + b \in X\}$.

When the operands in the rules are cones and the original SCBs are logarithmically homogeneous, so are the resulting barriers (in the case of **D**, provided that $b = 0$). All the statements remain true when, instead of SCBs, we are speaking about SC functions; in this case, the parameter-related parts should be skipped, and what remains become statements on preserving self-concordance.

Essentially all we need in addition to the outlined (and nearly evident) elementary calculus rules, are two more advanced rules, as follows.

E *Taking the conic hull.* Let $X \subset \mathbb{R}^n$ be a closed convex domain and let F be a ϑ-SCB for X. Then, with a properly chosen absolute constant κ, the function $F_+(x, t) = \kappa[F(x/t) - 2\vartheta \ln t]$ is a $2\kappa\vartheta$-SCB for the conic hull

$$X_+ := \mathrm{cl}\left\{(x, t) \in \mathbb{R}^n \times \mathbb{R} : t > 0, x/t \in \mathrm{int}\, X\right\}$$

of X.

To present the last calculus rule, which can be skipped on a first reading, we need to introduce the notion of compatibility, as follows. Let $K \subset \mathbb{R}^N$ and $G_- \subseteq \mathbb{R}^n$ be a closed convex cone and a closed convex domain, respectively, let $\beta \geq 1$, and let $\mathcal{A}(x) : \mathrm{int}\, G_- \to \mathbb{R}^N$ be a mapping. We say that $\mathcal{A}$

is β-*compatible with* K, if $\mathcal{A}$ is three times continuously differentiable on int G_-, is K-concave (that is, $D^2\mathcal{A}(x)[h,h] \in -K$ for all $x \in \text{int } G_-$ and all $h \in \mathbb{R}^n$) and

$$D^3\mathcal{A}(x)[h,h,h] \leq_K -3\beta D^2\mathcal{A}(x)[h,h]$$

for all $x \in \text{int } G_-$ and $h \in \mathbb{R}^n$ with $x \pm h \in G_-$, where $a \leq_K b$ means that $b - a \in K$. The calculus rule in question reads as follows.

F Let $G_- \subset \mathbb{R}^n$, $G_+ \subset \mathbb{R}^N$ be closed convex domains and $\mathcal{A} : \text{int } G_- \to \mathbb{R}^N$ be a mapping, β-compatible with the recession cone of G_+, whose image intersects int G_+. Given $\vartheta_\pm$-SCBs $F_\pm$ for G_+ and G_-, respectively, let us define $F : X^\circ := \{x \in \text{int } G_- : \mathcal{A}(x) \in \text{int } G_+\} \to \mathbb{R}$ by

$$F(x) = F_+(\mathcal{A}(x)) + \beta^2 F_-(x).$$

Then F is a $(\vartheta_+ + \beta^2\vartheta_-)$-SCB for $X = \text{cl } X^\circ$.

The most non-trivial and important example of a mapping which can be used in the context of rule **E** is the *fractional-quadratic substitution*. Specifically, let T, E, F be Euclidean spaces, let $Q(x,z) : E \times E \to F$ be a symmetric bilinear mapping, and let $A(t)$ be a symmetric linear operator on E, affinely depending on $t \in T$, and such that the bilinear form $Q(A(t)x,z)$ on $E \times E$ is symmetric in x, z for every $t \in T$. Further, let K be a closed convex cone in F such that $Q(x,x) \in K$ for all x, and let H be a closed convex domain in T such that $A(t)$ is positive definite for all $t \in \text{int } H$. It turns out that the mapping $\mathcal{A}(y,x,t) = y - Q([A(t)]^{-1}x,x)$ with the domain $F \times E \times \text{int } H$ is 1-compatible with K.

It turns out (see examples in Nesterov and Nemirovski (1994) and Nemirovski (2004)) that the combination rules **A–F** used 'from scratch' (from the sole observation that the function $-\ln x$ is a 1-LHSCB for the non-negative ray) permit one to build 'good' SCBs/LHSCBs almost without calculation for all interesting convex domains/cones, including epigraphs of numerous convex functions (*e.g.*, the elementary univariate functions such as powers and the exponential, and the multivariate p-norms), sets given by finite systems of convex linear and quadratic inequalities, and much more. This list includes, in particular, ϑ-LHSCBs underlying:

(a) the non-negative orthant $\mathbb{R}^n_+$ ($F(x) = -\sum_j \ln x_j$, $\vartheta = n$),

(b) Lorentz cones

$$\mathbf{L}^q = \{(\xi, x) \in \mathbb{R} \times \mathbb{R}^q : \xi \geq \|x\|_2\} \quad (F(\xi, x) = -\ln(\xi^2 - x^T x),\ \vartheta = 2),$$

(c) semidefinite cones $\mathbf{S}^p_+$ (the cone of all symmetric positive definite matrices of order p) ($F(x) = -\ln \det(x)$, $\vartheta = p$), and

(d) matrix norm cones $\{(\xi, x) : \xi \geq 0, x \in \mathbb{R}^{p \times q} : \xi^2 I_q \succeq x^T x\}$ (assuming without loss of generality $p \geq q$, $F(\xi, x) = -\ln \det(\xi I_q - \xi^{-1} x^T x) - \ln \xi$, $\vartheta = q + 1$).

With regard to (a)–(d), (a) is self-evident, and the remaining three barriers can be obtained from (a) and $\mathbf{F}$, *without calculations*, via the above result on fractional-quadratic substitution. For example, to get (b), we set $T = F = \mathbb{R}$, $E = \mathbb{R}^n = \mathbb{R}^{n \times 1}$, $Q(x, z) = x^T z$, $A(\xi) = \xi I_n$, $K = H = \mathbb{R}_+$, thus concluding that the mapping $\mathcal{A}(y, x, \xi) = y - \xi^{-1} x^T x$ with the domain $\operatorname{int} G_-$, $G_- = \{y \in \mathbb{R}\} \times \{x \in \mathbb{R}^n\} \times \{\xi \geq 0\}$, taking values in $\mathbb{R}$, is 1-compatible with $K = \mathbb{R}_+$. Applying $\mathbf{F}$ to G_- and with $G_+ = \mathbb{R}_+$, $F_-(y, x, \xi) = -\ln \xi$, $F_+(s) = -\ln s$, and using (a) and $\mathbf{D}$ to conclude that F_- and F_+ are SCBs for G_- and G_+ with the parameters $\vartheta_- = \vartheta_+ = 1$, we see that $-\ln(y - \xi^{-1} x^T x) - \ln \xi$ is a 2-SCB for the set $\{(y, x, \xi) : y\xi \geq x^T x, y, \xi \geq 0\}$. It remains to note that $\mathbf{L}_n$ is the inverse affine image of the latter set under the linear mapping $(\xi, x) \mapsto (\xi, x, \xi)$, and to apply $\mathbf{D}$.

Note that (a)–(c) combine with $\mathbf{B}$ to induce LHSCBs for the direct products K of non-negative rays, Lorentz and semidefinite cones. All cones K one can get in this fashion are self-dual, and the resulting barriers F turn out to be 'self-symmetric' $(F_*(\cdot) = F(\cdot) + \mathrm{const}_K)$, thus giving rise to primal–dual IPMs for linear, conic quadratic, and semidefinite programming. Moreover, it turns out that the barriers in question are 'optimal', with provably minimum possible values of the self-concordance parameter ϑ.

3. Conic optimization

Here we treat in more detail the case of the primal–dual conic problems in (2.19). We restate the primal problem:

$$(P) \quad \begin{aligned} \min_{x} \quad & c^T x \\ & Ax = b, \\ & x \in \mathbf{K}, \end{aligned}$$

where again $c \in \mathbb{R}^n$, $A \in \mathbb{R}^{m \times n}$, $b \in \mathbb{R}^m$, and $\mathbf{K}$ is a closed convex cone in $\mathbb{R}^n$. We call this the conic programming problem in primal or standard form, since when $\mathbf{K}$ is the non-negative orthant, it becomes the standard-form linear programming problem.

Recall the dual cone defined by

$$\mathbf{K}_* := \{s \in \mathbb{R}^n : s^T x \geq 0, \text{ for all } x \in \mathbf{K}\}. \tag{3.1}$$

Then we can construct the conic programming problem in dual form using the same data:

$$(D) \quad \begin{aligned} \max_{y, s} \quad & b^T y \\ & A^T y + s = c, \\ & s \in \mathbf{K}_*, \end{aligned}$$

with $y \in \mathbb{R}^m$, where we have introduced the dual slack variable s to make the later analysis cleaner. In terms of the variables y, we have the conic constraints $c - A^T y \in \mathbf{K}_*$, corresponding to the linear inequality constraints $c - A^T y \geq 0$ when (P) is the standard linear programming problem.

In fact, it is easy to see that (D) is the Lagrangian dual

$$\max_y \left\{ \min_{x \in \mathbf{K}} \{c^T x - (Ax - b)^T y\} \right\}$$

of (P), using the fact that $\min\{u^T x : x \in \mathbf{K}\}$ is 0 if $u \in \mathbf{K}_*$ and $-\infty$ otherwise. We can also easily check weak duality, as follows.

Proposition 3.1. If x is feasible in (P) and (y, s) in (D), then

$$c^T x \geq b^T y,$$

with equality if and only if $s^T x = 0$.

Proof. Indeed,

$$c^T x - b^T y = (A^T y + s)^T x - (Ax)^T y = s^T x \geq 0, \tag{3.2}$$

with the inequality following from the definition of the dual cone. $\square$

In the case of linear programming, when $\mathbf{K}$ (and then also $\mathbf{K}_*$) is the non-negative orthant, then whenever (P) or (D) is feasible, we have equality of their optimal values (possibly $\pm\infty$), and if both are feasible, we have strong duality: no duality gap, and both optimal values attained.

In the case of more general conic programming, these properties no longer hold (we will provide examples in the next subsection), and we need further regularity conditions. Nesterov and Nemirovski (1994, Theorem 4.2.1) derive the next result.

Theorem 3.1. If either (P) or (D) is bounded and has a strictly feasible solution (*i.e.*, a feasible solution where x (respectively, s) lies in the interior of $\mathbf{K}$ (respectively, $\mathbf{K}_*$)), then their optimal values are equal. If both have strictly feasible solutions, then strong duality holds.

The existence of an easily stated dual problem provides one motivation for considering problems in conic form (but its usefulness depends on having a closed form expression for the dual cone). We will also see that many important applications naturally lead to conic optimization problems. Finally, there are efficient primal–dual interior-point methods for this class of problems, or at least for important subclasses.

In Section 3.1, we consider several interesting special cases of (P) and (D). Section 3.2 discusses path-following interior-point methods. In Section 3.3, we consider a special class of conic optimization problems allowing symmetric primal–dual methods. Finally, Section 3.4 addresses recent extensions.

3.1. Examples of conic programming problems

First of all, it is worth pointing out that any convex programming problem can be put into conic form. Without loss of generality, after introducing a new variable if necessary to represent a convex nonlinear objective function, we can assume that the original problem is

$$\min_x \{c^T x : x \in X\},$$

with X a closed convex subset of $\mathbb{R}^n$. This is equivalent to the conic optimization problem, but for one dimension higher:

$$\min_{x,\xi} \{c^T x : \xi = 1, (x, \xi) \in \mathbf{K}\},$$

where $\mathbf{K} := \mathrm{cl}\{(x, \xi) \in \mathbb{R}^n \times \mathbb{R} : \xi > 0, x/\xi \in X\}$. However, this formal equivalence may not be very useful practically, partly because $\mathbf{K}$ and $\mathbf{K}_*$ may not be easy to work with. More importantly, even if we have a good self-concordant barrier for X, it may be hard to obtain an efficient self-concordant barrier for $\mathbf{K}$ (although general, if usually overconservative, procedures are available: see rule $\mathbf{E}$ in Section 2.4 and Freund, Jarre and Schaible (1996)).

Let us turn to examples with very concrete and useful cones. The first example is of course linear programming, where $\mathbf{K} = \mathbb{R}_+^n$. Then it is easy to see that $\mathbf{K}_*$ is also $\mathbb{R}_+^n$, and so the dual constraints are just $A^T y \leq c$. The significance and wide applicability of linear programming are well known. Our first case with a non-polyhedral cone is what is known as *second-order cone programming* (SOCP). Here $\mathbf{K}$ is a second-order, or Lorentz, or 'ice-cream' cone,

$$\mathbf{L}^q := \{(\xi, \bar{x}) \in \mathbb{R} \times \mathbb{R}^q : \xi \geq \|\bar{x}\|\},$$

or the product of such cones. It is not hard to see, using the Cauchy–Schwarz inequality, that such cones are also self-dual, *i.e.*, equal to their duals. We now provide an example showing the usefulness of SOCP problems (many more examples can be found in Lobo, Vandenberghe, Boyd and Lebret (1998) and in Ben-Tal and Nemirovski (2001)), and also a particular instance demonstrating that strong duality does not always hold for such problems.

Suppose we are interested in solving a linear programming problem $\max\{b^T y : A^T y \leq c\}$, but the constraints are not known exactly: for the jth constraint $a_j^T y \leq c_j$, we just know that $(c_j; a_j) \in \{(\bar{c}_j; \bar{a}_j) + P_j u_j : \|u_j\| \leq 1\}$, an ellipsoidal *uncertainty set* centred at the nominal values $(\bar{c}_j; \bar{a}_j)$. (We use the MATLAB-like notation $(u; v)$ to denote the concatenation of the vectors u and v.) Here P_j is a suitable matrix that determines the shape and size of this uncertainty set. We would like to choose our decision variable y so that it is feasible no matter what the constraint coefficients turn out to be, as long as they are in the corresponding uncertainty sets; with this limitation, we would like to maximize $b^T y$. This is (a particular case of) the so-called *robust* linear programming problem. Since the minimum of $c_j - a_j^T y = (c_j; a_j)^T (1; -y)$ over the jth uncertainty set is

$$(\bar{c}_j; \bar{a}_j)^T (1; -y) + \min\{(P_j u_j)^T (1; -y) : \|u_j\| \leq 1\}$$
$$= (\bar{c}_j; \bar{a}_j)^T (1; -y) - \|P_j^T (1; -y)\|,$$

this robust linear programming problem can be formulated as

$$\begin{aligned}
\max \quad & b^T y \\
& -\bar{c}_j + \bar{a}_j^T y + \quad s_{j1} = 0, \quad j = 1, \ldots, m, \\
& P_j^T (1; -y) + \quad \bar{s}_j = 0, \quad j = 1, \ldots, m, \\
& \qquad\qquad (s_{j1}; \bar{s}_j) \in \mathbf{K}_j, \quad j = 1, \ldots, m,
\end{aligned}$$

where each $\mathbf{K}_j$ is a second-order cone of appropriate dimension. This is a SOCP problem in dual form.

Next, consider the SOCP problem in dual form with data

$$A = \begin{pmatrix} -1 & 0 & -1 \\ -1 & 0 & 1 \end{pmatrix}, \quad b = \begin{pmatrix} -1 \\ 0 \end{pmatrix}, \quad c = \begin{pmatrix} 0 \\ 1 \\ 0 \end{pmatrix},$$

and $\mathbf{K}$ the second-order cone in $\mathbb{R}^3$. It can be checked that y is feasible in (D) if and only if y_1 and y_2 are positive, and $4y_1 y_2 \geq 1$. Subject to these constraints, we wish to maximize $-y_1$, so the problem is feasible, with objective function bounded above, but there is no optimal solution! In this case, the optimal values of primal and dual are equal: $(\xi; \bar{x}) = (1/2; 0; 1/2)$ is the unique feasible solution to (P), with zero objective function value.

The second class of non-polyhedral cones we consider gives rise to semi-definite programming problems. These correspond to the case when $\mathbf{K}$ is the cone of positive semidefinite matrices of a given order (or possibly a Cartesian product of such cones). Here we will restrict ourselves to the case of real symmetric matrices, and we use $\mathbf{S}^p$ to denote the space of all such matrices of order p. Of course, this can be identified with $\mathbb{R}^n$ for $n := p(p+1)/2$, by making a vector from the entries m_{ii} and $\sqrt{2} m_{ij}$, $i < j$. We use the factor $\sqrt{2}$ so that the usual scalar product of the vectors corresponding to two symmetric matrices U and V equals the Frobenius scalar product

$$U \bullet V := \text{Tr}(U^T V) = \sum_{i,j} u_{ij} v_{ij}$$

of the matrices. However, we will just state these problems in terms of the matrices for clarity. We write $\mathbf{S}_+^p$ for the cone of (real symmetric) positive semidefinite matrices, and sometimes write $X \succeq 0$ to denote that X lies in this cone for appropriate p. As in the case of the non-negative orthant and the second-order cone, $\mathbf{S}^p_+$ is self-dual. This can be shown using the spectral decomposition of a symmetric matrix. We note that the case of complex Hermitian positive semidefinite matrices can also be considered, and this is important in some applications.

In matrix form, the constraint $AX = b$ is defined using an operator A from $\mathbf{S}^p$ to $\mathbb{R}^m$, and we can find matrices $A_i \in \mathbf{S}^p$, $i = 1, \ldots, m$, so that $AX = (A_i \bullet X)^m_{i=1}$; A^T is then the adjoint operator from $\mathbb{R}^m$ to $\mathbf{S}^p$ defined by $A^T y = \sum_i y_i A_i$. The primal and dual semidefinite programming problems then become

$$\min C \bullet X, \quad A_i \bullet X = b_i, \, i = 1, \ldots, m, \quad X \succeq 0, \qquad (3.3)$$

and

$$\max b^T y, \quad \sum_i y_i A_i + S = C, \quad S \succeq 0. \qquad (3.4)$$

Once again, we give examples of the importance of this class of conic optimization problems, and also an instance demonstrating the failure of strong duality.

Let us first describe a very simple example that illustrates techniques used in optimal control. Suppose we have a linear dynamical system

$$\dot{z}(t) = A(t)z(t),$$

where the $p \times p$ matrices $A(t)$ are known to lie in the convex hull of a number $A_1, \ldots, A_k$ of given matrices. We want conditions that guarantee that the trajectories of this system stay bounded. Certainly a sufficient condition is that there is a positive definite matrix $Y \in \mathbf{S}^p$ so that the Lyapunov function $L(z(t)) := z(t)^T Y z(t)$ remains bounded. And this will hold as long as $\dot{L}(z(t)) \leq 0$. Now using the dynamical system, we find that

$$\dot{L}(z(t)) = z(t)^T (A(t)^T Y + Y A(t)) z(t),$$

and since we do not know where the current state might be, we want $-A(t)^T Y - Y A(t)$ to be positive semidefinite whatever $A(t)$ is, and so we are led to the constraints

$$-A_i^T Y - Y A_i \succeq 0, \, i = 1, \ldots, k, \quad Y - I_p \succeq 0,$$

where the last constraint ensures that Y is positive definite. (Here I_p denotes the identity matrix of order p. Since the first constraints are homogeneous in Y, we can assume that Y is scaled so its minimum eigenvalue is at least 1.) To make an optimization problem, we could for instance minimize the condition number of Y by adding the constraint $\eta I_p - Y \succeq 0$ and then maximizing $-\eta$. This is a semidefinite programming problem in dual form. Note that the variables y are the entries of the symmetric matrix Y and the scalar η, and the cone is the product of $k+2$ copies of $\mathbf{S}_+^p$. We can similarly find sufficient conditions for $z(t)$ to decay exponentially to zero.

Our second example is a relaxation of a quadratic optimization problem with quadratic constraints. Notice that we did not stipulate that the problem be convex, so we can include constraints like $x_j^2 = x_j$, which implies that x_j is 0 or 1, $i.e.$, we have included binary integer programming problems. Any quadratic function can be written as a linear function of a certain symmetric matrix (depending quadratically on x). Specifically, we see that

$$
\alpha + 2b^T x + x^T C x = \begin{pmatrix} 1 \\ x \end{pmatrix}^T \begin{pmatrix} \alpha & b^T \\ b & C \end{pmatrix} \begin{pmatrix} 1 \\ x \end{pmatrix}
$$

$$
= \begin{pmatrix} \alpha & b^T \\ b & C \end{pmatrix} \bullet \left(\begin{pmatrix} 1 \\ x \end{pmatrix} \begin{pmatrix} 1 \\ x \end{pmatrix}^T \right)
$$

$$
= \begin{pmatrix} \alpha & b^T \\ b & C \end{pmatrix} \bullet \begin{pmatrix} 1 & x^T \\ x & xx^T \end{pmatrix}.
$$

The set of all matrices $\begin{pmatrix} 1 & x^T \\ x & xx^T \end{pmatrix}$ is certainly a subset of the set of all positive semidefinite matrices with top left entry equal to 1, and so we can obtain a relaxation of the original hard problem in x by optimizing over a matrix X that is subject to the constraints defining this superset. This technique has been very successful in a number of combinatorial problems, and has led to worthwhile approximations to the stable set problem, various satisfiability problems, and notably the max-cut problem. Further details can be found, for example, in Goemans (1997) and Ben-Tal and Nemirovski (2001).

Let us give an example of two dual semidefinite programming problems where strong duality fails. The primal problem is

$$
\min_{X \succeq 0} \begin{pmatrix} 0 & 0 & 0 \\ 0 & 0 & 0 \\ 0 & 0 & 1 \end{pmatrix} \bullet X, \quad \begin{pmatrix} 1 & 0 & 0 \\ 0 & 0 & 0 \\ 0 & 0 & 0 \end{pmatrix} \bullet X = 0, \quad \begin{pmatrix} 0 & 1 & 0 \\ 1 & 0 & 0 \\ 0 & 0 & 2 \end{pmatrix} \bullet X = 2,
$$

where the first constraint implies that x_{11}, and hence x_{12} and x_{21}, are zero, and so the second constraint implies that x_{33} is 1. Hence one optimal solution is $X = \text{Diag}(0; 0; 1)$ with $optimal\ value$ 1. The dual problem is

$$
\max 2y_2, \quad S = \begin{pmatrix} 0 & 0 & 0 \\ 0 & 0 & 0 \\ 0 & 0 & 1 \end{pmatrix} - y_1 \begin{pmatrix} 1 & 0 & 0 \\ 0 & 0 & 0 \\ 0 & 0 & 0 \end{pmatrix} - y_2 \begin{pmatrix} 0 & 1 & 0 \\ 1 & 0 & 0 \\ 0 & 0 & 2 \end{pmatrix} \succeq 0,
$$

so the dual slack matrix S has $s_{22} = 0$, implying that s_{12} and s_{21} must be zero, so y_2 must be zero. So an optimal solution is $y = (0; 0)$ with $optimal\ value$ 0. Hence, while both problems have optimal solutions, their optimal values are not equal. Note that neither problem has a strictly feasible solution, and arbitrary small perturbations in the data can make the optimal values jump.

3.2. Basic interior-point methods for conic problems

Recall that, for conic problems, we want to use logarithmically homogeneous SCBs, those satisfying (2.16):

$$F(\tau x) = F(x) - \vartheta \ln \tau.$$

Examples of such ϑ-LHSCBs are

$$F(x) := -\sum_j \ln x_j, \qquad x \in \operatorname{int} \mathbb{R}^n_+,$$

$$F(\xi; \bar{x}) := -\ln(\xi^2 - \|\bar{x}\|^2), \quad (\xi; \bar{x}) \in \operatorname{int} \mathbf{L}^q,$$

$$F(X) := -\ln \det X, \qquad X \in \operatorname{int} \mathbf{S}^p_+,$$

as in Section 2.4, with values of ϑ equal to n, 2, and p respectively. Each of these cones is self-dual, and it is easy to check that the corresponding dual barriers are $F_*(s) = F(s) - n$, $F_*(\sigma; \bar{s}) = F(\sigma; \bar{s}) + 2\ln 2 - 2$, and $F_*(S) = F(S) - p$.

Henceforth, F and F_* are ϑ-LHSCBs for the cones $\mathbf{K}$ and $\mathbf{K}_*$ respectively. The key properties of such functions are listed after (2.16) and in Proposition 2.1, and from these we easily obtain the following result.

Proposition 3.2. For $x \in \operatorname{int} K$, $s \in \operatorname{int} K_*$, and positive t, we have

$$s + t^{-1}\nabla F(x) = 0 \quad \text{if and only if} \quad x + t^{-1}\nabla F_*(s) = 0,$$

and if these hold,

$$s^T x = t^{-1}\vartheta \quad \text{and} \quad t^{-1}\nabla^2 F_*(s) = [t^{-1}\nabla^2 F(x)]^{-1}. \tag{3.5}$$

Proof. If $ts = -\nabla F(x)$, $x = -\nabla F_*(ts)$ since $-\nabla F$ and $-\nabla F_*$ are inverse bijections. Using the homogeneity of ∇F_*, we obtain $x + t^{-1}\nabla F_*(s) = 0$. The reverse implication follows the same reasoning. If $ts = -\nabla F(x)$, then $s^T x = -t^{-1}\nabla F(x)^T x = t^{-1}\vartheta$ by (2.17) and $\nabla^2 F_*(ts) = [\nabla^2 F(x)]^{-1}$, and the final claim follows from the homogeneity of $\nabla^2 F_*$ of degree -2. □

We now examine in more detail the path-following methods described in Section 2.3, both to see the computation involved and to see how these basic methods can be modified in some cases for increased efficiency. We assume that both (P) and (D) have strictly feasible solutions available. As we noted, the basic primal path-following algorithm can be applied to the restriction of F to the relative interior of $\{x \in \mathbf{K} : Ax = b\}$, which amounts to tracing the path of solutions for positive t to the primal barrier problems:

$$\min_x \; tc^T x + F(x)$$

(PB_t)
$$Ax = b,$$

$$x \in \mathbf{K}.$$

If we associate Lagrange multipliers $\lambda \in \mathbb{R}^m$ with the constraints, and then define $y := -t^{-1}\lambda$, we see that the optimality conditions for (PB_t) are

$$t(c - A^T y) + \nabla F(x) = 0, \quad Ax = b, \quad x \in \text{int } \mathbf{K}.$$

Since $-\nabla F$ maps int $\mathbf{K}$ into int $\mathbf{K}_*$, we see that $s := c - A^T y$ lies in int $\mathbf{K}_*$, and so we have

$$
\begin{aligned}
A^T y \;+\; & s = c, \quad s \in \text{int } \mathbf{K}_*, \\
Ax \qquad & = b, \quad x \in \text{int } \mathbf{K}, \\
\nabla F(x) \qquad & + ts = 0.
\end{aligned}
\tag{3.6}
$$

These equations define the primal–dual central path $\{(x_*(t), s_*(t))\}$ as in Section 2.3. Note also that, using (3.2) and (3.5), the duality gap associated with $x_*(t)$ and $(y_*(t), s_*(t))$ is $s_*(t)^T x_*(t) = t^{-1}\vartheta$. In view of Proposition 3.2, the conditions above are remarkably symmetric. Indeed, let us consider the dual barrier problem

$$
(DB_t) \qquad
\begin{aligned}
\min_{y,s} \quad & -tb^T y \;+\; F_*(s) \\
& A^T y \;+\; s = c, \\
& s \;\in\; \text{int } \mathbf{K}_*.
\end{aligned}
$$

If we associate Lagrange multipliers $\mu \in \mathbb{R}^n$ with the constraints, and then define $x := t^{-1}\mu$, we see that the optimality conditions for (DB_t) are

$$-tb + tAx = 0, \quad \nabla F_*(s) + tx = 0, \quad A^T y + s = c, \quad s \in \text{int } \mathbf{K}_*.$$

We can now conclude that $x \in \text{int } \mathbf{K}$, and so the optimality conditions can be written as (3.6) again, where the last equation is replaced by its equivalent form $tx + \nabla F_*(s) = 0$.

This nice symmetry is not preserved at first sight when we consider Newton-like algorithms to trace the central path. Suppose we have strictly feasible solutions x and (y, s) to (P) and (D), approximating a point on the central path: $(x, y, s) \approx (x_*(t), y_*(t), s_*(t))$ for some $t > 0$. We wish to find strictly feasible points approximating a point further along the central path, say corresponding to $t_+ > t$. Let us make a quadratic approximation to the objective function in (PB_{t_+}); for future analysis, we use the Hessian of F at a point $v \in \text{int } \mathbf{K}$ which may or may not equal x. If we let the variable be $x_+ =: x + \Delta x$, we have

$$
(PQP) \qquad
\begin{aligned}
\min_{\Delta x} \quad & t_+ c^T \Delta x + \nabla F(x)^T \Delta x + \tfrac{1}{2}\Delta x^T \nabla^2 F(v) \Delta x \\
& A\Delta x = 0.
\end{aligned}
$$

Let $\bar{\lambda} \in \mathbb{R}^m$ be the Lagrange multipliers for this problem, and define $\bar{y}_+ := -t_+^{-1}\bar{\lambda}$. Then the optimality conditions for (PQP) can be written as

$$t_+(c - A^T \bar{y}_+) + \nabla F(x) + \nabla^2 F(v)\Delta x = 0, \quad A\Delta x = 0,$$

and if we define $\overline{\Delta y} := \bar{y}_+ - y$ and $\overline{\Delta s} := -A^T \overline{\Delta y}$, we obtain

$$A^T \overline{\Delta y} + \overline{\Delta s} = 0,$$

$$(PQPOC) \qquad A\Delta x \qquad\qquad = 0,$$

$$t_+^{-1} \nabla^2 F(v) \Delta x \qquad + \overline{\Delta s} = -s - t_+^{-1} \nabla F(x).$$

This system also arises as giving the Newton step for (3.6) with t_+ replacing t, where $\nabla^2 F(v)$ is used instead of $\nabla^2 F(x)$. We will discuss the solution of this system of equations after comparing it with the corresponding system for the dual problem.

Hence let us make a quadratic approximation to the objective function of (DB_{t_+}), again evaluating the Hessian of F_* at a point $u \in \text{int}\, \mathbf{K}_*$ which may or may not equal s for future analysis. If we make the variables of the problem $y_+ =: y + \Delta y$ and $s_+ =: s + \Delta s$, we obtain

$$\min_{\Delta y, \Delta s} \quad -t_+ b^T \Delta y + \nabla F_*(s)^T \Delta s + \tfrac{1}{2} \Delta s^T \nabla^2 F_*(u) \Delta s$$

$$(DQP) \qquad A^T \Delta y + \qquad\qquad\qquad\qquad \Delta s = 0.$$

Let $\bar{\mu} \in \mathbb{R}^n$ be the Lagrange multipliers for (DQP), and define $\bar{x}_+ := t_+^{-1} \bar{\mu}$. Then the optimality conditions become

$$-t_+(b - A\bar{x}_+) = 0, \quad \nabla F_*(s) + \nabla^2 F_*(u)\Delta s + t_+ \bar{x}_+ = 0, \quad A^T \Delta y + \Delta s = 0.$$

Writing $\overline{\Delta x} := \bar{x}_+ - x$, we obtain

$$A^T \Delta y + \qquad\qquad\qquad \Delta s = 0,$$

$$(DQPOC) \quad A\overline{\Delta x} \qquad\qquad = 0,$$

$$\overline{\Delta x} \qquad + t_+^{-1} \nabla^2 F_*(u) \Delta s = -x - t_+^{-1} \nabla F_*(s).$$

We note that this system can also be viewed as a Newton-like system for a modified form of (3.6), where $t_+ x + \nabla F_*(s) = 0$ replaces $ts + \nabla F(x) = 0$ as the final equation. From this viewpoint, a natural way to adapt the methods to the case where x or (y, s) is not a strictly feasible solution of (P) or (D) is apparent. As long as $x \in \text{int}\, \mathbf{K}$ and $s \in \text{int}\, \mathbf{K}_*$, we can define search directions using $(PQPOC)$ or $(DQPOC)$ where the zero right-hand sides in the first two equations are replaced by the appropriate residuals in the equality constraints. These so-called *infeasible-interior-point methods* are simple and much used in practice, although their analysis is hard. Polynomial-time complexity for linear programming was established by Zhang (1994) and Mizuno (1994). The other possibility to deal with infeasible iterates is to use a self-dual embedding: see the references in Section 2.3.

There is clearly a strong similarity between the conditions $(PQPOC)$ and $(DQPOC)$, but they will only define the same directions $(\overline{\Delta x} = \Delta x$ and

$(\overline{\Delta y}, \overline{\Delta s}) = (\Delta y, \Delta s))$ under rather strong hypotheses, for example, if

$$t_+^{-1}\nabla^2 F_*(u) = [t_+^{-1}\nabla^2 F(v)]^{-1}, \qquad (3.7)$$

$$t_+^{-1}\nabla^2 F(v)(-x - t_+^{-1}\nabla F_*(s)) = -s - t_+^{-1}\nabla F(x). \qquad (3.8)$$

Using Proposition 3.2, this holds if $v = x = x_*(t_+)$ and $u = s = s_*(t_+)$, but in this case all the directions are zero and it is pointless to solve the systems! (It also holds if $(t_+/t)^{1/2}v = x = x_*(t)$ and $(t_+/t)^{1/2}u = s = s_*(t)$, again a very special situation.) In the next subsection, we will describe situations where the equations above hold for any x and s by suitable choice of u and v.

The solution to $(PQPOC)$ can be obtained by solving for $\overline{\Delta s}$ in terms of $\overline{\Delta y}$ and then Δx in terms of $\overline{\Delta s}$. Substituting in the equation $A\Delta x = 0$, we see that we need to solve

$$(A[\nabla^2 F(v)]^{-1}A^T)\overline{\Delta y} = A[\nabla^2 F(v)]^{-1}(s + t_+^{-1}\nabla F(x). \qquad (3.9)$$

Let us examine the form of these equations in the cases of linear and semidefinite programming. (The analysis for the second-order cone is also straightforward, but the formulae are rather cumbersome.) In the first case, $\nabla^2 F(v)$ for the usual log barrier function becomes $[\text{Diag}(v)]^{-2}$ and $\nabla F(x)$ becomes $-[\text{Diag}(x)]^{-1}e$, with e a vector of ones. Hence (3.9) can be written

$$(A[\text{Diag}(v)]^2 A^T)\overline{\Delta y} = A[\text{Diag}(v)]^2 s - t_+^{-1}A[\text{Diag}(v)]^2[\text{Diag}(x)]^{-1}e.$$

In the large sparse case, the coefficient matrix in the equation above can be formed fairly cheaply and usually retains some of the sparsity of A; its Cholesky factorization can be obtained somewhat cheaply. The typically very low number of iterations required then compensates to a large extent for the iterations being considerably more expensive than pivots in the simplex method. (Indeed, for the primal–dual algorithms of the next subsection, 10 to 50 iterations almost always provide 8 digits of accuracy, even for very large LP problems.)

In the case of semidefinite programming, A can be thought of as an operator from symmetric matrices to $\mathbb{R}^m$ and A^T as the adjoint operator from $\mathbb{R}^m$ to the space of symmetric matrices; see the discussion preceding (3.3). With the usual log determinant barrier function, $\nabla^2 F(V)$ maps a symmetric matrix Z to $V^{-1}ZV^{-1}$ and $\nabla F(X)$ is $-X^{-1}$, so (3.9) becomes

$$A_i \bullet \sum_j (VA_jV)\overline{\Delta y}_j = A_i \bullet (VSV - t_+^{-1}VX^{-1}V), \quad i = 1, \ldots, m.$$

If we take $V = X$, as seems natural, then there is a large cost in even forming the $m \times m$ matrix with ijth entry $A_i \bullet (XA_jX)$: the A_is may well be sparse, but X is frequently not, and then we must compute the Cholesky factorization of the resulting usually dense matrix.

Let us return to the general case. Computing Δx in this way using $v = x$

gives the primal path-following algorithm. We could also use $\overline{\Delta y}$ and $\overline{\Delta s}$ to update the dual solution, but it is easily seen that in fact Δx is independent of y and s as long as $A^T y + s = c$, so that the 'true' iterates are in x-space. However, updating the dual solution (if feasible) does give an easy way to determine the quality of the primal points generated. If the Newton decrement for t (i.e., $\sqrt{\Delta x^T \nabla^2 F(x) \Delta x}$, where Δx is computed with $t_+ = t$) is small, then updating t_+ as in (2.14) and then using a damped Newton step will yield an updated primal point x_+ at which the Newton decrement for t_+ is also small (and the updated dual solution will be feasible). In practice, heuristics may be used to choose much longer steps and accept points whose Newton decrement is much larger.

A similar analysis for $(DQPOC)$ leads to the equations

$$(A\nabla^2 F_*(u) A^T)\Delta y = t_+ A x + A\nabla F_*(s).$$

Here it is more apparent that the dual direction $(\Delta y, \Delta s)$ is independent of x as long as $Ax = b$, so this is a pure dual path-following method, although again primal iterates can be carried along to assess the quality of the dual iterates. In the case of linear programming, the coefficient matrix takes the form $A[\mathrm{Diag}(u)]^{-2} A^T$, while for semidefinite programming it becomes $(A_i \bullet (U^{-1} A_j U^{-1}))_{i,j=1}^m$.

In the next subsection we consider the case that leads to a symmetric primal–dual path-following algorithm. This requires the notion of self-scaled barrier introduced by Nesterov; further details can be found in Nesterov and Todd (1997, 1998).

3.3. Self-scaled barriers and cones and symmetric primal–dual algorithms

Let us now consider barriers that satisfy a further property: a ϑ-LHSCB F for $\mathbf{K}$ is called a ϑ-self-scaled barrier (ϑ-SSB) if, for all $v \in \mathrm{int}\,\mathbf{K}$, $\nabla^2 F(v)$ maps $\mathrm{int}\,\mathbf{K}$ to $\mathrm{int}\,\mathbf{K}_*$ and

$$(\forall v, x \in \mathrm{int}\,\mathbf{K})\ F_*(\nabla^2 F(v)x) = F(x) - 2F(v) - \vartheta. \qquad (3.10)$$

If a cone admits such an SSB, we call it a self-scaled cone. It is easy to check that the three barriers we introduced above for the non-negative, Lorentz, and semidefinite cones are all self-scaled, and so these cones are self-scaled. Moreover, in these examples, v can be chosen so that $\nabla^2 F(v)$ is the identity, so (as we saw) F_* differs from F by a constant.

The condition above implies many other strong properties: the dual barrier F_* is also self-scaled; for all $v \in \mathrm{int}\,\mathbf{K}$, $\nabla^2 F(v)$ maps $\mathrm{int}\,\mathbf{K}$ *onto* $\mathrm{int}\,\mathbf{K}_*$; and we have the following result.

Theorem 3.2. If F is a ϑ-SSB for $\mathbf{K}$, then for every $x \in \mathrm{int}\,\mathbf{K}$ and $s \in \mathrm{int}\,\mathbf{K}_*$, there is a unique $w \in \mathrm{int}\,\mathbf{K}$ such that

$$\nabla^2 F(w)x = s.$$

Moreover,

$$\nabla^2 F(w)\nabla F_*(s) = \nabla F(x) \quad \text{and} \quad \nabla^2 F(w)\nabla^2 F_*(s)\nabla^2 F(w) = \nabla^2 F(x).$$

We call w the *scaling point* for x and s. Clearly, $-\nabla F(w)$ is the scaling point (using F_*) for s and x. Tunçel (1998) found a more symmetric form of the equation (3.10) defining self-scaled barriers: if $\nabla^2 F(v)x = -\nabla F(z)$ for $v, x, z \in \text{int } \mathbf{K}$, then $F(v) = (F(x) + F(z))/2$, and by the result above we also have $\nabla^2 F(v)z = -\nabla F(x)$.

The properties above imply that the cone $\mathbf{K}$ is *symmetric*: it is *self-dual*, since $\mathbf{K}$ and $\mathbf{K}_*$ are isomorphic by the non-singular linear mapping $\nabla^2 F(v)$ for any $v \in \text{int } \mathbf{K}$; and it is *homogeneous*, since there is an automorphism of $\mathbf{K}$ taking any point x_1 of int $\mathbf{K}$ into any other such point x_2. Indeed, we can choose the automorphism $[\nabla^2 F(w_2)]^{-1}\nabla^2 F(w_1)$, where w_i is the scaling point for x_i and some fixed $s \in \text{int } \mathbf{K}_*$, $i = 1, 2$. Symmetric cones have been much studied and even characterized: see the comprehensive book of Faraut and Koranyi (1994). They also coincide with cones of squares in Euclidean Jordan algebras. These connections were established by Güler (1996). Because of this connection, we know that self-scaled cones do not extend far beyond the cones we have considered: non-negative, Lorentz, and semidefinite cones, and Cartesian products of these.

Let us now return to the conditions (3.7) and (3.8) for $(PQPOC)$ and $(DQPOC)$ to define identical directions. If we set $\bar{u} := t_+^{1/2}u$ and $\bar{v} := t_+^{1/2}v$, these can be rewritten as

$$\nabla^2 F_*(\bar{u}) = [\nabla^2 F(\bar{v})]^{-1}, \quad \nabla^2 F(\bar{v})(-x - t_+^{-1}\nabla F_*(s)) = -s - t_+^{-1}\nabla F(x).$$

When F is self-scaled, these conditions can be satisfied by setting $\bar{v}$ to be the scaling point for x and s, and $\bar{u}$ (equal to $-\nabla F(\bar{v})$) to be the scaling point (for F_*) for s and x. (Notice that, if $(x, s) = (x_*(t), s_*(t))$, then these scaling points are $t^{-1/2}x$ and $t^{-1/2}s$ respectively, and, except for a scalar multiple, we come back to the primal (or dual) direction.)

Let us describe the resulting symmetric primal–dual short-step path-following algorithm. We need a symmetric measure of proximity to the central path. Hence, for x and (y, s) strictly feasible solutions to (P) and (D), define

$$t := t(x, s) := \frac{\vartheta}{s^T x} \quad \text{and} \quad \lambda_2(x, s) := \|ts + \nabla F(x)\|_{F,x}.$$

It can be shown (Nesterov and Todd 1998, Section 3) that $\lambda_2(x, s) = \|tx + \nabla F_*(s)\|_{F_*,s}$ also. Suppose x and s are such that

$$\lambda_2(x, s) \le 0.1,$$

and we choose

$$t_+ := (1 + 0.06\vartheta^{-1/2})t.$$

We compute the scaling point w for x and s, and let Δx, Δy, and Δs be the solution to $(PQPOC)$ with $v := t_+^{-1/2} w$ (or equivalently to $(DQPOC)$ with $u := -t_+^{-1/2} \nabla F(w)$). Finally, we set $x_+ := x + \Delta x$ and $(y_+, s_+) := (y + \Delta y, s + \Delta s)$. It can be shown (Nesterov and Todd 1998, Section 6) that

$$t(x_+, s_+) = t_+ \quad \text{and} \quad \lambda_2(x_+, s_+) \le 0.1,$$

so we can continue the process.

Theorem 3.3. Suppose (P) and (D) have strictly feasible solutions, and we have a ϑ-SSB F for $\mathbf{K}$. Suppose further we have a strictly feasible pair x_0, (y_0, s_0) for (P) and (D) with $\lambda_2(x_0, s_0) \le 0.1$. Then the algorithm described above (with x_{k+1} and (y_{k+1}, s_{k+1}) derived from x_k and (y_k, s_k) as are x_+ and (y_+, s_+) from x and (y, s)) is well-defined (all iterates are strictly feasible), maintains closeness to the path ($\lambda_2(x_k, s_k) \le 0.1$ for all k) and has the efficiency estimate

$$c^T x_k - b^T y_k = s_k^T x_k = \frac{\vartheta}{t_k} \le s_0^T x_0 \exp\left\{ -\frac{0.05}{\sqrt{\vartheta}} k \right\}.$$

Hence, for every $\epsilon > 0$, it takes at most

$$O(1) \sqrt{\vartheta} \ln\left(\frac{s_0^T x_0}{\epsilon} \right)$$

iterations to obtain strictly feasible solutions with duality gap at most ϵ.

Thus we have obtained an algorithm with complexity bounds of the same order as those for the primal path-following method in Theorem 2.1. In fact, the constants are a little worse than those for the primal method. However, it is important to realize that these are worst-case bounds, and that the primal–dual framework is much more conducive to allowing adaptive algorithms that can give much better results in practice: see, *e.g.*, Algorithms 6.2 and 6.3 in Nesterov and Todd (1998). Part of the reason that long-step algorithms are possible in this context is that approximations of F and of $\nabla^2 F$ hold for much larger perturbations of a point $x \in \text{int } \mathbf{K}$. Indeed, results like (2.5) hold true for any perturbation h with $x \pm h \in \text{int } \mathbf{K}$: see Theorems 4.1 and 4.2 of Nesterov and Todd (1997).

There are also symmetric primal–dual potential-reduction algorithms, using the Tanabe–Todd–Ye function (2.21). Note that

$$\nabla_x p(x, s) = \frac{\vartheta + \sqrt{\vartheta}}{s^T x} s + \nabla F(x), \quad \nabla_s p(x, s) = \frac{\vartheta + \sqrt{\vartheta}}{s^T x} x + \nabla F_*(s),$$

and the coefficient of s (or x) is $t_+ := (1 + 1/\sqrt{\vartheta}) t(x, s)$. Thus Newton-like steps to decrease the potential function (where the Hessian is replaced by $t_+\nabla^2 F(w))$ lead to exactly the same search directions as in the path-following algorithm above. Performing a line search on p in those directions leads to a guaranteed decrease of at least 0.24 (for details, see Section 8 of Nesterov and Todd (1997)), and again, this leads to an $O(\sqrt{\vartheta}\ln(s_0^T x_0/\epsilon))$-iteration algorithm from a well-centred initial pair to achieve an ϵ-optimal pair. The big advantage is that now there is no necessity to stay close to the central path, and indeed, the initial pair does not have to be well-centred – the only change is that the complexity bound is modified appropriately.

We now discuss how the scaling point w for x and s can be computed in the case of the non-negative orthant and the semidefinite cone; for the Lorentz cone, the computation is again straightforward but cumbersome. For the non-negative orthant $\mathbb{R}_+^n$, we have $\nabla^2 F(w) = [\mathrm{Diag}(w)]^{-2}$, so we find the scaling point w for positive vectors x and s is given by

$$w = \left(\sqrt{x_j/s_j}\right)_{j=1}^n,$$

so that the equation to be solved for Δy is

$$A\,\mathrm{Diag}(x)\,[\mathrm{Diag}(s)]^{-1}A^T\Delta y = A(x - t_+^{-1}[\mathrm{Diag}(s)]^{-1}e), \qquad (3.11)$$

leading to the usual LP primal–dual symmetric search direction. The computation required is of the same order as that for the primal or dual methods.

For the semidefinite cone $\mathbf{S}_+^p$, the defining relation $\nabla^2 F(w)x = s$ becomes $W^{-1}XW^{-1} = S$, or $WSW = X$, for positive definite X and S, from which we find

$$W = S^{-1/2}(S^{1/2}XS^{1/2})^{1/2}S^{-1/2},$$

where $V^{1/2}$ denotes the positive semidefinite square root of a positive semidefinite matrix V. Todd, Toh and Tütüncü (1998) show that W can be computed using two Cholesky factorizations ($X = L_X L_X^T$ and $S = R_S R_S^T$) and one eigenvalue (of $L_X^T S L_X$) or singular value (of $R_S^T L_X$) decomposition. (After W is obtained, Δy (and hence ΔS and ΔX) can be computed using a system like that for the primal or dual barrier method, but with W replacing V or U^{-1}.)

The need for an eigenvalue or singular value decomposition makes each iteration of a (path-following or potential-reduction) interior-point algorithm using the scaling point W quite expensive. While linear and second-order cone programming problems with hundreds of thousands of variables and constraints (with favourable sparsity patterns) can be solved in under 5 minutes on a fairly modest PC, semidefinite programming problems with matrices of order a thousand, even with very favourable structure, can take up to half an hour. When the matrices are of order two thousand, the times increase to an hour even for the simplest such problems.

Alternative methods greatly improve the computational time per iteration. The Jordan algebra approach (Faybusovich 1997, Schmieta and

Alizadeh 2001) replaces the last equation in (3.6) by one exhibiting more primal–dual symmetry. For linear programming, this is $x \circ s = t^{-1}e$, where $\circ$ denotes the Hadamard or componentwise product. A Newton step for this leads to the same direction as the self-scaled method. For semidefinite programming, it gives $XS + SX = 2t^{-1}I$. Unfortunately, linearizing this equation to get

$$(\Delta XS + S\Delta X) + (X\Delta S + \Delta SX) = -XS - SX + 2t_+^{-1}I,$$

as proposed by Alizadeh, Haeberly and Overton (1997), leads to a system that requires even more computation than the self-scaled approach, and does not enjoy scale-invariance properties (see, *e.g.*, Todd *et al.* (1998)). Suppose instead the iterates are first scaled (X by pre- and postmultiplying by $S^{1/2}$, and S by pre- and postmultiplying by $S^{-1/2}$) so that the current iterates are transformed into $\tilde{X} = S^{1/2}XS^{1/2}$ and $\tilde{S} = S^{-1/2}SS^{-1/2} = I$. If the Alizadeh–Haeberly–Overton approach is followed in the transformed space, the linearization becomes

$$2\Delta\tilde{X} + (\tilde{X}\Delta\tilde{S} + \Delta\tilde{S}\tilde{X}) = -2\tilde{X} + 2t_+^{-1}I,$$

or in terms of the original variables after transforming back,

$$\Delta X + \frac{1}{2}(X\Delta SS^{-1} + S^{-1}\Delta SX) = -X + t_+^{-1}S^{-1}.$$

Then the search directions can be obtained after solving the $m \times m$ system

$$A_i \bullet \left(\sum_j (XA_jS^{-1})\Delta y_j\right) = A_i \bullet (X - t_+^{-1}S^{-1}), \quad i = 1,\ldots,m.$$

This method was developed independently by Helmberg, Rendl, Vanderbei and Wolkowicz (1996) and Kojima, Shindoh and Hara (1997), and later derived from a different viewpoint by Monteiro (1997). This approach permits the solution of certain problems with matrices of order two thousand (and favourable structure) in under twenty minutes. A pure dual barrier method can also be used successfully on problems of this size with even faster results, but on some problems it seems not as successful as primal–dual methods.

For truly large semidefinite programming problems, either non-interior-point methods need to be used (see, *e.g.*, Section 6.3 in Todd (2001)), or iterative techniques employed to solve approximately the linear systems arising at each iteration (see, *e.g.*, Toh (2007) and Chai and Toh (2007)). For more information on semidefinite programming, the reader can consult Helmberg's page www-user.tu-chemnitz.de/~helmberg/semidef.html; software for linear, second-order cone, and semidefinite programming can be found at the NEOS solvers site neos.mcs.anl.gov/neos/solvers/index.html.

3.4. Recent developments

In this final subsection, we describe some recent developments in interior-point methods for conic optimization. We concentrate on classes of cones that are more general than self-scaled cones, but that have some structure that may help in developing efficient interior-point algorithms.

The first class of such cones consists of *hyperbolicity cones*. These cones arise in connection with hyperbolic polynomials: a homogeneous polynomial p on $\mathbb{R}^n$ is hyperbolic in direction $d \in \mathbb{R}^n$ if the univariate polynomial $t \mapsto p(x - td)$ has only real roots for every $x \in \mathbb{R}^n$. The associated hyperbolicity cone $\mathbf{K}(p, d)$ is the set of those x for which all these roots are non-negative. These objects were first studied in the context of PDEs, but were introduced to the interior-point community by Güler (1997) because of their generality and nice properties.

The polynomial $p(x) = x_1 x_2 \cdots x_n$ is hyperbolic in direction d for any positive vector $d \in \mathbb{R}^n$, and the associated hyperbolicity cone is the non-negative orthant. The Lorentz cone arises from $x_1^2 - \sum_{j=2}^n x_j^2$, hyperbolic in the direction $d = (1, 0, \ldots, 0)^T$. Finally, if $n = p(p+1)/2$ and we associate $\mathbb{R}^n$ with $\mathbf{S}^p$, the polynomial $\det(X)$ is hyperbolic in the direction of the identity and gives rise to the semidefinite cone. However, the range of hyperbolicity cones is much larger: Güler (1997) shows, for example, that it includes (properly) all homogeneous cones.

The significance of this class of cones for interior-point methods is that $F(x) := -\ln p(x)$ is an m-LHSCB for the cone $\mathbf{K}(p, d)$, where m is the degree of homogeneity of p. This function has very good properties: for any x, $\nabla^2 F(x)$ takes int $\mathbf{K}(p, d)$ into (but not necessarily onto) the interior of its dual cone; there is a unique scaling point for each $x \in$ int $\mathbf{K}(p, d)$ and s in the interior of its dual; and F has good 'long-step properties' like those hinted at below Theorem 3.3 for self-scaled barriers. These results were obtained by Güler (1997), who showed that long-step primal potential-reduction algorithms could be extended from self-scaled cones to hyperbolicity cones. However, the dual barrier of a hyperbolic barrier of this kind is itself a hyperbolic barrier only if the original barrier was self-scaled. Hence it seems unlikely that the primal–dual methods of the previous subsection can be extended to general hyperbolicity cones.

Bauschke, Güler, Lewis and Sendov (2001) study hyperbolic polynomials from the viewpoint of convex analysis and hence rederive some of Güler's results. Of more interest in optimization, Renegar (2006) makes use of an important property of hyperbolic polynomials, namely, that if p is hyperbolic in direction d, then so is the directional derivative $d^T \nabla p$, and the hyperbolicity cone of the latter contains that of p. In this way a hierarchy of relaxations of a hyperbolicity cone programming problem can be defined; Renegar suggests a homotopy method to solve the original problem by considering solutions to these relaxed problems. At present, there is no complexity analysis for this approach, but it seems promising.

The second class of cones we wish to mention arises in global polynomial optimization: that is, one seeks the global minimizer p_* of a polynomial function p of n variables, possibly subject to polynomial inequality constraints. Here the functions involved need not be convex, and the problem is NP-hard even for degree-four polynomials, but we would still like to be able to solve (even approximately) small-scale problems. We describe here briefly an approach, introduced by Parrilo (2003) and Lasserre (2001), that uses semidefinite programming problems as approximations.

Let us follow Lasserre in describing a convex formulation of such a polynomial optimization problem. Suppose p is a polynomial of degree $2m$ in n variables. Using the notation $x^\alpha := x_1^{\alpha_1} \cdots x_n^{\alpha_n}$ and $|\alpha| := \sum_j \alpha_j$, where α is a non-negative integer n-vector, we can associate p with its vector of coefficients $(p_\alpha)_{|\alpha| \leq 2m}$, where

$$p(x) = \sum_{|\alpha| \leq 2m} p_\alpha x^\alpha.$$

The key idea is to replace an optimization problem over the n-vector x with one over probability measures μ on $\mathbb{R}^n$. Then minimizing p over $\mathbb{R}^n$ can be replaced by minimizing $\int p(x)\,\mathrm{d}\mu(x)$, which is a convex (even linear!) function of the infinite-dimensional variable μ. Moreover, since p is a polynomial, we have

$$\int p(x)\,\mathrm{d}\mu(x) = \sum_{|\alpha| \leq 2m} p_\alpha y_\alpha,$$

where y_α is the α-moment of μ, $\int x^\alpha \,\mathrm{d}\mu(x)$. We now have a linear optimization problem over the finite-dimensional vector $(y_\alpha)_{|\alpha| \leq 2m}$, with the constraint that this vector be the vector of moments of some probability measure. The constraint can be separated: we need y to be the vector of moments of a Borel measure (this defines a convex cone, the *moment cone*), and $y_0 = 1$ (this requires the measure to be a probability measure).

Unfortunately (as we would expect from the NP-hardness result), this convex cone is hard to deal with: in particular, it is very unlikely that a computationally tractable barrier function for it exists. We would therefore like to approximate it. Here is one necessary condition, based on a large matrix whose entries are the components of y. Let us enumerate all $\binom{m+n}{n}$ monomials x^β with $|\beta| \leq m$ and use them to index the rows and columns of a matrix. Let $M_m(y)$ denote the symmetric matrix whose entry in the row corresponding to x^β and column corresponding to x^γ is $y_{\beta+\gamma}$. Then

$$M_m(y) \succeq 0.$$

Indeed, if $(q_\alpha)_{|\alpha| \leq m}$ is the vector of coefficients of a polynomial $q(x)$ of degree m, then $q^T M_m(y)q$ is $\int (q(x))^2 \, d\mu(x)$, which is non-negative. We can then minimize the linear function $\sum p_\alpha y_\alpha$ subject to $y_0 = 1$ and this semidefinite constraint. This is a relaxation of the original polynomial optimization problem and will provide a lower bound.

It turns out that this lower bound is tight exactly when $p(x) - p_*$ (a polynomial that is non-negative everywhere) can be written as a sum of squares. Indeed, finding the smallest $\bar{p}$ such that $p(x) - \bar{p}$ is a sum of squares can be formulated as a semidefinite programming problem, and it is precisely the dual of the problem above. The complication is that, except in very special cases, the set of non-negative polynomials is larger than the set of sums of squares (this is related to Hilbert's 17th problem), but there are results in semi-algebraic geometry that provide ways to attack the problem. Without going into details, we merely note that a sequence of semidefinite programming problems can be formulated, whose optimal values approach p_*, and frequently the value is attained in a finite (and small) number of steps. The disadvantage is that the sizes of these semidefinite problems grow very fast, so that only small-scale problems can be solved. Lasserre (2001) gives results for (constrained) problems with degree up to four and up to 10 variables; Parrilo and Sturmfels (2003) solve (unconstrained) degree four problems in 13 variables and degree six problems in 7 variables in under half an hour. A MATLAB package for solving sum of squares optimization problems using semidefinite programming is available at www.cds.caltech.edu/sostools/.

We described above the polynomial minimization problem as that of minimizing $\sum_{|\alpha| \leq 2m} p_\alpha y_\alpha$, subject to $y_0 = 1$ and $(y_\alpha)_{|\alpha| \leq 2m}$ belonging to the cone of moments (up to degree $2m$) of a Borel measure. It is not hard to see that the corresponding dual cone consists of the coefficients $(q_\alpha)_{|\alpha| \leq 2m}$ of polynomials q of degree at most $2m$ that are non-negative everywhere. These are two dual convex cones, easy to describe, but hard to deal with computationally, that are important in applications. Another such pair of cones arises in copositive programming.

Suppose we wish to minimize the quadratic function $x^T Q x$ over the standard simplex $\{x \in \mathbb{R}^n_+ : e^T x = 1\}$, where $e \in \mathbb{R}^n$ is the vector of ones. This *standard quadratic programming problem* includes the problem of computing a maximum stable set in a graph and can arise in general quadratic optimization as a test for global optimality (see Bomze (1998)). In fact, the standard quadratic programming problem can be written as the conic optimization problem of minimizing $Q \bullet X$ subject to $E \bullet X = 1$ and X lying in the cone of *completely positive* symmetric matrices: those that can be written as JJ^T for a non-negative (entrywise) matrix J. Here $E := ee^T$, the $n \times n$ matrix of ones. This equivalence can be seen by characterizing the extreme solutions of the latter problem, as in Bomze, de Klerk, Roos, Quist and Terlaky (2000). The dual of the completely positive cone is easily shown to be the cone of copositive matrices, *i.e.*, those that are positive semidefinite on the non-negative orthant. In turn, these are related to non-negative quartics: P is copositive if and only if the quartic $\sum_{i,j} P_{ij} z_i^2 z_j^2$ is everywhere non-negative. Hence copositive programming (and so also the standard quadratic programming problem) can be attacked using the techniques discussed above, introduced by Parrilo and Lasserre. This is a topic of considerable recent interest: see Bomze and de Klerk (2002) and the references therein.

4. IPMs for non-convex programming

In this short final section, we sketch the algorithms that have been proposed for general, not necessarily convex, nonlinear programming. For further details, see the survey papers of Forsgren *et al.* (2002) and Gould *et al.* (2005); the issues are also nicely treated in Nocedal and Wright (2006). These methods were inspired by the great success of interior-point methods for specially structured convex problems, and differ in many respects from the earlier barrier methods of the 1960s and 1970s. However, since they are designed for general problems, the motivating concerns are very different from those for convex optimization: global convergence (possibly to an infeasible point which is a local minimizer of some measure of infeasibility) replaces complexity analysis; superlinear convergence, and the resulting careful control of the parameter t, is of considerable interest; step-size control usually involves a merit function; and modifications to Newton systems are often employed to avoid convergence to stationary points that are not local minimizers. There are two families of interior-point methods for nonlinear programming: those based on line searches and those based on trust regions. Here we restrict ourselves to line-search methods as they are closer to what we have discussed for convex problems.

For simplicity, we concentrate on the inequality-constrained problem

$$(NLP) \quad \min f(y), \quad g(y) \le 0,$$

where $f : \mathbb{R}^m \to \mathbb{R}$ and $g : \mathbb{R}^m \to \mathbb{R}^n$ are twice continuously differentiable functions. Other forms of problem are discussed by many of the authors of the papers cited below, but the main ideas can be illustrated in this framework. The somewhat unconventional notation is chosen to facilitate comparison with the dual linear programming problem, where $f(y) = -b^T y$ and $g(y) = A^T y - c$.

The first step is to introduce slack variables to convert the inequality constraints to the form $g(y) + s = 0$, $s \ge 0$. A barrier method then tries to find approximate solutions to problems of the form

$$(NLB_t) \quad \min tf(y) - \sum_j \ln s_j, \quad g(y) + s = 0 \quad (s > 0),$$

for positive parameters t increasing to ∞. If we associate Lagrange multipliers $\lambda \in \mathbb{R}^n$ to the constraints, and then define $x := t^{-1}\lambda$, we find that the optimality conditions for (NLB_t) can be written as

$$
\begin{aligned}
\nabla f(y) \;+\; \nabla g(y)x \qquad\qquad &= 0, \\
g(y) \qquad\qquad\qquad\; + s &= 0, \qquad\qquad (4.1) \\
\mathrm{Diag}(x)\mathrm{Diag}(s)e \qquad &= t^{-1}e.
\end{aligned}
$$

Given a trial solution (y, x, s) with x and s positive, a Newton step towards a solution of (4.1) will move in the direction $(\Delta y, \Delta x, \Delta s)$ satisfying

$$
\begin{bmatrix}
K & \nabla g(y) & 0 \\
\nabla g(y)^T & 0 & I \\
0 & \mathrm{Diag}(s) & \mathrm{Diag}(x)
\end{bmatrix}
\begin{pmatrix}
\Delta y \\ \Delta x \\ \Delta s
\end{pmatrix}
=
\begin{pmatrix}
-\nabla f(y) - \nabla g(y)x \\
-g(y) - s \\
t^{-1}e - \mathrm{Diag}(x)\mathrm{Diag}(s)e
\end{pmatrix},
$$

$$(4.2)$$

where K denotes the Hessian of the Lagrangian function $L(y, x, s) := f(y) + x^T(g(y) + s)$ with respect to y. Using the last set of equations to solve for Δs, we arrive at

$$
\begin{bmatrix}
K & \nabla g(y) \\
\nabla g(y)^T & -[\mathrm{Diag}(x)]^{-1}\mathrm{Diag}(s)
\end{bmatrix}
\begin{pmatrix}
\Delta y \\ \Delta x
\end{pmatrix}
=
\begin{pmatrix}
-\nabla f(y) - \nabla g(y)x \\
-g(y) - t^{-1}[\mathrm{Diag}(x)]^{-1}e
\end{pmatrix};
$$

$$(4.3)$$

if we further eliminate Δx, we reach

$$
[K + \nabla g(y)\,\mathrm{Diag}(x)[\mathrm{Diag}(s)]^{-1}\nabla g(y)^T]\Delta y
$$
$$
= -\nabla f(y) - \nabla g(y)[x + t^{-1}s + \mathrm{Diag}(x)[\mathrm{Diag}(s)]^{-1}g(y)]. \qquad (4.4)
$$

This reduces to the primal–dual system (3.11) when (NLP) reduces to the linear programming problem $\min\{-b^T y : A^T y - c \leq 0\}$ and when $s = c - A^T y$.

Primal–dual line-search methods start by solving one of the three linear systems above. If the coefficient matrix in (4.4) is positive definite (this is guaranteed when sufficiently close to a local minimizer of (NLP) satisfying the strong second-order sufficient conditions), the resulting solution $(\Delta y, \Delta x, \Delta s)$ is taken as the search direction. Otherwise, most methods modify the system in some way: either a multiple of the identity matrix of order m is added to K, or a multiple of the identity matrix of order n is subtracted from the $(2, 2)$ block in (4.3), for example. The resulting direction Δy can then be shown to be a descent direction for a merit function such as

$$
f(y) - t^{-1}\sum_j \ln s_j + \rho\|g(y) + s\|, \qquad (4.5)
$$

possibly after increasing the positive penalty parameter ρ. A step is then taken along the direction $(\Delta y, \Delta x, \Delta s)$ to ensure 'sufficient' decrease in the merit function.

In feasible methods (called quasi-feasible methods if there are also equality constraints present that may not be satisfied exactly), s is reset after each iteration to $-g(y)$, so that $s > 0$ forces $g(y) < 0$ for all iterates. This requirement complicates and restricts the line search, but can avoid some undesirable convergence behaviour. Such methods include those of Gay, Overton and Wright (1998), Forsgren *et al.* (2002) (except in their Section 6.4), and the quasi-feasible method of Chen and Goldfarb (2006). The more common infeasible methods allow $g(y) + s$ to be non-zero, and control it implicitly through the merit function: see, *e.g.*, Vanderbei and Shanno (1999) (LOQO), Waltz, Morales, Nocedal and Orban (2006) (KNITRO/DIRECT), Wächter and Biegler (2006) (IPOPT), and the infeasible method of Chen and Goldfarb (2006). Important practical issues such as how the linear systems are modified and solved, how the line searches are performed, how the parameter t is adjusted, and what – if any – back-up techniques are employed if poor convergence is observed, are discussed further in these papers. For example, KNITRO (Byrd *et al.* 2006) reverts to a trust-region interior-point subproblem to ensure global convergence if negative curvature or slow convergence is detected, and IPOPT uses a filter approach instead of a traditional line search with a merit function, and also includes a feasibility restoration phase. Chen and Goldfarb (2006) modify the $(2, 2)$-block of (4.3) to correspond to the Newton system for moving to a local minimizer of the merit function in (4.5) and may also modify the $(1, 1)$-block; they prove strong global convergence properties for both quasi-feasible and infeasible algorithms.

Overall, these methods have proved strongly competitive for general nonlinear programming problems, and research remains very active. Our treatment has only scratched the surface; for details, consult the references cited and the comprehensive survey articles of Forsgren *et al.* (2002) and Gould *et al.* (2005). The software systems mentioned are available (sometimes free) from the NEOS solvers website neos.mcs.anl.gov/neos/solvers/index.html.

5. Summary

Interior-point methods have changed the way we look at optimization problems over the last twenty years. In this paper we have concentrated on convex problems, and in particular on the classes of structured convex problems for which interior-point methods provide provably efficient algorithms. We have highlighted the theory and motivation for these methods and their domains of applicability, and also pointed out new topics of research. Finally, we have sketched very briefly interior-point methods for general nonlinear programming.

Since the field is so active, we conclude by pointing out once more some sources for tracking current research and algorithms: the websites for Optimization Online at www.optimization-online.org/ and for the NEOS solvers at neos.mcs.anl.gov/neos/solvers/index.html, and, for semidefinite programming, Helmberg's page at www-user.tu-chemnitz.de/~helmberg/semidef.html.

Acknowledgements

We are very grateful to Renato Monteiro, Jorge Nocedal and Jim Renegar for help with this paper. The second author was partially supported by NSF through grant DMS-0513337 and ONR through grant N00014-02-1-0057.

REFERENCES

F. Alizadeh (1995), 'Interior point methods in semidefinite programming with applications to combinatorial optimization', *SIAM J. Optim.* **5**, 13–51.

F. Alizadeh, J.-P. A. Haeberly and M. L. Overton (1998), 'Primal–dual interior-point methods for semidefinite programming: Convergence rates, stability and numerical results', *SIAM J. Optim.* **8**, 746–768.

E. D. Andersen and Y. Ye (1999), 'On a homogeneous algorithm for monotone complementarity system', *Math. Program.* **84**, 375–399.

H. H. Bauschke, O. Güler, A. S. Lewis and H. S. Sendov (2001), 'Hyperbolic polynomials and convex analysis', *Canad. J. Math.* **53**, 470–488.

A. Ben-Tal and A. S. Nemirovski (2001), *Lectures on Modern Convex Optimization: Analysis, Algorithms, and Engineering Applications*, SIAM, Philadelphia.

R. E. Bixby (2002), 'Solving real-world linear programs: A decade and more of progress', *Oper. Res.* **50**, 3–15.

I. M. Bomze (1988), 'On standard quadratic optimization problems', *J. Global Optim.* **13**, 369–387.

I. M. Bomze and E. de Klerk (2002), 'Solving standard quadratic optimization problems via semidefinite and copositive programming', *J. Global Optim.* **24**, 163–185.

I. M. Bomze, M. Dür, E. de Klerk, C. Roos, A. J. Quist and T. Terlaky (2000), 'On copositive programming and standard quadratic optimization problems', *J. Global Optim.* **18**, 301–320.

S. Boyd and L. Vandenberghe (2004), *Convex Optimization*, Cambridge University Press.

S. Boyd, L. El Ghaoui, E. Feron and V. Balakrishnan (1994), *Linear Matrix Inequalities in System and Control Theory*, Vol. 15 of *Studies in Applied Mathematics*, SIAM.

R. H. Byrd, J. Nocedal and R. A. Waltz (2006), KNITRO: An integrated package for nonlinear optimization. In *Large-Scale Nonlinear Optimization* (G. di Pillo and M. Roma, eds), Springer, New York, pp. 35–59.

C. W. Carroll (1961), 'The created response surface technique for optimizing nonlinear, restrained systems', *Oper. Res.* **9**, 169–185.

J. S. Chai and K. C. Toh (2007), 'Preconditioning and iterative solution of symmetric indefinite linear systems arising from interior point methods for linear programming', *Comput. Optim. Appl.* **36**, 221–247.

L. Chen and D. Goldfarb (2006), 'Interior-point l_2-penalty methods for nonlinear programming with strong convergence properties', *Math. Program.* **108**, 1–36.

R. Courant (1943), 'Variational methods for the solution of problems of equilibrium and vibrations', *Bull. Amer. Math. Soc.* **49**, 1–23.

J. Faraut and A. Koranyi (1994), *Analysis on Symmetric Cones*, Clarendon Press, Oxford.

L. Faybusovich (1997), 'Linear systems in Jordan algebras and primal–dual interior-point algorithms', *J. Comput. Appl. Math.* **86**, 149–175.

A. V. Fiacco and G. P. McCormick (1968), *Nonlinear Programming: Sequential Unconstrained Minimization Techniques*, Wiley.

A. Forsgren, P. E. Gill and M. H. Wright (2002), 'Interior methods for nonlinear optimization', *SIAM Review* **44**, 525–597.

R. W. Freund, F. Jarre and S. Schaible (1996), 'On self-concordant barrier functions for conic hulls and fractional programming', *Math. Program.* **74**, 237–246.

K. R. Frisch (1955), The logarithmic potential method of convex programming. Memorandum of May 13, University Institute of Economics, Oslo, Norway.

D. M. Gay, M. L. Overton and M. H. Wright (1998), A primal–dual interior method for nonconvex nonlinear programming. In *Advances in Nonlinear Programming* (Y. Yuan, ed.), Kluwer, pp. 31–56.

P. E. Gill, W. Murray, M. A. Saunders, J. A. Tomlin and M. H. Wright (1986), 'On projected Newton barrier methods for linear programming and an equivalence to Karmarkar's projective method', *Math. Program.* **36**, 183–209.

M. X. Goemans (1997), 'Semidefinite programming in combinatorial optimization', *Math. Program.* **79**, 143–161.

C. C. Gonzaga (1989), An algorithm for solving linear programming problems in $O(n^3 L)$ operations. In *Progress in Mathematical Programming: Interior Point and Related Methods* (N. Megiddo, ed.), Springer, New York, pp. 1–28.

N. I. M. Gould, D. Orban and P. L. Toint (2005), Numerical methods for large-scale nonlinear optimization. In *Acta Numerica*, Vol. 14, Cambridge University Press, pp. 299–361.

C. Guéret, C. Prins and M. Sevaux (2002), *Applications of Optimization with XpressMP*, Dash Optimization. Translated and revised by Susanne Heipcke.

O. Güler (1996), 'Barrier functions in interior-point methods', *Math. Oper. Res.* **21**, 860–885.

O. Güler (1997), 'Hyperbolic polynomials and interior-point methods for convex programming', *Math. Oper. Res.* **22**, 350–377.

C. Helmberg, F. Rendl, R. Vanderbei and H. Wolkowicz (1996), 'An interior-point method for semidefinite programming', *SIAM J. Optim.* **6**, 342–361.

N. Karmarkar (1984), 'A new polynomial-time algorithm for linear programming', *Combinatorica* 4, 373–395.

E. de Klerk, C. Roos and T. Terlaky (1997), 'Initialization in semidefinite programming via a self-dual skew-symmetric embedding', *Oper. Res. Letters* **20**, 213–221.

M. Kojima, S. Mizuno and A. Yoshise (1989), 'A polynomial–time algorithm for a class of linear complementarity problems', *Math. Program.* **44**, 1–26.

M. Kojima, S. Shindoh and S. Hara (1997), 'Interior-point methods for the monotone semidefinite linear complementarity problem in symmetric matrices', *SIAM J. Optim.* **7**, 86–125.

J. B. Lasserre (2001), 'Global optimization with polynomials and the problem of moments', *SIAM J. Optim.* **11**, 796–817.

A. S. Lewis and M. L. Overton (1996), Eigenvalue optimization. In *Acta Numerica*, Vol. 5, Cambridge University Press, pp. 149–160.

M. S. Lobo, L. Vandenberghe, S. Boyd and H. Lebret (1998), 'Applications of second-order cone programming', *Linear Algebra Appl.* **284**, 193–228.

Z.-Q. Luo, J. F. Sturm and S. Zhang (2000), 'Conic convex programming and self-dual embedding', *Optim. Methods Software* **14**, 169–218.

N. Megiddo (1989), Pathways to the optimal set in linear programming. In *Progress in Mathematical Programming: Interior Point and Related Methods* (N. Megiddo, ed.), Springer, New York, pp. 131–158.

S. Mizuno (1994), 'Polynomiality of infeasible-interior-point algorithms for linear programming', *Math. Program.* **67**, 109–119.

R. D. C. Monteiro (1997), 'Primal–dual path-following algorithms for semidefinite programming', *SIAM J. Optim.* **7**, 663–678.

R. D. C. Monteiro and I. Adler (1989), 'Interior path following primal–dual algorithms I: Linear programming', *Math. Program.* **44**, 27–41.

S. G. Nash (1998), 'SUMT (revisited)', *Oper. Res.* **46**, 763–775.

A. Nemirovski (2004), *Interior Point Polynomial Time Methods in Convex Programming.* Lecture Notes: www2.isye.gatech.edu/~nemirovs/Lect_IPM.pdf.

A. Nemirovski and L. Tunçel (2005), ' " Cone-free" primal–dual path-following and potential-reduction polynomial time interior-point methods', *Math. Program.* **102**, 261–295.

Y. Nesterov (1997), 'Long-step strategies in interior-point primal–dual methods', *Math. Program.* **76**, 47–94.

Y. Nesterov (2003), *Introductory Lectures on Convex Optimization: A Basic Course*, Kluwer, Dordrecht.

Y. Nesterov and A. Nemirovski (1994), *Interior Point Polynomial Time Methods in Convex Programming*, SIAM, Philadelphia.

Y. E. Nesterov and M. J. Todd (1997), 'Self-scaled barriers and interior-point methods for convex programming', *Math. Oper. Res.* **22**, 1–42.

Y. E. Nesterov and M. J. Todd (1998), 'Primal–dual interior-point methods for self-scaled cones', *SIAM J. Optim.* **8**, 324–364.

Y. Nesterov, M. J. Todd and Y. Ye (1999), 'Infeasible-start primal–dual methods and infeasibility detectors for nonlinear programming problems', *Math. Program.* **84**, 227–267.

J. Nocedal and S. J. Wright (2006), *Numerical Optimization*, Springer, New York.

P. A. Parrilo (2003), 'Semidefinite programming relaxations for semialgebraic problems', *Math. Program.* **96**, 293–320.

P. A. Parrilo and B. Sturmfels (2003), Minimizing polynomials. In *Algorithmic and Quantitative Real Algebraic Geometry*, Vol. 60 of *DIMACS Series in Discrete Mathematics and Theoretical Computer Science*, AMS, pp. 83–99.

F. A. Potra and R. Sheng (1998), 'On homogeneous interior-point algorithms for semidefinite programming', *Optim. Methods Software* **9**, 161–184.

J. Renegar (1988), 'A polynomial-time algorithm, based on Newton's method, for linear programming', *Math. Program.* **40**, 59–93.

J. Renegar (2001), *A Mathematical View of Interior-Point Methods in Convex Optimization*, SIAM, Philadelphia.

J. Renegar (2006), 'Hyperbolic programs, and their derivative relaxations', *Found. Comput. Math.* **6**, 59–79.

C. Roos, T. Terlaky and J.-P. Vial (1997), *Theory and Algorithms for Linear Optimization: An Interior Point Approach*, Wiley.

S. H. Schmieta and F. Alizadeh (2001), 'Associative and Jordan algebras, and polynomial-time interior-point algorithms for symmetric cones', *Math. Oper. Res.* **26**, 543–564.

K. Tanabe (1988), Centered Newton method for mathematical programming. In *System Modeling and Optimization*, Springer, New York, pp. 197–206.

M. J. Todd (2001), Semidefinite optimization. In *Acta Numerica*, Vol. 10, Cambridge University Press, pp. 515–560.

M. J. Todd and Y. Ye (1990), 'A centered projective algorithm for linear programming', *Math. Oper. Res.* **15**, 508–529.

M. J. Todd, K.-C. Toh and R. H. Tütüncü (1998), 'On the Nesterov–Todd direction in semidefinite programming', *SIAM J. Optim.* **8**, 769–796.

K. C. Toh (2007), 'An inexact primal–dual path-following algorithm for convex quadratic SDP', *Math. Program.* **112**, 221–254.

L. Tunçel (1998), 'Primal–dual symmetry and scale-invariance of interior-point algorithms for convex programming', *Math. Oper. Res.* **23**, 708–718.

R. J. Vanderbei (2007), *Linear Programming: Foundations and Extensions*, Springer, New York.

R. J. Vanderbei and D. F. Shanno (1999), 'An interior-point algorithm for nonconvex nonlinear programming', *Comput. Optim. Appl.* **13**, 231–252.

R. A. Waltz, J. L. Morales, J. Nocedal and D. Orban (2006), 'An interior algorithm for nonlinear optimization that combines line search and trust region steps', *Math. Program.* **107**, 391–408.

A. Wächter and L. T. Biegler (2006), 'On the implementation of an interior-point filter line-search algorithm for large-scale nonlinear programming', *Math. Program.* **106**, 25–57.

H. Wolkowicz, R. Saigal and L. Vanderberghe, eds (2000), *Handbook of Semidefinite Programming: Theory, Algorithms, and Applications*, Kluwer, Boston.

M. H. Wright (1992), Interior methods for constrained optimization. In *Acta Numerica*, Vol. 1, Cambridge University Press, pp. 341–407.

S. J. Wright (1997), *Primal–Dual Interior-Point Methods*, SIAM, Philadelphia.

X. Xu, P. F. Hung and Y. Ye (1996), 'A simplified homogeneous self-dual linear programming algorithm and its implementation', *Ann. Oper. Res.* **62**, 151–171.

Y. Ye (1991), 'An $O(n^3L)$ potential reduction algorithm for linear programming', *Math. Program.* **50**, 239–258.

Y. Ye (1997), *Interior Point Algorithms: Theory and Analysis*, Wiley.

Y. Ye, M. J. Todd and S. Mizuno (1994), 'An $O(\sqrt{n}L)$-iteration homogeneous and self-dual linear programming algorithm', *Math. Oper. Res.* **19**, 53–67.

Y. Zhang (1994), 'On the convergence of a class of infeasible interior-point methods for the horizontal linear complementarity problem', *SIAM J. Optim.* **4**, 208–227.

Acta Numerica (2008), pp. 235–409
doi: 10.1017/S0962492906380014

Greedy approximation

V. N. Temlyakov
University of South Carolina,
Columbia, 29208, USA
E-mail: temlyak@math.sc.edu

In this survey we discuss properties of specific methods of approximation that belong to a family of greedy approximation methods (greedy algorithms). It is now well understood that we need to study nonlinear sparse representations in order to significantly increase our ability to process (compress, denoise, *etc.*) large data sets. Sparse representations of a function are not only a powerful analytic tool but they are utilized in many application areas such as image/signal processing and numerical computation. The key to finding sparse representations is the concept of m-term approximation of the target function by the elements of a given system of functions (dictionary). The fundamental question is how to construct good methods (algorithms) of approximation. Recent results have established that greedy-type algorithms are suitable methods of nonlinear approximation in both m-term approximation with regard to bases, and m-term approximation with regard to redundant systems. It turns out that there is one fundamental principle that allows us to build good algorithms, both for arbitrary redundant systems and for very simple well-structured bases, such as the Haar basis. This principle is the use of a greedy step in searching for a new element to be added to a given m-term approximant.

CONTENTS

Preface

This section provides a general introduction. Each chapter has its own more specific introduction. A generic problem of mathematical and numerical analysis is to approximate a function f from a Banach space X in the norm $\|\cdot\|$ of this space. There are two major approaches to this problem that lead to two different branches of mathematical analysis. In approach (I) we begin with an assumption of how an approximant should look. In other words, in approach (I) we specify the form of an approximant. In approach (II) we begin with an assumption on the information available about f. Here are typical settings that fall into approach (I).

(Ia) An approximant comes from a given linear subspace L_n of dimension n (algebraic polynomials of degree $n - 1$, trigonometric polynomials of appropriate order, splines with $n - 1$ fixed knots).

(Ib) An approximant comes from a nonlinear set (rational functions; m-term approximant with respect to a given system, splines with fixed number of free knots).

The following are typical settings for approach (II).

(IIa) Information on f is given by a vector $(f(x_1), \ldots, f(x_n))$, for some given set $(x_1, \ldots, x_n)$ of points, or some set that we can choose depending on the problem.

(IIb) The above setting (IIa) has a more general formulation with the functionals $f(x_j)$ replaced by arbitrary linear functionals $\lambda_j(f)$.

In this survey we mostly concentrate on approach (I). We will only touch upon approach (II) in Section 2.9 in a discussion of compressed sensing. Approach (II) is the main issue for information-based complexity (see Traub, Wasilkowski and Wozniakowski (1988)). A very important question in approach (I) is how to choose an appropriate form of approximants. This question has been intensely studied in approximation theory, and resulted in the invention of the concept of *width*. In 1936 A. N. Kolmogorov introduced the following quantity (known as Kolmogorov's width) for a compact $F \subset X$:

$$d_n(F, X) := \inf_{L_n} \sup_{f \in F} \inf_{a \in L_n} \|f - a\|,$$

where L_n is an n-dimensional linear subspace of X. The Kolmogorov width $d_n(F, X)$ of a compact F is an important characteristic of F that states that the best we can achieve in approximating functions from F by elements of linear subspaces of dimension n is $d_n(F, X)$. Therefore, if one can find an n-dimensional subspace L_n^* and an approximation method $A_n : F \to L_n^*$ such that, for any $f \in F$,

$$\|f - A_n(f)\| \le (1 + \epsilon)d_n(F, X),$$

then A_n is an almost ideal approximation method for F with respect to the Kolmogorov width. Thus, the concept of width provides a very nice theoretical way to compare optimal approximation methods. The major drawback of this approach from a practical point of view is that, in order to initialize a procedure of selection of L_n^* and A_n, we need to know the function class F. In many contemporary practical problems we have no idea which class to choose in place of F.

There are two ways to overcome the above problem. The first one is to return (in spirit) to the classical setting that goes back to Chebyshev and Weierstrass. In this setting, we fix *a priori* the form of the approximant (say, approximation by algebraic polynomials of degree n, as in the case of Chebyshev and Weierstrass) and look for an approximation method that is optimal, or near-optimal, for each individual function from X. For example, the approximation method that picks the algebraic polynomial of degree n of best approximation to f in X is an optimal method of approximation by algebraic polynomials of degree n. However, such an obvious optimal method of approximation may not be good from the perspective of practical implementation. This leads to the following natural setting. We specify not only the form of the approximant, but also choose a specific method of approximation (for instance, one known to be suitable for practical implementation). Now, we have a precise mathematical problem of studying the efficiency of our specific method of approximation. We discuss this problem in detail here. It turns out that a convenient and flexible way of measuring the efficiency of a specific approximation method is to derive the corresponding Lebesgue-type inequalities. Remember that we would like this method to work for all functions; therefore, it should at least converge for each $f \in X$, and hence convergence is a fundamental theoretical problem. In this survey we thoroughly discuss the problem of convergence for greedy algorithms.

The second way to overcome the above-mentioned drawback of a method based on the concept of width consists in weakening the *a priori* assumption that f is an element of F. Instead of looking for an approximation method that is optimal (or near-optimal) for a given single class F, we look for an approximation method that is near-optimal for each class from a given collection $\mathcal{F}$ of classes. Such a method is called *universal* for $\mathcal{F}$. Universal algorithms have been studied in approximation theory (see Temlyakov (1988, 2003a)) and in learning theory (see, for instance, Temlyakov (2005c)). In this survey we do not further discuss universal algorithms, but refer the reader to the survey of Temlyakov (2003a).

In this survey we discuss properties of specific methods of approximation that belong to a family of greedy approximation methods (greedy algorithms). Realizing approach (I) mentioned above, we need to specify the form of the approximant. We use a concept of *sparsity* in dealing with this problem. It is now well understood that we need to study nonlinear sparse representations in order to significantly increase our ability to process (compress, denoise, *etc.*) large data sets. Sparse representations are not only a powerful analytic tool but are used in many application areas, such as image/signal processing and numerical computation. The key to finding sparse representations is the concept of the m-term approximation of the target function by the elements of a given system of functions: a dictionary. Since the elements of the dictionary used in the m-term approximation are allowed to depend on the function being approximated, this type of approximation is very efficient when the approximants can be found. Thus, we specify the form of our approximant as an m-term approximant with regard to a given system of functions. It is clear that this method of approximation is a particular case of nonlinear methods of approximation.

The past decade has seen great success in studying nonlinear approximation, motivated by numerous applications: see the surveys by DeVore (1998) and Temlyakov (2003*a*). Nonlinear approximation is important in applications because of its concise representations and increased computational efficiency. Two types of nonlinear approximation are frequently employed in applications. Adaptive methods are used in PDE solvers, while m-term approximation, considered here, is used in image/signal/data processing, as well as in the design of neural networks. The fundamental question of nonlinear approximation is how to devise good constructive methods, or algorithms, of nonlinear approximation. This problem has two levels of nonlinearity. The first level of nonlinearity is m-term approximation with regard to bases. In this problem one can use the unique function expansion with regard to a given basis to build an approximant. Nonlinearity enters by looking for m-term approximants with terms (*i.e.*, basis elements in the approximant) allowed to depend on the given function. We discuss m-term approximation with regard to bases in detail in Chapter 1. On the second level of nonlinearity, we replace a basis by a more general system which is not necessarily minimal, for example, a redundant system, or dictionary. This setting is much more complicated than the bases case; however, there is a solid justification of the importance of redundant systems in both theoretical questions and in practical applications: see, for instance, Schmidt (1906), Huber (1985) and Donoho (2001). In Chapters 2 and 3 we discuss approximation by linear combinations of elements that are taken from a redundant (overcomplete) system of elements. We give a brief discussion of the question: Why do we need redundant systems? Answering this question, we first mention three classical redundant systems that are used in different areas of mathematics. Perhaps the first example of m-term approximation with regard to a redundant dictionary was considered by Schmidt (1906), who considered the approximation of functions $f(x, y)$ of two variables by bilinear forms $\sum_{i=1}^{m} u_i(x) v_i(y)$ in $L_2([0, 1]^2)$. This problem is closely connected to properties of the integral operator $J_f(g) := \int_0^1 f(x,y)g(y)\,dy$ with kernel $f(x,y)$.

Another example which is well known in statistics is the projection pursuit regression problem. We formulate the related setting in the language of function theory. Given a bounded domain $\Omega \subset \mathbb{R}^d$, the problem is to approximate a given function $f \in L_2(\Omega)$ by a sum of ridge functions, *i.e.*, by $\sum_{j=1}^m r_j(\omega_j \cdot x)$, for $x, \omega_j \in \mathbb{R}^d$, $j = 1, \ldots, m$, where r_j, $j = 1, \ldots, m$, are univariate functions.

The third example is from signal processing. In signal processing the most popular methods of approximation are wavelets and the system of Gabor functions $\{g_{a,b}(x - c) : g_{a,b}(x) := \mathrm{e}^{iax}\mathrm{e}^{-bx^2}, \ a, c \in \mathbb{R}, \ b \in \mathbb{R}_+\}$. The Gabor system gives more flexibility in constructing an approximant but it is a redundant, not minimal, system. It also seems natural to use redundant systems in modelling analysing elements for the visual system; see the discussion in Donoho (2001).

Thus, in order to address the contemporary needs of approximation theory and computational mathematics, a very general model of approximation with regard to a redundant system, or dictionary, has been considered in many recent papers. As such a model, we choose a Banach space X whose elements are our target functions, and an approximating system which can be any subset $\mathcal{D}$ of elements of this space such that the closure of span $\mathcal{D}$ coincides with X. We would like to have an algorithm to construct m-term approximants that, at each step, adds only one new element from $\mathcal{D}$ and keeps elements of $\mathcal{D}$ previously obtained. This requirement is an analogue of *on-line* computation that is very desirable in practical algorithms. Clearly, we are looking for good algorithms which, at least, converge for each target function. It is not obvious that such an algorithm exists in a setting at the above level of generality (X, $\mathcal{D}$ are arbitrary).

The fundamental question is how to construct good methods, or algorithms, of approximation. Recent results have established that greedy-type algorithms are suitable methods of nonlinear approximation, in both m-term approximation with regard to bases, and m-term approximation with regard to redundant systems. It turns out that there is one fundamental principle that allows us to build good algorithms both for arbitrary redundant systems and for very simple well-structured bases such as the Haar basis. This principle is the use of a greedy step in searching for a new element $g_m(f) \in \mathcal{D}$ to be added to a given m-term approximant, by which we mean that $g_m(f) \in \mathcal{D}$ should maximize a certain functional determined by information from the previous steps of the algorithm. We obtain different types of greedy algorithms by varying the above-mentioned functional and also by using different ways of constructing the m-term approximant (*i.e.*, choosing coefficients of the linear combination) from the previously found m elements of the dictionary. In Chapters 2 and 3 we present different greedy-type algorithms, beginning with a very simple and very natural Pure Greedy Algorithm in a Hilbert space, and ending with its rather complicated modifications in a Banach space. The general goal of different modifications is to prepare the corresponding greedy algorithms for practical implementation. We discuss this issue in detail in Chapters 2 and 3.

It is known that in many numerical problems, users are satisfied with a Hilbert space setting and do not consider a more general setting in a Banach space. We now give one remark that justifies our interest in Banach spaces. The first argument is an *a priori* argument that the spaces L_p are very natural, and should be studied along with the L_2-space. The second argument is an *a posteriori* argument. The study of greedy approximation in Banach spaces discovered that one very important characteristic of a Banach space X that governs the behaviour of greedy approximations is the *modulus of smoothness* $\rho(u)$ of X (see Section 3.1 for details). It is known that the spaces L_p, $2 \leq p < \infty$ have moduli of smoothness of the same order u^2. Thus, many results that are known for the Hilbert space L_2, and were proved using the special structure of a Hilbert space, can be generalized to the Banach spaces L_p, $2 \leq p < \infty$. The new proofs use only the geometry of the unit sphere of the space expressed in the form $\rho(u) \leq \gamma u^2$.

The theory of greedy approximation is developing rapidly and results are spread over hundreds of papers by different authors. There are several surveys that discuss greedy approximation: see DeVore (1998), Temlyakov (2003a), Konyagin and Temlyakov (2002), Wojtaszczyk (2002a) and Temlyakov (2006b). There are no books on greedy approximation at present. We decided to include in this survey proofs of the most important and typical results. In the majority of cases these proofs are not technically involved and allow the reader to understand a phenomenon much better than merely stating results. We have tried to make the presentation of ideas and techniques of greedy approximation sufficiently systematic to be used in a graduate course on greedy approximation.

We will use C, $C(p,d)$, $C_{p,d}$, *etc.*, to denote various positive constants, the indexes indicating dependence on other parameters. We will use the following symbols for brevity. For two non-negative sequences $a = \{a_n\}_{n=1}^{\infty}$ and $b = \{b_n\}_{n=1}^{\infty}$, the relation, or order inequality, $a_n \ll b_n$ means that there is a number $C(a,b)$ such that, for all n, we have $a_n \leq C(a,b)\, b_n$; and the relation $a_n \asymp b_n$ means that $a_n \ll b_n$ and $b_n \ll a_n$. Other notation is defined in the text itself.

CHAPTER ONE
Greedy approximation with respect to bases

1.1. Introduction

It is well known that in many problems it is very convenient to represent a function by a series with respect to a given system of functions. For example, in 1807 Fourier suggested representing a 2π-periodic function by its series (now known as the Fourier series) with respect to the trigonometric system. A very important feature of the trigonometric system that made it attractive for the representation of periodic functions is orthogonality. For an orthonormal system $\mathcal{B} := \{b_n\}_{n=1}^{\infty}$ of a Hilbert space H with an inner product $\langle \cdot, \cdot \rangle$, one can construct a Fourier series of an element f in the following way:

$$f \sim \sum_{n=1}^{\infty} \langle f, b_n \rangle b_n. \tag{1.1.1}$$

If the system $\mathcal{B}$ is a basis for H, then the series in (1.1.1) converges to f in H and (1.1.1) provides the unique representation

$$f = \sum_{n=1}^{\infty} \langle f, b_n \rangle b_n \tag{1.1.2}$$

of f with respect to $\mathcal{B}$. This representation has nice approximative properties. By Parseval's identity,

$$\|f\|^2 = \sum_{n=1}^{\infty} |\langle f, b_n \rangle|^2, \tag{1.1.3}$$

we obtain a convenient way to calculate, or estimate, the norm $\|f\|$.

It is known that the partial sums

$$S_m(f, \mathcal{B}) := \sum_{n=1}^{m} \langle f, b_n \rangle b_n \tag{1.1.4}$$

provide the best approximation, that is, defining

$$E_m(f, \mathcal{B}) := \inf_{\{c_n\}} \left\| f - \sum_{n=1}^{m} c_n b_n \right\| \tag{1.1.5}$$

to be the distance of f from the $\text{span}\{b_1, \ldots, b_m\}$, we have

$$\|f - S_m(f, \mathcal{B})\| = E_m(f, \mathcal{B}). \tag{1.1.6}$$

Identities (1.1.3) and (1.1.6) are fundamental properties of Hilbert spaces and their orthonormal bases. These properties make the theory of approximation in H from the $\text{span}\{b_1, \ldots, b_m\}$, or linear approximation theory, simple and convenient.

The situation becomes more complicated when we replace a Hilbert space H by a Banach space X. In a Banach space X we consider a Schauder basis Ψ instead of an orthonormal basis $\mathcal{B}$ in H. In Section 1.2 we discuss Schauder bases in detail. If $\Psi := \{\psi_n\}_{n=1}^{\infty}$ is a Schauder basis for X, then for any $f \in X$ there exists a unique representation

$$f = \sum_{n=1}^{\infty} a_n(f)\psi_n$$

that converges in X.

Theorem 1.2.3 states that the partial sum operators S_m, defined by

$$S_m(f, \Psi) := \sum_{n=1}^{m} a_n(f)\psi_n,$$

are uniformly bounded operators from X to X. In other words, there exists a constant B such that, for any $f \in X$ and any m, we have

$$\|S_m(f, \Psi)\| \leq B\|f\|.$$

This inequality implies the following analogue of (1.1.6): for any $f \in X$,

$$\|f - S_m(f, \Psi)\| \leq (B+1)E_m(f, \Psi), \tag{1.1.7}$$

where

$$E_m(f, \Psi) := \inf_{\{c_n\}} \left\| f - \sum_{n=1}^{m} c_n\psi_n \right\|.$$

Inequality (1.1.7) shows that the $S_m(f, \Psi)$ provides near-best approximation from $\text{span}\{\psi_1, \ldots, \psi_m\}$. Thus, if we are satisfied with near-best approximation instead of best approximation, then the linear approximation theory with respect to Schauder bases becomes simple and convenient. The partial sums $S_m(\cdot, \Psi)$ provide near-best approximation for any individual element of X.

Motivated by computational issues, researchers became interested in non-linear approximation with regard to a given system instead of linear approximation. For example, in the case of representation (1.1.2) in a Hilbert space, one can take an approximant of the form

$$S_\Lambda(f, \mathcal{B}) := \sum_{n \in \Lambda} \langle f, b_n \rangle b_n, \quad |\Lambda| = m,$$

instead of an approximant $S_m(f, \mathcal{B})$ from an m-dimensional linear subspace. Then the two approximants $S_m(f, \mathcal{B})$ and $S_\Lambda(f, \mathcal{B})$ have the same sparsity: both are linear combinations of m basis elements. However, we can achieve a better approximation error with $S_\Lambda(f, \mathcal{B})$ than with $S_m(f, \mathcal{B})$ if we choose Λ in the right way. In the case of a Hilbert space and an orthonormal basis $\mathcal{B}$, an optimal choice Λ_m of Λ is obvious: Λ_m is a set of m indices with the biggest (in absolute value) coefficients $\langle f, b_n \rangle$. Then, by Parseval's identity (1.1.3), we obtain

$$\|f - S_{\Lambda_m}(f, \mathcal{B})\| \leq \|f - S_m(f, \mathcal{B})\|.$$

Also, it is clear that the $S_{\Lambda_m}(f, \mathcal{B})$ realizes the best m-term approximation of f with regard to $\mathcal{B}$,

$$\|f - S_{\Lambda_m}(f, \mathcal{B})\| = \sigma_m(f, \mathcal{B}) := \inf_{\Lambda:|\Lambda|=m} \inf_{\{c_n\}} \left\| f - \sum_{n \in \Lambda} c_n b_n \right\|. \qquad (1.1.8)$$

The approximant $S_{\Lambda_m}(f, \mathcal{B})$ can be obtained as a realization of m iterations of the following *greedy approximation step*. For a given $f \in H$ we choose at a *greedy step* an index n_1 with the biggest $|\langle f, b_{n_1} \rangle|$. At a greedy approximation step we build a new element $f_1 := f - \langle f, b_{n_1} \rangle b_{n_1}$.

The identity (1.1.8) shows that the greedy approximation works perfectly in nonlinear approximation in a Hilbert space with regard to an orthonormal basis $\mathcal{B}$.

This chapter is devoted to a systematic study of greedy approximation in Banach spaces. In Section 1.2 we discuss the following natural question. Equation (1.1.8) proves the existence of the best m-term approximant in a Hilbert space with respect to an orthonormal basis. Further, we discuss existence of the best m-term approximant in a Banach space with respect to a Schauder basis. That discussion illustrates that the situation with existence theorems is much more complex in Banach spaces than in Hilbert spaces. We also give some sufficient conditions on a Schauder basis that guarantee existence of the best m-term approximant. However, the problem is far from being completely solved.

The central issue of this chapter is the following question. Which bases are suitable for greedy approximation? Greedy approximation with regard to a Schauder basis is defined in a similar way to the greedy approximation with regard to an orthonormal basis (see above). The greedy algorithm picks the terms with the biggest (in absolute value) coefficients from the expansion

$$f = \sum_{n=1}^{\infty} a_n(f) \psi_n, \qquad (1.1.9)$$

and gives a *greedy approximant*

$$G_m(f, \Psi) := S_{\Lambda_m}(f, \Psi) := \sum_{n \in \Lambda_m} a_n(f) \psi_n.$$

Here, Λ_m is such that $|\Lambda_m| = m$ and

$$\min_{n \in \Lambda_m} |a_n(f)| \geq \max_{n \notin \Lambda_m} |a_n(f)|.$$

We note that we need some restrictions on the basis Ψ (see Sections 1.3 and 1.4 for a detailed discussion) in order to be able to run the greedy algorithm for each $f \in X$. It is sufficient to assume that Ψ is normalized. We make this assumption for our further discussion in the Introduction.

An application of the greedy algorithm can also be seen as a rearrangement of the series from (1.1.9) in a special way: according to the size of coefficients. Let

$$|a_{n_1}| \geq |a_{n_2}| \geq \cdots .$$

Then

$$G_m(f, \Psi) = \sum_{j=1}^{m} a_{n_j}(f)\psi_{n_j}.$$

Thus, the greedy approximant $G_m(f, \Psi)$ is a partial sum of the rearranged series

$$\sum_{j=1}^{\infty} a_{n_j}(f)\psi_{n_j}. \qquad (1.1.10)$$

An immediate question with (1.1.10) is: When does this series converge? The theory of convergence of rearranged series is a classical topic in analysis. A series converges *unconditionally* if every rearrangement of this series converges. A basis Ψ of a Banach space X is said to be an *unconditional basis* if, for every $f \in X$, its expansion (1.1.9) converges unconditionally. For a set of indices Λ define

$$S_\Lambda(f, \Psi) := \sum_{n \in \Lambda} a_n(f)\psi_n.$$

It is well known that if Ψ is unconditional then there exists a constant K such that, for any Λ,

$$\|S_\Lambda(f, \Psi)\| \leq K\|f\|. \qquad (1.1.11)$$

This inequality is similar to $\|S_m(f, \Psi)\| \leq B\|f\|$ and implies an analogue of (1.1.7):

$$\|f - S_\Lambda(f, \Psi)\| \leq (K+1)E_\Lambda(f, \Psi), \qquad (1.1.12)$$

where

$$E_\Lambda(f, \Psi) := \inf_{\{c_n\}} \left\| f - \sum_{n \in \Lambda} c_n\psi_n \right\|.$$

Inequality (1.1.12) indicates that in the case of an unconditional basis Ψ it is sufficient for finding near-best m-term approximant to optimize only over the sets of indices Λ. The greedy algorithm $G_m(\cdot, \Psi)$ gives a simple recipe for building Λ_m: pick the indices with biggest coefficients. In Section 1.3 we discuss in detail when the above simple recipe provides a near-best m-term approximant. It turns out that the assumption that Ψ is merely unconditional does not guarantee that $G_m(\cdot, \Psi)$ provides a near-best m-term approximation. We also discuss a new class of bases (*greedy bases*) that has the property that $G_m(f, \Psi)$ provides a near-best m-term approximation for each $f \in X$. We show that the class of greedy bases is a proper subclass of the class of unconditional bases.

It follows from the definition of an unconditional basis that any rearrangement of the series in (1.1.9) converges. It is known that it converges to f. The rearrangement (1.1.10) is a specific rearrangement of (1.1.9). Clearly, for an unconditional basis Ψ, (1.1.10) converges to f. It turns out that unconditionality of Ψ is not a necessary condition for convergence of (1.1.10) for each $f \in X$. Bases that have the property of convergence of (1.1.10) for each $f \in X$ are exactly the *quasi-greedy bases* (see Section 1.4).

Let us summarize our discussion of bases in Banach spaces. Schauder bases are natural for convergence of $S_m(f, \Psi)$ and convenient for linear approximation theory. Other classical bases, namely, unconditional bases, are natural for convergence of all rearrangements of expansions. The needs of nonlinear approximation, or, more specifically, the needs of greedy approximation lead us to new concepts of bases: greedy bases and quasi-greedy bases. The relations between these bases are the following:

$$\{\text{greedy bases}\} \subset \{\text{unconditional bases}\} \subset$$
$$\{\text{quasi-greedy bases}\} \subset \{\text{Schauder bases}\}.$$

All the inclusions $\subset$ are proper inclusions. In this chapter we provide a justification of the importance of the new classes of bases. With a belief in the importance of greedy bases and quasi-greedy bases, we discuss here the following natural questions. Could we weaken a rule of building $G_m(f, \Psi)$ and still have good approximation and convergence properties? We answer this question in Sections 1.5 and 1.6. What can be said about classical systems, say, the Haar system and the trigonometric system, in this regard? We discuss this question in Sections 1.3 and 1.7. How to build the approximation theory (mostly, direct and inverse theorems) for m-term approximation with regard to greedy-type bases? Section 1.8 is devoted to this question.

1.2. Schauder bases in Banach spaces

Schauder bases in Banach spaces are used to associate a sequence of numbers with an element $f \in X$: these are the coefficients of f with respect to a basis. This helps in studying properties of a Banach space X. We begin with some classical results on Schauder bases: see, for instance, Lindenstrauss and Tzafriri (1977).

Definition 1.2.1. A sequence $\Psi := \{\psi_n\}_{n=1}^{\infty}$ in a Banach space X is called a Schauder basis of X (basis of X) if, for any $f \in X$, there exists a unique sequence $\{a_n\}_{n=1}^\infty := \{a_n(f)\}_{n=1}^\infty$ such that

$$f = \sum_{n=1}^\infty a_n \psi_n.$$

Let

$$S_0(f) := 0, \quad S_m(f) := S_m(f, \Psi) := \sum_{n=1}^m a_n(f) \psi_n.$$

For a fixed basis Ψ, consider the following quantity:

$$\|f\| := \sup_m \|S_m(f, \Psi)\|.$$

It is clear that, for any $f \in X$ we have

$$\|f\| \leq \|f\| < \infty. \tag{1.2.1}$$

It is easy to see that $\|\cdot\|$ provides a norm on the linear space X. Denote this new normed linear space by X^s.

Proposition 1.2.2. The space X^s is a Banach space.

Theorem 1.2.3. Let X be a Banach space with a Schauder basis Ψ. Then the operators $S_m : X \to X$ are bounded linear operators and

$$\sup_m \|S_m\| < \infty.$$

The proof of this theorem is based on the following fundamental theorem of Banach.

Theorem 1.2.4. Let U, V be Banach spaces and T be a bounded linear one-to-one operator from V to U. Then the inverse operator T^{-1} is a bounded linear operator from U to V.

We specify $U = X$, $V = X^s$, and let T be the identity map. It follows from (1.2.1) that T is a bounded operator from V to U. Thus, by Theorem 1.2.4, T^{-1} is also bounded. This means that there exists a constant C such that, for any $f \in X$, we have $\|f\| \leq C\|f\|$. This completes the proof of Theorem 1.2.3.

The operators $\{S_m\}_{m=1}^\infty$ are called the natural projections associated with a basis Ψ. The number $\sup_m \|S_m\|$ is called the basis constant of the basis Ψ. A basis whose basis constant is one is called a *monotone basis*. It is clear that an orthonormal basis in a Hilbert space is a monotone basis. Every Schauder basis Ψ is monotone with respect to the norm $\|f\| := \sup_m \|S_m(f, \Psi)\|$, which was already used above. Indeed, we have

$$\|S_m(f)\| = \sup_n \|S_n(S_m(f))\| = \sup_{1 \leq n \leq m} \|S_n(f)\| \leq \|f\|.$$

The above remark means that for any Schauder basis Ψ of X we can renorm X (take X^s) to make the basis Ψ monotone for a new norm.

Theorem 1.2.5. Let $\{x_n\}_{n=1}^\infty$ be a sequence of elements in a Banach space X. Then $\{x_n\}_{n=1}^\infty$ is a Schauder basis of X if and only if the following three conditions hold.

(a) $x_n \neq 0$ for all n.

(b) There is a constant K such that, for every choice of scalars $\{a_i\}_{i=1}^\infty$ and integers $n < m$, we have

$$\left\|\sum_{i=1}^n a_i x_i\right\| \leq K \left\|\sum_{i=1}^m a_i x_i\right\|.$$

(c) The closed linear span of $\{x_n\}_{n=1}^\infty$ coincides with X.

We note that for a basis Ψ with the basis constant K, we have for any $f \in X$

$$\|f - S_m(f, \Psi)\| \leq (K + 1) \inf_{\{c_k\}} \left\|f - \sum_{k=1}^m c_k \psi_k\right\|.$$

Thus, the partial sums $S_m(f, \Psi)$ provide near-best approximation from $\mathrm{span}\{\psi_1, \ldots, \psi_m\}$.

Let a Banach space X, with a basis $\Psi = \{\psi_k\}_{k=1}^\infty$, be given. In order to understand the efficiency of an algorithm providing an m-term approximation we compare its accuracy with the best-possible accuracy when an approximant is a linear combination of m terms from Ψ. We define the best m-term approximation with regard to Ψ as follows:

$$\sigma_m(f) := \sigma_m(f, \Psi)_X := \inf_{c_k, \Lambda} \left\|f - \sum_{k \in \Lambda} c_k \psi_k\right\|_X,$$

where the infimum is taken over coefficients c_k and sets of indices Λ with cardinality $|\Lambda| = m$. We note that in the above definition of $\sigma_m(f, \Psi)_X$ the system Ψ may be any system of elements from X, not necessarily a basis of X.

An immediate natural question is when the best m-term approximant exists. This question is a more difficult problem than the corresponding problem in the case of linear approximation. The problem of existence of best m-term approximant with regard to a basis has not been studied thoroughly. We present here some results in this direction.

Let us proceed to the approximation problem setting. Let a subset $A \subset X$ be given. For any $f \in X$, let

$$d(f, A) := d(f, A)_X := \inf_{a \in A} \|f - a\|$$

denote the distance from f to A, or in other words the best approximation error of f by elements from A in the norm of X. To illustrate some relevant techniques in this direction, let us prove existence theorems in the following two settings.

S1 Let $X = L_p(0, 2\pi)$, $1 \leq p < \infty$, or $X = L_\infty(0, 2\pi) := C(0, 2\pi)$ be the set of 2π-periodic functions. Consider A to be the set Σ_m of all complex trigonometric polynomials or $\Sigma_m(R)$ of all real trigonometric polynomials which have at most m non-zero coefficients:

$$\Sigma_m := \left\{ t : t = \sum_{k \in \Lambda} c_k e^{ikx}, \quad \#\Lambda \leq m \right\},$$

$$\Sigma_m(R) := \left\{ t : t = \sum_{k \in \Lambda_1} a_k \cos kx + \sum_{k \in \Lambda_2} b_k \sin kx, \quad \#\Lambda_1 + \#\Lambda_2 \leq m \right\}.$$

We will also use the following notation in this case:

$$\sigma_m(f, \mathcal{T})_X := d(f, \Sigma_m)_X.$$

S2 Let $X = L_p(0, 1)$, $1 \leq p < \infty$ and let A be the set Σ_m^S of piecewise constant functions with at most $m - 1$ break-points at $(0, 1)$.

In the setting S2 we prove here the following existence theorem (see De-Vore and Lorenz (1993), p. 363).

Theorem 1.2.6. For any $f \in L_p(0, 1)$, $1 \leq p < \infty$, there exists $g \in \Sigma_m^S$ such that

$$d(f, \Sigma_m^S)_p = \|f - g\|_p.$$

Proof. Fix the break-points $0 = y_0 \leq y_1 \leq \cdots \leq y_{m-1} \leq y_m = 1$, let $y := (y_0, \ldots, y_m)$, and let $S_0(y)$ be the set of piecewise constant functions with break-points $y_1, \ldots, y_{m-1}$. Further, let

$$e_m^y(f)_p := \inf_{a \in S_0(y)} \|f - a\|_p.$$

From the definition of $d(f, \Sigma_m^S)_p$, there exists a sequence y^i such that

$$e_m^{y^i}(f)_p \to d(f, \Sigma_m^S)_p$$

when $i \to \infty$. Considering a subsequence of $\{y^i\}$, if necessary we can assume that $y^i \to y^*$ for some $y^* \in \mathbb{R}^{m+1}$. Now we consider only those indices j for which $y_{j-1}^* \neq y_j^*$. Let Λ denote the corresponding set of indices. Take a positive number ϵ satisfying

$$\epsilon < \min_{j \in \Lambda}(y_j^* - y_{j-1}^*)/3,$$

and consider i such that

$$\|y^* - y^i\|_\infty < \epsilon, \quad \text{where } \|y\|_\infty := \max_k |y_k|. \tag{1.2.2}$$

By the existence theorem in the case of approximation by elements of a subspace of finite dimension, for each y^i there exists

$$g(f, y^i, c^i) := \sum_{j=1}^{m} c_j^i \chi_{[y_{j-1}^i, y_j^i]},$$

where χ_E denotes the characteristic function of a set E, with the property

$$\|f - g(f, y^i, c^i)\|_p = e_m^{y^i}(f)_p.$$

For i satisfying (1.2.2) and $j \in \Lambda$ we have $|c_j^i| \leq C(f, \epsilon)$, which allows us to assume (passing to a subsequence if necessary) the convergence

$$\lim_{i \to \infty} c_j^i = c_j, \quad j \in \Lambda.$$

Consider

$$g(f, c) := \sum_{j \in \Lambda} c_j \chi_{[y_{j-1}^*, y_j^*]}.$$

Let $U_\epsilon(y) := \cup_j (y_j - \epsilon, y_j + \epsilon)$ and introduce $G := [0, 1] \setminus U_\epsilon(y^*)$. Then we have

$$\int_G |f - g(f, c)|^p = \lim_{i \to \infty} \int_G |f - g(f, y^i, c^i)|^p \leq d(f, \Sigma_m^S)_p^p.$$

Letting $\epsilon \to 0$, we complete the proof. $\square$

We proceed now to the trigonometric case S1. We will give the proof in the general d-variable case for $\mathcal{T}^d := \mathcal{T} \times \cdots \times \mathcal{T}$ (d times) because this generality does not introduce any complication. The following theorem was essentially proved in Baishanski (1983). The presented proof is taken from Temlyakov (1998c).

Theorem 1.2.7. Let $1 \leq p \leq \infty$. For any $f \in L_p(\mathbb{T}^d)$ and any $m \in \mathbb{N}$, there exists a trigonometric polynomial t_m of the form

$$t_m(x) = \sum_{n=1}^{m} c_n e^{i(k^n, x)}, \tag{1.2.3}$$

such that

$$\sigma_m(f, \mathcal{T}^d)_p = \|f - t_m\|_p. \tag{1.2.4}$$

Proof. We prove this theorem by induction. Let us use the abbreviated notation $\sigma_m(f)_p := \sigma_m(f, \mathcal{T}^d)_p$.

First step. Let $m = 1$. We assume $\sigma_1(f)_p < \|f\|_p$, because in the case $\sigma_1(f)_p = \|f\|_p$ the proof is trivial: we take $t_1 = 0$. We now prove that polynomials of the form $c e^{i(k,x)}$ with big $|k|$ cannot provide approximation with error close to $\sigma_1(f)_p$. This will allow us to restrict the search for an optimal approximant $c_1 e^{i(k^1,x)}$ to a finite number of k^1, which in turn will imply the existence.

We introduce a parameter $N \in \mathbb{N}$, which will be specified later on, and consider the following polynomials:

$$\mathcal{K}_N(u) := \sum_{|k|<N} \left(1 - \frac{|k|}{N}\right) e^{iku}, \quad u \in \mathbb{T}, \tag{1.2.5}$$

and

$$\mathcal{K}_N(x) := \prod_{j=1}^{d} \mathcal{K}_N(x_j), \quad x = (x_1, \ldots, x_d) \in \mathbb{T}^d.$$

The functions $\mathcal{K}_N$ are the Fejér kernels. These polynomials have the following property (for (1.2.6) see Zygmund (1959, Chapter 3, Section 3)):

$$\|\mathcal{K}_N\|_1 = 1, \quad N = 1, 2, \ldots. \tag{1.2.6}$$

Consider the operator

$$(K_N(g))(x) = (2\pi)^{-d} \int_{\mathbb{T}^d} \mathcal{K}_N(x-y)g(y)\,dy. \tag{1.2.7}$$

Let

$$e_N(g) := \|g - K_N(g)\|_p. \tag{1.2.8}$$

It is known that for any $f \in L_p(\mathbb{T}^d)$ we have $e_N \to 0$ as $N \to \infty$. For fixed N take any $k \in \mathbb{Z}^d$ such that $\|k\|_\infty \geq N$. Consider $g(x) = f(x) - c e^{i(k,x)}$ with some c. Using (1.2.5) and (1.2.6), we get on the one hand

$$\|K_N(f)\|_p = \|K_N(g)\|_p \lesssim \|g\|_p. \tag{1.2.9}$$

On the other hand, we have

$$\|K_N(f)\|_p \geq \|f\|_p - \|f - K_N(f)\|_p \geq \|f\|_p - e_N(f). \tag{1.2.10}$$

Therefore, combining (1.2.9) and (1.2.10) we obtain, for all k, $\|k\|_\infty \geq N$, and any c,

$$\|f(x) - c e^{i(k,x)}\|_p \geq \|f\|_p - e_N(f). \tag{1.2.11}$$

Making N big enough, we get

$$\|f\|_p - e_N(f) \geq (\|f\|_p + \sigma_1(f)_p)/2. \tag{1.2.12}$$

Relations (1.2.11) and (1.2.12) imply

$$\sigma_1(f)_p = \inf_{c, \|k\|_\infty < N} \|f(x) - c e^{i(k,x)}\|_p,$$

which completes the proof for $m = 1$, by the existence theorem in the case of approximation by elements of a subspace of finite dimension.

General step. Assume that Theorem 1.2.7 has already been proved for $m - 1$. We prove it for m. If $\sigma_m(f)_p = \sigma_{m-1}(f)_p$, we are done, by the induction assumption. Let $\sigma_m(f)_p < \sigma_{m-1}(f)_p$. The idea of the proof in the general step is similar to that in the first step.

Take any $k^1, \ldots, k^m$. Assume $\|k^j\|_\infty \leq \|k^m\|_\infty$, $j = 1, \ldots, m - 1$, and $\|k^m\|_\infty > N$. We prove that a polynomial with frequencies $k^1, \ldots, k^m$ does not provide good approximation. Take any numbers $c_1, \ldots, c_m$, and consider

$$f_{m-1}(x) := f(x) - \sum_{j=1}^{m-1} c_j e^{i(k^j, x)},$$

$$g(x) := f_{m-1}(x) - c_m e^{i(k^m, x)}.$$

Then, replacing f by f_{m-1}, we get in the same way as above the estimate

$$\left\| f(x) - \sum_{j=1}^{m} c_j e^{i(k^j, x)} \right\|_p \geq \sigma_{m-1}(f)_p - e_N(f). \tag{1.2.13}$$

We remark here that the analogue to (1.2.10) looks as follows:

$$\|K_N(f_{m-1})\|_p \geq \sigma_{m-1}(K_N(f))_p$$
$$\geq \sigma_{m-1}(f)_p - \|f - K_N(f)\|_p$$
$$\geq \sigma_{m-1}(f)_p - e_N(f).$$

Making N big enough, we derive from (1.2.13) that

$$\sigma_m(f)_p = \inf \left(\inf_{c_j, j=1, \ldots, m} \left\| f(x) - \sum_{j=1}^{m} c_j e^{i(k^j, x)} \right\|_p \right),$$

where the infimum is taken over k^j satisfying the restriction $\|k^j\|_\infty \leq N$ for all $j = 1, \ldots, m$. In order to complete the proof of Theorem 1.2.7, it remains to remark that, by the existence theorem in the case of approximation by elements of a subspace of finite dimension, the inside infimum can always be replaced by minimum, and the outside infimum is taken over a finite set. This completes the proof. □

Concerning the problem of uniqueness of the best approximant, we will only make a remark that shows that in the m-term nonlinear approximation we can hardly expect the unicity. Let us consider problem S1 on best m-term trigonometric approximation in a particular case $X = L_2(0, 2\pi)$. Take

$$f(x) = \sum_{k=1}^{n} e^{ikx}.$$

Clearly, $\sigma_1(f)_2 = (n-1)^{1/2}$ and each e^{ikx}, $k = 1, \ldots, n$ may serve as a best approximant.

We can prove the following existence theorem (see Temlyakov (2001a)) in a similar way to the proof of Theorem 1.2.7.

Theorem 1.2.8. Let Ψ be a monotone basis of X. Then, for any $x \in X$ and any $m \in \mathbb{N}$, there exist Λ_m, $|\Lambda_m| \leq m$, and $\{c_i^* : i \in \Lambda_m\}$ such that

$$\left\| f - \sum_{i \in \Lambda_m} c_i^* \psi_i \right\| = \sigma_m(f, \Psi).$$

Here is one more existence theorem from Temlyakov (2001a).

Theorem 1.2.9. Let Ψ be a normalized ($\|\psi_k\| = 1$, $k = 1, \ldots$) Schauder basis of X with the additional property that ψ_k converges weakly to 0. Then, for any $f \in X$, and any $m \in \mathbb{N}$, there exist Λ_m, $|\Lambda_m| \leq m$, and $\{c_i^* : i \in \Lambda_m\}$ such that

$$\left\| f - \sum_{i \in \Lambda_m} c_i^* \psi_i \right\| = \sigma_m(f, \Psi).$$

Proof. The proof is a development of ideas from Baishanski (1983). In order to sketch the idea of the proof, let us consider first the case $m = 1$. Let

$$\| f - c_{k_n} \psi_{k_n} \| \to \sigma_1(f, \Psi), \quad n \to \infty. \tag{1.2.14}$$

If

$$\liminf_{n \to \infty} k_n < \infty,$$

then there exists k and a sequence $\{a_n\}$ such that

$$\| f - a_n \psi_k \| \to \sigma_1(f, \Psi), \quad n \to \infty. \tag{1.2.15}$$

Using the fact that Ψ is a Schauder basis, we infer from (1.2.15) that the sequence $\{a_n\}$ is bounded. Choosing a convergent subsequence of $\{a_n\}$, we construct an a such that

$$\| f - a \psi_k \| = \sigma_1(f, \Psi),$$

which proves existence in this case. Assume now that

$$\lim_{n \to \infty} k_n = \infty.$$

Let F_f be a norming (peak) functional for f: $F_f(f) = \|f\|$, $\|F_f\| = 1$. Then

$$\| f - c_{k_n} \psi_{k_n} \| \geq F_f(f - c_{k_n} \psi_{k_n}) = \|f\| - c_{k_n} F_f(\psi_{k_n}). \tag{1.2.16}$$

Relation (1.2.14) implies boundedness of $\{c_{k_n}\}$, and therefore, by weak convergence to 0 of $\{\psi_k\}$, we get from (1.2.16) and (1.2.14) that

$$\sigma_1(f, \Psi) = \|f\|.$$

Thus we can take 0 as a best approximant. Let us now consider the general case of m-term approximation. Let

$$f^n := \sum_{j=1}^{m} c_{k_j^n}^n \psi_{k_j^n}, \quad k_1^n < k_2^n < \cdots < k_m^n,$$

be such that

$$\|f - f^n\| \to \sigma_m(f, \Psi).$$

Then we have

$$|c_{k_j^n}^n| \leq M \qquad (1.2.17)$$

for all n, j with some constant M. Assume that we have

$$\liminf_{n \to \infty} k_j^n < \infty, \quad \text{for some (possibly none) } j = 1, \ldots, l \leq m,$$

$$\lim_{n \to \infty} k_j^n = \infty, \quad \text{for some (possibly none) } j = l+1, \ldots, m.$$

Then, as in the case of $m = 1$ we find Λ, $|\Lambda| \leq l$, and a subsequence $\{n_s\}_{s=1}^{\infty}$ such that

$$\sum_{k \in \Lambda} c_k^{n_s} \psi_k \to \sum_{k \in \Lambda} c_k \psi_k =: y. \qquad (1.2.18)$$

Consider the norming functional F_{f-y}. We have from (1.2.17), (1.2.18) and weak convergence of $\{\psi_k\}$ to 0 that

$$F_{f-y}(f^{n_s} - y) \to 0, \quad \text{as } s \to \infty.$$

Thus

$$\|f - y\| = F_{f-y}(f - y) = F_{f-y}(f - f^{n_s} + f^{n_s} - y)$$
$$\leq \|f - f^{n_s}\| + |F_{f-y}(f^{n_s} - y)| \to \sigma_m(f, \Psi),$$

as $s \to \infty$. This implies that

$$\|f - y\| = \sigma_m(f, \Psi),$$

which completes the proof of Theorem 1.2.9. $\qquad \qquad \square$

The following observation is from Wojtaszczyk (2002b).

Remark 1.2.10. It is clear from the proof of Theorem 1.2.9 that the condition of weak convergence of ψ_k to 0 can be replaced by the condition $y(\psi_k) \to 0$ for every $y \in Y$. Here, $Y \subset X^*$ is such that, for all $f \in X$,

$$\|f\| = \sup_{y \in Y, \|y\| \leq 1} |y(f)|.$$

Also, Wojtaszczyk (2002b) contains an example of an unconditional basis Ψ and an element f such that the best m-term approximation of f with regard to Ψ does not exist.

1.3. Greedy bases

Let a Banach space X, with a basis $\Psi = \{\psi_k\}_{k=1}^{\infty}$, be given. We assume that $\|\psi_k\| \geq C > 0$, $k = 1, 2, \ldots$, and consider the following theoretical greedy algorithm. For a given element $f \in X$ we consider the expansion

$$f = \sum_{k=1}^{\infty} c_k(f, \Psi)\psi_k. \tag{1.3.1}$$

For an element $f \in X$ we say that a permutation ρ of the positive integers is decreasing if

$$|c_{k_1}(f, \Psi)| \geq |c_{k_2}(f, \Psi)| \geq \cdots, \tag{1.3.2}$$

where $\rho(j) = k_j$, for $j = 1, 2, \ldots$, and write $\rho \in D(f)$. If the inequalities are strict in (1.3.2), then $D(f)$ consists of only one permutation. We define the mth greedy approximant of f, with respect to the basis Ψ corresponding to a permutation $\rho \in D(f)$, by the formula

$$G_m(f) := G_m(f, \Psi) := G_m(f, \Psi, \rho) := \sum_{j=1}^{m} c_{k_j}(f, \Psi)\psi_{k_j}.$$

We note that there is another natural greedy-type algorithm based on ordering $\|c_k(f, \Psi)\psi_k\|$ instead of ordering absolute values of coefficients. In this case we do not need the restriction $\|\psi_k\| \geq C > 0$, $k = 1, 2, \ldots$. Let $\Lambda_m(f)$ be a set of indices such that

$$\min_{k \in \Lambda_m(f)} \|c_k(f, \Psi)\psi_k\| \geq \max_{k \notin \Lambda_m(f)} \|c_k(f, \Psi)\psi_k\|.$$

We define $G_m^X(f, \Psi)$ by the formula

$$G_m^X(f, \Psi) := S_{\Lambda_m(f)}(f, \Psi), \quad \text{where } S_E(f) := S_E(f, \Psi) := \sum_{k \in E} c_k(f, \Psi)\psi_k.$$

It is clear that for a normalized basis ($\|\psi_k\| = 1$, $k = 1, 2, \ldots$) the above two greedy algorithms coincide. It is also clear that the above greedy algorithm $G_m^X(\cdot, \Psi)$ can be considered as a greedy algorithm $G_m(\cdot, \Psi')$, with $\Psi' := \{\psi_k/\|\psi_k\|\}_{k=1}^{\infty}$ being a normalized version of the Ψ. Thus, we will concentrate on studying the algorithm $G_m(\cdot, \Psi)$. In the above definition of $G_m(\cdot, \Psi)$ we impose an extra condition on a basis Ψ: $\inf_k \|\psi_k\| > 0$. This restriction allows us to define $G_m(f, \Psi)$ for all $f \in X$. For the sake of completeness we will also discuss the case

$$\inf_k \|\psi_k\| = 0. \tag{1.3.3}$$

In this case we define the $G_m(f, \Psi)$ in the same way as above, but only for f of a special form:

$$f = \sum_{k \in Y} c_k(f, \Psi) \psi_k, \quad |Y| < \infty. \tag{1.3.4}$$

The above algorithm $G_m(\cdot, \Psi)$ is a simple algorithm which describes the theoretical scheme for m-term approximation of an element f. We call this algorithm the Greedy Algorithm (GA). In order to understand the efficiency of this algorithm we compare its accuracy with the best-possible accuracy when an approximant is a linear combination of m terms from Ψ. We define the best m-term approximation error with respect to Ψ as follows:

$$\sigma_m(f) := \sigma_m(f, \Psi)_X := \inf_{c_k, \Lambda} \left\| f - \sum_{k \in \Lambda} c_k \psi_k \right\|_X,$$

where the infimum is taken over coefficients c_k and sets of indices Λ with cardinality $|\Lambda| = m$. The best we can achieve with the algorithm G_m is

$$\| f - G_m(f, \Psi, \rho) \|_X = \sigma_m(f, \Psi)_X,$$

or the slightly weaker

$$\| f - G_m(f, \Psi, \rho) \|_X \le G \sigma_m(f, \Psi)_X, \tag{1.3.5}$$

for all elements $f \in X$, and with a constant $G = C(X, \Psi)$ independent of f and m. It was mentioned in the Introduction (see (1.1.8)) that, when $X = H$ is a Hilbert space and $\mathcal{B}$ is an orthonormal basis, we have

$$\| f - G_m(f, \mathcal{B}, \rho) \|_H = \sigma_m(f, \mathcal{B})_H.$$

Let us begin our discussion with an important class of bases: wavelet-type bases. For $X = L_p$, we will write p instead of L_p. Let $\mathcal{H} := \{H_k\}_{k=1}^{\infty}$ denote the Haar basis on $[0, 1)$ normalized in $L_2(0, 1)$. We denote by $\mathcal{H}_p := \{H_{k,p}\}_{k=1}^{\infty}$ the Haar basis $\mathcal{H}$ renormalized in $L_p(0, 1)$, which is defined as follows: $H_{1,p} = 1$ on $[0, 1)$ and, for $k = 2^n + l$, $l = 1, 2, \ldots, 2^n$, $n = 0, 1, \ldots,$

$$H_{k,p} = \begin{cases} 2^{n/p}, & x \in [(2l - 2)2^{-n-1}, (2l - 1)2^{-n-1}), \\ -2^{n/p}, & x \in [(2l - 1)2^{-n-1}, 2l2^{-n-1}), \\ 0, & \text{otherwise.} \end{cases}$$

We will use the following definition of the L_p-equivalence of bases. We say that $\Psi = \{\psi_k\}_{k=1}^{\infty}$ is L_p-equivalent to $\Phi = \{\phi_k\}_{k=1}^{\infty}$ if for any finite set Λ and any coefficients c_k, $k \in \Lambda$, we have

$$C_1(p, \Psi, \Phi) \left\| \sum_{k \in \Lambda} c_k \phi_k \right\|_p \le \left\| \sum_{k \in \Lambda} c_k \psi_k \right\|_p \le C_2(p, \Psi, \Phi) \left\| \sum_{k \in \Lambda} c_k \phi_k \right\|_p$$

with two positive constants $C_1(p, \Psi, \Phi), C_2(p, \Psi, \Phi)$ which may depend on p, Ψ, and Φ. For sufficient conditions on Ψ to be L_p-equivalent to $\mathcal{H}$, see Frazier and Jawerth (1990) and DeVore, Konyagin and Temlyakov (1998). In particular, it is known that all reasonable univariate wavelet-type bases are L_p-equivalent to $\mathcal{H}$ for $1 < p < \infty$. We proved the following theorem in Temlyakov (1998a).

Theorem 1.3.1. Let $1 < p < \infty$ and let a basis Ψ be L_p-equivalent to the Haar basis $\mathcal{H}$. Then, for any $f \in L_p(0,1)$, we have

$$\|f - G_m^p(f, \Psi)\|_p \leq C(p, \Psi) \sigma_m(f, \Psi)_p$$

with a constant $C(p, \Psi)$ independent of f and m.

By a simple renormalization argument we obtain the following version of Theorem 1.3.1.

Theorem 1.3.1A. Let $1 < p < \infty$ and let a basis Ψ be L_p-equivalent to the Haar basis $\mathcal{H}_p$. Then, for any $f \in L_p(0,1)$ and any $\rho \in D(f)$, we have

$$\|f - G_m(f, \Psi, \rho)\|_p \leq C(p, \Psi) \sigma_m(f, \Psi)_p$$

with a constant $C(p, \Psi)$ independent of f, ρ, and m.

We note that Temlyakov (1998a) also contains a generalization of Theorem 1.3.1 to the multivariate Haar basis obtained by the multi-resolution analysis procedure. These theorems motivated us to consider the general setting of greedy approximation in Banach spaces. We concentrated on studying bases which satisfy (1.3.5) for all individual functions. Definitions 1.3.2–1.3.4, below, are from Konyagin and Temlyakov (1999a).

Definition 1.3.2. We call a basis Ψ a greedy basis if, for every $f \in X$ (in the case $\inf_k \|\psi_k\| > 0$) and for f of the form (1.3.4) (in the case $\inf_k \|\psi_k\| = 0$), there exists a permutation $\rho \in D(f)$ such that the inequality

$$\|f - G_m(f, \Psi, \rho)\|_X \leq G \sigma_m(f, \Psi)_X \qquad (1.3.6)$$

holds with a constant independent of f, m.

Theorem 1.3.1A shows that each basis Ψ which is L_p-equivalent to the univariate Haar basis $\mathcal{H}_p$ is a greedy basis for $L_p(0,1)$, $1 < p < \infty$. We note that for a Hilbert space each orthonormal basis is a greedy basis with a constant $G = 1$ (see (1.3.6)).

We now give the definitions of unconditional and democratic bases.

Definition 1.3.3. A basis $\Psi = \{\psi_k\}_{k=1}^{\infty}$ of a Banach space X is said to be unconditional if, for every choice of signs $\theta = \{\theta_k\}_{k=1}^{\infty}$, $\theta_k = 1$ or -1,

$k = 1, 2, \ldots$, the linear operator M_θ defined by

$$M_\theta \left(\sum_{k=1}^{\infty} a_k \psi_k \right) = \sum_{k=1}^{\infty} a_k \theta_k \psi_k$$

is a bounded operator from X into X.

Definition 1.3.4. We say that a basis $\Psi = \{\psi_k\}_{k=1}^{\infty}$ is a democratic basis for X if there exists a constant $D := D(X, \Psi)$ such that, for any two finite sets of indices P and Q with the same cardinality $|P| = |Q|$, we have

$$\left\| \sum_{k \in P} \psi_k \right\| \leq D \left\| \sum_{k \in Q} \psi_k \right\|.$$

We proved in Konyagin and Temlyakov (1999a) the following theorem.

Theorem 1.3.5. A basis is greedy if and only if it is unconditional and democratic.

This theorem gives a characterization of greedy bases. Further investigations (Temlyakov 1998b, Cohen, DeVore and Hochmuth 2000, Kerkyacharian and Picard 2004, Gribonval and Nielsen 2001b, Kamont and Temlyakov 2004) showed that the concept of greedy bases is very useful in direct and inverse theorems of nonlinear approximation and also in applications in statistics.

Let us make a remark on bases Ψ that satisfy condition (1.3.3). In this case the greedy algorithm $G_m(\cdot, \Psi)$ is defined only for f of the form (1.3.4). However, if Ψ is a greedy basis, then by Theorem 1.3.5 it is democratic, and therefore satisfies the condition $\inf_k \|\psi_k\| > 0$. Thus, there are no greedy bases satisfying (1.3.3).

An interesting generalization of m-term approximation was considered in Cohen *et al.* (2000). Let $\Psi = \{\psi_I\}_I$ be a basis indexed by dyadic intervals. Take an α and assign to each index set Λ the following measure:

$$\Phi_\alpha(\Lambda) := \sum_{I \in \Lambda} |I|^\alpha.$$

In the case $\alpha = 0$ we get $\Phi_0(\Lambda) = |\Lambda|$. An analogue of best m-term approximation is as follows:

$$\inf_{\Lambda : \Phi_\alpha(\Lambda) \leq m} \inf_{c_I, I \in \Lambda} \left\| f - \sum_{I \in \Lambda} c_I \psi_I \right\|_p.$$

A detailed study of this type of approximation (restricted approximation) can be found in Cohen *et al.* (2000).

We now elaborate on the idea of assigning to each basis element ψ_k a non-negative weight w_k. We discuss weight-greedy bases and prove a criterion for weight-greedy bases similar to that for greedy bases.

Let Ψ be a basis for X. As above, if $\inf_n \|\psi_n\| > 0$ then $c_n(f) \to 0$ as $n \to \infty$, where

$$f = \sum_{n=1}^{\infty} c_n(f)\psi_n.$$

Then we can rearrange the coefficients $\{c_n(f)\}$ in the decreasing order

$$|c_{n_1}(f)| \geq |c_{n_2}(f)| \geq \cdots,$$

and define the mth greedy approximant as

$$G_m(f, \Psi) := \sum_{k=1}^{m} c_{n_k}(f)\psi_{n_k}. \tag{1.3.7}$$

In the case $\inf_n \|\psi_n\| = 0$ we define $G_m(f, \Psi)$ by (1.3.7) for f of the form

$$f = \sum_{n \in Y} c_n(f)\psi_n, \quad |Y| < \infty. \tag{1.3.8}$$

Let a weight sequence $w = \{w_n\}_{n=1}^{\infty}$, $w_n > 0$, be given. For $\Lambda \subset \mathbb{N}$, denote $w(\Lambda) := \sum_{n \in \Lambda} w_n$. For a positive real number $v > 0$ define

$$\sigma_v^w(f, \Psi) := \inf_{\{b_n\}, \Lambda: w(\Lambda) \leq v} \left\| f - \sum_{n \in \Lambda} b_n \psi_n \right\|,$$

where Λ are finite.

We present results from Kerkyacharian, Picard and Temlyakov (2006).

Definition 1.3.6. We call a basis Ψ a weight-greedy basis (w-greedy basis) if for any $f \in X$ in the case $\inf_n \|\psi_n\| > 0$ or for any $f \in X$ of the form (1.3.8) in the case $\inf_n \|\psi_n\| = 0$, we have

$$\|f - G_m(f, \Psi)\| \leq C_G \sigma_{w(\Lambda_m)}^w(f, \Psi),$$

where Λ_m is obtained from the representation

$$G_m(f, \Psi) = \sum_{n \in \Lambda_m} c_n(f)\psi_n, \quad |\Lambda_m| = m.$$

Definition 1.3.7. We call a basis Ψ a weight-democratic basis (w-democratic basis) if, for any finite $A, B \subset \mathbb{N}$ such that $w(A) \leq w(B)$, we have

$$\left\| \sum_{n \in A} \psi_n \right\| \leq C_D \left\| \sum_{n \in B} \psi_n \right\|.$$

Theorem 1.3.8. A basis Ψ is a w-greedy basis if and only if it is unconditional and w-democratic.

Proof. **I** We first prove the implication

$$\text{unconditional} + w\text{-democratic} \Rightarrow w\text{-greedy}.$$

Let f be any function, or a function of the form (1.3.8) if $\inf_n \|\psi_n\| = 0$. Consider

$$G_m(f, \Psi) = \sum_{n \in Q} c_n(f)\psi_n =: S_Q(f).$$

We take any finite set $P \subset \mathbb{N}$ satisfying $w(P) \leq w(Q)$. Then our assumption $w_n > 0$, $n \in \mathbb{N}$ implies that either $P = Q$ or $Q \setminus P$ is non-empty. As in the Introduction, let

$$E_P(f, \Psi) := \inf_{\{b_n\}} \left\| f - \sum_{n \in P} b_n \psi_n \right\|.$$

Then, by unconditionality of Ψ, we have (see (1.1.12))

$$\|f - S_P(f)\| \leq (K + 1) E_P(f, \Psi). \tag{1.3.9}$$

This (with $P = Q$) completes the proof in the case $\sigma^w_{w(Q)}(f, \Psi) = E_Q(f, \Psi)$. Suppose that $\sigma^w_{w(Q)}(f, \Psi) < E_Q(f, \Psi)$. Clearly, we may now consider only those P that satisfy the following two conditions:

$$w(P) \leq w(Q) \quad \text{and} \quad E_P(f, \Psi) < E_Q(f, \Psi).$$

For P satisfying the above conditions we have $Q \setminus P \neq \emptyset$. We estimate

$$\|f - S_Q(f)\| \leq \|f - S_P(f)\| + \|S_P(f) - S_Q(f)\|. \tag{1.3.10}$$

We have

$$S_P(f) - S_Q(f) = S_{P \setminus Q}(f) - S_{Q \setminus P}(f). \tag{1.3.11}$$

As for (1.3.9) we get

$$\|S_{Q \setminus P}(f)\| \leq K E_P(f, \Psi). \tag{1.3.12}$$

It remains to estimate $\|S_{P \setminus Q}(f)\|$. By unconditionality and w-democracy in the case of a real Banach space X, we have

$$\|S_{P \setminus Q}(f)\| \leq 2K \max_{n \in P \setminus Q} |c_n(f)| \left\| \sum_{n \in P \setminus Q} \psi_n \right\| \tag{1.3.13}$$

$$\leq 2K C_D \min_{n \in Q \setminus P} |c_n(f)| \left\| \sum_{n \in Q \setminus P} \psi_n \right\| \leq C(K) C_D \|S_{Q \setminus P}(f)\|.$$

In the case of a complex Banach space X the above inequalities hold with $2K$ replaced by $4K$. Combining (1.3.9)–(1.3.13), we complete the proof of part I. $\qquad\square$

II We now prove the implication

$$w\text{-greedy} \quad \Rightarrow \quad \text{unconditional} + w\text{-democratic}.$$

IIa We begin with the following one:

$$w\text{-greedy} \quad \Rightarrow \quad \text{unconditional}.$$

We will prove a slightly stronger statement.

Lemma 1.3.9. Let Ψ be a basis such that, for any f of the form (1.3.8), we have

$$\|f - G_m(f, \Psi)\| \le CE_\Lambda(f, \Psi),$$

where

$$G_m(f, \Psi) = \sum_{n \in \Lambda} c_n(f)\psi_n.$$

Then Ψ is unconditional.

Proof. It is clear that it is sufficient to prove that there exists a constant C_0 such that, for any finite Λ and any f of the form (1.3.8), we have

$$\|S_\Lambda(f)\| \le C_0\|f\|.$$

Let f and Λ be given and $\Lambda \subset [1, M]$. Consider

$$f_M := S_{[1,M]}(f).$$

Then $\|f_M\| \le C_B\|f\|$. We take a $b > \max_{1 \le n \le M} |c_n(f)|$ and define a new function

$$g := f_M - S_\Lambda(f_M) + b \sum_{n \in \Lambda} \psi_n.$$

Then

$$G_m(g, \Psi) = b \sum_{n \in \Lambda} \psi_n, \quad m := |\Lambda|,$$

and

$$E_\Lambda(g, \Psi) \le \|f_M\|.$$

Thus,

$$\|f_M - S_\Lambda(f_M)\| = \|g - G_m(g, \Psi)\| \le CE_\Lambda(g, \Psi) \le C\|f_M\|.$$

Therefore,

$$\|S_\Lambda(f)\| = \|S_\Lambda(f_M)\| \le C_0\|f\|. \qquad \square$$

IIb It remains to prove the implication

$$w\text{-greedy} \quad \Rightarrow \quad w\text{-democratic}.$$

First, let $A, B \subset \mathbb{N}$, $w(A) \le w(B)$, be such that $A \cap B = \emptyset$. Consider

$$f := \sum_{n \in A} \psi_n + (1 + \epsilon) \sum_{n \in B} \psi_n, \quad \epsilon > 0.$$

Then

$$G_m(f, \Psi) = (1 + \epsilon) \sum_{n \in B} \psi_n, \quad m := |B|,$$

and

$$E_A(f, \Psi) \leq \left\| \sum_{n \in B} \psi_n \right\| (1 + \epsilon).$$

Therefore, by the w-greedy assumption we get

$$\left\| \sum_{n \in A} \psi_n \right\| \leq C(1 + \epsilon) \left\| \sum_{n \in B} \psi_n \right\|.$$

Now let A, B be any finite subsets of $\mathbb{N}$ for which $w(A) \leq w(B)$. Then, using the unconditionality of Ψ proved in IIa and the above part of IIb, we obtain

$$\left\| \sum_{n \in A} \psi_n \right\| \leq \left\| \sum_{n \in A \setminus B} \psi_n \right\| + \left\| \sum_{n \in A \cap B} \psi_n \right\|$$

$$\leq C \left\| \sum_{n \in B \setminus A} \psi_n \right\| + K \left\| \sum_{n \in B} \psi_n \right\| \leq C_1 \left\| \sum_{n \in B} \psi_n \right\|.$$

This completes the proof of Theorem 1.3.8.   □

Theorems 1.3.5 and 1.3.8 show that *greedy = unconditional + democratic*. We now show that unconditionality does not imply democracy, and *vice versa*.

Unconditionality does not imply democracy. This follows from properties of the multivariate Haar system $\mathcal{H}^2 = \mathcal{H} \times \mathcal{H}$ defined as the tensor product of the univariate Haar systems $\mathcal{H}$ (see (1.3.14) below).

Democracy does not imply unconditionality. Let X be the set of all real sequences $x = (x_1, x_2, \ldots)$ such that

$$\|x\|_X = \sup_{N \in \mathbb{N}} \left| \sum_{n=1}^{N} x_n \right|$$

is finite. Clearly, X equipped with the norm $\|\cdot\|_X$ is a Banach space. Let $\psi_k \in X$, $k = 1, 2, \ldots$, be defined as $(\psi_k)_n = 1$ if $n = k$ and $(\psi_k)_n = 0$ otherwise. Let X_0 denote the subspace of X generated by the elements ψ_k. It is easy to see that $\{\psi_k\}$ is a democratic basis in X_0. However, it is not an unconditional basis, since

$$\left\| \sum_{k=1}^{m} \psi_k \right\|_X = m,$$

but

$$\left\|\sum_{k=1}^{m}(-1)^k\psi_k\right\|_X = 1.$$

We let $\mathcal{H}_p := \{H_{k,p}\}_{k=1}^{\infty}$ be the Haar basis $\mathcal{H}$ renormalized in $L_p([0,1))$. We define the multivariate Haar basis $\mathcal{H}_p^d$ to be the tensor product of the univariate Haar bases: $\mathcal{H}_p^d := \mathcal{H}_p \times \cdots \times \mathcal{H}_p$;

$$H_{\mathbf{n},p}(x) := H_{n_1,p}(x_1)\cdots H_{n_d,p}(x_d), \quad x = (x_1,\ldots,x_d), \quad \mathbf{n} = (n_1,\ldots,n_d).$$

Supports of functions $H_{\mathbf{n},p}$ are arbitrary dyadic parallelepipeds (intervals). It is known (see Temlyakov (2002a)) that the tensor product structure of the multivariate wavelet bases makes them universal for approximation of anisotropic smoothness classes with different anisotropy. It is also known that the study of such bases is more difficult than the study of the univariate bases. In many cases we need to develop new techniques and in some cases we encounter new phenomena. For instance, it turns out that the democratic property does not hold for the multivariate Haar basis $\mathcal{H}_p^d$ for $p \neq 2$. The following relation is known for $1 < p < \infty$:

$$\sup_{f \in L_p} \|f - G_m(f, \mathcal{H}_p^d)\|_p / \sigma_m(f, \mathcal{H}_p^d) \asymp (\log m)^{(d-1)|1/2-1/p|}. \qquad (1.3.14)$$

The lower bound in (1.3.14) was proved by R. Hochmuth; the upper bound in (1.3.14) was proved in the case $d = 2$, $4/3 \leq p \leq 4$, and was conjectured for all d, $1 < p < \infty$, in Temlyakov (1998b). The conjecture was proved in Wojtaszczyk (2000).

Let us return to the problem of finding a near-best m-term approximant of $f \in X$ with regard to a basis Ψ. This problem consists of two subproblems. First, we need to identify a set Λ_m of m indices that can be used in achieving near-best m-term approximation of f. Second, we need to find the coefficients $\{c_k\}$, $k \in \Lambda_m$, such that the approximant $\sum_{k\in\Lambda} c_k\psi_k$ provides near-best approximation of f. It is clear from the properties of an unconditional basis Ψ that, for any $f \in X$ and any Λ, we have (see (1.1.12))

$$\left\|f - \sum_{k\in\Lambda} c_k(f,\Psi)\psi_k\right\| \leq C \inf_{\{c_k\}} \left\|f - \sum_{k\in\Lambda} c_k\psi_k\right\|.$$

Therefore, in the case of an unconditional basis Ψ the second subproblem is easy: we can always choose the expansion coefficients $c_k(f, \Psi)$, $k \in \Lambda$. Theorem 1.3.5 shows that if a basis Ψ is simultaneously unconditional and democratic then the first subproblem is also easy: it follows from the definition of greedy basis that the algorithm of choosing the m biggest in absolute-value coefficients gives the set Λ_m.

It would be very interesting to understand how we can find Λ_m in the case when we only know that Ψ is unconditional. The following special case of the above problem is of great interest: $X = L_p([0,1]^d)$, $d \geq 2$, Ψ is the multivariate Haar basis $\mathcal{H}_p^d$, $1 < p < \infty$. It is known from Temlyakov (1998b), Wojtaszczyk (2000) and Kamont and Temlyakov (2004) that the function

$$\mu(m, \mathcal{H}_p^d) := \sup_{k \leq m} \left(\sup_{\Lambda:|\Lambda|=k} \left\| \sum_{\mathbf{n} \in \Lambda} H_{\mathbf{n},p} \right\|_p \Big/ \inf_{\Lambda:|\Lambda|=k} \left\| \sum_{\mathbf{n} \in \Lambda} H_{\mathbf{n},p} \right\|_p \right)$$

plays a very important role in estimates of the m-term greedy approximation in terms of the best m-term approximation. For instance (see Temlyakov (1998b)),

$$\| f - G_m^{L_p}(f, \mathcal{H}_p^d) \|_p \leq C(p,d) \mu(m, \mathcal{H}_p^d) \sigma_m(f, \mathcal{H}_p^d)_p, \quad 1 < p < \infty. \quad (1.3.15)$$

The following theorem gives, in particular, upper bounds for $\mu(m, \mathcal{H}_p^d)$.

Theorem 1.3.10. Let $1 < p < \infty$. Then, for any Λ, $|\Lambda| = m$, we have for $2 \leq p < \infty$

$$C_{p,d}^1 m^{1/p} \min_{\mathbf{n} \in \Lambda} |c_{\mathbf{n}}| \leq \left\| \sum_{\mathbf{n} \in \Lambda} c_{\mathbf{n}} H_{\mathbf{n},p} \right\|_p \leq C_{p,d}^2 m^{1/p} (\log m)^{h(p,d)} \max_{\mathbf{n} \in \Lambda} |c_{\mathbf{n}}|,$$

and for $1 < p \leq 2$

$$C_{p,d}^3 m^{1/p} (\log m)^{-h(p,d)} \min_{\mathbf{n} \in \Lambda} |c_{\mathbf{n}}| \leq \left\| \sum_{\mathbf{n} \in \Lambda} c_{\mathbf{n}} H_{\mathbf{n},p} \right\|_p \leq C_{p,d}^4 m^{1/p} \max_{\mathbf{n} \in \Lambda} |c_{\mathbf{n}}|,$$

where $h(p,d) := (d-1)|1/2 - 1/p|$.

Theorem 1.3.10 for $d = 1$, $1 < p < \infty$ was proved in Temlyakov (1998a), and for $d = 2$, $4/3 \leq p \leq 4$ it was proved in Temlyakov (1998b). Theorem 1.3.10 in the general case was proved in Wojtaszczyk (2000). It is known (Temlyakov 2002c) that the extra log factors in Theorem 1.3.10 are sharp.

Let Ψ be a normalized basis for $L_p([0,1))$. For the space $L_p([0,1)^d)$ we define $\Psi^d := \Psi \times \cdots \times \Psi$ (d times), and

$$\psi_{\mathbf{n}}(x) := \psi_{n_1}(x_1) \cdots \psi_{n_d}(x_d), \quad \text{for } x = (x_1, \dots, x_d), \quad \mathbf{n} = (n_1, \dots, n_d).$$

In Kerkyacharian *et al.* (2006) we proved the following theorem using a proof whose structure is similar to that from Wojtaszczyk (2000).

Theorem 1.3.11. Let $1 < p < \infty$ and let Ψ be a greedy basis for $L_p([0,1))$. Then, for any Λ, $|\Lambda| = m$, we have for $2 \leq p < \infty$

$$C_{p,d}^5 m^{1/p} \min_{\mathbf{n} \in \Lambda} |c_{\mathbf{n}}| \leq \left\| \sum_{\mathbf{n} \in \Lambda} c_{\mathbf{n}} \psi_{\mathbf{n}} \right\|_p \leq C_{p,d}^6 m^{1/p} (\log m)^{h(p,d)} \max_{\mathbf{n} \in \Lambda} |c_{\mathbf{n}}|,$$

and for $1 < p \leq 2$

$$C_{p,d}^7 m^{1/p} (\log m)^{-h(p,d)} \min_{\mathbf{n} \in \Lambda} |c_{\mathbf{n}}| \leq \left\| \sum_{\mathbf{n} \in \Lambda} c_{\mathbf{n}} \psi_{\mathbf{n}} \right\|_p \leq C_{p,d}^8 m^{1/p} \max_{\mathbf{n} \in \Lambda} |c_{\mathbf{n}}|,$$

where $h(p,d) := (d-1)|1/2 - 1/p|$.

Inequality (1.3.15) was extended in Wojtaszczyk (2000) to a normalized unconditional basis Ψ for X instead of $\mathcal{H}_p^d$ for $L_p([0,1)^d)$. Therefore, as a corollary of Theorem 1.3.11 we obtain the following inequality for a greedy basis Ψ (for $L_p([0,1))$)

$$\|f - G_m^{L_p}(f, \Psi^d)\|_p \leq C(\Psi, d, p)(\log m)^{h(p,d)} \sigma_m(f, \Psi^d)_p, \quad 1 < p < \infty. \tag{1.3.16}$$

1.4. Quasi-greedy and almost greedy bases

In Section 1.3 we imposed the condition

$$\inf_k \|\psi_k\| > 0 \tag{1.4.1}$$

on a basis Ψ, to define $G_m(f, \Psi)$ for all $f \in X$. We noticed that in the case of a greedy basis this condition is always satisfied. In this section we assume that (1.4.1) is satisfied.

Let us discuss the question of weakening the requirement that a basis be a greedy basis. We begin with a concept of quasi-greedy basis that was introduced in Konyagin and Temlyakov (1999a).

Definition 1.4.1. We call a basis Ψ a quasi-greedy basis if, for every $f \in X$ and every permutation $\rho \in D(f)$, we have

$$\|G_m(f, \Psi, \rho)\|_X \leq C\|f\|_X \tag{1.4.2}$$

with a constant C independent of f, m, and ρ.

It is clear that (1.4.2) is weaker then (1.3.6). Wojtaszczyk (2000) proved the following theorem.

Theorem 1.4.2. A basis Ψ is quasi-greedy if and only if, for any $f \in X$ and any $\rho \in D(f)$, we have

$$\|f - G_m(f, \Psi, \rho)\| \to 0 \quad \text{as } m \to \infty. \tag{1.4.3}$$

Theorem 1.4.2 allows us to use (1.4.3) as an equivalent definition of a quasi-greedy basis. We give one more equivalent definition of a quasi-greedy basis.

Definition 1.4.3. We say that a basis Ψ is quasi-greedy if there exists a constant C_Q such that, for any $f \in X$ and any finite set of indices Λ having the property

$$\min_{k \in \Lambda} |c_k(f)| \geq \max_{k \notin \Lambda} |c_k(f)|, \tag{1.4.4}$$

we have

$$\|S_\Lambda(f, \Psi)\| \leq C_Q \|f\|. \tag{1.4.5}$$

It is clear that for elements f with the unique decreasing rearrangement of coefficients ($\#D(f) = 1$), inequalities (1.4.2) and (1.4.5) are equivalent. By slightly modifying the coefficients and using the continuity argument we deduce that (1.4.2) and (1.4.5) are equivalent for general f.

We now continue a discussion from Section 1.3 of relations between the following concepts: greedy basis, unconditional basis, democratic basis, and quasi-greedy basis. Theorem 1.3.5 states that *greedy* = *unconditional* + *democratic*. It is clear from the definition of quasi-greedy basis that an unconditional basis is always a quasi-greedy basis. We now give an example from Konyagin and Temlyakov (1999a) of a basis that is quasi-greedy and democratic (even superdemocratic) and is not an unconditional basis.

It is clear that an unconditional and democratic basis Ψ satisfies the following inequality:

$$\left\| \sum_{k \in P} \theta_k \psi_k \right\| \leq D_S \left\| \sum_{k \in Q} \epsilon_k \psi_k \right\|, \tag{1.4.6}$$

for any two finite sets P and Q, $|P| = |Q|$, and any choices of signs $\theta_k = \pm 1$, $k \in P$, and $\epsilon_k = \pm 1$, $k \in Q$.

Definition 1.4.4. We say that a basis Ψ is a superdemocratic basis if it satisfies (1.4.6).

Theorem 1.3.5 implies that a greedy basis is a superdemocratic one. Now we will construct an example of a superdemocratic quasi-greedy basis which is not an unconditional basis, and therefore, by Theorem 1.3.5, is not a greedy basis.

Let X be the set of all real sequences $x = (x_1, x_2, \ldots) \in l_2$ such that

$$\|x\|_1 = \sup_{N \in \mathbb{N}} \left| \sum_{n=1}^{N} x_n / \sqrt{n} \right|$$

is finite. Clearly, X equipped with the norm

$$\| \cdot \| = \max(\| \cdot \|_{l_2}, \| \cdot \|_1)$$

is a Banach space. Let $\psi_k \in X$, $k = 1, 2, \ldots$, be defined as $(\psi_k)_n = 1$ if $n = k$ and $(\psi_k)_n = 0$ otherwise. Let X_0 denote the subspace of X generated by the elements ψ_k. It is easy to see that $\Psi = \{\psi_k\}$ is a democratic basis in X_0. Moreover, it is superdemocratic: for any $k_1, \ldots, k_m$ and for any choice of signs,

$$\sqrt{m} \leq \left\| \sum_{j=1}^{m} \pm \psi_{k_j} \right\| < 2\sqrt{m}. \tag{1.4.7}$$

Indeed, we have

$$\left\| \sum_{j=1}^{m} \pm \psi_{k_j} \right\|_{l_2} = \sqrt{m},$$

$$\left\| \sum_{j=1}^{m} \pm \psi_{k_j} \right\|_1 \leq \sum_{j=1}^{m} 1/\sqrt{j} < 2\sqrt{m},$$

and (1.4.7) follows. However, Ψ is not an unconditional basis since, for $m \geq 2$,

$$\left\| \sum_{k=1}^{m} \psi_k/\sqrt{k} \right\| \geq \sum_{k=1}^{m} 1/k \asymp \log m,$$

but

$$\left\| \sum_{k=1}^{m} (-1)^k \psi_k/\sqrt{k} \right\| \asymp \sqrt{\log m}.$$

We now prove that the basis Ψ constructed above is a quasi-greedy basis. Assume $\|f\| = 1$. Then, by definition of $\| \cdot \|$ we have

$$\sum_{k=1}^{\infty} |c_k(f)|^2 \leq 1, \tag{1.4.8}$$

and for any M

$$\left| \sum_{k=1}^{M} c_k(f) k^{-1/2} \right| \leq 1. \tag{1.4.9}$$

It is clear that for any Λ we have

$$\|S_\Lambda(f, \Psi)\|_{l_2} \leq \|f\|_{l_2} \leq 1. \tag{1.4.10}$$

We now estimate $\|S_\Lambda(f, \Psi)\|_1$. Let Λ be any finite set of indices satisfying (1.4.4), and let

$$\alpha := \min_{k \in \Lambda} |c_k(f)|.$$

If $\alpha = 0$, then $S_\Lambda(f, \Psi) = f$ and (1.4.5) holds. Therefore consider $\alpha > 0$, and, for any N, let

$$\Lambda^+(N) := \{k \in \Lambda : k > N\}, \qquad \Lambda^-(N) := \{k \in \Lambda : k \leq N\}.$$

By Hölder's inequality we have, for any N,

$$\sum_{k \in \Lambda^+(N)} |c_k(f)| k^{-1/2} \leq \left(\sum_{k \in \Lambda^+(N)} |c_k(f)|^{3/2} \right)^{2/3} \left(\sum_{k > N} k^{-3/2} \right)^{1/3} \quad (1.4.11)$$

$$\ll N^{-1/6} \left(\sum_{k \in \Lambda^+(N)} |c_k(f)|^{3/2} (|c_k(f)|/\alpha)^{1/2} \right)^{2/3} \ll (\alpha^2 N)^{-1/6}.$$

Choose $N_\alpha := [\alpha^{-2}] + 1$. Then, for any $M \leq N_\alpha$ we have by (1.4.9) that

$$\left| \sum_{k \in \Lambda^-(M)} c_k(f) k^{-1/2} \right| \leq \left| \sum_{k=1}^{M} c_k(f) k^{-1/2} \right| + \left| \sum_{k \notin \Lambda^-(M), k \leq M} c_k(f) k^{-1/2} \right|$$

$$\leq 1 + \alpha \sum_{k=1}^{M} k^{-1/2} \leq 1 + 2\alpha M^{1/2} \ll 1. \quad (1.4.12)$$

For $M > N_\alpha$, we get using (1.4.11) and (1.4.12)

$$\left| \sum_{k \in \Lambda^-(M)} c_k(f) k^{-1/2} \right| \leq \left| \sum_{k \in \Lambda^-(N_\alpha)} c_k(f) k^{-1/2} \right| + \sum_{k \in \Lambda^+(N_\alpha)} |c_k(f)| k^{-1/2} \ll 1.$$

Thus

$$\|S_\Lambda(f, \Psi)\|_1 \leq C,$$

which completes the proof.

The above example and Theorem 1.3.5 show that a quasi-greedy basis is not necessarily a greedy basis. Further results on quasi-greedy bases can be found in Wojtaszczyk (2000) and Dilworth, Kalton, Kutzarova and Temlyakov (2003).

The above discussion shows that a quasi-greedy basis is not necessarily an unconditional basis. However, quasi-greedy bases have some properties that are close to those of unconditional bases. We formulate two of them (see, for instance, Konyagin and Temlyakov (2002)).

Lemma 1.4.5. Let Ψ be a quasi-greedy basis. Then, for any two finite sets of indices $A \subseteq B$ and coefficients $0 < t \leq |a_j| \leq 1$, $j \in B$, we have

$$\left\| \sum_{j \in A} a_j \psi_j \right\| \leq C(X, \Psi, t) \left\| \sum_{j \in B} a_j \psi_j \right\|.$$

It will be convenient to define the quasi-greedy constant K to be the least constant such that

$$\|G_m(f)\| \leq K\|f\| \quad \text{and} \quad \|f - G_m(f)\| \leq K\|f\|, \quad f \in X.$$

Lemma 1.4.6. Suppose Ψ is a quasi-greedy basis with a quasi-greedy constant K. Then, for any real numbers a_j and any finite set of indices P, we have

$$(4K^2)^{-1} \min_{j \in P} |a_j| \left\| \sum_{j \in P} \psi_j \right\| \leq \left\| \sum_{j \in P} a_j \psi_j \right\| \leq 2K \max_{j \in P} |a_j| \left\| \sum_{j \in P} \psi_j \right\|.$$

We note that the mth greedy approximant $G_m(x, \Psi)$ changes if we renormalize the basis Ψ (replace it by a basis $\{\lambda_n \psi_n\}$). This gives us more flexibility in adjusting a given basis Ψ for greedy approximation. Let us make one observation from Konyagin and Temlyakov (2003a) along these lines.

Proposition 1.4.7. Let $\Psi = \{\psi_n\}_{n=1}^\infty$ be a normalized basis for a Banach space X. Then the basis $\{e_n\}_{n=1}^\infty$, $e_n := 2^n \psi_n$, $n = 1, 2, \ldots$ is a quasi-greedy basis in X.

We proceed to an intermediate concept of *almost greedy basis*. This concept was introduced and studied in Dilworth *et al.* (2003). Let

$$f = \sum_{k=1}^\infty c_k(f) \psi_k.$$

We define the following expansional best m-term approximation of f:

$$\tilde{\sigma}_m(f) := \tilde{\sigma}_m(f, \Psi) := \inf_{\Lambda, |\Lambda| = m} \left\| f - \sum_{k \in \Lambda} c_k(f) \psi_k \right\|.$$

It is clear that

$$\sigma_m(f, \Psi) \leq \tilde{\sigma}_m(f, \Psi).$$

It is also clear that for an unconditional basis Ψ we have

$$\tilde{\sigma}_m(f, \Psi) \leq C \sigma_m(f, \Psi).$$

Definition 1.4.8. We call a basis Ψ an almost greedy basis if, for every $f \in X$, there exists a permutation $\rho \in D(f)$ such that we have the inequality

$$\|f - G_m(f, \Psi, \rho)\|_X \leq C \tilde{\sigma}_m(f, \Psi)_X, \tag{1.4.13}$$

with a constant independent of f and m.

The following proposition follows from the proof of Theorem 3.3 of Dilworth *et al.* (2003) (see Theorem 1.4.10 below).

Proposition 1.4.9. If Ψ is an almost greedy basis then (1.4.13) holds for any permutation $\rho \in D(f)$.

The following characterization of almost greedy bases was obtained in Dilworth *et al.* (2003).

Theorem 1.4.10. Suppose Ψ is a basis of a Banach space. The following are equivalent.

A Ψ is almost greedy.

B Ψ is quasi-greedy and democratic.

C For any (respectively, every) $\lambda > 1$ there is a constant $C = C_\lambda$ such that

$$\|f - G_{[\lambda m]}(f, \Psi)\| \le C_\lambda \sigma_m(f, \Psi).$$

In order to give the reader an idea of relations between $\tilde{\sigma}$ and σ we present an estimate for $\tilde{\sigma}_n(f, \Psi)$ in terms of $\sigma_m(f, \Psi)$ for a quasi-greedy basis Ψ. For a basis Ψ we define the fundamental function

$$\varphi(m) := \sup_{|A| \le m} \left\| \sum_{k \in A} \psi_k \right\|.$$

We also need the following function:

$$\phi(m) := \inf_{|A| = m} \left\| \sum_{k \in A} \psi_k \right\|.$$

The following inequality was obtained in Dilworth *et al.* (2003).

Theorem 1.4.11. Let Ψ be a quasi-greedy basis. Then, for any m and r there exists a set E, $|E| \le m + r$ such that

$$\|f - S_E(f, \Psi)\| \le C \left(1 + \frac{\varphi(m)}{\phi(r+1)} \right) \sigma_m(f, \Psi).$$

In Section 1.3, in addition to bases Ψ satisfying (1.4.1), we discussed a more general case that included bases satisfying (1.3.3). In the latter case we defined the greedy algorithm $G_m(f, \Psi)$ for functions f of the form (1.3.4). We gave a definition of a greedy basis in the general case, which included those bases satisfying (1.3.3). However, the characterization of greedy bases given by Theorem 1.3.5 excluded bases satisfying (1.3.3). We note that a similar attempt to include bases Ψ satisfying (1.3.3) into the consideration of quasi-greedy bases does not work. Indeed, let Ψ be a normalized unconditional basis and consider a renormalized basis $\Psi' := \{\psi'_k := k^{-3}\psi_k\}$. Clearly, Ψ' is also an unconditional basis, and therefore inequality (1.4.2) is satisfied for any f of the form (1.3.4). However, for the function

$$f := \sum_{k=1}^{\infty} k^{-2} \psi_k = \sum_{k=1}^{\infty} k \psi'_k,$$

we cannot apply the algorithm $G_m(\cdot, \Psi')$ because the expansion coefficients are not bounded.

1.5. Weak Greedy Algorithms with respect to bases

The greedy approximant $G_m(f, \Psi)$ considered in Sections 1.3 and 1.4 was defined to be the sum

$$\sum_{j=1}^{m} c_{k_j}(f, \Psi)\psi_{k_j}$$

of the expansion terms with the m biggest coefficients in absolute value (see (1.3.2)). In this section we discuss a more flexible way to construct a greedy approximant. The rule for choosing the expansion terms for approximation will be weaker than in the greedy algorithm $G_m(\cdot, \Psi)$. Instead of taking m terms with the biggest coefficients we now take m terms with near-biggest coefficients. We proceed to a formal definition of the Weak Greedy Algorithm with regard to a basis Ψ. We assume here that Ψ satisfies (1.4.1).

Let $t \in (0, 1]$ be a fixed parameter. For a given basis Ψ and a given $f \in X$, let $\Lambda_m(t)$ be any set of m indices such that

$$\min_{k \in \Lambda_m(t)} |c_k(f, \Psi)| \geq t \max_{k \notin \Lambda_m(t)} |c_k(f, \Psi)|, \qquad (1.5.1)$$

and define

$$G_m^t(f) := G_m^t(f, \Psi) := \sum_{k \in \Lambda_m(t)} c_k(f, \Psi)\psi_k.$$

We call it the Weak Greedy Algorithm (WGA) with the weakness sequence $\{t\}$ (the weakness parameter t). We note that the WGA with regard to a basis was introduced in the very first paper (see Temlyakov (1998a)) on greedy bases. It is clear that $G_m^1(f, \Psi) = G_m(f, \Psi)$. It is also clear that, in the case $t < 1$, we have more flexibility in building a weak greedy approximant $G_m^t(f, \Psi)$ than in building $G_m(f, \Psi)$: it is one advantage of a weak greedy approximant $G_m^t(f, \Psi)$. The question is: How much does this flexibility affect efficiency of the algorithm? Surprisingly, it turns out that the effect is minimal: it is only reflected in a multiplicative constant (see below).

We begin our discussion with the case when Ψ is a greedy basis. It was proved in Temlyakov (1998a) that, when $X = L_p$, $1 < p < \infty$, and Ψ is the Haar system $\mathcal{H}_p$ normalized in L_p, we have

$$\|f - G_m^t(f, \mathcal{H}_p)\|_{L_p} \leq C(p, t)\sigma_m(f, \mathcal{H}_p)_{L_p}, \qquad (1.5.2)$$

for any $f \in L_p$. It was noted in Konyagin and Temlyakov (2002) that the proof of (1.5.2) from Temlyakov (1998a) works for any greedy basis, not merely the Haar system $\mathcal{H}_p$. Thus, we have the following result.

Theorem 1.5.1. For any greedy basis Ψ of a Banach space X, and any $t \in (0, 1]$, we have

$$\|f - G_m^t(f, \Psi)\|_X \leq C(\Psi, t)\sigma_m(f, \Psi)_X, \qquad (1.5.3)$$

for each $f \in X$.

We now consider the Weak Greedy Algorithm with regard to a quasi-greedy basis Ψ. It was proved in Konyagin and Temlyakov (2002) that the weak greedy approximant has properties similar to the greedy approximant.

Theorem 1.5.2. Let Ψ be a quasi-greedy basis. Then, for a fixed $t \in (0,1]$ and any m, we have for any $f \in X$

$$\|G_m^t(f, \Psi)\| \leq C(t)\|f\|. \tag{1.5.4}$$

The following theorem from Konyagin and Temlyakov (2002) is essentially due to Wojtaszczyk (2000).

Theorem 1.5.3. Let Ψ be a quasi-greedy basis for a Banach space X. Then, for any fixed $t \in (0,1]$, we have for each $f \in X$ that

$$G_m^t(f, \Psi) \to f \quad \text{as } m \to \infty.$$

Let us now proceed to an almost greedy basis Ψ. The following result was established in Konyagin and Temlyakov (2002).

Theorem 1.5.4. Let Ψ be an almost greedy basis. Then, for $t \in (0,1]$ we have for any m

$$\|f - G_m^t(f, \Psi)\| \leq C(t)\tilde{\sigma}_m(f, \Psi). \tag{1.5.5}$$

Proof. We drop Ψ from the notation for the sake of brevity. Take any $\epsilon > 0$ and find P, $|P| = m$ such that

$$\|f - S_P(f)\| \leq \tilde{\sigma}_m(f) + \epsilon.$$

Let $Q := \Lambda_m(t)$ with $\Lambda_m(t)$ from the definition of $G_m^t(f)$. Then

$$\|f - G_m^t(f)\| \leq \|f - S_P(f)\| + \|S_P(f) - S_Q(f)\|. \tag{1.5.6}$$

We have

$$S_P(f) - S_Q(f) = S_{P \setminus Q}(f) - S_{Q \setminus P}(f). \tag{1.5.7}$$

Let us first estimate $\|S_{Q \setminus P}(f)\|$. Denote $f_1 := f - S_P(f)$. Then

$$S_{Q \setminus P}(f) = S_{Q \setminus P}(f_1).$$

Next,

$$\min_{k \in Q \setminus P} |c_k(f_1)| = \min_{k \in Q \setminus P} |c_k(f)| \geq \min_{k \in Q} |c_k(f)|$$

$$\geq t \max_{k \notin Q} |c_k(f)| \geq t \max_{k \notin Q} |c_k(f_1)| = t \max_{k \notin Q \setminus P} |c_k(f_1)|.$$

Thus $Q \setminus P = \Lambda_n(t)$ for f_1 with $n := |Q \setminus P|$. By Theorem 1.5.2 we have

$$\|S_{Q \setminus P}(f)\| \leq C_1(t)\|f_1\|. \tag{1.5.8}$$

We now estimate $\|S_{P\setminus Q}(f)\|$. From the definition of Q we easily derive

$$at \leq b, \quad \text{where} \quad a := \max_{k \in P\setminus Q} |c_k(f)|, \quad b := \min_{k \in Q\setminus P} |c_k(f)|. \tag{1.5.9}$$

By Lemma 1.4.6 (see Lemma 2.1 from Dilworth $et\ al.$ (2003)),

$$\|S_{P\setminus Q}(f)\| \leq 2Ka \left\| \sum_{k \in P\setminus Q} \psi_k \right\| \tag{1.5.10}$$

and (see Lemma 2.2 from Dilworth $et\ al.$ (2003))

$$\|S_{Q\setminus P}(f)\| \geq (4K^2)^{-1} b \left\| \sum_{k \in Q\setminus P} \psi_k \right\|. \tag{1.5.11}$$

By Theorem 1.4.10 an almost greedy basis is a democratic basis. Thus we obtain

$$\left\| \sum_{k \in P\setminus Q} \psi_k \right\| \leq D \left\| \sum_{k \in Q\setminus P} \psi_k \right\|. \tag{1.5.12}$$

Combining (1.5.6)–(1.5.12) we obtain (1.5.5). Theorem 1.5.4 is proved. $\square$

We now discuss the stability of the greedy-type property of a basis. Let $0 < a \leq \lambda_k \leq b < \infty$, $k = 1, 2, \ldots$ and for a basis $\Psi = \{\psi_k\}$ consider $\Psi^\lambda := \{\lambda_k \psi_k\}$. The following theorem is from Konyagin and Temlyakov (2002). We note that the case for quasi-greedy bases was proved in Wojtaszczyk (2000).

Theorem 1.5.5.   Let a basis Ψ have one of the following properties:

(1) greedy,
(2) almost greedy,
(3) quasi-greedy.

Then the basis Ψ^λ has the same property.

Proof.   Let $f \in X$ and

$$f = \sum_k c_k(f)\psi_k = \sum_k c_k(f)\lambda_k^{-1}\lambda_k\psi_k.$$

Consider

$$G_m(f, \Psi^\lambda) = \sum_{k \in \Lambda_m} (c_k(f)\lambda_k^{-1})\lambda_k\psi_k.$$

Then, using $\lambda_k \in [a, b]$ and the definition of the $G_m(f, \Psi^\lambda)$, we obtain

$$\min_{k \in \Lambda_m} |c_k(f)| \geq a \min_{k \in \Lambda_m} |c_k(f)|\lambda_k^{-1} \geq a \max_{k \notin \Lambda_m} |c_k(f)|\lambda_k^{-1} \geq \frac{a}{b} \max_{k \notin \Lambda_m} |c_k(f)|.$$

Therefore, the set Λ_m can be interpreted as a $\Lambda_m(t)$ with $t = a/b$ with regard to the basis Ψ. It remains to apply the corresponding results for $G_m^t(f, \Psi)$: (1.5.3) in case (1), (1.5.4) in case (3), and (1.5.5) in case (2). This completes the proof of Theorem 1.5.5. $\square$

Kamont and Temlyakov (2004) studied the following modification of the above weak-type greedy algorithm as a way to further weaken restriction (1.5.1). We call this modification the Weak Greedy Algorithm (WGA) with a weakness sequence $\tau = \{t_k\}$. Let a weakness sequence $\tau := \{t_k\}_{k=1}^{\infty}$, $t_k \in [0, 1]$, $k = 1, \dots$ be given. We define the WGA by induction. We take an element $f \in X$, and at the first step we let

$$\Lambda_1(\tau) := \{n_1\}, \qquad G_1^{\tau}(f, \Psi) := c_{n_1} \psi_{n_1},$$

with any n_1 satisfying

$$|c_{n_1}| \geq t_1 \max_n |c_n|,$$

where we write $c_n := c_n(f, \Psi)$ for brevity. Assume we have already defined

$$G_{m-1}^{\tau}(f, \Psi) := G_{m-1}^{X,\tau}(f, \Psi) := \sum_{n \in \Lambda_{m-1}(\tau)} c_n \psi_n.$$

Then, at the mth step we define

$$\Lambda_m(\tau) := \Lambda_{m-1}(\tau) \cup \{n_m\}, \qquad G_m^{\tau}(f, \Psi) := G_m^{X,\tau}(f, \Psi) := \sum_{n \in \Lambda_m(\tau)} c_n \psi_n,$$

with any $n_m \notin \Lambda_{m-1}(\tau)$ satisfying

$$|c_{n_m}| \geq t_m \max_{n \notin \Lambda_{m-1}(\tau)} |c_n|.$$

Thus, for $f \in X$ the WGA builds a rearrangement of a subsequence of the expansion (1.3.1). If Ψ is an unconditional basis then we also have the limit $G_m^{\tau}(f, \Psi) \to f^*$. It is clear that in this case $f^* = f$ if and only if the sequence $\{n_k\}_{k=1}^{\infty}$ contains indices of all non-zero $c_n(f, \Psi)$. We say that the WGA corresponding to Ψ and τ is convergent if, for any realization $G_m^{\tau}(f, \Psi)$, we have

$$\|f - G_m^{\tau}(f, \Psi)\| \to 0 \quad \text{as } m \to \infty,$$

for all $f \in X$.

We formulate here only one theorem from Kamont and Temlyakov (2004).

Theorem 1.5.6. Let $2 \leq p < \infty$, $d \geq 1$ and let Ψ be a normalized unconditional basis in $L_p([0, 1]^d)$. Let $\tau = \{t_n : n \geq 1\}$ be a weakness sequence. Then the WGA corresponding to Ψ and τ converges if and only if $\tau \notin l_p$.

1.6. Thresholding and minimal systems

In this section we briefly discuss some further generalizations. Here, we assume that X is a quasi-Banach space and replace a basis by a complete minimal system. In addition, we consider the Weak Thresholding Algorithm and prove that its convergence is equivalent to convergence of the Weak Greedy Algorithm (see Proposition 1.6.3). Thresholding algorithms are very useful in statistics (see, for instance, Donoho and Johnstone (1994)).

Let X be a quasi-Banach space (real or complex) with the quasi-norm $\|\cdot\|$ such that for all $x, y \in X$ we have $\|x+y\| \leq \alpha(\|x\| + \|y\|)$ and $\|tx\| = |t|\|x\|$. It is well known (see Kalton, Beck and Roberts (1984, Lemma 1.1)) that there is a p, $0 < p \leq 1$, such that

$$\left\| \sum_n x_n \right\| \leq 4^{1/p} \left(\sum_n \|x_n\|^p \right)^{1/p}. \tag{1.6.1}$$

Let $\{e_n\} \subset X$ be a complete minimal system in X with the conjugate (dual) system $\{e_n^*\} \subset X^*$ ($e_n^*(e_n) = 1$, $e_n^*(e_k) = 0$, $k \neq n$). We assume that $\sup_n \|e_n^*\| < \infty$. This implies that for each $x \in X$ we have

$$\lim_{n \to \infty} e_n^*(x) = 0. \tag{1.6.2}$$

Any element $x \in X$ has a formal expansion

$$x \sim \sum_n e_n^*(x) e_n, \tag{1.6.3}$$

and various types of convergence of the series (1.6.3) can be studied. In this section we deal with greedy-type approximations with regard to the system $\{e_n\}$. We note that in this section we use the notation x and $\{e_n\}$ for an element and for a system, respectively, differing from the notation f and Ψ used in previous sections, to emphasize that we are now in a more general setting. It will be convenient for us to define a unique 'greedy ordering' in this section. For any $x \in X$ we define the greedy ordering for x as the map $\rho : \mathbb{N} \to \mathbb{N}$ for which $\{j : e_j^*(x) \neq 0\} \subset \rho(\mathbb{N})$, and such that, if $j < k$, then either $|e_{\rho(j)}^*(x)| > |e_{\rho(k)}^*(x)|$ or $|e_{\rho(j)}^*(x)| = |e_{\rho(k)}^*(x)|$ and $\rho(j) < \rho(k)$. The mth greedy approximation is given by

$$G_m(x) := G_m(x, \{e_n\}) := \sum_{j=1}^m e_{\rho(j)}^*(x) e_{\rho(j)}.$$

The system $\{e_n\}$ is a quasi-greedy system (Konyagin and Temlyakov 1999a) if there exists a constant C such that $\|G_m(x)\| \leq C\|x\|$ for all $x \in X$ and $m \in \mathbb{N}$. Wojtaszczyk (2000) proved that these are precisely the systems for which $\lim_{m \to \infty} G_m(x) = x$ for all x. If, as in Section 1.4, a quasi-greedy system $\{e_n\}$ is a basis, then we say that $\{e_n\}$ is a quasi-greedy basis. As we mentioned above, it is clear that any unconditional basis is a quasi-greedy basis. We note that there are conditional quasi-greedy bases $\{e_n\}$ in some Banach spaces. Hence, for such a basis $\{e_n\}$ there exists a permutation of $\{e_n\}$ which forms a quasi-greedy system but not a basis. This remark justifies the study of the class of quasi-greedy systems rather than the class of quasi-greedy bases.

Greedy approximations are close to thresholding approximations (sometimes they are called *thresholding greedy approximations*). Thresholding approximations are defined by

$$T_\epsilon(x) := \sum_{|e_j^*(x)| \geq \epsilon} e_j^*(x) e_j, \quad \epsilon > 0.$$

Clearly, for any $\epsilon > 0$ there exists an m such that $T_\epsilon(x) = G_m(x)$. Therefore, if $\{e_n\}$ is a quasi-greedy system then

$$\forall x \in X \quad \lim_{\epsilon \to 0} T_\epsilon(x) = x. \tag{1.6.4}$$

Conversely, following the Remark from Wojtaszczyk (2000, pp. 296–297), it is easy to show that condition (1.6.4) implies that $\{e_n\}$ is a quasi-greedy system.

As in Section 1.5, one can define the Weak Thresholding Approximation. Fix $t \in (0, 1)$. For $\epsilon > 0$ let

$$D_{t,\epsilon}(x) := \{j : t\epsilon \leq |e_j^*(x)| < \epsilon\}.$$

The Weak Thresholding Approximations are defined as all possible sums

$$T_{\epsilon,D}(x) = \sum_{|e_j^*(x)| \geq \epsilon} e_j^*(x) e_j + \sum_{j \in D} e_j^*(x) e_j,$$

where $D \subseteq D_{t,\epsilon}(x)$. We say that the Weak Thresholding Algorithm converges for $x \in X$, and write $x \in \mathrm{WT}\{e_n\}(t)$ if, for any $D(\epsilon) \subseteq D_{t,\epsilon}$,

$$\lim_{\epsilon \to 0} T_{\epsilon,D(\epsilon)}(x) = x.$$

It is clear that the above relation is equivalent to

$$\lim_{\epsilon \to 0} \sup_{D \subseteq D_{t,\epsilon}(x)} \|x - T_{\epsilon,D}(x)\| = 0.$$

We proved in Konyagin and Temlyakov (2003a) (see Theorem 1.6.1 below) that the set $\mathrm{WT}\{e_n\}(t)$ does not depend on $t \in (0, 1)$. Therefore, we can drop t from the notation: $\mathrm{WT}\{e_n\} = \mathrm{WT}\{e_n\}(t)$.

It turns out that the Weak Thresholding Algorithm has more regularity than the Thresholding Algorithm: we will see that the set $\mathrm{WT}\{e_n\}$ is linear. On the other hand, by 'weakening' the Thresholding Algorithm (making convergence stronger), we do not narrow the convergence set too much. It is known that for many natural classes of sets $Y \subseteq X$ the convergence of $T_\epsilon(x)$ to x for all $x \in Y$ is equivalent to the condition $Y \subseteq \mathrm{WT}\{e_n\}$. In particular, it can be derived from Wojtaszczyk (2000, Proposition 3) that the above two conditions are equivalent for $Y = X$.

We suppose that X and $\{e_n\}$ satisfy the conditions stated in the beginning of this section. The following two theorems were proved in Konyagin and Temlyakov (2003 a).

Theorem 1.6.1. Let $t, t' \in (0, 1)$, $x \in X$. Then the following conditions are equivalent.

(1) $\lim_{\epsilon \to 0} \sup_{D \subseteq D_{t,\epsilon}(x)} \|T_{\epsilon,D}(x) - x\| = 0$.

(2) $\lim_{\epsilon \to 0} T_\epsilon(x) = x$ and

$$\lim_{\epsilon \to 0} \sup_{D \subseteq D_{t,\epsilon}(x)} \left\| \sum_{j \in D} e_j^*(x) e_j \right\| = 0. \qquad (1.6.5)$$

(3) $\lim_{\epsilon \to 0} T_\epsilon(x) = x$ and

$$\lim_{\epsilon \to 0} \sup_{|a_j| \leq 1 (j \in D_{t,\epsilon}(x))} \left\| \sum_{j \in D_{t,\epsilon}(x)} a_j e_j^*(x) e_j \right\| = 0. \qquad (1.6.6)$$

(4) $\lim_{\epsilon \to 0} T_\epsilon(x) = x$ and

$$\lim_{\epsilon \to 0} \sup_{|b_j| < \epsilon (j : |e_j^*(x)| \geq \epsilon)} \left\| \sum_{j : |e_j^*(x)| \geq \epsilon} b_j e_j \right\| = 0. \qquad (1.6.7)$$

(5) $\lim_{\epsilon \to 0} \sup_{D \subseteq D_{t',\epsilon}(x)} \|T_{\epsilon,D}(x) - x\| = 0$.

So, the set $\mathrm{WT}\{e_n\}(t)$ defined above is indeed independent of $t \in (0, 1)$.

Theorem 1.6.2. The set $\mathrm{WT}\{e_n\}$ is linear.

Let us discuss relations between the Weak Thresholding Algorithm $T_{\epsilon,D}(x)$ and the Weak Greedy Algorithm $G_m^t(x)$. We define $G_m^t(x)$ with regard to a minimal system $\{e_n\}$ in the same way as it was defined for a basis Ψ. For a given system $\{e_n\}$ and $t \in (0, 1]$, we denote for $x \in X$ and $m \in \mathbb{N}$ by $W_m(t)$ any set of m indices such that

$$\min_{j \in W_m(t)} |e_j^*(x)| \geq t \max_{j \notin W_m(t)} |c_j^*(x)|, \qquad (1.6.8)$$

and define

$$G_m^t(x) := G_m^t(x, \{e_n\}) := S_{W_m(t)}(x) := \sum_{j \in W_m(t)} e_j^*(x) e_j.$$

It is clear that for any $t \in (0, 1]$ and any $D \subseteq D_{t,\epsilon}(x)$ there exist m and $W_m(t)$ satisfying (1.6.8) such that

$$T_{\epsilon,D}(x) = S_{W_m(t)}(x). \qquad (1.6.9)$$

Thus the convergence $G_m^t(x) \to x$ as $m \to \infty$ implies the convergence $T_{\epsilon,D}(x) \to x$ as $\epsilon \to 0$ for any $t \in (0,1]$. We will now prove (see Konyagin and Temlyakov (2003a, Proposition 2.2)) that for $t \in (0,1)$ the inverse is also true.

Proposition 1.6.3. Let $t \in (0,1)$ and $x \in X$. Then the following two conditions are equivalent:

$$\lim_{\epsilon \to 0} \sup_{D \subseteq D_{t,\epsilon}(x)} \|T_{\epsilon,D}(x) - x\| = 0, \qquad (1.6.10)$$

$$\lim_{m \to \infty} \|G_m^t(x) - x\| = 0, \qquad (1.6.11)$$

for any realization $G_m^t(x)$.

Proof. The implication $(1.6.11) \Rightarrow (1.6.10)$ is simple and follows from the remark following (1.6.9). We prove that $(1.6.10) \Rightarrow (1.6.11)$. Let

$$\epsilon_m := \max_{j \notin W_m(t)} |e_j^*(x)|.$$

Clearly $\epsilon_m \to 0$ as $m \to \infty$. We have

$$G_m^t(x) = T_{2\epsilon_m}(x) + \sum_{j \in D_m} e_j^*(x) e_j, \qquad (1.6.12)$$

with D_m having the following property: for any $j \in D_m$,

$$t\epsilon_m \le |e_j^*(x)| < 2\epsilon_m.$$

Thus, by condition (5) from Theorem 1.6.1, for $t' = t/2$ we obtain (1.6.11).
    Proposition 1.6.3 is now proved. $\qquad \square$

Proposition 1.6.3 and Theorem 1.6.1 imply that the convergence set of the Weak Greedy Algorithm $G_m^t(\cdot)$ does not depend on $t \in (0,1)$ and coincides with WT$\{e_n\}$. By Theorem 1.6.2 this set is a linear set.
    Let us make a comment on the case $t = 1$ that is not covered by Proposition 1.6.3. It is clear that $T_\epsilon(x) = G_m(x)$ with some m, and therefore $G_m(x) \to x$ as $m \to \infty$ implies $T_\epsilon(x) \to x$ as $\epsilon \to 0$. It is also not difficult to understand that, in general, $T_\epsilon(x) \to x$ as $\epsilon \to 0$ does not imply $G_m(x) \to x$ as $m \to \infty$. This can be done, for instance, by considering the trigonometric system in the space L_p, $p \ne 2$, and using the Rudin–Shapiro polynomials (see Temlyakov (1998c)). However, if, for the trigonometric system, we put the Fourier coefficients with equal absolute values in a natural order (say, lexicographic), then, in the case $1 < p < \infty$, by Riesz's theorem we obtain convergence of $G_m(f)$ from convergence of $T_\epsilon(f)$. Results from Konyagin and Skopina (2001) show that the situation is different for $p = 1$. In this case the natural order does not help to derive convergence of $G_m(f)$ from convergence of $T_\epsilon(f)$.

1.7. Greedy approximation with respect to the trigonometric system

The first results (see Theorem 1.3.1) on greedy approximation with regard to bases showed that the Haar basis and other bases similar to it are very well designed for greedy approximation. In this section we discuss another classical system, namely, the trigonometric system from the point of view of greedy approximation. It is well known that the trigonometric system is not an unconditional basis for L_p, $p \neq 2$. Therefore, by Theorem 1.3.5 it is not a greedy basis for L_p, $p \neq 2$. In this section we mostly discuss convergence properties of the Weak Greedy Algorithm with regard to the trigonometric system. It is a non-trivial problem. We will demonstrate how it relates to some deep results in harmonic and functional analysis.

Consider a periodic function $f \in L_p(\mathbb{T}^d)$, $1 \leq p \leq \infty$, $(L_\infty(\mathbb{T}^d) = \mathcal{C}(\mathbb{T}^d))$, defined on the d-dimensional torus $\mathbb{T}^d$. Let a number $m \in \mathbb{N}$ and a number $t \in (0, 1]$ be given, and let Λ_m be a set of $k \in \mathbb{Z}^d$ with the properties

$$\min_{k \in \Lambda_m} |\hat{f}(k)| \geq t \max_{k \notin \Lambda_m} |\hat{f}(k)|, \quad |\Lambda_m| = m, \tag{1.7.1}$$

where

$$\hat{f}(k) := (2\pi)^{-d} \int_{\mathbb{T}^d} f(x) e^{-i(k,x)} \, dx$$

is a Fourier coefficient of f. We define

$$G_m^t(f) := G_m^t(f, \mathcal{T}^d) := S_{\Lambda_m}(f) := \sum_{k \in \Lambda_m} \hat{f}(k) e^{i(k,x)},$$

and call it an mth weak greedy approximant of f with regard to the trigonometric system $\mathcal{T}^d := \{e^{i(k,x)}\}_{k \in \mathbb{Z}^d}$, $\mathcal{T} := \mathcal{T}^1$. We write $G_m(f) = G_m^1(f)$ and call it an mth greedy approximant. Clearly, an mth weak greedy approximant and even an mth greedy approximant may not be unique. In this section we do not impose any extra restrictions on Λ_m in addition to (1.7.1). Thus, theorems formulated below hold for any choice of Λ_m satisfying (1.7.1), or, in other words, for any realization $G_m^t(f)$ of the weak greedy approximation.

We will discuss in detail only results concerning convergence of the WGA with regard to the trigonometric system. T. W. Körner (1996), answering a question raised by Carleson and Coifman, constructed a function from $L_2(\mathbb{T})$ and then, in Körner (1999), a continuous function such that $\{G_m(f, \mathcal{T})\}$ diverges almost everywhere. It was proved in Temlyakov (1998c) for $p \neq 2$, and in Cordoba and Fernandez (1998) for $p < 2$, that there exists an $f \in L_p(\mathbb{T})$ such that $\{G_m(f, \mathcal{T})\}$ does not converge in L_p. It was remarked in Temlyakov (2003a) that the method from Temlyakov (1998c) gives a little more.

(1) There exists a continuous function f such that $\{G_m(f, \mathcal{T})\}$ does not converge in $L_p(\mathbb{T})$ for any $p > 2$.

(2) There exists a function f that belongs to any $L_p(\mathbb{T})$, $p < 2$, such that $\{G_m(f, \mathcal{T})\}$ does not converge in measure.

Thus the above negative results show that the condition $f \in L_p(\mathbb{T}^d)$, $p \neq 2$, does not guarantee convergence of $\{G_m(f, \mathcal{T})\}$ in the L_p-norm. The main goal of this section is to complement the survey of Temlyakov (2003a) by recent results in the following setting: find an additional (to $f \in L_p$) condition on f to guarantee that $\|f - G_m(f, \mathcal{T})\|_p \to 0$ as $m \to \infty$. In Konyagin and Temlyakov (2003b) we proved the following theorem.

Theorem 1.7.1. Let $f \in L_p(\mathbb{T}^d)$, $2 < p \leq \infty$, and let $q > p' := p/(p-1)$. Assume that f satisfies the condition

$$\sum_{|k|>n} |\hat{f}(k)|^q = o(n^{d(1-q/p')})$$

where $|k| := \max_{1 \leq j \leq d} |k_j|$. Then we have

$$\lim_{m \to \infty} \|f - G_m^t(f, \mathcal{T}^d)\|_p = 0.$$

It was proved in Konyagin and Temlyakov (2003b) that Theorem 1.7.1 is sharp.

Proposition 1.7.2. For each $2 < p \leq \infty$ there exists $f \in L_p(\mathbb{T}^d)$ such that

$$|\hat{f}(k)| = O(|k|^{-d(1-1/p)}),$$

and the sequence $\{G_m(f)\}$ diverges in L_p.

Let us make some comments. For a given set Λ denote

$$E_\Lambda(f)_p := \inf_{c_k, k \in \Lambda} \left\| f - \sum_{k \in \Lambda} c_k e^{i(k,x)} \right\|_p, \quad S_\Lambda(f) := \sum_{k \in \Lambda} \hat{f}(k) e^{i(k,x)}.$$

Define a special domain

$$Q(n) := \{k : |k| \leq n^{1/d}\}.$$

Remark 1.7.3. Theorem 1.7.1 implies that if $f \in L_p$, $2 < p \leq \infty$, and

$$E_{Q(n)}(f)_2 = o(n^{1/p-1/2}),$$

then $G_m^t(f) \to f$ in L_p.

Remark 1.7.4. The proof of Proposition 1.7.2 (see Konyagin and Temlyakov (2003b)) implies that there is an $f \in L_p(\mathbb{T}^d)$ such that

$$E_{Q(n)}(f)_\infty = O(n^{1/p-1/2})$$

and $\{G_m(f)\}$ diverges in L_p, $2 < p \leq \infty$.

We note that Remark 1.7.3 can also be obtained from some general inequalities for $\|f - G_m^t(f)\|_p$. As in the above general definition of best m-term approximation, we define the best m-term approximation with regard to $\mathcal{T}^d$:

$$\sigma_m(f)_p := \sigma_m(f, \mathcal{T}^d)_p := \inf_{k^j \in \mathbb{Z}^d, c_j} \left\| f - \sum_{j=1}^{m} c_j e^{i(k^j, x)} \right\|_p.$$

The following inequality was proved in Temlyakov (1998c) for $t = 1$ and in Konyagin and Temlyakov (2003b) for general t.

Theorem 1.7.5. For each $f \in L_p(\mathbb{T}^d)$ and any $0 < t \le 1$ we have

$$\|f - G_m^t(f)\|_p \le (1 + (2 + 1/t)m^{h(p)})\sigma_m(f)_p, \quad 1 \le p \le \infty, \qquad (1.7.2)$$

where $h(p) := |1/2 - 1/p|$.

It was proved in Temlyakov (1998c) that the inequality (1.7.2) is sharp: there is a positive absolute constant C such that, for each m and $1 \le p \le \infty$, there exists a function $f \ne 0$ with the property

$$\|G_m(f)\|_p \ge C m^{h(p)} \|f\|_p. \qquad (1.7.3)$$

The above inequality (1.7.3) shows that the trigonometric system is not a quasi-greedy basis for L_p, $p \ne 2$. We formulate one more inequality from Konyagin and Temlyakov (2003b).

Theorem 1.7.6. Let $2 \le p \le \infty$. Then, for any $f \in L_p(\mathbb{T}^d)$ and any Q, $|Q| \le m$, we have

$$\|f - G_m^t(f)\|_p \le \|f - S_Q(f)\|_p + (3 + 1/t)(2m)^{h(p)} E_Q(f)_2.$$

We present some results from Konyagin and Temlyakov (2003b) that are formulated in terms of the Fourier coefficients. For $f \in L_1(\mathbb{T}^d)$ let $\{\hat{f}(k(l))\}_{l=1}^{\infty}$ denote the decreasing rearrangement of $\{\hat{f}(k)\}_{k \in \mathbb{Z}^d}$, i.e.,

$$|\hat{f}(k(1))| \ge |\hat{f}(k(2))| \ge \cdots.$$

Let $a_n(f) := |\hat{f}(k(n))|$.

Theorem 1.7.7. Let $2 < p < \infty$ and let a decreasing sequence $\{A_n\}_{n=1}^{\infty}$ satisfy the condition

$$A_n = o(n^{1/p-1}) \quad \text{as } n \to \infty.$$

Then, for any $f \in L_p(\mathbb{T}^d)$ with the property $a_n(f) \le A_n$, $n = 1, 2, \ldots$, we have

$$\lim_{m \to \infty} \|f - G_m^t(f)\|_p = 0.$$

We also proved in Konyagin and Temlyakov (2003b) that, for any decreasing sequence $\{A_n\}$ satisfying

$$\limsup_{n\to\infty} A_n n^{1-1/p} > 0,$$

there exists a function $f \in L_p$ such that $a_n(f) \le A_n$, $n = 1, \ldots$, whose sequence $\{G_m(f)\}$ of greedy approximants is divergent in L_p.

In Konyagin and Temlyakov (2003b) we proved a necessary and sufficient condition on the majorant $\{A_n\}$ to guarantee, under the assumption that f is a continuous function, the uniform convergence of greedy approximants to a function f.

Theorem 1.7.8. Let a decreasing sequence $\{A_n\}_{n=1}^{\infty}$ satisfy the condition ($\mathcal{A}_\infty$):

$$\sum_{M<n\le e^M} A_n = o(1) \quad \text{as } M \to \infty.$$

Then, for any $f \in C(\mathbb{T})$ with the property $a_n(f) \le A_n$, $n = 1, 2, \ldots$, we have

$$\lim_{m\to\infty} \|f - G_m^t(f, \mathcal{T})\|_\infty = 0.$$

The condition ($\mathcal{A}_\infty$) is very close to the convergence of the series $\sum_n A_n$; if the condition ($\mathcal{A}_\infty$) holds then we have

$$\sum_{n=1}^{N} A_n = o(\log_*(N)), \quad \text{as } N \to \infty,$$

where a function $\log_*(u)$ is defined to be bounded for $u \le 0$ and to satisfy $\log_*(u) = \log_*(\log u) + 1$ for $u > 0$. The function $\log_*(u)$ grows more slowly than any iterated logarithmic function.

The condition ($\mathcal{A}_\infty$) in Theorem 1.7.8 is sharp.

Theorem 1.7.9. Assume that a decreasing sequence $\{A_n\}_{n=1}^{\infty}$ does not satisfy the condition ($\mathcal{A}_\infty$). Then there exists a function $f \in C(\mathbb{T})$ with the property $a_n(f) \le A_n$, $n = 1, 2, \ldots$, and such that we have

$$\limsup_{m\to\infty} \|f - G_m(f, \mathcal{T})\|_\infty > 0$$

for some realization $G_m(f, \mathcal{T})$.

In Konyagin and Temlyakov (2005) we concentrated on imposing extra conditions in the following form. We assume that for some sequence $\{M(m)\}$, $M(m) > m$, we have

$$\|G_{M(m)}(f) - G_m(f)\|_p \to 0 \quad \text{as } m \to \infty.$$

When p is an even number, or $p = \infty$, we found, in Konyagin and Temlyakov

(2005), necessary and sufficient conditions on the growth of the sequence $\{M(m)\}$ to provide convergence $\|f - G_m(f)\|_p \to 0$ as $m \to \infty$. We proved the following theorem in Konyagin and Temlyakov (2005).

Theorem 1.7.10. Let $p = 2q$, $q \in \mathbb{N}$, be an even integer, $\delta > 0$. Assume that $f \in L_p(\mathbb{T})$ and there exists a sequence of positive integers $M(m) > m^{1+\delta}$ such that

$$\|G_m(f) - G_{M(m)}(f)\|_p \to 0 \quad \text{as } m \to \infty.$$

Then we have

$$\|G_m(f) - f\|_p \to 0 \quad \text{as } m \to \infty.$$

In Konyagin and Temlyakov (2005) we proved that the condition $M(m) > m^{1+\delta}$ cannot be replaced by the condition $M(m) > m^{1+o(1)}$.

Theorem 1.7.11. For any $p \in (2, \infty)$ there exists a function $f \in L_p(\mathbb{T})$ with an $L_p(\mathbb{T})$-divergent sequence $\{G_m(f)\}$ of greedy approximations with the following property. For any sequence $\{M(m)\}$ such that $m \leq M(m) \leq m^{1+o(1)}$, we have

$$\|G_{M(m)}(f) - G_m(f)\|_p \to 0, \quad \text{as } m \to \infty.$$

In Konyagin and Temlyakov (2005) we also considered the case $p = \infty$, and proved necessary and sufficient conditions for convergence of greedy approximations in the uniform norm. For a mapping $\alpha : W \to W$ we let α_k denote its k-fold iteration: $\alpha_k := \alpha \circ \alpha_{k-1}$.

Theorem 1.7.12. Let $\alpha : \mathbb{N} \to \mathbb{N}$ be strictly increasing. Then the following conditions are equivalent.

(a) For some $k \in \mathbb{N}$ and for any sufficiently large $m \in \mathbb{N}$, we have the inequality $\alpha_k(m) > e^m$.

(b) If $f \in C(\mathbb{T})$ and

$$\|G_{\alpha(m)}(f) - G_m(f)\|_\infty \to 0, \quad \text{as } m \to \infty,$$

then

$$\|f - G_m(f)\|_\infty \to 0 \quad \text{as } m \to \infty.$$

In order to illustrate the techniques used in the proofs of the above results we discuss some inequalities that were used in proving Theorems 1.7.10 and 1.7.12. The reader will also see from the further discussion a connection to some deep results in harmonic analysis. The general style of these inequalities is as follows. A function that has a sparse representation with regard to the trigonometric system cannot be approximated in L_p by functions with small Fourier coefficients. We begin our discussion with some concepts introduced in Konyagin and Temlyakov (2005) that are useful in proving such inequalities. The following new characteristic of a Banach space L_p plays an important role in such inequalities. We introduce some more notation. Let Λ be a finite subset of $\mathbb{Z}^d$. We let $|\Lambda|$ denote its cardinality and let $\mathcal{T}(\Lambda)$ be the span of $\{e^{i(k,x)}\}_{k\in\Lambda}$. Denote

$$\Sigma_m(\mathcal{T}) = \cup_{\Lambda:|\Lambda|\le m}\mathcal{T}(\Lambda).$$

For $f \in L_p$, $F \in L_{p'}$, $1 \le p \le \infty$, $p' = p/(p-1)$, we write

$$\langle F, f\rangle := \int_{\mathbb{T}^d} F\bar{f}\,d\mu, \quad d\mu := (2\pi)^{-d}\,dx.$$

Definition 1.7.13. Let Λ be a finite subset of $\mathbb{Z}^d$ and $1 \le p \le \infty$. We call a set $\Lambda' := \Lambda'(p,\gamma)$, $\gamma \in (0,1]$, a (p,γ)-dual to Λ if, for any $f \in \mathcal{T}(\Lambda)$, there exists $F \in \mathcal{T}(\Lambda')$ such that $\|F\|_{p'} = 1$ and $\langle F, f\rangle \ge \gamma\|f\|_p$.

Let $D(\Lambda, p, \gamma)$ denote the set of all (p,γ)-dual sets Λ'. The following function is important for us:

$$v(m,p,\gamma) := \sup_{\Lambda:|\Lambda|=m}\ \inf_{\Lambda'\in D(\Lambda,p,\gamma)}\ |\Lambda'|.$$

We note that in a particular case $p = 2q$, $q \in \mathbb{N}$ we have

$$v(m,p,1) \le m^{p-1}. \tag{1.7.4}$$

This follows immediately from the form of the norming functional F for $f \in L_p$:

$$F = f^{q-1}(\bar{f})^q\|f\|_p^{1-p}. \tag{1.7.5}$$

In Konyagin and Temlyakov (2005) we used the quantity $v(m,p,\gamma)$ in greedy approximation. We first prove a lemma.

Lemma 1.7.14. Let $2 \le p \le \infty$. For any $h \in \Sigma_m(\mathcal{T})$ and any $g \in L_p$, we have

$$\|h + g\|_p \ge \gamma\|h\|_p - v(m,p,\gamma)^{1-1/p}\|\{\hat{g}(k)\}\|_{\ell_\infty}.$$

Proof. Let $h \in \mathcal{T}(\Lambda)$ with $|\Lambda| = m$ and let $\Lambda' \in D(\Lambda,p,\gamma)$. Then, using the Definition 1.7.13 we find $F(h,\gamma) \in \mathcal{T}(\Lambda')$ such that

$$\|F(h,\gamma)\|_{p'} = 1 \quad \text{and} \quad \langle F(h,\gamma), h\rangle \ge \gamma\|h\|_p.$$

We have

$$\langle F(h,\gamma), h\rangle = \langle F(h,\gamma), h+g\rangle - \langle F(h,\gamma), g\rangle \le \|h+g\|_p + |\langle F(h,\gamma), g\rangle|.$$

Next,

$$|\langle F(h,\gamma), g\rangle| \le \|\{\hat{F}(h,\gamma)(k)\}\|_{\ell_1}\|\{\hat{g}(k)\}\|_{\ell_\infty}.$$

Using $F(h,\gamma) \in \mathcal{T}(\Lambda')$ and the Hausdorff–Young theorem (Zygmund 1959,

Chapter 12, Section 1.2), we obtain

$$\|\{\hat{F}(h,\gamma)(k)\}\|_{\ell_1} \leq |\Lambda'|^{1-1/p}\|\{\hat{F}(h,\gamma)(k)\}\|_{\ell_p}$$
$$\leq |\Lambda'|^{1-1/p}\|F(h,\gamma)\|_{p'} = |\Lambda'|^{1-1/p}.$$

We now combine the above inequalities and use the definition of $v(m,p,\gamma)$.

$\square$

Definition 1.7.15. Let X be a finite-dimensional subspace of L_p, $1 \leq p \leq \infty$. We call a subspace $Y \subset L_{p'}$ a (p,γ)-dual to X, $\gamma \in (0,1]$, if for any $f \in X$ there exists $F \in Y$ such that $\|F\|_{p'} = 1$ and $\langle F, f \rangle \geq \gamma\|f\|_p$.

As above, let $D(X,p,\gamma)$ denote the set of all (p,γ)-dual subspaces Y. Consider the following function:

$$w(m,p,\gamma) := \sup_{X:\dim X=m} \inf_{Y\in D(X,p,\gamma)} \dim Y.$$

We begin our discussion with a particular case: $p = 2q$, $q \in \mathbb{N}$. Let X be given and let $e_1, \ldots, e_m$ form a basis of X. Using the Hölder inequality for n functions $f_1, \ldots, f_n \in L_n$, we have

$$\int |f_1 \cdots f_n| \, d\mu \leq \|f_1\|_n \cdots \|f_n\|_n.$$

Setting $f_i = |e_j|^{p'}$, $n = p - 1$, we deduce that any function of the form

$$\prod_{i=1}^m |e_i|^{k_i}, \quad k_i \in \mathbb{N}, \quad \sum_{i=1}^m k_i = p - 1,$$

belongs to $L_{p'}$. It now follows from (1.7.5) that

$$w(m,p,1) \leq m^{p-1}, \quad p = 2q, \quad q \in \mathbb{N}. \tag{1.7.6}$$

There is a general theory of the uniform approximation property (UAP) which provides some estimates for $w(m,p,\gamma)$ and $v(m,p,\gamma)$. We give some definitions from this theory. For a given subspace X of L_p, $\dim X = m$, and a constant $K > 1$, let $k_p(X,K)$ be the smallest k such that there is an operator $I_X : L_p \to L_p$, with $I_X(f) = f$ for $f \in X$, $\|I_X\|_{L_p \to L_p} \leq K$, and rank $I_X \leq k$. Define

$$k_p(m,K) := \sup_{X:\dim X=m} k_p(X,K),$$

and let us discuss how $k_p(m,K)$ can be used in estimating $w(m,p,\gamma)$. Consider the dual operator I_X^* to I_X. Then $\|I_X^*\|_{L_{p'} \to L_{p'}} \leq K$ and rank $I_X^* \leq k_p(m,K)$. Let $f \in X$, $\dim X = m$, and let F_f be the norming functional for f. Define

$$F := I_X^*(F_f)/\|I_X^*(F_f)\|_{p'}.$$

Then, for any $f \in X$,

$$\langle f, I_X^*(F_f) \rangle = \langle I_X(f), F_f \rangle = \langle f, F_f \rangle = \|f\|_p$$

and

$$\|I_X^*(F_f)\|_{p'} \leq K$$

imply

$$\langle f, F \rangle \geq K^{-1}\|f\|_p.$$

Therefore

$$w(m, p, K^{-1}) \leq k_p(m, K). \qquad (1.7.7)$$

We note that the behaviour of functions $w(m, p, \gamma)$ and $k_p(m, K)$ may be very different. Bourgain (1992) proved that, for any $p \in (1, \infty)$, $p \neq 2$, the function $k_p(m, K)$ grows faster than any polynomial in m. The estimate (1.7.6) shows that, in the particular case $p = 2q$, $q \in \mathbb{N}$, the growth of $w(m, p, \gamma)$ is at most polynomial. This means that we cannot expect to obtain accurate estimates for $w(m, p, K^{-1})$ using inequality (1.7.7). We give one more application of the UAP in the style of Lemma 1.7.14.

Lemma 1.7.16. Let $2 \leq p \leq \infty$. For any $h \in \Sigma_m(\mathcal{T})$ and any $g \in L_p$ we have

$$\|h + g\|_p \geq K^{-1}\|h\|_p - k_p(m, K)^{1/2}\|g\|_2, \qquad (1.7.8)$$

$$\|h + g\|_p \geq K^{-2}\|h\|_p - k_p(m, K)\|\{\hat{g}(k)\}\|_{\ell_\infty}. \qquad (1.7.9)$$

Proof. Let $h \in \mathcal{T}(\Lambda)$, $|\Lambda| = m$. Take $X = \mathcal{T}(\Lambda)$ and consider the operator I_X provided by the UAP. Let $\psi_1, \ldots, \psi_M$ form an orthonormal basis for the range Y of the operator I_X. Then $M \leq k_p(m, K)$. Let

$$I_X(e^{i(k,x)}) = \sum_{j=1}^{M} c_j^k \psi_j.$$

Then the property $\|I_X\|_{L_p \to L_p} \leq K$ implies

$$\left(\sum_{j=1}^{M} |c_j^k|^2\right)^{1/2} = \|I_X(e^{i(k,x)})\|_2 \leq \|I_X(e^{i(k,x)})\|_p \leq K.$$

Consider along with the operator I_X, the new operator,

$$A := (2\pi)^{-d} \int_{\mathbb{T}^d} T_t I_X T_{-t} \, dt,$$

where T_t is the shift operator: $T_t(f) = f(\cdot + t)$. Then

$$A(e^{i(k,x)}) = \sum_{j=1}^{M} c_j^k (2\pi)^{-d} \int_{\mathbb{T}^d} e^{-i(k,t)} \psi_j(x+t) \, dt = \left(\sum_{j=1}^{M} c_j^k \hat{\psi}_j(k)\right) e^{i(k,x)}.$$

Let

$$\lambda_k := \sum_{j=1}^{M} c_j^k \hat{\psi}_j(k).$$

We have

$$\sum_k |\lambda_k|^2 \le \sum_k \left(\sum_{j=1}^{M} |c_j^k|^2 \right) \left(\sum_{j=1}^{M} |\hat{\psi}(k)|^2 \right) \le K^2 M.$$

Also, $\lambda_k = 1$ for $k \in \Lambda$. For the operator A we have

$$\|A\|_{L_p \to L_p} \le K \quad \text{and} \quad \|A\|_{L_2 \to L_\infty} \le K M^{1/2}.$$

Therefore

$$\|A(h+g)\|_p \le K \|h+g\|_p$$

and

$$\|A(h+g)\|_p \ge \|h\|_p - K M^{1/2} \|g\|_2.$$

This proves the first inequality.

Consider the operator $B := A^2$. Then

$$B(h) = h, \quad h \in \mathcal{T}(\Lambda), \quad \|B\|_{L_p \to L_p} \le K^2,$$

and

$$\|B(f)\|_\infty \le K^2 M \|\{\hat{f}(k)\}\|_{\ell_\infty}.$$

Now, on the one hand

$$\|B(h+g)\|_p \le K^2 \|h+g\|_p,$$

and on the other hand

$$\|B(h+g)\|_p = \|h + B(g)\|_p \ge \|h\|_p - K^2 M \|\{\hat{g}(k)\}\|_{\ell_\infty}.$$

This proves inequality (1.7.9).   □

Theorem 1.7.17.   For any $h \in \Sigma_m(\mathcal{T})$ and any $g \in L_\infty$ we have

$$\|h+g\|_\infty \ge K^{-1} \|h\|_\infty - e^{C(K)m/2} \|g\|_2,$$
$$\|h+g\|_\infty \ge K^{-2} \|h\|_\infty - e^{C(K)m} \|\{\hat{g}(k)\}\|_{\ell_\infty}.$$

Proof.   This theorem is a direct corollary of Lemma 1.7.16 and the known estimate

$$k_\infty(m, K) \le e^{C(K)m}$$

(see Figiel, Johnson and Schechtman (1988)).   □

As we have already mentioned, $k_p(m, K)$ increases faster than any polynomial. In Konyagin and Temlyakov (2005) we improved inequality (1.7.8) by using other arguments.

Lemma 1.7.18. Let $2 \leq p \leq \infty$. For any $h \in \Sigma_m(\mathcal{T})$ and any $g \in L_p$, we have

$$\|h + g\|_p^p \geq 2^{-p-1}\|h\|_p^p - 2m^{p/2}\|h\|_p^{p-2}\|g\|_2^2. \qquad (1.7.10)$$

We mention two inequalities from Konyagin and Temlyakov (2003b) in the style of the inequalities in Lemmas 1.7.14–1.7.18.

Lemma 1.7.19. Let $2 \leq p < \infty$ and $h \in L_p$, $\|h\|_p \neq 0$. Then, for any $g \in L_p$ we have

$$\|h\|_p \leq \|h + g\|_p + (\|h\|_{2p-2}/\|h\|_p)^{p-1}\|g\|_2.$$

Lemma 1.7.20. Let $h \in \Sigma_m(\mathcal{T})$, $\|h\|_\infty = 1$. Then, for any function g such that $\|g\|_2 \leq \frac{1}{4}(4\pi m)^{-m/2}$, we have

$$\|h + g\|_\infty \geq 1/4.$$

We proceed to estimate $v(m, p, \gamma)$ and $w(m, p, \gamma)$ for $p \in [2, \infty)$. In the special case of even p, we have by (1.7.4) and (1.7.6) that

$$v(m, p, 1) \leq m^{p-1}, \quad w(m, p, 1) \leq m^{p-1}.$$

The following bound was proved in Konyagin and Temlyakov (2005).

Lemma 1.7.21. Let $2 \leq p < \infty$, and let $\alpha := p/2 - [p/2]$. Then we have

$$v(m, p, \gamma) \leq m^{c(\alpha, \gamma)m^{1/2}+p-1}.$$

1.8. Greedy-type bases; direct and inverse theorems

Theorem 1.3.1 points out the importance of bases L_p-equivalent to the Haar basis. We will now discuss necessary and sufficient conditions for f to have a prescribed decay of $\{\sigma_m(f, \Psi)\}$ under the assumption that Ψ is L_p-equivalent to the Haar basis $\mathcal{H}$, $1 < p < \infty$. We will express these conditions in terms of coefficients $\{c_n(f)\}$ of the expansion

$$f = \sum_{n=1}^{\infty} c_n(f)\psi_n.$$

The direct theorems of approximation theory provide bounds of approximation error (in our case $\sigma_m(f, \Psi)$) in terms of smoothness properties of a function f. These theorems are also known under the name of Jackson-type inequalities. The inverse theorems of approximation theory (also known as Bernstein-type inequalities) provide some smoothness properties of a function f from the sequence of approximation errors (in our case $\{\sigma_m(f, \Psi)\}$). It is well understood in approximation theory (see Petrushev (1988), DeVore and Lorenz (1993) and DeVore (1998)) how the Jackson-type and Bernstein-type inequalities can be used in order to characterize the corresponding approximation spaces. In the case of our interest, when we study best m-term approximation with regard to bases that are L_p-equivalent to the Haar basis, the theory of Jackson and Bernstein inequalities has been developed in Cohen *et al.* (2000). It was used in Cohen *et al.* (2000) for a description of approximation spaces defined in terms of $\{\sigma_m(f, \Psi)\}$. We want to point out that in the special case of bases that are L_p-equivalent to the Haar basis (and also for some more general bases) there exists a simple direct way to describe the approximation spaces defined in terms of $\{\sigma_m(f, \Psi)\}$ (Temlyakov 1998b, Kamont and Temlyakov 2004, Kerkyacharian and Picard 2004). We present results from Temlyakov (1998b) here. The following lemma from Temlyakov (1998a) (see Lemmas 3.1 and 3.2) plays the key role in this consideration.

Lemma 1.8.1. Let a basis Ψ be L_p-equivalent to $\mathcal{H}_p$, $1 < p < \infty$. Then, for any finite Λ and $a \leq |c_n| \leq b$, $n \in \Lambda$, we have

$$C_1(p, \Psi)a(|\Lambda|)^{1/p} \leq \left\| \sum_{n \in \Lambda} c_n \psi_n \right\|_p \leq C_2(p, \Psi)b(|\Lambda|)^{1/p}. \qquad (1.8.1)$$

We note that the results that follow use only the assumption that Ψ is a greedy basis satisfying (1.8.1). We formulate a general statement and then consider several important particular examples of the rate of decrease of $\{\sigma_m(f, \Psi)_p\}$. We begin by introducing some notation. For a sequence $\mathcal{E} = \{\epsilon_k\}_{k=0}^{\infty}$ of positive numbers monotonically decreasing to zero (we write $\mathcal{E} \in$ MDP), we define inductively a sequence $\{N_s\}_{s=0}^{\infty}$ of non-negative integers:

$$N_0 = 0, \quad \text{and } N_s \text{ is the smallest integer satisfying} \qquad (1.8.2)$$

$$\epsilon_{N_s} < 2^{-s}, \qquad d_s := \max(N_{s+1} - N_s, 1).$$

We are going to consider the following examples of sequences.

Example 1.8.1. Take $\epsilon_0 = 1$ and $\epsilon_k = k^{-r}$, $r > 0$, $k = 1, 2, \ldots$. Then

$$N_s \asymp 2^{s/r} \quad \text{and} \quad d_s \asymp 2^{s/r}.$$

Example 1.8.2. Fix $0 < b < 1$ and take $\epsilon_k = 2^{-k^b}$, $k = 0, 1, 2, \ldots$. Then

$$N_s = s^{1/b} + O(1) \quad \text{and} \quad d_s \asymp s^{1/b-1}.$$

Let $f \in L_p$. Rearrange the sequence $\|c_n(f)\psi_n\|_p$ in decreasing order,

$$\|c_{n_1}(f)\psi_{n_1}\|_p \geq \|c_{n_2}(f)\psi_{n_2}\|_p \geq \cdots,$$

and define

$$a_k(f, p) := \|c_{n_k}(f)\psi_{n_k}\|_p.$$

We now give some inequalities for $a_k(f, p)$ and $\sigma_m(f, \Psi)_p$. We will use the abbreviations $\sigma_m(f)_p := \sigma_m(f, \Psi)_p$ and $\sigma_0(f)_p := \|f\|_p$.

Lemma 1.8.2. For any two positive integers $N < M$ we have

$$a_M(f, p) \leq C(p, \Psi)\sigma_N(f)_p(M - N)^{-1/p}.$$

Proof. By Theorem 1.3.1 we have, for all m,

$$\|f - G_m^p(f, \Psi)\|_p \leq C(p, \Psi)\sigma_m(f)_p.$$

Hence, and by definition of G_m^p, we get

$$J := \left\| \sum_{k=N+1}^{M} c_{n_k}(f)\psi_{n_k} \right\|_p \leq C(p, \Psi)(\sigma_N(f)_p + \sigma_M(f)_p). \qquad (1.8.3)$$

Next, we have for $k \in (N, M]$,

$$\|c_{n_k}(f)\psi_{n_k}\|_p \geq \|c_{n_M}(f)\psi_{n_M}\|_p = a_M(f, p),$$

and by Lemma 1.8.1 we obtain

$$a_M(f, p)(M - N)^{1/p} \leq C(p, \Psi)J. \qquad (1.8.4)$$

Relations (1.8.3) and (1.8.4) imply the conclusion of Lemma 1.8.2. $\qquad \square$

Lemma 1.8.3. For any sequence $m_0 < m_1 < m_2 < \cdots$ of non-negative integers we have

$$\sigma_{m_s}(f)_p \leq C(p, \Psi) \sum_{l=s}^{\infty} a_{m_l}(f, p)(m_{l+1} - m_l)^{1/p}.$$

Proof. We have

$$\sigma_{m_s}(f)_p \leq \left\| \sum_{k>m_s} c_{n_k}(f)\psi_{n_k} \right\|_p \leq \sum_{l=s}^{\infty} \left\| \sum_{k \in (m_l, m_{l+1}]} c_{n_k}(f)\psi_{n_k} \right\|_p.$$

Hence, using Lemma 1.8.1,

$$\sigma_{m_s}(f)_p \leq C(p, \Psi) \sum_{l=s}^{\infty} a_{m_l}(f, p)(m_{l+1} - m_l)^{1/p},$$

as required. $\qquad \square$

Theorem 1.8.4. Assume a given sequence $\mathcal{E} \in \text{MDP}$ satisfies the conditions

$$\epsilon_{N_s} \geq C_1 2^{-s}, \qquad d_{s+1} \leq C_2 d_s, \qquad s = 0, 1, 2, \ldots.$$

Then we have the equivalence

$$\sigma_n(f)_p \ll \epsilon_n \iff a_{N_s}(f, p) \ll 2^{-s} d_s^{-1/p}.$$

Proof. We first prove $\Rightarrow$. If $N_{s+1} > N_s$, then we use Lemma 1.8.2 with $M = N_{s+1}$ and $N = N_s$,

$$a_{N_{s+1}}(f, p) \leq C(p, \Psi)\sigma_{N_s}(f)_p d_s^{-1/p} \leq C(p, \Psi)2^{-s-1}(d_{s+1}/C_2)^{-1/p},$$

which implies the statement of Theorem 1.8.4 in this case. Let $N_{s+1} = N_s = \cdots = N_{s-j} > N_{s-j-1}$. The assumption $\epsilon_{N_s} \geq C_1 2^{-s}$ combined with the definition of N_s: $\epsilon_{N_s} < 2^{-s}$ imply that $j \leq C_3$. Then, from the above case we get

$$a_{N_{s-j}}(f,p) \ll 2^{-s+j}(d_{s-j})^{-1/p},$$

and therefore

$$a_{N_{s+1}}(f,p) \ll 2^{-s-1}(d_{s+1})^{-1/p}.$$

The implication $\Rightarrow$ has been proved.

We now prove the inverse statement $\Leftarrow$. Using Lemma 1.8.3, we get

$$\sigma_{N_s}(f)_p \ll \sum_{l=s}^{\infty} a_{N_l}(f,p)(N_{l+1} - N_l)^{1/p} \ll \sum_{l=s}^{\infty} 2^{-l} \ll 2^{-s} \ll \epsilon_{N_s},$$

and for $n \in [N_s, N_{s+1})$

$$\sigma_n(f)_p \leq \sigma_{N_s}(f)_p \ll \epsilon_{N_s}(f)_p \ll 2^{-s} \ll \epsilon_{N_{s+1}}(f)_p \leq \epsilon_n(f)_p. \qquad \square$$

Corollary 1.8.5. Theorem 1.8.4 applied to Examples 1.8.1 and 1.8.2 gives the following relations:

$$\sigma_m(f)_p \ll (m+1)^{-r} \iff a_n(f,p) \ll n^{-r-1/p}, \qquad (1.8.5)$$

$$\sigma_m(f)_p \ll 2^{-m^b} \iff a_n(f,p) \ll 2^{-n^b} n^{(1-1/b)/p}. \qquad (1.8.6)$$

Remark 1.8.6. Making use of Lemmas 1.8.2 and 1.8.3 we can prove a version of Corollary 1.8.5 with the sign $\ll$ replaced by $\asymp$.

Theorem 1.8.4 and Corollary 1.8.5 are in the spirit of the classical Jackson–Bernstein direct and inverse theorems in linear approximation theory, where conditions of the form

$$E_n(f)_p \ll \epsilon_n, \quad \text{or} \quad \|E_n(f)_p/\epsilon_n\|_{l_\infty} < \infty \qquad (1.8.7)$$

are imposed on the corresponding sequences of approximating characteristics. It is well known (see DeVore (1998)) that, in studying many questions of approximation theory, it is convenient to consider, along with the restriction (1.8.7), its following generalization:

$$\|E_n(f)_p/\epsilon_n\|_{l_q} < \infty. \qquad (1.8.8)$$

Lemmas 1.8.2 and 1.8.3 are also useful in handling this more general case. For instance, in the particular case of Example 1.8.1 we get the following statement.

Theorem 1.8.7. Let $1 < p < \infty$ and $0 < q < \infty$. Then, for any positive r we have the equivalence relation

$$\sum_m \sigma_m(f)_p^q m^{rq-1} < \infty \iff \sum_n a_n(f,p)^q n^{rq-1+q/p} < \infty.$$

Remark 1.8.8. The condition

$$\sum_n a_n(f,p)^q n^{rq-1+q/p} < \infty$$

with $q = \beta := (r+1/p)^{-1}$ takes a very simple form:

$$\sum_n a_n(f,p)^\beta = \sum_n \|c_n(f)\psi_n\|_p^\beta < \infty. \qquad (1.8.9)$$

In the case $\Psi = \mathcal{H}_p$, condition (1.8.9) is equivalent to f being in Besov space $B_\beta^r(L_\beta)$.

Corollary 1.8.9. Theorem 1.8.7 implies the following relation:

$$\sum_m \sigma_m(f,\mathcal{H})_p^\beta m^{r\beta-1} < \infty \iff f \in B_\beta^r(L_\beta),$$

where $\beta := (r+1/p)^{-1}$.

The statement similar to Corollary 1.8.9 for free-knot spline approximation was proved in Petrushev (1988). Corollary 1.8.9 and further results in this direction can be found in DeVore and Popov (1988) and DeVore, Jawerth and Popov (1992). We want to remark here that conditions in terms of $a_n(f,p)$ are convenient in applications. For instance, relation (1.8.5) can be rewritten using the idea of thresholding. For a given $f \in L_p$ denote

$$T(\epsilon) := \#\{a_k(f,p) : a_k(f,p) \geq \epsilon\}.$$

Then (1.8.5) is equivalent to

$$\sigma_m(f)_p \ll (m+1)^{-r} \iff T(\epsilon) \ll \epsilon^{-(r+1/p)^{-1}}.$$

For further results in this direction see DeVore (1998), Cohen *et al.* (2000) and Oswald (2001).

The above direct and inverse Theorem 1.8.7 that holds for greedy bases satisfying (1.8.1) was extended in Kerkyacharian and Picard (2004) to the case of quasi-greedy bases satisfying (1.8.1). Kerkyacharian and Picard (2004) say that a basis Ψ of a Banach space X has the p-Temlyakov property if there exists $0 < C < \infty$ such that, for any finite set of indices Λ, we have

$$C^{-1}\left(\min_{n\in\Lambda}|c_n|\right)|\Lambda|^{1/p} \leq \left\|\sum_{n\in\Lambda} c_n\psi_n\right\|_X \leq C\left(\max_{n\in\Lambda}|c_n|\right)|\Lambda|^{1/p}. \qquad (1.8.10)$$

Now let

$$f = \sum_{k=1}^\infty c_k(f)\psi_k$$

and

$$|c_{k_1}| \geq |c_{k_2}| \geq \cdots$$

be a decreasing reordering of the coefficients. The following result is from Kerkyacharian and Picard (2004).

Theorem 1.8.10. Let Ψ be a quasi-greedy basis.

(1) If Ψ has the p-Temlyakov property (1.8.10), then for any $0 < r < \infty$, $0 < q < \infty$ we have

$$\sum_m \sigma_m(f, \Psi)_X^q m^{rq-1} < \infty \iff \sum_n |c_{k_n}(f)|^q n^{rq-1+q/p} < \infty.$$

(1.8.11)

(2) If (1.8.11) holds with some $r > 0$, then Ψ has the p-Temlyakov property (1.8.10).

We note that property (1.8.10) implies that Ψ is democratic. Therefore, by Theorem 1.4.10 a quasi-greedy basis satisfying (1.8.10) is an almost greedy basis. The basis $\mathcal{H}_p^d$ is not a democratic basis for L_p, $p \neq 2$, $d > 1$. So, we cannot apply the above results in this case. Some direct and inverse theorems for $\mathcal{H}_p^d$ are obtained in Kamont and Temlyakov (2004).

1.9. Some further results

We begin our discussion with the case of $X = L_p$, $p = 1$ or $p = \infty$ and $\Psi = \mathcal{H}_p^d$. It turns out that efficiency of greedy algorithms $G_m(\cdot, \mathcal{H}_p^d)$, $p = 1, \infty$, drops down dramatically compared with the case $1 < p < \infty$. We formulate a result from Temlyakov (1998b).

Theorem 1.9.1. Let $p = 1$ or $p = \infty$. Then we have for each $f \in L_p$

$$\|f - G_m(f, \mathcal{H}_p^d)\|_p \leq (3m + 1)\sigma_m(f, \mathcal{H}^d)_p.$$

The extra factor $(3m + 1)$ cannot be replaced by a factor $c(m)$ such that $c(m)/m \to 0$ as $m \to \infty$.

This particular result indicates that there are problems with greedy approximation in L_1 and in $\mathcal{C}$ with regard to the Haar basis. We note that, as is proved in Oswald (2001), the extra factor $3m+1$ is the best-possible extra factor in Theorem 1.9.1. The greedy-type bases have nice properties and they are important in nonlinear m-term approximation. Therefore, one of the new directions of research in functional analysis and in approximation theory is to understand which Banach spaces may have such bases. Another direction is to understand in which Banach spaces some classical bases are of greedy type. Some results in this direction can be derived immediately from known results on Banach spaces that have unconditional bases, and from the characterization Theorem 1.3.5. For instance, it is well known that the spaces L_1 and $\mathcal{C}$ do not have unconditional bases. Therefore, Theorem 1.3.5 implies that there is no greedy basis in L_1 and in $\mathcal{C}$.

It was proved in Dilworth, Kutzarova and Wojtaszczyk (2002) that the Haar basis $\mathcal{H}_1$ is not a quasi-greedy basis for L_1. We saw in Section 1.6 that the use of the Weak Greedy Algorithm has some advantages over the Greedy Algorithm. Theorem 1.6.2 states that the convergence set $\mathrm{WT}\{e_n\}$ of the WGA is linear for any $t \in (0,1)$, while the convergence set may not be linear for the Greedy Algorithm. Recently, Gogyan (2006) proved that, for any $t \in (0,1)$ and for any $f \in L_1(0,1)$, there exists a realization of the WGA with respect to the Haar basis that converges to f in L_1.

It was proved in Dilworth *et al.* (2002) that there exists an increasing sequence of integers $\{n_j\}$ such that the lacunary Haar system $\{H^1_{2^{n_j}+l}; l = 1, \ldots, 2^{n_j}, j = 1, 2, \ldots\}$ is a quasi-greedy basis for its linear span in L_1. Gogyan (2005) proved that the above property holds if either $\{n_j\}$ is a sequence of all even numbers or $\{n_j\}$ is a sequence of all odd numbers. We also note that the space $L_1(0,1)$ has a quasi-greedy basis (Dilworth, Kalton and Kutzarova 2003). The reader can find further results on existence (and non-existence) of quasi-greedy and almost greedy bases in Dilworth *et al.* (2003). In particular, it is proved in Dilworth *et al.* (2003) that $\mathcal{C}[0,1]$ does not have quasi-greedy bases.

We pointed out in Section 1.7 that the trigonometric system is not a quasi-greedy basis for L_p, $p \neq 2$. The question of when (and for which weights w) the trigonometric system forms a quasi-greedy basis for a weighted space $L_p(w)$ was studied in Nielsen (2006). The author proved that this can happen only for $p = 2$ and, whenever the system forms a quasi-greedy basis, the basis must be a Riesz basis.

Theorem 1.3.1A shows that, in the case when a basis Ψ is L_p-equivalent to the Haar basis $\mathcal{H}_p$, $1 < p < \infty$, the Greedy Algorithm $G_m(f, \Psi)$ provides near-best approximation for each individual function $f \in L_p$. For a function class $F \subset X$, let

$$\sigma_m(F, \Psi)_X := \sup_{f \in F} \sigma_m(F, \Psi)_X,$$

$$G_m(F, \Psi)_X := \sup_{f \in F} \|f - G_m(f, \Psi)\|_X.$$

Obviously, if $G_m(\cdot, \Psi)$ provides near-best approximation for each individual function, then it provides near-best approximation for each function class F:

$$G_m(F, \Psi)_X \leq C\sigma_m(F, \Psi)_X.$$

In Section 1.7 we pointed out that the trigonometric system is not a quasi-greedy basis for L_p, $p \neq 2$ (see (1.7.3)). Thus, the trigonometric system is not a greedy basis for L_p, $p \neq 2$, and for some functions $f \in L_p$, $p \neq 2$, $G_m(f, \mathcal{T})$, does not provide near-best approximation. However, it was proved in Temlyakov (1998c) that in many cases the algorithm $G_m(\cdot, \mathcal{T})$ is optimal for a given class of functions. The reader can find further results on $\sigma_m(F, \mathcal{T}^d)_p$ and $G_m(F, \mathcal{T}^d)_p$ for different classes F in DeVore and Temlyakov (1995) and Temlyakov (1998 c, 2000 a, 2002 a).

Consideration of approximation in a function class leads to a concept of the *optimal* (*best*) *basis* for a given class. The first results for best basis approximation were given by Kashin (1985), who showed that, for any orthonormal basis Ψ and any $0 < \alpha \leq 1$, we have

$$\sigma_m(\mathrm{Lip}\,\alpha, \Psi)_{L_2} \geq cm^{-\alpha}, \qquad (1.9.1)$$

where the constant c depends only on α. It follows from this that any of the standard wavelet or Fourier bases are best for the Lipschitz classes, when the approximation is carried out in L_2 and the competition is held over all orthonormal bases. The estimate (1.9.1) rests on some fundamental estimates for the best basis approximation of finite-dimensional hypercubes using orthonormal bases.

The problem of best basis selection was studied in Coifman and Wickerhauser (1992). Donoho (1993, 1997) also studied the problem of best bases for a function class F. He calls a basis Ψ from a collection $\mathbb{B}$ best for F if

$$\sigma_m(F, \Psi)_X = O(m^{-\alpha}), \qquad m \to \infty,$$

and no other basis Ψ' from $\mathbb{B}$ satisfies

$$\sigma_n(F, \Psi')_X = O(n^{-\beta}), \qquad n \to \infty,$$

for a value of $\beta > \alpha$. Donoho has shown that in some cases it is possible to determine a best basis (in the above sense) for the class F by intrinsic properties of how the class gets represented with respect to the basis. In Donoho's analysis (as was the case for Kashin as well) the space X is L_2 (or equivalently any Hilbert space), and the competition for a best basis takes place over all complete orthonormal systems (*i.e.*, $\mathbb{B}$ consists of all complete orthonormal bases for L_2).

In DeVore, Petrova and Temlyakov (2003) we continued to study the problem of optimal basis selection with regard to natural collections of bases. We worked on the following problem in this direction. We say that a function class F is aligned to the basis Ψ if, whenever $f = \sum a_k \psi_k$ is in F, then

$$\sum a'_k \psi_k \in F \quad \text{for any } |a'_k| \leq c|a_k|, \qquad k = 1, 2, \dots,$$

where $c > 0$ is a fixed constant. We pointed out in DeVore *et al.* (2003) that the results from Kashin (1985) and Donoho (1993) imply the following result.

Theorem 1.9.2. Let Φ be an orthonormal basis for a Hilbert space H and let F be a function class aligned with Φ such that, for some $\alpha > 0$, $\beta \in \mathbb{R}$, we have

$$\limsup_{m \to \infty} m^\alpha (\log m)^\beta \sigma_m(F, \Phi) > 0.$$

Then, for any orthonormal basis B we have

$$\limsup_{m\to\infty} m^\alpha (\log m)^\beta \sigma_m(F, B) > 0.$$

We have obtained in DeVore *et al.* (2003) a generalization of this important result in the following direction. We replaced the Hilbert space with the Banach space and also widened the search for optimal basis selection from the collection of orthonormal bases to the collection of unconditional bases. Here is the corresponding theorem from DeVore *et al.* (2003).

Theorem 1.9.3. Let Ψ be a normalized unconditional basis for X with the property

$$\left\| \sum_{j\in A} \psi_j \right\|_X \asymp (\#A)^\mu,$$

for some $\mu > 0$. Assume that the function class F is aligned with Ψ, and for some $\alpha > 0$, $\beta \in \mathbb{R}$ we have

$$\limsup_{m\to\infty} m^\alpha (\log m)^\beta \sigma_m(F, \Psi) > 0.$$

Then, for any unconditional basis B we have

$$\limsup_{m\to\infty} m^\alpha (\log m)^{\alpha+\beta} \sigma_m(F, B) > 0. \tag{1.9.2}$$

Theorem 1.9.3 is weaker than Theorem 1.9.2 in the sense that we have an extra factor $(\log m)^\alpha$ in (1.9.2). Recently, Bednorz (2006) proved Theorem 1.9.3 with (1.9.2) replaced by (1.9.3):

$$\limsup_{m\to\infty} m^\alpha (\log m)^\beta \sigma_m(F, B) > 0. \tag{1.9.3}$$

The following nonlinear analogues of the Kolmogorov widths and the ortho-widths (see, for instance, Temlyakov (1989a)) were considered in Temlyakov (2000a, 2002a, 2003a). Let a function class F and a Banach space X be given. Assume that, on the basis of some additional information, we know that our basis for m-term approximation should satisfy some structural properties, for instance, it has to be orthogonal. Let $\mathbb{B}$ be a collection of bases satisfying a given property.

I Define an analogue of the Kolmogorov width

$$\sigma_m(F, \mathbb{B})_X := \inf_{\Psi\in\mathbb{B}} \sup_{f\in F} \sigma_m(f, \Psi)_X.$$

II Define an analogue of the orthowidth

$$\gamma_m(F, \mathbb{B})_X := \inf_{\Psi\in\mathbb{B}} \sup_{f\in F} \|f - G_m(f, \Psi)\|_X.$$

In the papers cited above some results were obtained when $\mathbb{B} = \mathbb{O}$, the set of orthonormal bases, and F is either a multivariate smoothness class of an anisotropic Sobolev–Nikol'skii kind, or a class of functions with bounded mixed derivatives.

We conclude this section with a very recent result from Wojtaszczyk (2006). Theorem 1.3.1 says that the univariate Haar basis $\mathcal{H}$ is a greedy basis for $L_p := L_p([0,1])$, $1 < p < \infty$. The spaces L_p are examples of rearrangement-invariant spaces. Let us recall that a rearrangement-invariant space of functions defined on $[0,1]$ is a Banach space X with norm $\|\cdot\|$ whose elements are measurable (in the sense of Lebesgue) functions defined on $[0,1]$ satisfying the following conditions.

(1) If $f \in X$ and g is a measurable function such that $|g(x)| \leq |f(x)|$ almost everywhere, then $g \in X$ and $\|g\| \leq \|f\|$.

(2) If $f \in X$ and g has the same distribution as f, i.e., for all λ,

$$\text{measure}(\{x \in [0,1] : f(x) \leq \lambda\}) = \text{measure}(\{x \in [0,1] : g(x) \leq \lambda\}),$$

then $g \in X$ and $\|g\| = \|f\|$.

The following result was proved in Wojtaszczyk (2006).

Theorem 1.9.4. Let X be a rearrangement-invariant space on $[0,1]$. If the Haar system normalized in X is a greedy basis for X, then $X = L_p([0,1])$ with some $1 < p < \infty$.

It is a very interesting result that singles out the L_p-spaces with $1 < p < \infty$ from the collection of rearrangement-invariant spaces. Theorem 1.9.4 emphasizes the importance of the L_p-spaces in the theory of greedy approximation.

1.10. Systems L_p-equivalent to $\mathcal{H}$

In the previous sections of this chapter we have presented elements of a general theory of greedy-type bases. In this section we concentrate on construction of greedy bases and related bases that are useful in approximation of functions in the L_p-norm. Theorem 1.3.1 indicates importance of bases that are L_p-equivalent to the Haar basis $\mathcal{H}$. It says that such bases are greedy bases for L_p, $1 < p < \infty$. Theorem 1.3.1 addresses the case of $L_p([0,1])$. The same proof works for the $L_p(\mathbb{R})$. In this section we will give some sufficient conditions on a system of functions in order to be L_p-equivalent to the Haar basis. It is more convenient to give these conditions in the case of $L_p(\mathbb{R})$. These results are a part of general Littlewood–Paley theory. We begin in this section by introducing various forms of Littlewood–Paley theory for systems of functions. From the univariate wavelet ψ, we can construct efficient bases for $L_2(\mathbb{R})$ and other function spaces by dilation and shifts (see, for instance, DeVore (1998)). For example, the functions

$$\psi_{j,k} := 2^{k/2}\psi(2^k \cdot -j), \quad j, k \in \mathbb{Z},$$

form a stable basis (orthogonal basis in the case of an orthogonal wavelet ψ) for $L_2(\mathbb{R})$.

It is convenient to use a different indexing for the functions $\psi_{j,k}$. Let $D := D(\mathbb{R})$ denote the set of dyadic intervals. Each such interval I is of the form $I = [j2^{-k}, (j+1)2^{-k}]$. We define

$$\psi_I := \psi_{j,k}, \quad I = [j2^{-k}, (j+1)2^{-k}]. \tag{1.10.1}$$

Thus the basis $\{\psi_{j,k}\}_{j,k\in\mathbb{Z}}$ is the same as $\{\psi_I\}_{I\in D(\mathbb{R})}$.

We consider in this section systems of functions $\{\eta(I, \cdot)\}_{I\in D}$ defined on $\mathbb{R}$. If $1 < p < \infty$, we say that a family of real-valued functions $\eta(I, \cdot)$, $I \in D$, satisfies the *strong Littlewood–Paley property* for p if, for any finite sequence (c_I) of real numbers, we have

$$\left\| \sum_{I\in D} c_I \eta(I, \cdot) \right\|_p \asymp \left\| \left(\sum_{I\in D} [c_I \eta(I, \cdot)]^2 \right)^{1/2} \right\|_p \tag{1.10.2}$$

with constants of equivalency depending at most on p. Here and later we use the notation $A \asymp B$ to mean that there are two constants $C_1, C_2 > 0$ such that

$$C_1 A \leq B \leq C_2 A.$$

We shall indicate what the constants depend on (in the case of (1.10.2) they may depend on p).

Here is a useful remark concerning (1.10.2). From the validity of (1.10.2) for finite sequences, we can deduce its validity for infinite sequences by a limiting argument. For example, if $(c_I)_{I\in D}$ is an infinite sequence for which the sum on the left-hand side of (1.10.2) converges in $L_p(\mathbb{R})$ with respect to some ordering of the $I \in D$, then the right-hand side of (1.10.2) will converge with respect to the same ordering and the right-hand side of (1.10.2) will be less than a multiple of the left. Likewise, we can reverse the roles of the left- and right-hand sides. Similar remarks hold for other statements such as (1.10.2).

We use *strong Littlewood–Paley inequality* to differentiate (1.10.2) from other possible forms of Littlewood–Paley inequalities. For example, the Littlewood–Paley inequalities for the complex exponentials take a different form (see Zygmund (1959, Chapter XV)). Another point of interest in our considerations is the following:

$$\left\| \sum_{I\in D} c_I \eta(I, \cdot) \right\|_p \asymp \left\| \left(\sum_{I\in D} [c_I \chi_I]^2 \right)^{1/2} \right\|_p. \tag{1.10.3}$$

We use the notation χ for the characteristic function of $[0, 1]$ and χ_I for its $L_2(\mathbb{R})$-normalized, shifted dilates given by (1.10.1) (with $\psi = \chi$).

The two forms (1.10.2) and (1.10.3) are equivalent under very mild conditions on the functions $\eta(I, \cdot)$. To see this, we shall use the Hardy–Littlewood maximal operator, which is defined for a locally integrable function g on $\mathbb{R}$ by

$$Mg(x) := \sup_{J \ni x} \frac{1}{|J|} \int_J |g(y)| \, dy$$

with the supremum taken over all intervals J that contain x. It is well known that M is a bounded operator on $L_p(\mathbb{R})$ for all $1 < p \le \infty$. The Fefferman–Stein inequality (Fefferman and Stein 1972) bounds the mapping M on sequences of functions. We shall only need the following special case of this inequality, which says that for any functions $\eta(I, \cdot)$ and constants c_I, $I \in D$, we have for $1 < p \le \infty$,

$$\left\| \left(\sum_{I \in D} (c_I M\eta(I, \cdot))^2 \right)^{1/2} \right\|_p \le A \left\| \left(\sum_{I \in D} (c_I \eta(I, \cdot))^2 \right)^{1/2} \right\|_p, \qquad (1.10.4)$$

with an absolute constant A.

Consider now as an example the equivalence of (1.10.2). If the functions $\eta(I, \cdot)$, $I \in D$, satisfy

$$|\eta(I, x)| \le CM\chi_I(x), \quad \chi_I(x) \le CM\eta(I, x), \quad \text{for almost all } x \in \mathbb{R},$$
$$(1.10.5)$$

then, using (1.10.4), we see that (1.10.2) holds if and only if (1.10.3) holds. The first inequality in (1.10.5) is a decay condition on $\eta(I, \cdot)$. For example, if $\eta(I, \cdot)$ is given by the normalized, shifted dilates of the function ψ, $\eta(I, \cdot) = \psi_I$, then the first inequality in (1.10.5) holds whenever

$$|\psi(x)| \le C[\max(1, |x|)]^{-\lambda}, \quad \text{for almost all } x \in \mathbb{R},$$

with $\lambda \ge 1$. The second condition in (1.10.5) is extremely mild. For example, it is always satisfied when the family $\eta(I, \cdot)$ is generated by the shifted dilates of a non-zero function ψ.

Suppose that we have in hand two families $\eta(I, \cdot), \mu(I, \cdot)$, $I \in D(\mathbb{R})$. We shall use the notation $\{\eta(I, \cdot)\}_{I \in D} \prec \{\mu(I, \cdot)\}_{I \in D}$ if there is a constant $C > 0$ such that

$$\left\| \sum_{I \in D} c_I \eta(I, \cdot) \right\|_p \le C \left\| \sum_{I \in D} c_I \mu(I, \cdot) \right\|_p \qquad (1.10.6)$$

holds for all finite sequences $(c_I)_{I \in D}$ with C independent of the sequence. If $\{\eta(I, \cdot)\}_{I \in D} \prec \{\mu(I, \cdot)\}_{I \in D}$ and $\{\mu(I, \cdot)\}_{I \in D} \prec \{\eta(I, \cdot)\}_{I \in D}$, then we write $\{\eta(I, \cdot)\}_{I \in D} \approx \{\mu(I, \cdot)\}_{I \in D}$ and say that these systems are L_p-equivalent.

Given two families $\eta(I,\cdot), \mu(I,\cdot)$, $I \in D(\mathbb{R})$, we define the operator T which maps $\mu(I,\cdot)$ into $\eta(I,\cdot)$ for all $I \in D$, and we extend T to finite linear combinations of the $\mu(I,\cdot)$ by linearity. Then (1.10.6) holds if and only if T is a bounded operator with respect to the L_p-norm, and $\{\mu(I,\cdot)\}_{I\in D} \prec \{\eta(I,\cdot)\}_{I\in D}$ holds if and only if T has a bounded inverse with respect to the L_p-norm.

The strong Littlewood–Paley inequalities (1.10.3) are the same as the L_p-equivalence $\{\eta(I,\cdot)\} \approx \{H_I\}$. We begin with a presentation of sufficient conditions in order that $\{\eta(I,\cdot)\} \prec \{H_I\}$. Let ξ_I, $I \in D$, denote the centre of the dyadic interval I. We shall assume in this section that $\eta(I,\cdot)$, $I \in D$, is a family of univariate functions that satisfy the following assumptions.

A1 There is an $\epsilon > 0$, and a constant C_1 such that, for all $t \in \mathbb{R}$ and all $J \in D$, we have

$$|\eta(J, \xi_J + t|J|)| \leq C_1 |J|^{-1/2}(1 + |t|)^{-1-\epsilon}.$$

A2 There is an $\epsilon > 0$ and a constant C_2 and a partition of $[-1/2, 1/2]$ into intervals $J_1, \ldots, J_m$ that are dyadic with respect to $[-1/2, 1/2]$, such that, for any $J \in D$, any $j \in \mathbb{Z}$, and any t_1, t_2 in the interior of the same interval J_k, $k = 1, \ldots, m$, we have

$$|\eta(J, \xi_J + j|J| + t_1|J|) - \eta(J, \xi_J + j|J| + t_2|J|)|$$
$$\leq C_2 |J|^{-1/2}(1 + |j|)^{-1-\epsilon}|t_2 - t_1|^\epsilon,$$

A3 For any $J \in D$, we have

$$\int_\mathbb{R} \eta(J, x)\, dx = 0.$$

When $\eta(J,\cdot) = \psi_J$ for a function ψ, it is enough to check these assumptions for $J = [0,1]$, i.e., for the function ψ alone. They follow for all other dyadic intervals J by dilation and translation.

Condition A1 is a standard decay assumption and A3 is the zero moment condition. Condition A2 requires that the functions $\eta(I,\cdot)$ be piecewise in Lip ϵ.

Let T be the linear operator which satisfies

$$T\left(\sum_{I\in D} c_I H_I\right) = \sum_{I\in D} c_I \eta(I,\cdot) \qquad (1.10.7)$$

for each finite linear combination $\sum_{I\in D} c_I H_I$ of the H_I. We wish to show that

$$\left\|T\left(\sum_{I\in D} c_I H_I\right)\right\|_p \leq C \left\|\sum_{I\in D} c_I H_I\right\|_p$$

for each such sum. From this it would follow that T extends (by continuity) to a bounded operator on all of $L_p(\mathbb{R})$ and therefore $\{\eta(I,\cdot)\} \prec \{H_I\}$.

We can expand $\eta(J, \cdot)$ into its Haar decomposition. Let

$$\lambda(I, J) := \int_{\mathbb{R}} \eta(J, x) H_I(x) \, dx, \tag{1.10.8}$$

so that

$$\eta(J, \cdot) = \sum_{I \in D} \lambda(I, J) H_I.$$

It follows that

$$T\left(\sum_{J \in D} c_J H_J\right) = \sum_{I \in D} \sum_{J \in D} \lambda(I, J) c_J H_I. \tag{1.10.9}$$

Thus the mapping T is tied to the bi-infinite matrix $\Lambda := (\lambda(I, J))_{I, J \in D}$, which maps the sequence $c := (c_J)$ to the sequence

$$(c_I') := \Lambda c.$$

One approach to proving Littlewood–Paley inequalities is to show that the matrix Λ decays sufficiently fast away from the diagonal (see Frazier and Jawerth (1990, Section 3)). Following Frazier and Jawerth (1990), we say that a matrix $A = (a(I, J))_{I, J \in D}$ is *almost diagonal* if, for some $\epsilon > 0$, we have

$$|a(I, J)| \le C\omega(I, J), \tag{1.10.10}$$

with

$$\omega(I, J) := \left(1 + \frac{|\xi_I - \xi_J|}{\max(|I|, |J|)}\right)^{-1-\epsilon} \left(\min\left(\frac{|I|}{|J|}, \frac{|J|}{|I|}\right)\right)^{(1+\epsilon)/2}. \tag{1.10.11}$$

In DeVore *et al.* (1998) we used the following special case of a theorem of Frazier and Jawerth (1990, Theorem 3.3), concerning almost diagonal operators.

Theorem 1.10.1. If $(a(I, J))_{I, J \in \mathcal{D}}$ is an almost diagonal matrix, then the operator A defined by

$$A\left(\sum_{J \in \mathcal{D}} c_J H_J\right) := \sum_{I \in \mathcal{D}} \sum_{J \in \mathcal{D}} a(I, J) c_J H_I \tag{1.10.12}$$

is bounded on $L_p(\mathbb{R})$ for each $1 < p < \infty$.

In DeVore *et al.* (1998) we proved the following theorems.

Theorem 1.10.2. If $\eta(I, \cdot)$, $I \in D$, satisfy assumptions A1–A3, then the operator T defined by (1.10.7) is bounded from $L_p(\mathbb{R})$ into itself for each $1 < p < \infty$.

Corollary 1.10.3. If $\eta(I, \cdot)$, $I \in \mathcal{D}$, satisfy assumptions A1–A3, then $\{\eta(I, \cdot)\}_{I \in D} \prec \{H_I\}_{I \in D}$.

We can use a duality argument to give sufficient conditions that the operator T of (1.10.7) is boundedly invertible. For this, we assume that $\eta(I, \cdot)$, $I \in D$, is a family of functions for which there is a dual family $\eta^*(I, \cdot)$, $I \in D$, that satisfies

$$\langle \eta(I, \cdot), \eta^*(J, \cdot) \rangle = \delta(I, J), \quad I, J \in D.$$

Theorem 1.10.4. If the functions $\eta^*(I, \cdot)$, $I \in D$, satisfy assumptions A1–A3, then $\{H_I\}_{I \in D} \prec \{\eta(I, \cdot)\}_{I \in D}$.

Theorem 1.10.5. If the systems of functions $\{\eta(I, \cdot)\}_{I \in D}$, $\{\eta^*(I, \cdot)\}_{I \in D}$, satisfy assumptions A1–A3, then the system $\{\eta(I, \cdot)\}_{I \in D}$ is L_p-equivalent to the Haar system $\{H_I\}_{I \in D}$ for $1 < p < \infty$.

It is known from different results (see DeVore *et al.* (1992), DeVore (1998), Temlyakov (2003*a*)) that wavelets are well designed for nonlinear approximation. We present here one general result in this direction. We fix $p \in (1, \infty)$ and consider in $L_p([0, 1]^d)$ a basis $\Psi := \{\psi_I\}_{I \in D}$ indexed by dyadic intervals I of $[0, 1]^d$, $I = I_1 \times \cdots \times I_d$, I_j is a dyadic interval of $[0, 1]$, $j = 1, \ldots, d$, which satisfies certain properties. Set $L_p := L_p(\Omega)$ with a normalized Lebesgue measure on Ω, $|\Omega| = 1$. First of all we assume that, for all $1 < q, p < \infty$ and $I \in D$, where $D := D([0, 1]^d)$ is the set of all dyadic intervals of $[0, 1]^d$, we have

$$\|\psi_I\|_p \asymp \|\psi_I\|_q |I|^{1/p - 1/q}, \tag{1.10.13}$$

with constants independent of I. This property can be easily checked for a given basis.

Next, assume that for any $s = (s_1, \ldots, s_d) \in \mathbb{Z}^d$, $s_j \geq 0$, $j = 1, \ldots, d$, and any $\{c_I\}$, we have for $1 < p < \infty$

$$\left\| \sum_{I \in D_s} c_I \psi_I \right\|_p^p \asymp \sum_{I \in D_s} \|c_I \psi_I\|_p^p, \tag{1.10.14}$$

where

$$D_s := \{I = I_1 \times \cdots \times I_d \in D : |I_j| = 2^{-s_j}, \quad j = 1, \ldots, d\}.$$

This assumption allows us to estimate the L_p-norm of a dyadic block in terms of Fourier coefficients.

The third assumption is that Ψ is a basis satisfying the following version (weak form) of the Littlewood–Paley inequality, as follows. Let $1 < p < \infty$ and let $f \in L_p$ have the expansion

$$f = \sum_I f_I \psi_I.$$

We assume that

$$\lim_{\min_j \mu_j \to \infty} \left\| f - \sum_{s_j \leq \mu_j, j=1,\ldots,d} \sum_{I \in D_s} f_I \psi_I \right\|_p = 0, \tag{1.10.15}$$

and

$$\|f\|_p \asymp \left\| \left(\sum_s \left| \sum_{I \in D_s} f_I \psi_I \right|^2 \right)^{1/2} \right\|_p. \tag{1.10.16}$$

Let $\mu \in \mathbb{Z}^d$, $\mu_j \geq 0$, $j = 1, \ldots, d$. Denote by $\Psi(\mu)$ the subspace of polynomials of the form

$$\psi = \sum_{s_j \leq \mu_j, j=1,\ldots,d} \sum_{I \in D_s} c_I \psi_I.$$

We now define a function class. Let $R = (R_1, \ldots, R_d)$, $R_j > 0$, $j = 1, \ldots, d$, and

$$g(R) := \left(\sum_{j=1}^d R_j^{-1} \right)^{-1}.$$

For any natural number l, define

$$\Psi(R, l) := \Psi(\mu), \qquad \mu_j = [g(R)l/R_j], \quad j = 1, \ldots, d.$$

We define the class $H_q^R(\Psi)$ as the set of functions $f \in L_q$ representable in the form

$$f = \sum_{l=1}^{\infty} t_l, \quad t_l \in \Psi(R, l), \quad \|t_l\|_q \leq 2^{-g(R)l}.$$

We proved in Temlyakov (2002a) the following theorem.

Theorem 1.10.6. Let $1 < q, p < \infty$ and $g(R) > (1/q - 1/p)_+$. Then, for Ψ satisfying (1.10.13)–(1.10.16), we have

$$\sup_{f \in H_q^R(\Psi)} \|f - G_m^{L_p}(f, \Psi)\|_p \ll m^{-g(R)}.$$

In the periodic case the basis $U^d := U \times \cdots \times U$ can be used in place of Ψ in Theorem 1.10.6. We define the system $U := \{U_I\}$ in the univariate case. Denote

$$U_n^+(x) := \sum_{k=0}^{2^n-1} e^{ikx} = \frac{e^{i2^n x} - 1}{e^{ix} - 1}, \quad n = 0, 1, 2, \ldots,$$

$$U_{n,k}^+(x) := e^{i2^n x} U_n^+(x - 2\pi k 2^{-n}), \quad k = 0, 1, \ldots, 2^n - 1,$$

$$U_{n,k}^-(x) := e^{-i2^n x} U_n^+(-x + 2\pi k 2^{-n}), \quad k = 0, 1, \ldots, 2^n - 1.$$

We normalize the system of functions $\{U_{n,k}^+, U_{n,k}^-\}$ in L_2 and enumerate it by dyadic intervals. We write

$$U_I(x) := 2^{-n/2} U_{n,k}^+(x) \quad \text{with} \quad I = [(k+1/2)2^{-n}, (k+1)2^{-n}),$$

$$U_I(x) := 2^{-n/2} U_{n,k}^-(x) \quad \text{with} \quad I = [k2^{-n}, (k+1/2)2^{-n}),$$

and

$$U_{[0,1)}(x) := 1.$$

Wojtaszczyk (1997) proved that U is an unconditional basis of L_p, $1 < p < \infty$. It is well known that $H_q^R(U^d)$ is equivalent to the standard anisotropic multivariate periodic Hölder–Nikol'skii classes NH_p^R. We define these classes in the following way (see Nikol'skii (1975)). The class NH_p^R, $R = (R_1, \ldots, R_d)$ and $1 \le p \le \infty$, is the set of periodic functions $f \in L_p([0, 2\pi]^d)$ such that, for each $l_j = [R_j] + 1$, $j = 1, \ldots, d$, the following relations hold:

$$\|f\|_p \le 1, \qquad \|\Delta_t^{l_j, j} f\|_p \le |t|^{R_j}, \quad j = 1, \ldots, d, \tag{1.10.17}$$

where $\Delta_t^{l,j}$ is the lth difference with step t in the variable x_j. For $d = 1$, NH_p^R coincides with the standard Hölder class H_p^R. Theorem 1.10.6 gives the following result.

Theorem 1.10.7. Let $1 < q, p < \infty$; then for R such that $g(R) > (1/q - 1/p)_+$, we have

$$\sup_{f \in NH_q^R} \|f - G_m^{L_p}(f, U^d)\|_p \ll m^{-g(R)}.$$

We also proved in Temlyakov (2002a) that the basis U^d is an optimal orthonormal basis for approximation of classes NH_q^R in L_p:

$$\sigma_m(NH_q^R, \mathbb{O})_p \asymp \sigma_m(NH_q^R, U^d)_p \asymp m^{-g(R)} \tag{1.10.18}$$

for $1 < q < \infty$, $2 \le p < \infty$, $g(R) > (1/q - 1/p)_+$. Here $\mathbb{O}$ is a collection of orthonormal bases. It is important to note that Theorem 1.10.7 guarantees that the estimate in (1.10.18) can be realized by greedy algorithm $G_m^{L_p}(\cdot, U^d)$ with regard to U^d. Another important feature of (1.10.18) is that the basis U^d is optimal (in the sense of order) for each class NH_q^R independently of $R = (R_1, \ldots, R_d)$ and q. This property is known as universality for a collection of classes (in the above case, the collection $\{NH_q^R\}$). Further discussion of this important issue can be found in Temlyakov (2002a, 2003a).

CHAPTER TWO
Greedy approximation with respect to dictionaries: Hilbert spaces

2.1. Introduction

We discuss greedy approximation with regard to redundant systems in this chapter. Greedy approximation is a special form of nonlinear approximation. The basic idea behind nonlinear approximation is that the elements used in the approximation do not come from a fixed linear space but are allowed to depend on the function being approximated. The standard problem in this regard is the problem of m-term approximation, where one fixes a basis and aims to approximate a target function f by a linear combination of m terms of the basis. We discussed this problem in detail in Chapter 1. When the basis is a wavelet basis or a basis of other waveforms, then this type of approximation is the starting point for compression algorithms. An important feature of approximation using a basis

$$\Psi := \{\psi_k\}_{k=1}^{\infty}$$

of a Banach space X is that each function $f \in X$ has a unique representation

$$f = \sum_{k=1}^{\infty} c_k(f)\psi_k, \tag{2.1.1}$$

and we can identify f with the set of its coefficients $\{c_k(f)\}_{k=1}^{\infty}$. The problem of m-term approximation with regard to a basis has been studied thoroughly and rather complete results have been established (see Chapter 1). In particular, it was established that the greedy-type algorithm which forms a sum of m terms with the largest $\|c_k(f)\psi_k\|_X$ out of expansion (2.1.1) realizes in many cases near-best m-term approximation for function classes (DeVore *et al.* 1992) and even for individual functions (see Chapter 1).

Recently, there has emerged another more complicated form of nonlinear approximation, which we call highly nonlinear approximation. It takes many forms but has the basic ingredient that a basis is replaced by a larger system of functions that is usually redundant. We call such systems dictionaries. On the one hand, redundancy offers much promise for greater efficiency in terms of the approximation rate, but on the other hand gives rise to highly non-trivial theoretical and practical problems. The problem of characterizing approximation rate for a given function or function class is now much more substantial and results are quite fragmentary. However, such results are very important for understanding what this new type of approximation offers. Perhaps the first example of this type was considered by Schmidt (1906), who studied the approximation of functions $f(x,y)$ of two variables by bilinear forms,

$$\sum_{i=1}^{m} u_i(x)v_i(y),$$

in $L_2([0,1]^2)$. This problem is closely connected with properties of the integral operator

$$J_f(g) := \int_0^1 f(x,y)g(y)\,dy$$

with kernel $f(x,y)$. Schmidt (1906) gave an expansion (known as the Schmidt expansion)

$$f(x,y) = \sum_{j=1}^{\infty} s_j(J_f)\phi_j(x)\psi_j(y),$$

where $\{s_j(J_f)\}$ is a non-increasing sequence of singular numbers of J_f, i.e., $s_j(J_f) := \lambda_j(J_f^* J_f)^{1/2}$, where $\{\lambda_j(A)\}$ is the sequence of eigenvalues of an operator A, and J_f^* is the adjoint operator to J_f. The two sequences $\{\phi_j(x)\}$ and $\{\psi_j(y)\}$ form orthonormal sequences of eigenfunctions of the operators $J_f J_f^*$ and $J_f^* J_f$, respectively. He also proved that

$$\left\| f(x,y) - \sum_{j=1}^{m} s_j(J_f)\phi_j(x)\psi_j(y) \right\|_{L_2}$$

$$= \inf_{u_j,v_j \in L_2,\ j=1,\dots,m} \left\| f(x,y) - \sum_{j=1}^{m} u_j(x)v_j(y) \right\|_{L_2}.$$

It was understood later that the above best bilinear approximation can be realized by the following greedy algorithm. Assume c_j, $u_j(x)$, $v_j(y)$, $\|u_j\|_{L_2} = \|v_j\|_{L_2} = 1$, $j = 1, \dots, m-1$, have been constructed after $m-1$ steps of the algorithm. At the mth step we choose c_m, $u_m(x)$, $v_m(y)$, $\|u_m\|_{L_2} = \|v_m\|_{L_2} = 1$, to minimize

$$\left\| f(x,y) - \sum_{j=1}^{m} c_j u_j(x)v_j(y) \right\|_{L_2}.$$

We call this type of algorithm the Pure Greedy Algorithm (PGA) (see the general definition below).

Another problem of this type which is well known in statistics is the projection pursuit regression problem, mentioned in the Preface. The problem is to approximate in L_2 a given function $f \in L_2$ by a sum of ridge functions, i.e., by

$$\sum_{j=1}^{m} r_j(\omega_j \cdot x), \quad x, \omega_j \in \mathbb{R}^d, \quad j = 1, \dots, m,$$

where r_j, $j = 1, \ldots, m$, are univariate functions. The following greedy-type algorithm (projection pursuit) was proposed in Friedman and Stuetzle (1981) to solve this problem. Assume functions $r_1, \ldots, r_{m-1}$ and vectors $\omega_1, \ldots, \omega_{m-1}$ have been determined after $m - 1$ steps of algorithm. Choose at the mth step a unit vector ω_m and a function r_m to minimize the error

$$\left\| f(x) - \sum_{j=1}^{m} r_j(\omega_j \cdot x) \right\|_{L_2}.$$

This is one more example of a Pure Greedy Algorithm. The Pure Greedy Algorithm and some other versions of greedy-type algorithms have recently been intensively studied: see Barron (1993), Donahue, Gurvits, Darken and Sontag (1997), Davis, Mallat and Avellaneda (1997), DeVore and Temlyakov (1996, 1997), Dubinin (1997), Huber (1985), Jones (1987, 1992), Konyagin and Temlyakov (1999b), Livshitz (2006, 2007a, 2007b), Livshitz and Temlyakov (2001, 2003) and Temlyakov (1999, 2000b, 2002b, 2003b). There are several survey papers that discuss greedy approximation with regard to redundant systems: see DeVore (1998) and Temlyakov (2003a, 2006a). In this chapter we discuss along with the PGA some of its modifications which are more suitable for implementation. This new type of greedy algorithms will be termed Weak Greedy Algorithms.

In order to orient the reader we recall some notation and definitions from the theory of greedy algorithms. Let H be a real Hilbert space with an inner product $\langle \cdot, \cdot \rangle$ and the norm $\|x\| := \langle x, x \rangle^{1/2}$. We say a set $\mathcal{D}$ of functions (elements) from H is a dictionary if each $g \in \mathcal{D}$ has norm one ($\|g\| = 1$) and the closure of span $\mathcal{D}$ is equal to H. Sometimes it will be convenient for us also to consider the symmetrized dictionary $\mathcal{D}^{\pm} := \{\pm g : g \in \mathcal{D}\}$. In DeVore and Temlyakov (1996) we studied the following two greedy algorithms. If $f \in H$, we let $g = g(f) \in \mathcal{D}$ be the element from $\mathcal{D}$ which maximizes $|\langle f, g \rangle|$ (we make an additional assumption that a maximizer exists) and define

$$G(f) := G(f, \mathcal{D}) := \langle f, g \rangle g \qquad (2.1.2)$$

and

$$R(f) := R(f, \mathcal{D}) := f - G(f).$$

Pure Greedy Algorithm (PGA). We define $f_0 := R_0(f) := R_0(f, \mathcal{D}) := f$ and $G_0(f) := G_0(f, \mathcal{D}) := 0$. Then, for each $m \geq 1$, we inductively define

$$G_m(f) := G_m(f, \mathcal{D}) := G_{m-1}(f) + G(R_{m-1}(f)),$$

$$f_m := R_m(f) := R_m(f, \mathcal{D}) := f - G_m(f) = R(R_{m-1}(f)).$$

We note that the Pure Greedy Algorithm is known under the name Matching Pursuit in signal processing (see, for instance, Mallat and Zhang (1993)).

If H_0 is a finite-dimensional subspace of H, we let P_{H_0} be the orthogonal projector from H onto H_0. That is, $P_{H_0}(f)$ is the best approximation to f from H_0.

Orthogonal Greedy Algorithm (OGA). We define $f_0^o := R_0^o(f) := R_0^o(f, \mathcal{D}) := f$ and $G_0^o(f) := G_0^o(f, \mathcal{D}) := 0$. Then, for each $m \geq 1$, we inductively define

$$H_m := H_m(f) := \operatorname{span}\{g(R_0^o(f)), \dots, g(R_{m-1}^o(f))\},$$

$$G_m^o(f) := G_m^o(f, \mathcal{D}) := P_{H_m}(f),$$

$$f_m^o := R_m^o(f) := R_m^o(f, \mathcal{D}) := f - G_m^o(f).$$

We remark that for each f we have

$$\|f_m^o\| \leq \|f_{m-1}^o - G_1(f_{m-1}^o, \mathcal{D})\|. \tag{2.1.3}$$

In Section 1.5 we realized that the Weak Greedy Algorithms with regard to bases work as well as the corresponding Greedy Algorithms. In this chapter we study similar modifications of the Pure Greedy Algorithm (PGA) and the Orthogonal Greedy Algorithm (OGA), which we call, respectively, the Weak Greedy Algorithm (WGA) and the Weak Orthogonal Greedy Algorithm (WOGA). We now give the corresponding definitions from Temlyakov (2000*b*). Let a sequence $\tau = \{t_k\}_{k=1}^{\infty}$, $0 \leq t_k \leq 1$, be given.

Weak Greedy Algorithm (WGA). We define $f_0^\tau := f$. Then, for each $m \geq 1$ we have the following inductive definition.

(1) $\varphi_m^\tau \in \mathcal{D}$ is any element satisfying

$$|\langle f_{m-1}^\tau, \varphi_m^\tau \rangle| \geq t_m \sup_{g \in \mathcal{D}} |\langle f_{m-1}^\tau, g \rangle|.$$

(2)
$$f_m^\tau := f_{m-1}^\tau - \langle f_{m-1}^\tau, \varphi_m^\tau \rangle \varphi_m^\tau.$$

(3)
$$G_m^\tau(f, \mathcal{D}) := \sum_{j=1}^{m} \langle f_{j-1}^\tau, \varphi_j^\tau \rangle \varphi_j^\tau.$$

We note that, for a particular case $t_k = t$, $k = 1, 2, \dots$, this algorithm was considered in Jones (1987). Thus, the WGA is a generalization of the PGA making it easier to construct an element φ_m^τ at the mth greedy step. We point out that the WGA contains, in addition to the first (greedy) step, the second step (see (2) and (3) in the above definition) where we update an approximant by adding an orthogonal projection of the residual f_{m-1}^τ onto φ_m^τ. Therefore, the WGA provides for each $f \in H$ an expansion into a series (greedy expansion)

$$f \sim \sum_{j=1}^{\infty} c_j(f) \varphi_j^\tau, \quad c_j(f) := \langle f_{j-1}^\tau, \varphi_j^\tau \rangle.$$

In general it is not an orthogonal expansion but it has some similar properties. The coefficients $c_j(f)$ of an expansion are obtained by the Fourier formulas with f replaced by the residuals f_{j-1}^τ. It is easy to see that

$$\|f_m^\tau\|^2 = \|f_{m-1}^\tau\|^2 - |c_m(f)|^2.$$

We prove convergence of greedy expansion (see, for instance, Theorem 2.2.4 below), and therefore, from the above equality, we get for this expansion an analogue of the Parseval formula for orthogonal expansions:

$$\|f\|^2 = \sum_{j=1}^\infty |c_j(f)|^2.$$

Weak Orthogonal Greedy Algorithm (WOGA). We define $f_0^{o,\tau} := f$, $f_1^{o,\tau} := f_1^\tau$, and $\varphi_1^{o,\tau} := \varphi_1^\tau$, where f_1^τ, φ_1^τ are are given in the above definition of the WGA. Then, for each $m \geq 2$ we have the following inductive definition.

(1) $\varphi_m^{o,\tau} \in \mathcal{D}$ is any element satisfying

$$|\langle f_{m-1}^{o,\tau}, \varphi_m^{o,\tau}\rangle| \geq t_m \sup_{g \in \mathcal{D}} |\langle f_{m-1}^{o,\tau}, g\rangle|.$$

(2) $G_m^{o,\tau}(f, \mathcal{D}) := P_{H_m^\tau}(f), \quad$ where $\quad H_m^\tau := \operatorname{span}(\varphi_1^{o,\tau}, \ldots, \varphi_m^{o,\tau}).$

(3) $f_m^{o,\tau} := f - G_m^{o,\tau}(f, \mathcal{D}).$

It is clear that G_m^τ and $G_m^{o,\tau}$ in the case $t_k = 1$, $k = 1, 2, \ldots$, coincide with the PGA G_m and the OGA G_m^o, respectively. It is also clear that the WGA and the WOGA are more ready for implementation than the PGA and the OGA. The WOGA has the same greedy step as the WGA and differs in the construction of a linear combination of $\varphi_1, \ldots, \varphi_m$. In the WOGA we do our best to construct an approximant out of $H_m := \operatorname{span}(\varphi_1, \ldots, \varphi_m)$: we take an orthogonal projection onto H_m. Clearly, in this way we lose a property of the WGA to build an expansion into a series in the case of the WOGA. However, this modification pays off in the sense of improving the convergence rate of approximation. To see this, compare Theorems 2.3.5 and 2.3.6.

There is one more greedy-type algorithm that works well for functions from the convex hull of $\mathcal{D}^\pm$, where $\mathcal{D}^\pm := \{\pm g : g \in \mathcal{D}\}$.

For a general dictionary $\mathcal{D}$ we define the class of functions

$$\mathcal{A}_1^o(\mathcal{D}, M) := \left\{ f \in H : f = \sum_{k \in \Lambda} c_k w_k, \ w_k \in \mathcal{D}, \ \#\Lambda < \infty, \ \sum_{k \in \Lambda} |c_k| \leq M \right\}$$

and we define $\mathcal{A}_1(\mathcal{D}, M)$ to be the closure (in H) of $\mathcal{A}_1^o(\mathcal{D}, M)$. Furthermore, we define $\mathcal{A}_1(\mathcal{D})$ to be the union of the classes $\mathcal{A}_1(\mathcal{D}, M)$ over all $M > 0$.

For $f \in \mathcal{A}_1(\mathcal{D})$, we define the norm

$$|f|_{\mathcal{A}_1(\mathcal{D})}$$

to be the smallest M such that $f \in \mathcal{A}_1(\mathcal{D}, M)$.

For $M = 1$ we denote $A_1(\mathcal{D}) := \mathcal{A}_1(\mathcal{D}, 1)$. We proceed to discuss the relaxed type of greedy algorithms. We begin with the simplest one.

Relaxed Greedy Algorithm (RGA). Let $f_o^r := R_0^r(f) := R_0^r(f, \mathcal{D}) := f$ and $G_0^r(f) := G_0^r(f, \mathcal{D}) := 0$. For $m = 1$, we define $G_1^r(f) := G_1^r(f, \mathcal{D}) := G_1(f)$ and $f_1^r := R_1^r(f) := R_1^r(f, \mathcal{D}) := R_1(f)$. For a function $h \in H$, let $g = g(h)$ denote the function from $\mathcal{D}^{\pm}$ which maximizes $\langle h, g \rangle$ (we assume the existence of such an element). Then, for each $m \geq 2$ we inductively define

$$G_m^r(f) := G_m^r(f, \mathcal{D}) := \left(1 - \frac{1}{m}\right) G_{m-1}^r(f) + \frac{1}{m} g(R_{m-1}^r(f)),$$

$$f_m^r := R_m^r(f) := R_m^r(f, \mathcal{D}) := f - G_m^r(f).$$

There are several modifications of the Relaxed Greedy Algorithm (see, for instance, Barron (1993) and DeVore and Temlyakov (1996)). Before giving the definition of the Weak Relaxed Greedy Algorithm (WRGA), we make one remark which helps to motivate the corresponding definition. Assume $G_{m-1} \in A_1(\mathcal{D})$ is an approximant to $f \in A_1(\mathcal{D})$ obtained at the $(m-1)$th step. The major idea of relaxation in greedy algorithms is to look for an approximant at the mth step of the form $G_m := (1-a)G_{m-1} + ag$, $g \in \mathcal{D}^{\pm}$, $0 \leq a \leq 1$. This form guarantees that $G_m \in A_1(\mathcal{D})$. Thus we are looking for co-convex approximants. The best we can do at the mth step is to achieve

$$\delta_m := \inf_{g \in \mathcal{D}^{\pm}, 0 \leq a \leq 1} \|f - ((1-a)G_{m-1} + ag)\|.$$

Let $f_n := f - G_n$, $n = 1, \ldots, m$. It is clear that for a given $g \in \mathcal{D}^{\pm}$ we have

$$\inf_a \|f_{m-1} - a(g - G_{m-1})\|^2 = \|f_{m-1}\|^2 - \langle f_{m-1}, g - G_{m-1}\rangle^2 \|g - G_{m-1}\|^{-2},$$

and this infimum is attained for

$$a(g) = \langle f_{m-1}, g - G_{m-1}\rangle \|g - G_{m-1}\|^{-2}.$$

Next, it is not difficult to derive from the definition of $A_1(\mathcal{D})$ and from our assumption on existence of a maximizer that, for any $h \in H$ and $u \in A_1(\mathcal{D})$, there exists $g \in \mathcal{D}^{\pm}$ such that

$$\langle h, g \rangle \geq \langle h, u \rangle. \tag{2.1.4}$$

Taking $h = f_{m-1}$ and $u = f$, we get from (2.1.4) that there exists $g_m \in \mathcal{D}^{\pm}$ such that

$$\langle f_{m-1}, g_m - G_{m-1}\rangle \geq \langle f_{m-1}, f - G_{m-1}\rangle = \|f_{m-1}\|^2. \tag{2.1.5}$$

This implies in particular that we get for g_m

$$\|g_m - G_{m-1}\| \geq \|f_{m-1}\| \tag{2.1.6}$$

and $0 \leq a(g_m) \leq 1$. Thus,

$$\delta_m^2 \leq \|f_{m-1}\|^2 - \frac{1}{4} \sup_{g \in \mathcal{D}^\pm} \langle f_{m-1}, g - G_{m-1}\rangle^2.$$

We now give the definition of the WRGA for $f \in A_1(\mathcal{D})$.

Weak Relaxed Greedy Algorithm (WRGA). We define $f_0 := f$ and $G_0 := 0$. Then, for each $m \geq 1$ we have the following inductive definition.

(1) $\varphi_m \in \mathcal{D}^\pm$ is any element satisfying

$$\langle f_{m-1}, \varphi_m - G_{m-1}\rangle \geq t_m \|f_{m-1}\|^2. \tag{2.1.7}$$

(2) $$G_m := G_m(f, \mathcal{D}) := (1 - \beta_m)G_{m-1} + \beta_m \varphi_m,$$

$$\beta_m := t_m \left(1 + \sum_{k=1}^{m} t_k^2\right)^{-1} \quad \text{for} \ \ m \geq 1.$$

(3) $$f_m := f - G_m.$$

2.2. Convergence

We begin this section with convergence of the Weak Orthogonal Greedy Algorithm (WOGA). The following theorem was proved in Temlyakov (2000b).

Theorem 2.2.1.   Assume

$$\sum_{k=1}^{\infty} t_k^2 = \infty. \tag{2.2.1}$$

Then, for any dictionary $\mathcal{D}$ and any $f \in H$, we have for the WOGA

$$\lim_{m \to \infty} \|f_m^{o,\tau}\| = 0. \tag{2.2.2}$$

Remark 2.2.2.   It is easy to see that if $\mathcal{D} = \mathcal{B}$, an orthonormal basis, the assumption (2.2.1) is also necessary for convergence (2.2.2) for all f.

Proof of Theorem 2.2.1.   Let $f \in H$ and let $\varphi_1^{o,\tau}, \varphi_2^{o,\tau}, \ldots$ be as given in the definition of the WOGA. Let

$$H_n := H_n^\tau = \operatorname{span}(\varphi_1^{o,\tau}, \ldots, \varphi_n^{o,\tau}).$$

It is clear that $H_n \subseteq H_{n+1}$, and therefore $\{P_{H_n}(f)\}$ converges to some function v. The following Lemma 2.2.3 says that $v = f$ and completes the proof of Theorem 2.2.1. $\qquad\qquad\square$

Lemma 2.2.3. Assume that (2.2.1) is satisfied. Then, if $\{f_m^\tau\}_{m=1}^\infty$ or $\{f_m^{o,\tau}\}_{m=1}^\infty$ converges, it converges to zero.

Proof of Lemma 2.2.3. We prove this lemma by contradiction. Let us consider first the case of $\{f_m^\tau\}_{m=1}^\infty$. Assume $f_m^\tau \to u \neq 0$ as $m \to \infty$. It is clear that

$$\sup_{g \in \mathcal{D}} |\langle u, g \rangle| \geq 2\delta$$

with some $\delta > 0$. Therefore, there exists N such that, for all $m \geq N$, we have

$$\sup_{g \in \mathcal{D}} |\langle f_m^\tau, g \rangle| \geq \delta.$$

From the definition of the WGA we get for all $m > N$

$$\|f_m^\tau\|^2 = \|f_{m-1}^\tau\|^2 - |\langle f_{m-1}^\tau, \varphi_m^\tau \rangle|^2 \leq \|f_N^\tau\|^2 - \delta^2 \sum_{k=N+1}^m t_k^2,$$

which contradicts (2.2.1).

We now proceed to the case $\{f_m^{o,\tau}\}_{m=1}^\infty$. Assume $f_m^{o,\tau} \to u \neq 0$ as $m \to \infty$. Then, as in the above proof, there exist $\delta > 0$ and N such that, for all $m \geq N$, we have

$$\sup_{g \in \mathcal{D}} |\langle f_m^{o,\tau}, g \rangle| \geq \delta.$$

Next, as in (2.1.3) we have

$$\|f_m^{o,\tau}\|^2 \leq \|f_{m-1}^{o,\tau}\|^2 - t_m^2 \left(\sup_{g \in \mathcal{D}} |\langle f_{m-1}^{o,\tau}, g \rangle| \right)^2 \leq \|f_N^{o,\tau}\|^2 - \delta^2 \sum_{k=N+1}^m t_k^2,$$

which contradicts the divergence of $\sum_k t_k^2$. $\qquad\square$

Theorem 2.2.1 and Remark 2.2.2 show that (2.2.1) is a necessary and sufficient condition on weakness sequence $\tau = \{t_k\}$ in order that the WOGA converges for each f and all $\mathcal{D}$. Condition (2.2.1) can be rewritten as $\tau \notin \ell_2$. It turns out that the convergence of the PGA is more delicate. We now proceed to the corresponding results. The following theorem gives a criterion of convergence in a special case of monotone weakness sequences $\{t_k\}$. Sufficiency was proved in Temlyakov (2000b) and necessity in Livshitz and Temlyakov (2001).

Theorem 2.2.4. In the class of monotone sequences $\tau = \{t_k\}_{k=1}^\infty$, $1 \geq t_1 \geq t_2 \geq \cdots \geq 0$, the condition

$$\sum_{k=1}^\infty \frac{t_k}{k} = \infty \tag{2.2.3}$$

is necessary and sufficient for convergence of the Weak Greedy Algorithm for each f and all Hilbert spaces H and dictionaries $\mathcal{D}$.

Remark 2.2.5. We note that the sufficiency part of Theorem 2.2.4 (see Temlyakov (2000b)) does not need the monotonicity of τ.

Proof of sufficiency condition in Theorem 2.2.4. This proof (see Temlyakov (2000b)) is a refinement of the original proof of Jones (1987). The following lemma, Lemma 2.2.6, combined with Lemma 2.2.3, implies sufficiency in Theorem 2.2.4. $\square$

Lemma 2.2.6. Assume (2.2.3) is satisfied. Then $\{f_m^\tau\}_{m=1}^\infty$ converges.

Proof of Lemma 2.2.6. It is easy to derive from the definition of the WGA the following two relations:

$$f_m^\tau = f - \sum_{j=1}^m \langle f_{j-1}^\tau, \varphi_j^\tau \rangle \varphi_j^\tau, \tag{2.2.4}$$

$$\|f_m^\tau\|^2 = \|f\|^2 - \sum_{j=1}^m |\langle f_{j-1}^\tau, \varphi_j^\tau \rangle|^2. \tag{2.2.5}$$

Let $a_j := |\langle f_{j-1}^\tau, \varphi_j^\tau \rangle|$. We get from (2.2.5) that

$$\sum_{j=1}^\infty a_j^2 \le \|f\|^2. \tag{2.2.6}$$

We take any two indices $n < m$ and consider

$$\|f_n^\tau - f_m^\tau\|^2 = \|f_n^\tau\|^2 - \|f_m^\tau\|^2 - 2\langle f_n^\tau - f_m^\tau, f_m^\tau \rangle.$$

Let

$$\theta_{n,m}^\tau := |\langle f_n^\tau - f_m^\tau, f_m^\tau \rangle|.$$

Using (2.2.4) and the definition of the WGA, we obtain, for all $n < m$ and all m such that $t_{m+1} \ne 0$,

$$\theta_{n,m}^\tau \le \sum_{j=n+1}^m |\langle f_{j-1}^\tau, \varphi_j^\tau \rangle| |\langle f_m^\tau, \varphi_j^\tau \rangle| \le \frac{a_{m+1}}{t_{m+1}} \sum_{j=1}^{m+1} a_j. \tag{2.2.7}$$

$\square$

We now need a property of ℓ_2-sequences.

Lemma 2.2.7. Assume $y_j \ge 0$, $j = 1, 2, \dots$, and

$$\sum_{k=1}^\infty \frac{t_k}{k} = \infty, \qquad \sum_{j=1}^\infty y_j^2 < \infty.$$

Then

$$\lim_{n\to\infty} \frac{y_n}{t_n} \sum_{j=1}^{n} y_j = 0.$$

Proof. Let $P(\tau) := \{n \in \mathbb{N} : t_n \neq 0\}$. Consider a series

$$\sum_{n \in P(\tau)} \frac{t_n}{n} \frac{y_n}{t_n} \sum_{j=1}^{n} y_j. \qquad (2.2.8)$$

We shall prove that this series converges. It is clear that convergence of this series together with the assumption $\sum_{k=1}^{\infty} t_k/k = \infty$ imply the statement of Lemma 2.2.7.

We use the following known fact. If $\{y_j\}_{j=1}^{\infty} \in \ell_2$ then $\{n^{-1}\sum_{j=1}^{n} y_j\}_{n=1}^{\infty} \in \ell_2$ (see Zygmund (1959, Chapter 1, Section 9)). By the Cauchy inequality, we have

$$\sum_{n \in P(\tau)} \frac{t_n}{n} \frac{y_n}{t_n} \sum_{j=1}^{n} y_j \leq \left(\sum_{n=1}^{\infty} y_n^2\right)^{1/2} \left(\sum_{n=1}^{\infty} \left(n^{-1} \sum_{j=1}^{n} y_j\right)^2\right)^{1/2} < \infty.$$

This completes the proof of Lemma 2.2.7. $\qquad\qquad\qquad\qquad\qquad\qquad \square$

Relation (2.2.7) and Lemma 2.2.7 imply that

$$\lim_{m\to\infty} \max_{n<m} \theta_{n,m}^{\tau} = 0.$$

It remains to use the following simple lemma.

Lemma 2.2.8. In a Banach space X, let a sequence $\{x_n\}_{n=1}^{\infty}$ be given, such that, for any k, l, we have

$$\|x_k - x_l\|^2 = y_k - y_l + \vartheta_{k,l},$$

where $\{y_n\}_{n=1}^{\infty}$ is a convergent sequence of real numbers and the real sequence $\vartheta_{k,l}$ satisfies the property

$$\lim_{l\to\infty} \max_{k<l} |\vartheta_{k,l}| = 0.$$

Then $\{x_n\}_{n=1}^{\infty}$ converges.

The necessary condition in Theorem 2.2.4 was proved in Livshitz and Temlyakov (2001). We do not present it here.

Theorem 2.2.4 solves the problem of convergence of the WGA in the case of monotone weakness sequences. We now consider the case of general weakness sequences. In Theorem 2.2.4 we reduced the proof of convergence of the WGA with weakness sequence τ to some properties of ℓ_2-sequences with regard to τ. The sufficiency part of Theorem 2.2.4 was derived from the following two statements.

Proposition 2.2.9. Let τ be such that, for any $\{a_j\}_{j=1}^{\infty} \in \ell_2$, $a_j \geq 0$, $j = 1, 2, \ldots$, we have

$$\liminf_{n \to \infty} a_n \sum_{j=1}^{n} a_j / t_n = 0.$$

Then, for any H, $\mathcal{D}$, and $f \in H$ we have

$$\lim_{m \to \infty} \|f_m^{\tau}\| = 0.$$

Proposition 2.2.10. If τ satisfies condition (2.2.3) then τ satisfies the assumption of Proposition 2.2.9.

We proved in Temlyakov (2002b) a criterion on τ for convergence of the WGA. Let us introduce some notation. We define by $\mathcal{V}$ the class of sequences $x = \{x_k\}_{k=1}^{\infty}$, $x_k \geq 0$, $k = 1, 2, \ldots$, with the following property: there exists a sequence $0 = q_0 < q_1 < \cdots$ that may depend on x such that

$$\sum_{s=1}^{\infty} \frac{2^s}{\Delta q_s} < \infty \qquad (2.2.9)$$

and

$$\sum_{s=1}^{\infty} 2^{-s} \sum_{k=1}^{q_s} x_k^2 < \infty, \qquad (2.2.10)$$

where $\Delta q_s := q_s - q_{s-1}$.

Remark 2.2.11. It is clear from this definition that, if $x \in \mathcal{V}$ and for some N and c, we have $0 \leq y_k \leq cx_k$, $k \geq N$, then $y \in \mathcal{V}$.

Theorem 2.2.12. The condition $\tau \notin \mathcal{V}$ is necessary and sufficient for convergence of all realizations of the Weak Greedy Algorithm with weakness sequence τ for each f and all Hilbert spaces H and dictionaries $\mathcal{D}$.

The proof of the sufficiency part of Theorem 2.2.12 is a refinement of the corresponding proof of Theorem 2.2.4. The study of the behaviour of sequences $a_n \sum_{j=1}^{n} a_j$ for $\{a_j\}_{j=1}^{\infty} \in \ell_2$, $a_j \geq 0$, $j = 1, 2, \ldots$, plays an important role in both proofs. It turns out that the class $\mathcal{V}$ appears naturally in the study of the above-mentioned sequences. We proved the following theorem in Temlyakov (2002b).

Theorem 2.2.13. The following two conditions are equivalent:

$$\tau \notin \mathcal{V}, \qquad (2.2.11)$$

$$\forall \{a_j\}_{j=1}^{\infty} \in \ell_2, \quad a_j \geq 0, \quad \liminf_{n \to \infty} a_n \sum_{j=1}^{n} a_j / t_n = 0. \qquad (2.2.12)$$

Theorem 2.2.12 solves the problem of convergence of the WGA in a very general situation. The sufficiency part of Theorem 2.2.12 guarantees that, whenever $\tau \notin \mathcal{V}$, the WGA converges for each f and all $\mathcal{D}$. The necessity part of Theorem 2.2.12 states that, if $\tau \in \mathcal{V}$, then there exist an element f and a dictionary $\mathcal{D}$ such that some realization $G_m^\tau(f, \mathcal{D})$ of the WGA does not converge to f. However, Theorem 2.2.12 leaves open the following interesting and important problem. Let a dictionary $\mathcal{D} \subset H$ be given. Find necessary and sufficient conditions on a weakness sequence τ in order that $G_m^\tau(f, \mathcal{D}) \to f$ for each $f \in H$. The corresponding open problems for special dictionaries are formulated in Temlyakov (2003a, pp. 78, 81). They concern the following two classical dictionaries:

$$\Pi_2 := \{u(x)v(y) : u, v \in L_2([0,1]), \quad \|u\|_2 = \|v\|_2 = 1\},$$

and

$$\mathcal{R}_2 := \{g(x) = r(\omega \cdot x) : \|g\|_2 = 1\},$$

where r is a univariate function and $\omega \cdot x$ is the scalar product of x, $\|x\|_{\ell_2} \leq 1$, and a unit vector $\omega \in \mathbb{R}^2$.

2.3. Rate of convergence

2.3.1. Upper bounds for approximation by general dictionaries

We shall discuss here approximation from a general dictionary $\mathcal{D}$. We begin with a discussion of the approximation properties of the Relaxed Greedy Algorithm. The result we give below in Theorem 2.3.2 is from DeVore and Temlyakov (1996), and can be found in the paper of Jones (1992) in a different form. We begin with the following elementary lemma about numerical sequences.

Lemma 2.3.1. If $A > 0$ and $\{a_n\}_{n=1}^\infty$ is a sequence of non-negative numbers satisfying $a_1 \leq A$ and

$$a_m \leq a_{m-1} - \frac{2}{m} a_{m-1} + \frac{A}{m^2}, \quad m = 2, 3, \ldots, \tag{2.3.1}$$

then

$$a_m \leq \frac{A}{m}. \tag{2.3.2}$$

Proof. The proof is by induction. Suppose we have

$$a_{m-1} \leq \frac{A}{m-1}$$

for some $m \geq 2$. Then, from our assumption (2.3.1) we have

$$a_m \leq \frac{A}{m-1}\left(1 - \frac{2}{m}\right) + \frac{A}{m^2} = A\left(\frac{1}{m} - \frac{1}{(m-1)m} + \frac{1}{m^2}\right) \leq \frac{A}{m}. \quad \square$$

If $f \in \mathcal{A}_1^o(\mathcal{D})$, then $f = \sum_j c_j g_j$, for some $g_j \in \mathcal{D}$ and with $\sum_j |c_j| \leq 1$. Since the functions g_j all have norm one, it follows that

$$\|f\| \leq \sum_j |c_j| \|g_j\| \leq 1.$$

Since the functions $g \in \mathcal{D}$ have norm one, it follows that $G_1^r(f) = G_1(f)$ also has norm at most one. By induction, we find that $\|G_m^r(f)\| \leq 1$, $m \geq 1$.

Theorem 2.3.2. For the Relaxed Greedy Algorithm we have, for each $f \in A_1(\mathcal{D})$, the estimate

$$\|f - G_m^r(f)\| \leq \frac{2}{\sqrt{m}}, \quad m \geq 1. \tag{2.3.3}$$

Proof. We use the abbreviation $r_m := G_m^r(f)$ and $g_m := g(R_{m-1}^r(f))$. From the definition of r_m, we have

$$\|f - r_m\|^2 = \|f - r_{m-1}\|^2 + \frac{2}{m}\langle f - r_{m-1}, r_{m-1} - g_m \rangle + \frac{1}{m^2}\|r_{m-1} - g_m\|^2. \tag{2.3.4}$$

The last term on the right-hand side of (2.3.4) does not exceed $4/m^2$. For the middle term, we have

$$\langle f - r_{m-1}, r_{m-1} - g_m \rangle = \inf_{g \in \mathcal{D}^\pm} \langle f - r_{m-1}, r_{m-1} - g \rangle$$

$$= \inf_{\phi \in A_1(\mathcal{D})} \langle f - r_{m-1}, r_{m-1} - \phi \rangle$$

$$\leq \langle f - r_{m-1}, r_{m-1} - f \rangle = -\|f - r_{m-1}\|^2.$$

We substitute this in (2.3.4) to obtain

$$\|f - r_m\|^2 \leq \left(1 - \frac{2}{m}\right)\|f - r_{m-1}\|^2 + \frac{4}{m^2}. \tag{2.3.5}$$

Thus the theorem follows from Lemma 2.3.1 with $A = 4$ and $a_m := \|f - r_m\|^2$. $\square$

We now turn our discussion to the approximation properties of the Pure Greedy Algorithm and the Orthogonal Greedy Algorithm.

We shall need the following simple known lemma (see, for example, De-Vore and Temlyakov (1996)).

Lemma 2.3.3. Let $\{a_m\}_{m=1}^\infty$ be a sequence of non-negative numbers satisfying the inequalities

$$a_1 \leq A, \quad a_{m+1} \leq a_m(1 - a_m/A), \quad m = 1, 2, \ldots.$$

Then we have for each m

$$a_m \leq A/m.$$

Proof. The proof is by induction on m. For $m = 1$ the statement is true by assumption. We assume $a_m \leq A/m$ and prove that $a_{m+1} \leq A/(m+1)$. If $a_{m+1} = 0$ this statement is obvious. Assume therefore that $a_{m+1} > 0$. Then we have

$$a_{m+1}^{-1} \geq a_m^{-1}(1 - a_m/A)^{-1} \geq a_m^{-1}(1 + a_m/A) = a_m^{-1} + A^{-1} \geq (m+1)A^{-1},$$

which implies $a_{m+1} \leq A/(m+1)$.   □

We now want to estimate the decrease in error provided by one step of the Pure Greedy Algorithm. Let $\mathcal{D}$ be an arbitrary dictionary. If $f \in H$ and

$$\rho(f) := \langle f, g(f) \rangle / \|f\|, \tag{2.3.6}$$

where as before $g(f) \in \mathcal{D}^{\pm}$ satisfies

$$\langle f, g(f) \rangle = \sup_{g \in \mathcal{D}^{\pm}} \langle f, g \rangle,$$

then

$$R(f)^2 = \|f - G(f)\|^2 = \|f\|^2 (1 - \rho(f)^2). \tag{2.3.7}$$

The larger $\rho(f)$, the better the decrease of the error in the step of the Pure Greedy Algorithm. The following lemma estimates $\rho(f)$ from below.

Lemma 2.3.4. *If $f \in \mathcal{A}_1(\mathcal{D}, M)$, then*

$$\rho(f) \geq \|f\|/M. \tag{2.3.8}$$

Proof. It is sufficient to prove (2.3.8) for $f \in \mathcal{A}_1^o(\mathcal{D}, M)$ since the general result follows from this by taking limits. We can write $f = \sum c_k g_k$, where this sum has a finite number of terms and $g_k \in \mathcal{D}$ and $\sum |c_k| \leq M$. Hence,

$$\|f\|^2 = \langle f, f \rangle = \left\langle f, \sum c_k g_k \right\rangle = \sum c_k \langle f, g_k \rangle \leq M\rho(f)\|f\|,$$

and (2.3.8) follows.   □

The following theorem was proved in DeVore and Temlyakov (1996).

Theorem 2.3.5. *Let $\mathcal{D}$ be an arbitrary dictionary in H. Then, for each $f \in \mathcal{A}_1(\mathcal{D}, M)$ we have*

$$\|f - G_m(f, \mathcal{D})\| \leq Mm^{-1/6}.$$

Proof. It is enough to prove the theorem for $f \in \mathcal{A}_1(\mathcal{D}, 1)$; the general result then follows by rescaling. We shall use the abbreviated notation $f_m := R_m(f)$ for the residual. Let

$$a_m := \|f_m\|^2 = \|f - G_m(f, \mathcal{D})\|^2, \quad m = 0, 1, \ldots, \quad f_0 := f,$$

and define the sequence $\{b_m\}_{m=0}^{\infty}$ by

$$b_0 := 1, \quad b_{m+1} := b_m + \rho(f_m)\|f_m\|, \quad m = 0, 1, \ldots.$$

Since $f_{m+1} := f_m + \rho(f_m)\|f_m\|g(f_m)$, we obtain by induction that

$$f_m \in \mathcal{A}_1(\mathcal{D}, b_m), \quad m = 0, 1, \dots,$$

and consequently we have the following relations for $m = 0, 1, \dots$:

$$a_{m+1} = a_m(1 - \rho(f_m)^2), \tag{2.3.9}$$

$$b_{m+1} = b_m + \rho(f_m)a_m^{1/2}, \tag{2.3.10}$$

$$\rho(f_m) \geq a_m^{1/2}b_m^{-1}. \tag{2.3.11}$$

The last two relations give

$$b_{m+1} = b_m(1 + \rho(f_m)a_m^{1/2}b_m^{-1}) \leq b_m(1 + \rho(f_m)^2). \tag{2.3.12}$$

Combining this inequality with (2.3.9) we find

$$a_{m+1}b_{m+1} \leq a_m b_m(1 - \rho(f_m)^4),$$

which in turn implies for all m

$$a_m b_m \leq a_0 b_0 = \|f\|^2 \leq 1. \tag{2.3.13}$$

Further, using (2.3.9) and (2.3.11) we get

$$a_{m+1} = a_m(1 - \rho(f_m)^2) \leq a_m(1 - a_m/b_m^2).$$

Since $b_n \leq b_{n+1}$, this gives

$$a_{n+1}b_{n+1}^{-2} \leq a_n b_n^{-2}(1 - a_n b_n^{-2}).$$

Applying Lemma 2.3.3 to the sequence $(a_m b_m^{-2})$ we obtain

$$a_m b_m^{-2} \leq m^{-1}. \tag{2.3.14}$$

Relations (2.3.13) and (2.3.14) imply

$$a_m^3 = (a_m b_m)^2 a_m b_m^{-2} \leq m^{-1}.$$

In other words,

$$\|f_m\| = a_m^{1/2} \leq m^{-1/6},$$

which proves the theorem. □

The next theorem (DeVore and Temlyakov 1996) estimates the error in approximation by the Orthogonal Greedy Algorithm.

Theorem 2.3.6. Let $\mathcal{D}$ be an arbitrary dictionary in H. Then, for each $f \in \mathcal{A}_1(\mathcal{D}, M)$ we have

$$\|f - G_m^o(f, \mathcal{D})\| \leq Mm^{-1/2}.$$

Proof. The proof of this theorem is similar to the proof of Theorem 2.3.5 but technically even simpler. We can again assume that $M = 1$. We let

$f_m^o := R_m^o(f)$ be the residual in the Orthogonal Greedy Algorithm. Then, from the definition of Orthogonal Greedy Algorithm, we have

$$\|f_{m+1}^o\| \le \|f_m^o - G_1(f_m^o, \mathcal{D})\|. \tag{2.3.15}$$

From (2.3.7), we obtain

$$\|f_{m+1}^o\|^2 \le \|f_m^o\|^2(1 - \rho(f_m^o)^2). \tag{2.3.16}$$

By the definition of the Orthogonal Greedy Algorithm, $G_m^o(f) = P_{H_m}f$ and hence $f_m^o = f - G_m^o(f)$ is orthogonal to $G_m^o(f)$. Using this as in the proof of Lemma 2.3.4, we obtain

$$\|f_m^o\|^2 = \langle f_m^o, f \rangle \le \rho(f_m^o)\|f_m^o\|.$$

Hence,

$$\rho(f_m^o) \ge \|f_m^o\|.$$

Using this inequality in (2.3.16), we find

$$\|f_{m+1}^o\|^2 \le \|f_m^o\|^2(1 - \|f_m^o\|^2).$$

In order to complete the proof it remains to apply Lemma 2.3.3 with $A = 1$ and $a_m = \|f_m^o\|^2$. $\qquad\square$

2.3.2. Upper estimates for weak-type greedy algorithms

We begin this subsection with an error estimate for the Weak Orthogonal Greedy Algorithm. The following theorem from Temlyakov (2000b) is a generalization of Theorem 2.3.6.

Theorem 2.3.7. Let $\mathcal{D}$ be an arbitrary dictionary in H. Then, for each $f \in \mathcal{A}_1(\mathcal{D}, M)$ we have

$$\|f - G_m^{o,\tau}(f, \mathcal{D})\| \le M\left(1 + \sum_{k=1}^m t_k^2\right)^{-1/2}.$$

We now turn to the Weak Relaxed Greedy Algorithms. The following theorem from Temlyakov (2000b) shows that the WRGA performs on the $\mathcal{A}_1(\mathcal{D}, M)$ similar to the WOGA. We note that the approximation step of building the $G_m(f, \mathcal{D})$ in the WRGA is simpler than the corresponding step of building the $G_m^{o,\tau}(f, \mathcal{D})$ in the WOGA.

Theorem 2.3.8. Let $\mathcal{D}$ be an arbitrary dictionary in H. Then, for each $f \in \mathcal{A}_1(\mathcal{D}, M)$ we have for the Weak Relaxed Greedy Algorithm

$$\|f - G_m(f, \mathcal{D})\| \le 2M\left(1 + \sum_{k=1}^m t_k^2\right)^{-1/2}.$$

We now proceed to the Weak Greedy Algorithm. The construction of an approximant $G_m^\tau(f, \mathcal{D})$ in the WGA is the simplest out of the three types of algorithms (WGA, WOGA, WRGA) discussed here. We pointed out above that the WGA provides for each $f \in H$ an expansion into a series that satisfies an analogue of the Parseval formula. The following theorem from Temlyakov (2000b) gives the upper bounds for the residual $\|f_m^\tau\|$ of the WGA that are not as good as in Theorems 2.3.7 and 2.3.8 for the WOGA and WRGA, respectively. The next theorem, Theorem 2.3.10, shows that the bound (2.3.17) is sharp in a certain sense.

Theorem 2.3.9. Let $\mathcal{D}$ be an arbitrary dictionary in H. Assume $\tau := \{t_k\}_{k=1}^\infty$ is a non-increasing sequence. Then, for $f \in \mathcal{A}_1(\mathcal{D}, M)$ we have

$$\|f - G_m^\tau(f, \mathcal{D})\| \le M\left(1 + \sum_{k=1}^m t_k^2\right)^{-t_m/2(2+t_m)}. \tag{2.3.17}$$

In a particular case $\tau = \{t\}$, $(t_k = t, \ k = 1, 2, \ldots)$, (2.3.17) gives

$$\|f - G_m^t(f, \mathcal{D})\| \le M(1 + mt^2)^{-t/(4+2t)}, \quad 0 < t \le 1. \tag{2.3.18}$$

This estimate implies the inequality

$$\|f - G_m^t(f, \mathcal{D})\| \le C_1(t)m^{-at}|f|_{\mathcal{A}_1(\mathcal{D})}, \tag{2.3.19}$$

with the exponent at approaching 0 linearly in t. We proved in Livshitz and Temlyakov (2003) that this exponent cannot decrease to 0 at a slower rate than linear.

Theorem 2.3.10. There exists an absolute constant $b > 0$ such that, for any $t > 0$, we can find a dictionary $\mathcal{D}_t$ and a function $f_t \in \mathcal{A}_1(\mathcal{D}_t)$ such that, for some realization $G_m^t(f_t, \mathcal{D}_t)$ of the Weak Greedy Algorithm, we have

$$\liminf_{m \to \infty} \|f_t - G_m^t(f_t, \mathcal{D}_t)\| m^{bt} / |f_t|_{\mathcal{A}_1(\mathcal{D}_t)} > 0. \tag{2.3.20}$$

Remark 2.3.11. The estimate (2.3.18) implies that for small t the parameter a in (2.3.19) can be taken close to $1/4$. The proof from Livshitz and Temlyakov (2003) implies that the parameter b in (2.3.20) can be taken close to $(\ln 2)^{-1}$.

We now discuss some further results on the rate of convergence of the PGA and related results on greedy expansions. Theorem 2.3.5 states that for a general dictionary $\mathcal{D}$ the Pure Greedy Algorithm provides the estimate

$$\|f - G_m(f, \mathcal{D})\| \le |f|_{\mathcal{A}_1(\mathcal{D})} m^{-1/6}.$$

The above estimate was improved a little in Konyagin and Temlyakov (1999b) to

$$\|f - G_m(f, \mathcal{D})\| \le 4|f|_{\mathcal{A}_1(\mathcal{D})} m^{-11/62}.$$

We now discuss recent progress on the following open problem (Temlyakov 2003*a*, p. 65, Open Problem 3.1). This problem is a central theoretical problem in greedy approximation in Hilbert spaces.

Open problem. Find the order of decay of the sequence

$$\gamma(m) := \sup_{f,\mathcal{D},\{G_m\}} \left(\|f - G_m(f,\mathcal{D})\| |f|^{-1}_{\mathcal{A}_1(\mathcal{D})} \right),$$

where the supremum is taken over all dictionaries $\mathcal{D}$, all elements $f \in \mathcal{A}_1(\mathcal{D}) \setminus \{0\}$ and all possible choices of $\{G_m\}$.

Recently, the known upper bounds in approximation by the Pure Greedy Algorithm were improved in Sil'nichenko (2004). Sil'nichenko proved the estimate

$$\gamma(m) \le Cm^{-\frac{s}{2(2+s)}},$$

where s is a solution from $[1, 1.5]$ of the equation

$$(1+x)^{\frac{1}{2+x}} \left(\frac{2+x}{1+x} \right) - \frac{1+x}{x} = 0.$$

Numerical calculations of s (see Sil'nichenko (2004)) give

$$\frac{s}{2(2+s)} = 0.182 \cdots > 11/62.$$

The technique used in Sil'nichenko (2004) is a further development of a method from Konyagin and Temlyakov (1999*b*).

There is also some progress in the lower estimates. The estimate

$$\gamma(m) \ge Cm^{-0.27},$$

with a positive constant C, was proved in Livshitz and Temlyakov (2003). For previous lower estimates see Temlyakov (2003*a*, p. 59). Very recently Livshitz (2007*b*), using the technique from Livshitz and Temlyakov (2003), proved the following lower estimate:

$$\gamma(m) \ge Cm^{-0.1898}. \qquad (2.3.21)$$

We mentioned above that the PGA and its generalization the Weak Greedy Algorithm (WGA) give, for every element $f \in H$, a convergent expansion in a series with respect to a dictionary $\mathcal{D}$. We discuss a further generalization of the WGA that also provides a convergent expansion. We consider here a generalization of the WGA obtained by introducing to it a tuning parameter $b \in (0, 1]$ (see Temlyakov (2007*a*)). Let a sequence $\tau = \{t_k\}_{k=1}^{\infty}$, $0 \le t_k \le 1$, and a parameter $b \in (0, 1]$ be given. We define the Weak Greedy Algorithm with parameter b as follows.

Weak Greedy Algorithm with parameter b (WGA(b)). We define $f_0^{\tau,b} := f$. Then, for each $m \ge 1$ we have the following inductive definition.

(1) $\varphi_m^{\tau,b} \in \mathcal{D}$ is any element satisfying

$$|\langle f_{m-1}^{\tau,b}, \varphi_m^{\tau,b}\rangle| \geq t_m \sup_{g \in \mathcal{D}} |\langle f_{m-1}^{\tau,b}, g\rangle|.$$

(2) $$f_m^{\tau,b} := f_{m-1}^{\tau,b} - b\langle f_{m-1}^{\tau,b}, \varphi_m^{\tau,b}\rangle \varphi_m^{\tau,b}.$$

(3) $$G_m^{\tau,b}(f, \mathcal{D}) := b \sum_{j=1}^{m} \langle f_{j-1}^{\tau,b}, \varphi_j^{\tau,b}\rangle \varphi_j^{\tau,b}.$$

We note that the WGA(b) can be seen as a realization of the Approximate Greedy Algorithm studied in Gribonval and Nielsen (2001a) and Galatenko and Livshitz (2003, 2005).

We point out that the WGA(b), like the WGA, contains, in addition to the first (greedy) step, the second step (see (2) and (3) in the above definition) where we update an approximant by adding an orthogonal projection of the residual $f_{m-1}^{\tau,b}$ onto $\varphi_m^{\tau,b}$ multiplied by b. The WGA(b), therefore, provides, for each $f \in H$, an expansion into a series (greedy expansion):

$$f \sim \sum_{j=1}^{\infty} c_j(f)\varphi_j^{\tau,b}, \quad c_j(f) := b\langle f_{j-1}^{\tau,b}, \varphi_j^{\tau,b}\rangle.$$

We begin with a convergence result from Temlyakov (2007a).

Theorem 2.3.12. Let $\tau \notin \mathcal{V}$. Then the WGA(b) with $b \in (0,1]$ converges for each f and all Hilbert spaces H and dictionaries $\mathcal{D}$.

Theorem 2.3.12 is an extension of the corresponding result for the WGA (see Theorem 2.2.12).

We proved in Temlyakov (2007a) the following convergence rate of the WGA(b).

Theorem 2.3.13. Let $\mathcal{D}$ be an arbitrary dictionary in H. Assume $\tau := \{t_k\}_{k=1}^{\infty}$ is a non-increasing sequence and $b \in (0,1]$. Then, for $f \in A_1(\mathcal{D})$ we have

$$\|f - G_m^{\tau,b}(f, \mathcal{D})\| \leq \left(1 + b(2-b)\sum_{k=1}^{m} t_k^2\right)^{-(2-b)t_m/2(2+(2-b)t_m)}. \qquad (2.3.22)$$

This theorem is an extension of the corresponding result for the WGA (see Theorem 2.3.9). In the particular case $t_k = 1$, $k = 1, 2, \ldots$, we get the following rate of convergence:

$$\|f - G_m^{1,b}(f, \mathcal{D})\| \leq Cm^{-r(b)}, \quad r(b) := \frac{2-b}{2(4-b)}.$$

We note that $r(1) = 1/6$ and $r(b) \to 1/4$ as $b \to 0$. Thus we can offer the following observation. At each step of the Pure Greedy Algorithm we can choose a fixed fraction of the optimal coefficient for that step instead of the optimal coefficient itself. Surprisingly, this leads to better upper estimates than those known for the Pure Greedy Algorithm.

2.4. Greedy algorithms for systems that are not dictionaries

In this section we discuss greedy algorithms with regard to a system $\mathcal{G}$ that is not a dictionary. Here, we will discuss a variant of the RGA that is a generalization of the version of the RGA suggested by Barron (1993). Let H be a real Hilbert space and let $\mathcal{G} := \{g\}$ be a system of elements $g \in H$ such that $\|g\| \leq C_0$. Usually, in the theory of greedy algorithms we consider approximation with regard to a dictionary $\mathcal{D}$. One of the properties of a dictionary $\mathcal{D}$ is that the closure of span $\mathcal{D}$ is equal to H. In this section we do not assume that the system $\mathcal{G}$ is a dictionary. In particular, we do not assume that the closure of span $\mathcal{G}$ is H. This setting is motivated by applications in learning theory (see Section 2.8). We present here results from Temlyakov (2005d). Let $\mathcal{G}^{\pm} := \{\pm g : g \in \mathcal{G}\}$ denote the symmetrized system $\mathcal{G}$, and let $\theta > 0$.

RGA(θ) with respect to $\mathcal{G}$. For $f \in H$ we define $f_0 := f$, $G_0 := G_0(f) := 0$. Then, for each $n \geq 1$ we have the following inductive definition.

(1) $\varphi_n \in \mathcal{G}^{\pm}$ is an element satisfying (we assume existence)

$$\langle f_{n-1}, \varphi_n \rangle = \max_{g \in \mathcal{G}^{\pm}} \langle f_{n-1}, g \rangle.$$

(2) $\qquad G_n := G_n(f) := \left(1 - \frac{\theta}{n+\theta}\right) G_{n-1} + \frac{\theta}{n+\theta} \varphi_n, \quad f_n := f - G_n.$

Let $A_1(\mathcal{G})$ denote the closure in H of the convex hull of $\mathcal{G}^{\pm}$. Then, for $f \in H$ there exists a unique element $f' \in A_1(\mathcal{G})$ such that

$$d(f, A_1(\mathcal{G}))_H = \|f - f'\| \leq \|f - \phi\|, \quad \phi \in A_1(\mathcal{G}). \tag{2.4.1}$$

In analysis of the RGA(θ) we will use the following simple lemma (see DeVore and Temlyakov (1996) and Lemma 2.3.1 above for a variant of this lemma). Our analysis is similar to that of DeVore and Temlyakov (1996) and Lee, Bartlett and Williamson (1996).

Lemma 2.4.1. Let a sequence $\{a_n\}_{n=0}^{\infty}$ of non-negative numbers satisfy the relations (with $\beta > 1$, $B > 0$)

$$a_n \leq \frac{n}{n+\beta} a_{n-1} + \frac{B}{(n+\beta)^2}, \quad n = 1, 2, \ldots, \qquad a_0 \leq \frac{B}{(\beta-1)\beta}.$$

Then, for all n,

$$a_n \leq \frac{B}{(\beta-1)(n+\beta)}.$$

Proof. Setting $A := B/(\beta - 1)$, we obtain by induction

$$a_n \le \frac{A}{n-1+\beta}\frac{n}{n+\beta} + \frac{B}{(n+\beta)^2} = \frac{A}{n+\beta} - \frac{A(\beta-1)}{(n+\beta)(n-1+\beta)} + \frac{B}{(n+\beta)^2}.$$

Taking into account the inequality

$$\frac{A(\beta-1)}{(n+\beta)(n-1+\beta)} \ge \frac{A(\beta-1)}{(n+\beta)^2} = \frac{B}{(n+\beta)^2},$$

we complete the proof. □

Theorem 2.4.2. For $\theta > 1$ there exists a constant $C(\theta)$ such that, for any $f \in H$, we have

$$\|f_n\|^2 \le d(f, A_1(\mathcal{G}))_H^2 + C(\theta)(\|f\| + C_0)^2 n^{-1}.$$

Proof. From the definition of G_n and f_n we get, setting $\alpha := \frac{\theta}{n+\theta}$,

$$f_n = f - G_n = (1-\alpha)f_{n-1} + \alpha(f - \varphi_n)$$

and

$$\|f_n\|^2 = (1-\alpha)^2\|f_{n-1}\|^2 + 2\alpha(1-\alpha)\langle f_{n-1}, f - \varphi_n\rangle + \alpha^2\|f - \varphi_n\|^2. \quad (2.4.2)$$

It is known and easy to check that for any $h \in H$ we have

$$\sup_{g \in \mathcal{G}^\pm} \langle h, g\rangle = \sup_{\phi \in A_1(\mathcal{G})} \langle h, \phi\rangle. \quad (2.4.3)$$

Denote f' as above and set $f^* := f - f'$. Using (2.4.3) and the definition of φ_n, we obtain from (2.4.2)

$$\|f_n\|^2 \le (1-\alpha)^2\|f_{n-1}\|^2 + 2\alpha(1-\alpha)\langle f_{n-1}, f - f'\rangle + \alpha^2\|f - \varphi_n\|^2$$
$$= (1-\alpha)(\|f_{n-1}\|^2 - \alpha\|f_{n-1}\|^2 + 2\alpha\langle f_{n-1}, f^*\rangle - \alpha\|f^*\|^2)$$
$$\qquad + \alpha(1-\alpha)\|f^*\|^2 + \alpha^2\|f - \varphi_n\|^2$$
$$\le (1-\alpha)\|f_{n-1}\|^2 + \alpha\|f^*\|^2 + \alpha^2\|f - \varphi_n\|^2.$$

This implies

$$\|f_n\|^2 - \|f^*\|^2 \le (1-\alpha)(\|f_{n-1}\|^2 - \|f^*\|^2) + \alpha^2(\|f\| + C_0)^2.$$

Setting $a_n := \|f_n\|^2 - \|f^*\|^2$, $\beta := \theta$, and applying Lemma 2.4.1, we complete the proof. □

Theorem 2.4.3. For $\theta > 1/2$ there exists a constant $C := C(\theta, C_0)$ such that, for any $f \in H$, we have

$$\|f' - G_n(f)\|^2 \le C/n.$$

Proof. If $f \in A_1(\mathcal{G})$ then the statement of Theorem 2.4.3 follows from known results (see Barron (1993) and Theorem 2.3.2). If $d(f, A_1(\mathcal{G})) > 0$, then property (2.4.1) implies that, for any $\phi \in A_1(\mathcal{G})$, we have

$$\langle f^*, \phi - f' \rangle \leq 0. \tag{2.4.4}$$

It follows from the definition of f_n that

$$f_n = \left(1 - \frac{\theta}{n+\theta}\right) f_{n-1} + \frac{\theta}{n+\theta}(f - \varphi_n).$$

We set $f'_n := f_n - f^*$. Then, we get from the above representation

$$f'_n = \left(1 - \frac{\theta}{n+\theta}\right) f'_{n-1} + \frac{\theta}{n+\theta}(f' - \varphi_n).$$

We note that $f'_n = f' - G_n(f)$. Let us estimate

$$\|f'_n\|^2 = \|f'_{n-1}\|^2 \left(1 - \frac{\theta}{n+\theta}\right)^2 \tag{2.4.5}$$

$$+ \frac{2\theta}{n+\theta}\left(1 - \frac{\theta}{n+\theta}\right) \langle f'_{n-1}, f' - \varphi_n \rangle + \frac{\theta^2}{(n+\theta)^2}\|f' - \varphi_n\|^2.$$

Next,

$$\langle f'_{n-1}, f' - \varphi_n \rangle = \langle f'_{n-1} + f^*, f' - \varphi_n \rangle - \langle f^*, f' - \varphi_n \rangle$$

$$= \langle f_{n-1}, f' - \varphi_n \rangle + \langle f^*, \varphi_n - f' \rangle. \tag{2.4.6}$$

First, we prove that

$$\langle f_{n-1}, f' - \varphi_n \rangle \leq 0. \tag{2.4.7}$$

It easily follows from $f' \in A_1(\mathcal{G})$ that

$$\langle f_{n-1}, f' \rangle \leq \max_{g \in \mathcal{G}^\pm} \langle f_{n-1}, g \rangle. \tag{2.4.8}$$

By the definition of φ_n we get

$$\max_{g \in \mathcal{G}^\pm} \langle f_{n-1}, g \rangle = \langle f_{n-1}, \varphi_n \rangle. \tag{2.4.9}$$

Thus, (2.4.7) follows from (2.4.8) and (2.4.9).

Secondly, we note that (2.4.4) implies

$$\langle f^*, \varphi_n - f' \rangle \leq 0. \tag{2.4.10}$$

Therefore, by (2.4.6), (2.4.7), and (2.4.10) we obtain

$$\langle f'_{n-1}, f' - \varphi_n \rangle \leq 0. \tag{2.4.11}$$

Substitution of (2.4.11) in (2.4.5) gives

$$\|f'_n\|^2 \leq \|f'_{n-1}\|^2 \left(1 - \frac{2\theta}{n+\theta}\right) + \frac{\theta^2}{(n+\theta)^2}(\|f'_{n-1}\|^2 + \|f' - \varphi_n\|^2). \tag{2.4.12}$$

Using bounds $\|f'_{n-1}\| \le C_0$ and $\|f' - \varphi_n\| \le 2C_0$, we find

$$\|f'_n\|^2 \le \|f'_{n-1}\|^2 \left(1 - \frac{2\theta}{n + \theta}\right) + 5C_0^2\theta^2/(n + \theta)^2.$$

We note that

$$1 - \frac{2\theta}{n + \theta} < 1 - \frac{2\theta}{n + 2\theta}.$$

We now apply Lemma 2.4.1 with $a_n = \|f'_n\|^2$, $\beta = 2\theta$, and get

$$\|f'_n\|^2 \le C(\theta, C_0)/n. \tag{2.4.13}$$

This completes the proof. $\qquad\qquad\qquad\qquad\qquad\qquad\qquad\qquad\square$

2.5. Saturation property of greedy-type algorithms

In this section we shall give an example from DeVore and Temlyakov (1996) which shows that replacing a dictionary $\mathcal{B}$ given by an orthogonal basis by a non-orthogonal redundant dictionary $\mathcal{D}$ may damage the efficiency of the Pure Greedy Algorithm. The dictionary $\mathcal{D}$ in our example differs from the dictionary $\mathcal{B}$ by the one addition of the element g for a certain suitably chosen function g.

Let $\mathcal{B} := \{h_k\}_{k=1}^\infty$ be an orthonormal basis in a Hilbert space H. Consider the following element:

$$g := Ah_1 + Ah_2 + aA\sum_{k \ge 3}(k(k + 1))^{-1/2}h_k, \tag{2.5.1}$$

with

$$A := (33/89)^{1/2} \quad \text{and} \quad a := (23/11)^{1/2}.$$

Then $\|g\| = 1$. We define the dictionary $\mathcal{D} := \mathcal{B} \cup \{g\}$.

Theorem 2.5.1. For the function

$$f := h_1 + h_2,$$

we have

$$\|f - G_m(f)\| \ge m^{-1/2}, \quad m \ge 4.$$

Proof. We shall examine the steps of the Pure Greedy Algorithm applied to the function $f = h_1 + h_2$. We shall use the abbreviated notation $f_m := R_m(f) := f - G_m(f)$ for the residual at step m.

First step. We have

$$\langle f, g \rangle = 2A > 1, \quad |\langle f, h_k \rangle| \le 1, \quad k = 1, 2, \ldots,$$

This implies

$$G_1(f, D) = \langle f, g \rangle g,$$

and

$$f_1 = f - \langle f, g \rangle g = (1 - 2A^2)(h_1 + h_2) - 2aA^2 \sum_{k \geq 3} (k(k+1))^{-1/2} h_k.$$

Second step. We have

$$\langle f_1, g \rangle = 0, \quad \langle f_1, h_k \rangle = (1 - 2A^2), \quad k = 1, 2, \quad \langle f_1, h_3 \rangle = -aA^2 3^{-1/2}.$$

Comparing $\langle f_1, h_1 \rangle$ and $|\langle f_1, h_3 \rangle|$ we get

$$|\langle f_1, h_3 \rangle| = (23/89)(33/23)^{1/2} > 23/89 = 1 - 2A^2 = \langle f_1, h_1 \rangle.$$

This implies that the second approximation $G_1(f_1, D)$ is $\langle f_1, h_3 \rangle h_3$ and

$$f_2 = f_1 - \langle f_1, h_3 \rangle h_3 = (1 - 2A^2)(h_1 + h_2) - 2aA^2 \sum_{k \geq 4} (k(k+1))^{-1/2} h_k.$$

Third step. We have

$$\langle f_2, g \rangle = -\langle f_1, h_3 \rangle \langle h_3, g \rangle = (A/2)(23/89),$$
$$\langle f_2, h_1 \rangle = \langle f_2, h_2 \rangle = 1 - 2A^2 = 23/89,$$
$$\langle f_2, h_4 \rangle = -aA^2 5^{-1/2} = -(23/89)(99/115)^{1/2}.$$

Therefore, the third approximation should be $\langle f_2, h_1 \rangle h_1$ or $\langle f_2, h_2 \rangle h_2$. Let us take the first of these so that

$$f_3 = f_2 - \langle f_2, h_1 \rangle h_1.$$

Fourth step. It is clear that for all $k \neq 1$ we have

$$\langle f_3, h_k \rangle = \langle f_2, h_k \rangle.$$

This equality and the calculations from the third step show that it is sufficient to compare $\langle f_3, h_2 \rangle$ and $\langle f_3, g \rangle$. We have

$$\langle f_3, g \rangle = \langle f_2, g \rangle - \langle f_2, h_1 \rangle \langle h_1, g \rangle = -(23/89)(A/2).$$

This means that

$$f_4 = f_3 - \langle f_3, h_2 \rangle h_2 = -2aA^2 \sum_{k \geq 4} (k(k+1))^{-1/2} h_k. \tag{2.5.2}$$

mth step ($m > 4$). We prove by induction that for all $m \geq 4$ we have

$$f_m = -2aA^2 \sum_{k \geq m} (k(k+1))^{-1/2} h_k. \tag{2.5.3}$$

For $m = 4$ this relation follows from (2.5.2). We assume we have proved (2.5.3) for some m and derive that (2.5.3) also holds true for $m + 1$. To find f_{m+1}, we only have to compare the two inner products: $\langle f_m, h_m \rangle$ and $\langle f_m, g \rangle$. We have

$$|\langle f_m, h_m \rangle| = 2aA^2 (m(m+1))^{-1/2},$$

and

$$|\langle f_m, g \rangle| = 2a^2 A^3 \sum_{k \geq m} (k(k+1))^{-1} = 2a^2 A^3 m^{-1}.$$

Since

$$(|\langle f_m, g \rangle|/|\langle f_m, h_m \rangle|)^2 = (aA)^2(1 + 1/m) \leq 345/356 < 1,$$

we have that

$$|\langle f_m, g \rangle| < |\langle f_m, h_m \rangle|, \quad m \geq 4.$$

This proves (2.5.3) with m replaced by $m+1$.

From (2.5.3), we obtain

$$\|f - G_m(f, D)\| = \|f_m\| = 2aA^2 m^{-1/2} > m^{-1/2}, \quad m \geq 4. \qquad \square$$

2.6. Lebesgue-type inequalities for greedy approximation

Lebesgue proved the following inequality: for any 2π-periodic continuous function f we have

$$\|f - S_n(f)\|_\infty \leq \left(4 + \frac{4}{\pi^2} \ln n\right) E_n(f)_\infty, \qquad (2.6.1)$$

where $S_n(f)$ is the nth partial sum of the Fourier series of f and $E_n(f)_\infty$ is the error of the best approximation of f by the trigonometric polynomials of order n in the uniform norm $\| \cdot \|_\infty$. The inequality (2.6.1) relates the error of a particular method (S_n) of approximation by the trigonometric polynomials of order n to the best-possible error $E_n(f)_\infty$ of approximation by the trigonometric polynomials of order n. By a Lebesgue-type inequality we mean an inequality that provides an upper estimate for the error of a particular method of approximation of f by elements of a special form, say, form $\mathcal{A}$, by the best-possible approximation of f by elements of the form $\mathcal{A}$. In the case of approximation with regard to bases (or minimal systems), the Lebesgue-type inequalities are known both in linear and in nonlinear settings (see Chapter 1 and surveys by Konyagin and Temlyakov (2002) and Temlyakov (2003a)). It would be very interesting to prove the Lebesgue-type inequalities for redundant systems (dictionaries). However, there are substantial difficulties.

We begin our discussion with the Pure Greedy Algorithm (PGA). It is natural to compare performance of the PGA with the best m-term approximation with regard to a dictionary $\mathcal{D}$. We let $\Sigma_m(\mathcal{D})$ denote the collection of all functions (elements) in H which can be expressed as a linear combination of at most m elements of $\mathcal{D}$. Thus, each function $s \in \Sigma_m(\mathcal{D})$ can be written in the form

$$s = \sum_{g \in \Lambda} c_g g, \quad \Lambda \subset \mathcal{D}, \quad \#\Lambda \leq m,$$

where the c_g are real or complex numbers. In some cases, it may be possible to write an element from $\Sigma_m(\mathcal{D})$ in this form in more than one way. The space $\Sigma_m(\mathcal{D})$ is not linear: the sum of two functions from $\Sigma_m(\mathcal{D})$ is generally not in $\Sigma_m(\mathcal{D})$.

For a function $f \in H$ we define its best m-term approximation error:

$$\sigma_m(f) := \sigma_m(f, \mathcal{D}) := \inf_{s \in \Sigma_m(\mathcal{D})} \|f - s\|.$$

It seems there is no hope of proving a non-trivial Lebesgue-type inequality for the PGA in the case of an arbitrary dictionary $\mathcal{D}$. This pessimism is based on the following result from DeVore and Temlyakov (1996) (see Section 2.5).

Let $\mathcal{B} := \{h_k\}_{k=1}^{\infty}$ be an orthonormal basis in a Hilbert space H. Consider the following element:

$$g := Ah_1 + Ah_2 + aA \sum_{k \geq 3} (k(k+1))^{-1/2} h_k,$$

with

$$A := (33/89)^{1/2} \quad \text{and} \quad a := (23/11)^{1/2}.$$

Then $\|g\| = 1$. We define the dictionary $\mathcal{D} = \mathcal{B} \cup \{g\}$. It was proved in DeVore and Temlyakov (1996) (see Section 2.5 above) that, for the function

$$f = h_1 + h_2,$$

we have

$$\|f - G_m(f, \mathcal{D})\| \geq m^{-1/2}, \quad m \geq 4.$$

It is clear that $\sigma_2(f, \mathcal{D}) = 0$.

Therefore, we look for conditions on a dictionary $\mathcal{D}$ that allow us to prove Lebesgue-type inequalities. The condition $\mathcal{D} = \mathcal{B}$, an orthonormal basis for H, guarantees that

$$\|R_m(f, \mathcal{B})\| = \sigma_m(f, \mathcal{B}).$$

This is an ideal situation. The results that we will discuss here concern the case when we replace an orthonormal basis $\mathcal{B}$ by a dictionary that is, in a certain sense, not far from an orthonormal basis.

Let us begin with results from Donoho, Elad and Temlyakov (2007) that are close to results from Temlyakov (1999). We give a definition of a λ-quasi-orthogonal dictionary with depth D. When $D = \infty$ this definition coincides with the definition of a λ-quasi-orthogonal dictionary from Temlyakov (1999).

Definition 2.6.1. We say $\mathcal{D}$ is a λ-quasi-orthogonal dictionary with depth D if, for any $n \in [1, D]$ and any $g_i \in \mathcal{D}$, $i = 1, \ldots, n$, there exists a collection

$\varphi_j \in \mathcal{D}$, $j = 1, \ldots, J$, $J \leq N := \lambda n$, with the properties

$$g_i \in X_J := \text{span}(\varphi_1, \ldots, \varphi_J), \quad i = 1, \ldots, n,$$

and for any $f \in X_J$ we have

$$\max_{1 \leq j \leq J} |\langle f, \varphi_j \rangle| \geq N^{-1/2} \|f\|.$$

Remark 2.6.2. It is clear that an orthonormal dictionary is a 1-quasi-orthogonal dictionary.

The following theorem for $D = \infty$ was established in Temlyakov (1999). It is pointed out in Donoho *et al.* (2007) that the proof from Temlyakov (1999) also works in the case $D < \infty$, and gives the following result.

Theorem 2.6.3. Let a given dictionary $\mathcal{D}$ be λ-quasi-orthogonal with depth D, and let $0 < r < (2\lambda)^{-1}$ be a real number. Then, for any f such that

$$\sigma_m(f, \mathcal{D}) \leq m^{-r}, \quad m = 1, 2, \ldots, D,$$

we have

$$\|f_m\| = \|f - G_m(f, \mathcal{D})\| \leq C(r, \lambda) m^{-r}, \quad m \in [1, D/2].$$

In this section we consider dictionaries that have become popular in signal processing. Denote

$$M(\mathcal{D}) := \sup_{g \neq h; g, h \in \mathcal{D}} |\langle g, h \rangle|,$$

the coherence parameter of a dictionary $\mathcal{D}$. For an orthonormal basis $\mathcal{B}$ we have $M(\mathcal{B}) = 0$. It is clear that the smaller $M(\mathcal{D})$, the more $\mathcal{D}$ resembles an orthonormal basis. However, we should note that in the case $M(\mathcal{D}) > 0$ the $\mathcal{D}$ can be a redundant dictionary. We showed in Donoho *et al.* (2007) (see Proposition 2.1) that a dictionary with coherence $M := M(\mathcal{D})$ is a $(1 + 4\delta)$-quasi-orthogonal dictionary with depth δ/M, for any $\delta \in (0, 1/7]$. Therefore, Theorem 2.6.3 applies to M-coherent dictionaries. We proved in Donoho *et al.* (2007) a general Lebesgue-type inequality for the PGA with regard to an M-coherent dictionary.

Theorem 2.6.4. Let a dictionary $\mathcal{D}$ have the mutual coherence $M = M(\mathcal{D})$. Then, for any $S \leq 1/(2M)$ we have the following inequality:

$$\|f_S\|^2 \leq 2\|f\|(\sigma_S(f, \mathcal{D}) + 5MS\|f\|). \tag{2.6.2}$$

As a direct corollary of this theorem we obtain the following inequality for functions f that allow an S-sparse representation in $\mathcal{D}$ ($\sigma_S(f) = 0$):

$$\|f_S\| \leq (10MS)^{1/2} \|f\|.$$

Inequality (2.6.2) is the first Lebesgue-type inequality for the PGA in the case of incoherent dictionary $\mathcal{D}$.

We now proceed to a discussion of the Orthogonal Greedy Algorithm (OGA). It is clear from the definition of the OGA that at each step we have (see (2.1.3))

$$\|f_m^o\|^2 \leq \|f_{m-1}^o\|^2 - |\langle f_{m-1}^o, g(f_{m-1}^o)\rangle|^2.$$

We noted in Donoho *et al.* (2007) that the use of this inequality instead of the equality

$$\|f_m\|^2 = \|f_{m-1}\|^2 - |\langle f_{m-1}, g(f_{m-1})\rangle|^2,$$

which holds for the PGA, allows us to prove an analogue of Theorem 2.6.3 for the OGA. The proof repeats the corresponding proof from Temlyakov (1999). We formulate this as a remark.

Remark 2.6.5. Theorem 2.6.3 holds for the OGA instead of the PGA (for $\|f_m^o\|$ instead of $\|f_m\|$).

The first general Lebesgue-type inequality for the OGA for the M-coherent dictionary was obtained in Gilbert, Muthukrishnan and Strauss (2003). They proved that

$$\|f_m^o\| \leq 8m^{1/2}\sigma_m(f) \quad \text{for } m < 1/(32M).$$

The constants in this inequality were improved in Tropp (2004) (see also Donoho, Elad and Temlyakov (2006)):

$$\|f_m^o\| \leq (1 + 6m)^{1/2}\sigma_m(f) \quad \text{for } m < 1/(3M). \qquad (2.6.3)$$

We proved in Donoho *et al.* (2007) an analogue of (2.6.2) for the OGA.

Theorem 2.6.6. Let a dictionary $\mathcal{D}$ have the mutual coherence $M = M(\mathcal{D})$. Then, for any $S \leq 1/(2M)$ we have the following inequalities:

$$\|f_S^o\|^2 \leq 2\|f_k^o\|(\sigma_{S-k}(f_k^o) + 3MS\|f_k^o\|), \quad 0 \leq k \leq S. \qquad (2.6.4)$$

Inequality (2.6.4) can be used to improve (2.6.3) for small m. We proved in Donoho *et al.* (2007) the following inequality.

Theorem 2.6.7. Let a dictionary $\mathcal{D}$ have the mutual coherence $M = M(\mathcal{D})$. Assume $m \leq 0.05M^{-2/3}$. Then, for $l \geq 1$ satisfying $2^l \leq \log m$ we have

$$\|f_{m(2^l-1)}^o\| \leq 6m^{2^{-l}}\sigma_m(f).$$

Corollary 2.6.8. Let a dictionary $\mathcal{D}$ have the mutual coherence $M = M(\mathcal{D})$. Assume $m \leq 0.05M^{-2/3}$. Then we have

$$\|f_{[m \log m]}^o\| \leq 24\sigma_m(f). \qquad (2.6.5)$$

Inequality (2.6.5) is an almost perfect Lebesgue-type inequality. It has the following two deficiencies. First, clearly we would like to replace $[m \log m]$ by m or Cm in the number of iterations of the OGA. However, this is a minor drawback of (2.6.5). Second, as is stated in Corollary 2.6.8, inequality (2.6.5) holds for only a small number of iterations: $m \leq 0.05 M^{-2/3}$. It would be interesting to know if we can push the limit from $M^{-2/3}$ to a natural limit of M^{-1}.

The above results show that the smaller the coherence parameter $M(\mathcal{D})$, the better the performance of the OGA. In particular, (2.6.3) implies that if f is S-sparse with respect to $\mathcal{D}$ ($\sigma_S(f, \mathcal{D}) = 0$), then $f_S^o = 0$, provided that $S < 1/(3M)$. This means that the OGA exactly recovers S-sparse elements with respect to the M-coherent dictionary $\mathcal{D}$ if $S < 1/(3M)$. This is a very nice property of M-coherent dictionaries, which is important in applications (see Donoho *et al.* (2006) for a discussion). Therefore, it is very desirable to build dictionaries with a small coherent parameter $M(\mathcal{D})$. A rigorous setting in this regard is the following. Let $H = \mathbb{R}^d$ and let the cardinality of $\mathcal{D}$ be equal to N ($|\mathcal{D}| = N$). Find

$$c(N, d) := \inf_{\mathcal{D}, |\mathcal{D}|=N} M(\mathcal{D})$$

and describe the *Grassmannian dictionaries* for which $M(\mathcal{D}) = c(N, d)$, $|\mathcal{D}| = N$.

In a special case when $\mathcal{D}$ is assumed to be a frame, the dictionaries (frames) described above are known as Grassmannian frames. The theory of Grassmannian frames is a beautiful mathematical theory that has connections to areas such as spherical codes, algebraic geometry, graph theory, and sphere packings (see Stromberg and Heath (2003)). Some fundamental problems of this theory are still open. For instance, it is known that in the case of frames we have

$$c^{\mathrm{frame}}(N, d) \geq \left(\frac{N - d}{d(N - 1)} \right)^{1/2}. \tag{2.6.6}$$

However, it is not known for which pairs (N, d) we have equality in (2.6.6).

2.7. Some further remarks

In the Preface we mentioned two classical examples of redundant dictionaries:

$$\Pi_2 := \left\{ u(x_1)v(x_2) : (x_1, x_2) \in [0, 1]^2, \|u\|_{L_2([0,1])} = \|v\|_{L_2([0,1])} = 1 \right\},$$

$$\mathcal{R}_2 := \left\{ r(\omega \cdot x) : x, \omega \in \mathbb{R}^d, \|x\|_{\ell_2} \leq 1, \|\omega\|_{\ell_2} = 1, \|r(\omega \cdot x)\|_{L_2(B_2^d)} = 1 \right\}.$$

The reader can find detailed discussion of the m-term approximation with regard to these dictionaries in the survey of Temlyakov (2003*a*). The dictionary Π_2 is a very interesting example from the point of view of greedy approximation. As mentioned in the Preface, we have for each $f \in L_2([0, 1]^2)$

$$\|f - G_m(f, \Pi_2)\|_{L_2([0,1]^2)} = \sigma_m(f, \Pi_2)_{L_2([0,1]^2)}, \tag{2.7.1}$$

for the PGA. This means that the Π_2 is ideally designed for greedy approximation. Clearly, all the general results of this chapter apply in the case $\mathcal{D} = \Pi_2$. Surprisingly, we do not quite understand how to use specific properties of Π_2 in greedy approximation. For instance (see the end of Section 2.2), there has been no progress on the following open problem (see Temlyakov (2003a), p.78). Find the necessary and sufficient conditions on a weakness sequence τ to guarantee convergence of the WGA with regard to Π_2 for each $f \in L_2([0,1]^2)$.

We note that the Schmidt expansion formula for $f \in L_2([0,1]^2)$,

$$f(x_1, x_2) = \sum_{j=1}^{\infty} s_j(J_f)\phi_j(x_1)\psi_j(x_2),$$

points out the importance of the sequence $\{s_j(J_f)\}$ of singular numbers of the integral operator J_f associated with f. There is an extensive literature devoted to estimating $s_j(J_f)$ and $\sigma_m(f, \Pi_2)$ (in different norms) in terms of smoothness of f. We mention some of the papers: Fredholm (1903), Weyl (1911), Hille and Tamarkin (1931), Smithies (1937), Birman and Solomyak (1977), Cochran (1977) and Temlyakov (1989b, 1990, 1992a, 1992b, 1993b). For a further discussion see the survey of Temlyakov (2003a).

The dictionary $\mathcal{R}_2$ is not as good as Π_2 for greedy approximation. There are some weaker analogues of (2.7.1) for greedy approximation with regard to $\mathcal{R}_2$ (see Maiorov, Oskolkov and Temlyakov (2002)). However, as in the case of Π_2, we do not know how to use specific features of $\mathcal{R}_2$ in greedy approximation. There has also been no progress on an open problem (see Temlyakov (2003a, p. 81)), similar to the one mentioned above, on convergence of the WGA with regard to $\mathcal{R}_2$.

We now proceed to a discussion of some recent results on simultaneous greedy approximation. A new ingredient of the papers of Lutoborski and Temlyakov (2003), Leviatan and Temlyakov (2005, 2006) and Temlyakov (2004) is the move from approximating a single element f, to simultaneous approximation of a set of elements $f^1, \ldots, f^N$. We will give a description of some results from Lutoborski and Temlyakov (2003), Leviatan and Temlyakov (2006) and Temlyakov (2004). The main purpose of the above papers is to construct greedy-type expansions,

$$f^i \sim \sum_{j=1}^{\infty} c_j^i(f)\varphi_j, \quad c_j^i(f) := \langle f_{j-1}^i, \varphi_j \rangle, \qquad (2.7.2)$$

for a given finite set of elements $f^1, \ldots, f^N$, simultaneously with the same sequence $\{\varphi_j\}$ for all f^i, $i = 1, \ldots, N$. The first result in this direction was obtained in Lutoborski and Temlyakov (2003). The Vector Greedy Algorithms that are designed for the purpose of constructing mth greedy approximants, simultaneously for a given finite number of elements, were introduced and studied in Lutoborski and Temlyakov (2003).

Vector Weak Greedy Algorithm (VWGA). Let a vector of elements $f^i \in H$, $i = 1, \ldots, N$ be given. We write $f_0^{i,v,\tau} := f^i$. Then, for each $m \geq 1$ we have the following inductive definition.

(1) Let $\varphi_m^{v,\tau} \in \mathcal{D}$ be any element satisfying

$$\max_i |\langle f_{m-1}^{i,v,\tau}, \varphi_m^{v,\tau} \rangle| \geq t_m \max_i \sup_{g \in \mathcal{D}} |\langle f_{m-1}^{i,v,\tau}, g \rangle|. \tag{2.7.3}$$

(2)
$$f_m^{i,v,\tau} := f_{m-1}^{i,v,\tau} - \langle f_{m-1}^{i,v,\tau}, \varphi_m^{v,\tau} \rangle \varphi_m^{v,\tau}, \quad i = 1, \ldots, N.$$

(3)
$$G_m^{v,\tau}(f^i, \mathcal{D}) := \sum_{j=1}^{m} \langle f_{j-1}^{i,v,\tau}, \varphi_j^{v,\tau} \rangle \varphi_j^{v,\tau}, \quad i = 1, \ldots, N.$$

It was proved in Lutoborski and Temlyakov (2003) that the VWGA converges for $\tau \notin \mathcal{V}$. Therefore the VWGA with $\tau \notin \mathcal{V}$ provides the convergent expansions

$$f^i = \sum_{j=1}^{\infty} b_j^i g_j, \qquad g_j \in \mathcal{D},$$

with the property

$$\|f^i\|^2 = \sum_{j=1}^{\infty} |b_j^i|^2, \qquad i = 1, \ldots, N.$$

The following estimate of the rate of convergence of the VWGA was obtained in Lutoborski and Temlyakov (2003).

Theorem 2.7.1. Let $\mathcal{D}$ be an arbitrary dictionary in H. Assume $\tau := \{t_k\}_{k=1}^{\infty}$, $t_k = t$, $k = 1, \ldots$, $0 < t < 1$. Then, for any vector of elements $f^1, \ldots, f^N$, $f^i \in A_1(\mathcal{D})$, $i = 1, \ldots, N$, we have

$$\sum_{i=1}^{N} \|f_m^{i,v,\tau}\|^2 \leq \left(1 + \frac{mt^2}{N}\right)^{-t/(2N+t)} N^{\frac{2N+2t}{2N+t}}.$$

Comparing Theorem 2.3.9 with $\tau = \{t\}$ with Theorem 2.7.1, we see that the exponent $t/(2N + t)$ of decay is seriously affected by the number N of simultaneously approximated elements. Also, simultaneous approximation brings an extra factor, $N^{\frac{2N+2t}{2N+t}} \asymp N$. In Leviatan and Temlyakov (2006) we improve the exponent of decay, replacing $t/(2N + t)$ by $t/(2N^{1/2} + t)$, and we get the worse constant N^2 instead of N. Here is the corresponding theorem from Leviatan and Temlyakov (2006).

Theorem 2.7.2. Let $\mathcal{D}$ be an arbitrary dictionary in H. Assume $\tau :=$ $\{t_k\}_{k=1}^{\infty}$ is a non-increasing sequence. Then, for any vector of elements $f^1, \ldots, f^N$, $f^i \in A_1(\mathcal{D})$, $i = 1, \ldots, N$, we have

$$\sum_{i=1}^{N} \|f_m^{i,v,\tau}\|^2 \leq N^2 \left(1 + \frac{1}{N}\sum_{k=1}^{m} t_k^2\right)^{\frac{-t_m}{2N^{1/2}+t_m}}.$$

In addition to the VWGA the following two modifications of the VWGA were considered in Leviatan and Temlyakov (2006). The modifications differ from the VWGA only in the first step. In the first step of the Simultaneous Weak Greedy Algorithm 1 (SWGA1), we have the following.

(1)[SWGA1]   We look for any $\varphi_m^{s1,\tau} \in \mathcal{D}$ satisfying

$$\left(\sum_{i=1}^{N} |\langle f_{m-1}^i, \varphi_m^{s1,\tau}\rangle|^2\right)^{1/2} \geq t_m \max_i \sup_{g \in \mathcal{D}} |\langle f_{m-1}^i, g\rangle|, \quad f_{m-1}^i := f_{m-1}^{i,s1,\tau}.$$

(2.7.4)

The first step of the Simultaneous Weak Greedy Algorithm 2 (SWGA2) is then as follows.

(1)[SWGA2]   We look for any $\varphi_m^{s2,\tau} \in \mathcal{D}$ satisfying

$$\left(\sum_{i=1}^{N} |\langle f_{m-1}^i, \varphi_m^{s2,\tau}\rangle|^2\right)^{1/2} \geq t_m \sup_{g \in \mathcal{D}} \left(\sum_{i=1}^{N} |\langle f_{m-1}^i, g\rangle|^2\right)^{1/2}, \quad f_{m-1}^i := f_{m-1}^{i,s2,\tau}.$$

(2.7.5)

Clearly, any φ_m satisfying (2.7.3) or (2.7.5) also satisfies (2.7.4). Thus, any upper estimate for the SWGA1 yields an upper estimate for both the VWGA and the SWGA2. It was proved in Leviatan and Temlyakov (2006) that Theorem 2.7.2 holds for both variants of the Simultaneous Weak Greedy Algorithm.

We proved in Temlyakov (2004) the following estimate that improves the estimates in Theorems 2.7.1 and 2.7.2. It combines good features of estimates from Theorems 2.7.1 and 2.7.2. We proved in Temlyakov (2004) an estimate with the exponent $t/(2N^{1/2}+t)$ from Theorem 2.7.2 and with the constant N as in Theorem 2.7.1. Let s stand for either v or $s1$ or $s2$.

Theorem 2.7.3. Let $\mathcal{D}$ be an arbitrary dictionary in H. Assume $\tau :=$ $\{t_k\}_{k=1}^{\infty}$, $t_k = t \in (0,1]$, $k = 1, 2, \ldots$. Then, for any vector of elements $f^1, \ldots, f^N$, $f^i \in A_1(\mathcal{D})$, $i = 1, \ldots, N$, we have

$$\sum_{i=1}^{N} \|f_m^{i,s,\tau}\|^2 \leq N\left(1 + \frac{1}{N}mt^2\right)^{\frac{-t}{2N^{1/2}+t}}.$$

Theorem 2.7.4. Let $\mathcal{D}$ be an arbitrary dictionary in H. Assume $\tau :=$ $\{t_k\}_{k=1}^{\infty}$ is a non-increasing sequence. Then, for any vector of elements

$f^1, \ldots, f^N$, $f^i \in A_1(\mathcal{D})$, $i = 1, \ldots, N$, we have

$$\sum_{i=1}^{N} \|f_m^{i,s,\tau}\|^2 \le CN \left(N + \sum_{k=1}^{m} t_k^2 \right)^{\frac{-t_m}{2N^{1/2} + t_m}},$$

with an absolute constant $C = e^{2/e} < 3$.

Let us make some comments on proofs of Theorems 2.7.1–2.7.4. The proof of Theorem 2.7.1 from Lutoborski and Temlyakov (2003) is a modification of the proof of Theorem 2.3.9 from Temlyakov (2000b) to the vector case. This proof does not use Theorem 2.3.9. The proof of Theorem 2.7.2 from Leviatan and Temlyakov (2006) directly uses Theorem 2.3.9. In Leviatan and Temlyakov (2006) we interpret a simultaneous approximation of $f^1, \ldots, f^N$ in H with respect to $\mathcal{D}$ as an approximation of $F = (f^1, \ldots, f^N)$ in $H_N := H \times \cdots \times H$ with respect to a special dictionary $\mathcal{D}_N \subset H_N$ built from $\mathcal{D}$. The proof of Theorems 2.7.3 and 2.7.4 is more like the proof of Theorem 2.7.1. It is a modification of the proof of Theorem 2.3.9.

2.8. Application of greedy algorithms in learning theory

We discuss in this section some mathematical aspects of supervised learning theory. Supervised learning, or learning from examples, refers to a process that builds on the base of available data of inputs x_i and outputs y_i, $i = 1, \ldots, m$, a function that best represents the relation between the inputs $x \in X$ and the corresponding outputs $y \in Y$. The central question is how well this function estimates the outputs for general inputs. This is a big area of research both in non-parametric statistics and in learning theory.

A standard mathematical framework for the setting of the above learning problem is the following (Cucker and Smale 2001, Poggio and Smale 2003, DeVore, Kerkyacharian, Picard and Temlyakov 2004, 2006, Konyagin and Temlyakov 2004, 2007, Temlyakov 2005c, 2005d, 2006d). Let $X \subset \mathbb{R}^d$, $Y \subset \mathbb{R}$ be Borel sets, and let ρ be a Borel probability measure on $Z = X \times Y$. For $f : X \to Y$ define *the error*

$$\mathcal{E}(f) := \int_Z (f(x) - y)^2 \, d\rho.$$

Consider $\rho(y|x)$, the conditional (with respect to x) probability measure on Y, and ρ_X, the marginal probability measure on X (for $S \subset X$, $\rho_X(S) = \rho(S \times Y)$). Here we consider only bounded sets Y and, therefore, there exists a regular conditional probability $\rho(\cdot|x)$. Define $f_\rho(x)$ to be the conditional expectation of y with respect to measure $\rho(\cdot|x)$. The function f_ρ is known in statistics as the *regression function* of ρ. It is clear that if $f_\rho \in L_2(\rho_X)$ then it minimizes the error $\mathcal{E}(f)$ over all $f \in L_2(\rho_X)$: $\mathcal{E}(f_\rho) \le \mathcal{E}(f)$, $f \in L_2(\rho_X)$. Thus, in the sense of error $\mathcal{E}(\cdot)$, the regression function f_ρ is optimal to describe the relation between inputs $x \in X$ and outputs $y \in Y$. Now, our goal is to find an estimator $f_{\mathbf{z}}$, given data $\mathbf{z} = ((x_1, y_1), \ldots, (x_m, y_m))$ that approximates f_ρ well with high probability. We assume that (x_i, y_i), $i = 1, \ldots, m$ are independent and distributed according to ρ. We note that it is easy to see that, for any $f \in L_2(\rho_X)$,

$$\mathcal{E}(f) - \mathcal{E}(f_\rho) = \|f - f_\rho\|_{L_2(\rho_X)}^2.$$

The fundamental problem of learning theory is how to build a good estimator. It is well known in statistics that the following way of building $f_{\mathbf{z}}$ provides a near-optimal estimator in many cases. First, choose the right hypothesis space $\mathcal{H}$. Second, construct $f_{\mathbf{z}, \mathcal{H}} \in \mathcal{H}$ as the empirical optimum (least squares estimator). We explain this in more detail. We define

$$f_{\mathbf{z}, \mathcal{H}} = \arg\min_{f \in \mathcal{H}} \mathcal{E}_{\mathbf{z}}(f),$$

where

$$\mathcal{E}_{\mathbf{z}}(f) := \frac{1}{m} \sum_{i=1}^{m} (f(x_i) - y_i)^2$$

is the *empirical error (risk)* of f. This $f_{\mathbf{z}, \mathcal{H}}$ is called the *empirical optimum* or the *Least Squares Estimator* (LSE). Clearly, a crucial role in this approach is played by a choice of the hypothesis space $\mathcal{H}$. In other words, we need to begin our construction of an estimator with a decision on what should be the form of the estimator. In this section we discuss only the case relevant to the use of nonlinear approximation, in particular, greedy approximation in such a construction. We want to construct a good estimator that will provide high accuracy and that will be practically implementable. We will discuss a realization of this plan in several stages. We begin with results on accuracy. We will give a presentation in a rather general form of nonlinear approximation.

Let $\mathcal{D}(n, q) := \{g_l^n\}_{l=1}^{N_n}$, $n \in \mathbb{N}$, $N_n \leq n^q$, $q \geq 1$, be a system of bounded functions defined on X. We will consider a sequence $\{\mathcal{D}(n, q)\}_{n=1}^{\infty}$ of such systems. In building an estimator, based on $\mathcal{D}(n, q)$, we are going to use n-term approximations with regard to $\mathcal{D}(n, q)$:

$$G_\Lambda := \sum_{l \in \Lambda} c_l g_l^n, \quad |\Lambda| = n. \tag{2.8.1}$$

A standard assumption that we make in supervised learning theory is that $|y| \leq M$ almost surely. This implies that we always assume that $|f_\rho| \leq M$. Denoting $\|f\|_{B(X)} := \sup_{x \in X} |f(x)|$, we rewrite the above assumption in the form $\|f_\rho\|_{B(X)} \leq M$. It is clear that with such an assumption it is natural to restrict our search to estimators $f_{\mathbf{z}}$ satisfying the same inequality $\|f_{\mathbf{z}}\|_{B(X)} \leq M$. Now, in learning theory there are two standard ways to go. In the first approach, (I), we are looking for an estimator of the form (2.8.1) with an extra condition

$$\|G_\Lambda\|_{B(X)} \le M. \tag{2.8.2}$$

In the second approach, (II), we take an approximant G_Λ of the form (2.8.1) and truncate it, *i.e.*, consider $T_M(G_\Lambda)$, where T_M is a truncation operator: $T_M(u) = u$ if $|u| \le M$ and $T_M(u) = M \operatorname{sign} u$ if $|u| \ge M$. Then automatically $\|T_M(G_\Lambda)\|_{B(X)} \le M$.

Let us look in more detail at the hypothesis spaces generated in the above two cases. In case (I) we use the following compacts in $B(X)$ as a source of estimators:

$$F_n(q) := \left\{ f : \exists \Lambda \subset [1, N_n], |\Lambda| = n, f = \sum_{l \in \Lambda} c_l g_l^n, \|f\|_{B(X)} \le M \right\}.$$

An important feature of $F_n(q)$ is that it is a collection of sparse (at most n terms) estimators. An important drawback is that it may not be easy to check if (2.8.2) is satisfied for a particular G_Λ of the form (2.8.1).

In case (II) we use the following sets in $B(X)$ as a source of estimators:

$$F_n^T(q) := \left\{ f : \exists \Lambda \subset [1, N_n], |\Lambda| = n, f = T_M \left(\sum_{l \in \Lambda} c_l g_l^n \right) \right\}.$$

An obvious good feature of $F_n^T(q)$ is that by definition we have $\|f\|_{B(X)} \le M$ for any f from $F_n^T(q)$. An important drawback is that $F_n^T(q)$ has (in general) a rather complex structure. In particular, applying the truncation operator T_M to G_Λ we lose (in general) the sparseness property of G_Λ.

Now, when we have specified our hypothesis spaces, we can look for an existing theory that provides the corresponding error bounds. The general theory is well developed in case (I). We will use a variant of such a general theory developed in Temlyakov (2005c). This theory is based on the following property of compacts $F_n(q)$, formulated in terms of covering numbers:

$$N(F_n(q), \epsilon, B(X)) \le (1 + 2M/\epsilon)^n n^{qn}. \tag{2.8.3}$$

We now formulate the corresponding results from Temlyakov (2005c). For a compact Θ in a Banach space B we let $N(\Theta, \epsilon, B)$ denote the covering number, that is, the minimal number of balls of radius ϵ, with centres in Θ, needed to cover Θ. Let a and b be two positive numbers. Consider a collection $\mathcal{K}(a, b)$ of compact subsets K_n in $B(X)$ that are contained in the M-ball of $B(X)$ and satisfy the following covering numbers condition:

$$N(K_n, \epsilon, B(X)) \le (a(1 + 1/\epsilon))^n n^{bn}, \quad n = 1, 2, \dots. \tag{2.8.4}$$

The following theorem was proved in Temlyakov (2005c). We begin with the definition of our estimator. As above, let $\mathcal{K} := \mathcal{K}(a, b)$ be a collection of compacts K_n in $B(X)$ satisfying (2.8.4).

We take a parameter $A \geq 1$ and consider the following Penalized Least Squares Estimator (PLSE):

$$f_{\mathbf{z}}^A := f_{\mathbf{z}}^A(\mathcal{K}) := f_{\mathbf{z}, K_{n(\mathbf{z})}},$$

with

$$n(\mathbf{z}) := \arg \min_{1 \leq j \leq m} \left(\mathcal{E}_{\mathbf{z}}(f_{\mathbf{z}, K_j}) + \frac{Aj \ln m}{m} \right).$$

For a set L of a Banach space B, let

$$d(\Theta, L)_B := \sup_{f \in \Theta} \inf_{g \in L} \|f - g\|_B.$$

Theorem 2.8.1. For $\mathcal{K} := \{K_n\}_{n=1}^{\infty}$ satisfying (2.8.4) and $M > 0$, there exists $A_0 := A_0(a, b, M)$ such that, for any $A \geq A_0$ and any ρ such that $|y| \leq M$ a.s., we have

$$\|f_{\mathbf{z}}^A - f_\rho\|_{L_2(\rho_X)}^2 \leq \min_{1 \leq j \leq m} \left(3d(f_\rho, K_j)_{L_2(\rho_X)}^2 + \frac{4Aj \ln m}{m} \right),$$

with probability $\geq 1 - m^{-c(M)A}$.

It is clear from (2.8.3) and from the definition of $F_n(q)$ that we can apply Theorem 2.8.1 to the sequence of compacts $\{F_n(q)\}$ and obtain the following error bound with probability $\geq 1 - m^{-c(M)A}$:

$$\|f_{\mathbf{z}}^A - f_\rho\|_{L_2(\rho_X)}^2 \leq \min_{1 \leq j \leq m} \left(3d(f_\rho, F_j(q))_{L_2(\rho_X)}^2 + \frac{4Aj \ln m}{m} \right). \qquad (2.8.5)$$

We note that inequality (2.8.5) is the Lebesgue-type inequality (see Section 2.6). Indeed, on the left-hand side of (2.8.5) we have an error of a particular estimator $f_{\mathbf{z}}^A$ built as the PLSE and on the right-hand side of (2.8.5) we have $d(f_\rho, F_j(q))_{L_2(\rho_X)}$: the best error that we can get using estimators from $F_j(q)$, $j = 1, 2, \ldots$. We recall that by construction $f_{\mathbf{z}}^A \in F_{n(\mathbf{z})}(q)$.

Let us now discuss an application of the theory from Temlyakov (2005c) in case (II). We cannot apply that theory directly to the sequence of sets $\{F_n^T(q)\}$ because we do not know if these sets satisfy the covering number condition (2.8.4). However, we can modify the sets $F_n^T(q)$ to make them satisfy condition (2.8.4). Let $c \geq 0$ and define

$$F_n^T(q, c) := \Big\{ f : \exists G_\Lambda := \sum_{l \in \Lambda} c_l g_l^n, \Lambda \subset [1, N_n], |\Lambda| = n,$$

$$\|G_\Lambda\|_{B(X)} \leq C_2 n^c, f = T_M(G_\Lambda) \Big\}$$

with some fixed $C_2 \geq 1$. Then, using the inequality

$$|T_M(f_1(x)) - T_M(f_2(x))| \leq |f_1(x) - f_2(x)|, \quad \text{for } x \in X,$$

it is easy to get that

$$N(F_n^T(q, c), \epsilon, B(X)) \leq (2C_2(1 + 1/\epsilon))^n n^{(q+c)n}.$$

Therefore, (2.8.4) is satisfied with $a = 2C_2$ and $b = q + c$. We note that, from a practical point of view, an extra restriction $\|G_\Lambda\|_{B(X)} \leq C_2 n^c$ is not a big constraint.

The above estimators (built as the PLSE) are very good from the theoretical point of view. Their error bounds satisfy Lebesgue-type inequalities. However, they are not good from the point of view of implementation. For example, there is no simple algorithm to find $f_{\mathbf{z}, F_n(q)}$ because $F_n(q)$ is a union of $\binom{N_n}{n}$ M-balls of n-dimensional subspaces. Thus, finding an exact LSE $f_{\mathbf{z}, F_n(q)}$ is practically impossible. We now use a remark from Temlyakov (2005c) that allows us to build an approximate LSE with good approximation error. We proceed to the definition of the Penalized Approximate Least Squares Estimator (PALSE) (see Temlyakov (2005c)). Let $\delta := \{\delta_{j,m}\}_{j=1}^m$ be a sequence of non-negative numbers. We define $f_{\mathbf{z}, \delta, K_j}$ as an estimator satisfying the relation

$$\mathcal{E}_{\mathbf{z}}(f_{\mathbf{z}, \delta, K_j}) \leq \mathcal{E}_{\mathbf{z}}(f_{\mathbf{z}, K_j}) + \delta_{j,m}. \tag{2.8.6}$$

In other words, $f_{\mathbf{z}, \delta, K_j}$ is an approximation to the least squares estimator $f_{\mathbf{z}, K_j}$.

Next, we take a parameter $A \geq 1$ and define the Penalized Approximate Least Squares Estimator (PALSE)

$$f_{\mathbf{z}, \delta}^A := f_{\mathbf{z}, \delta}^A(\mathcal{K}) := f_{\mathbf{z}, \delta, K_{n(\mathbf{z})}},$$

with

$$n(\mathbf{z}) := \arg \min_{1 \leq j \leq m} \left(\mathcal{E}_{\mathbf{z}}(f_{\mathbf{z}, \delta, K_j}) + \frac{Aj \ln m}{m} \right).$$

The theory developed in Temlyakov (2005c) gives the following error estimate.

Theorem 2.8.2. Under the assumptions of Theorem 2.8.1 we have

$$\|f_{\mathbf{z}, \delta}^A - f_\rho\|_{L_2(\rho_X)}^2 \leq \min_{1 \leq j \leq m} \left(3d(f_\rho, K_j)_{L_2(\rho_X)}^2 + \frac{4Aj \ln m}{m} + 2\delta_{j,m} \right),$$

with probability $\geq 1 - m^{-c(M)A}$.

We point out here that the approximate least squares estimator $f_{\mathbf{z}, \delta, K_j}$ approximates the least squares estimator $f_{\mathbf{z}, K_j}$ in the sense that $\mathcal{E}_{\mathbf{z}}(f_{\mathbf{z}, \delta, K_j}) - \mathcal{E}_{\mathbf{z}}(f_{\mathbf{z}, K_j})$ is small, and not in the sense that $\|f_{\mathbf{z}, \delta, K_j} - f_{\mathbf{z}, K_j}\|$ is small. Theorem 2.8.2 guarantees a good error bound for any penalized estimator built from $\{f_{\mathbf{z}, \delta, K_j}\}$ satisfying (2.8.6). We will use greedy algorithms in building an approximate estimator. We now present results from Temlyakov (2005d).

We will need more specific compacts $F(n, q)$ and will impose some restrictions on g_l^n. We assume that $\|g_l^n\|_{B(X)} \leq C_1$ for all n and l. We consider the following compacts instead of $F_n(q)$:

$$F(n, q) := \left\{ f : \exists \Lambda \subset [1, N_n], |\Lambda| = n, f = \sum_{l \in \Lambda} c_l g_l^n, \sum_{l \in \Lambda} |c_l| \leq 1 \right\}.$$

Then we have $\|f\|_{B(X)} \leq C_1$ for any $f \in F(n, q)$ and $\|f\|_{B(X)} \leq M$ if $M \geq C_1$. Let $\mathbf{z} = (z_1, \ldots, z_m)$, $z_i = (x_i, y_i)$, be given. Consider the following system of vectors in $\mathbb{R}^m$:

$$v^{j,l} := (g_l^j(x_1), \ldots, g_l^j(x_m)), \quad l \in [1, N_j].$$

We equip the $\mathbb{R}^m$ with the norm $\|v\| := (m^{-1} \sum_{i=1}^{m} v_i^2)^{1/2}$. Then

$$\|v^{j,l}\| \leq \|g_l^j\|_{B(X)} \leq C_1.$$

Consider the system $\mathcal{G} := \{v^{j,l}\}_{l=1}^{N_j}$ in $H = \mathbb{R}^m$ with the norm $\|\cdot\|$ defined above. Finding the estimator

$$f_{\mathbf{z}, F(j,q)} = \sum_{l \in \Lambda} c_l g_l^j, \quad \sum_{l \in \Lambda} |c_l| \leq 1, \quad |\Lambda| = j, \quad \Lambda \subset [1, N_j],$$

is equivalent to finding best j-term approximant of $y \in \mathbb{R}^m$ from the $A_1(\mathcal{G})$ in the space H. We apply the RGA(θ) from Section 2.4 with $\theta = 2$ with respect to $\mathcal{G}$ to y and find, after j steps, an approximant

$$v^j := \sum_{l \in \Lambda'} a_l v^{j,l}, \quad \sum_{l \in \Lambda'} |a_l| \leq 1, \quad |\Lambda'| = j, \quad \Lambda' \subset [1, N_j],$$

such that

$$\|y - v^j\|^2 \leq d(y, A_1(\mathcal{G}))^2 + Cj^{-1}, \quad C = C(M, C_1).$$

We define an estimator

$$\hat{f}_{\mathbf{z}} := \hat{f}_{\mathbf{z}, F(j,q)} := \sum_{l \in \Lambda'} a_l g_l^j.$$

Then $\hat{f}_{\mathbf{z}} \in F(j, q)$ and

$$\mathcal{E}_{\mathbf{z}}(\hat{f}_{\mathbf{z}, F(j,q)}) \leq \mathcal{E}_{\mathbf{z}}(f_{\mathbf{z}, F(j,q)}) + Cj^{-1}.$$

We let $\delta := \{Cj^{-1}\}_{j=1}^{m}$, and define for $A \geq 1$

$$f_{\mathbf{z}, \delta}^A := \hat{f}_{\mathbf{z}, F(n(\mathbf{z}),q)},$$

with

$$n(\mathbf{z}) := \arg \min_{1 \leq j \leq m} \left(\mathcal{E}_{\mathbf{z}}(\hat{f}_{\mathbf{z}, F(j,q)}) + \frac{Aj \ln m}{m} \right).$$

By Theorem 2.8.2 we have for $A \geq A_0(M)$

$$\|f_{\mathbf{z},\delta}^A - f_\rho\|_{L_2(\rho_X)}^2 \leq \min_{1 \leq j \leq m} \left(3d(f_\rho, F(j,q))^2 + \frac{4Aj \ln m}{m} + 2Cj^{-1} \right) \quad (2.8.7)$$

with probability $\geq 1 - m^{-c(M)A}$.

In particular, (2.8.7) means that the estimator $f_{\mathbf{z},\delta}^A$ is an estimator that provides the error

$$\|f_{\mathbf{z},\delta}^A - f_\rho\|_{L_2(\rho_X)}^2 \ll \left(\frac{\ln m}{m} \right)^{\frac{2r}{1+2r}}$$

for f_ρ such that $d(f_\rho, F(j,q))_{L_2(\rho_X)} \ll j^{-r}$, $r \leq 1/2$. We note that the estimator $f_{\mathbf{z},\delta}^A$ is based on the greedy algorithm and it can easily be implemented.

We now describe an application of greedy algorithm in learning theory from Barron, Cohen, Dahmen and DeVore (2005). In this application one can use the Orthogonal Greedy Algorithm or the following variant of the Relaxed Greedy Algorithm.

Let $\alpha_1 := 0$ and $\alpha_m := 1 - 2/m$, $m \geq 2$. We set $f_0 := f$, $G_0 := 0$ and inductively define two sequences $\{\beta_m\}_{m=1}^\infty$, $\{\varphi_m\}_{m=1}^\infty$ as follows:

$$(\beta_m, \varphi_m) := \arg \min_{\beta \in \mathbb{R}, g \in \mathcal{D}} \|f - (\alpha_m G_{m-1} + \beta g)\|.$$

Then we set

$$f_m := f_{m-1} - \beta_m \varphi_m, \quad G_m := G_{m-1} + \beta_m \varphi_m.$$

For systems $\mathcal{D}(n,q)$ the following estimator is considered in Barron *et al.* (2005). As above, let $\mathbf{z} = (z_1, \ldots, z_m)$, $z_i = (x_i, y_i)$, be given. Consider the following system of vectors in $\mathbb{R}^m$:

$$v^{j,l} := (g_l^j(x_1), \ldots, g_l^j(x_m)), \quad l \in [1, N_j].$$

We equip the $\mathbb{R}^m$ with the norm $\|v\| := (m^{-1} \sum_{i=1}^m v_i^2)^{1/2}$ and normalize the above system of vectors. Denote the new system of vectors by $\mathcal{G}_j$. Now we apply either the OGA or the version of the RGA defined above to the vector $y \in \mathbb{R}$ with respect to the system $\mathcal{G}_j$. As in the case discussed above of the system $\mathcal{G}$, we obtain an estimator $\hat{f}_j$. Next, we look for the penalized estimator built from the estimators $\{\hat{f}_j\}$ in the following way. Let

$$n(\mathbf{z}) := \arg \min_{1 \leq j \leq m} \left(\mathcal{E}_{\mathbf{z}}(T_M(\hat{f}_j)) + \frac{Aj \log m}{m} \right).$$

Define

$$\hat{f} := T_M(\hat{f}_{n(\mathbf{z})}).$$

Assuming that the systems $\mathcal{D}(n, q)$ are normalized in $L_2(\rho_X)$, Barron *et al.* (2005) proved the following error estimate.

Theorem 2.8.3. There exists $A_0(M)$ such that for $A \geq A_0$ we have the following bound for the expectation of the error:

$$E(\|f_\rho - \hat{f}\|^2_{L_2(\rho_X)}) \leq \min_{1 \leq j \leq m} (C(A, M, q) j \log m/m \tag{2.8.8}$$

$$+ \inf_{h \in \mathrm{span}\, \mathcal{D}(j,q)} (2\|f_\rho - h\|^2_{L_2(\rho_X)} + 8\|h\|^2_{\mathcal{A}_1(\mathcal{D}(j,q))}/j)).$$

Let us make a comparison of (2.8.8) with (2.8.7). First of all, (2.8.8) gives an error bound for the expectation and (2.8.7) gives an error bound with high probability. In this sense (2.8.7) is better than (2.8.8). However, the condition $\|g_l^n\|_{B(X)} \leq C_1$ imposed on the systems $\mathcal{D}(n, q)$ in order to obtain (2.8.7) is more restrictive than the corresponding assumption for (2.8.8).

2.9. A remark on compressed sensing

Recently, compressed sensing (compressive sampling) has attracted a lot of attention from both mathematicians and computer scientists. Compressed sensing refers to a problem of *economical* recovery of an unknown vector $u \in \mathbb{R}^m$ from the information provided by linear measurements $\langle u, \varphi_j \rangle$, $\varphi_j \in \mathbb{R}^m$, $j = 1, \ldots, n$. The goal is to design an algorithm that finds (approximates) u from the information $y = (\langle u, \varphi_1 \rangle, \ldots, \langle u, \varphi_n \rangle) \in \mathbb{R}^n$. The crucial step here is to build a *sensing* set of vectors $\varphi_j \in \mathbb{R}^m$, $j = 1, \ldots, n$ that is *good* for all vectors $u \in \mathbb{R}^m$. Clearly, the terms *economical* and *good* should be clarified in a mathematical setting of the problem. A natural variant of such a setting, which we discuss here, uses the concept of *sparsity*. We call a vector $u \in \mathbb{R}^m$ k-sparse if it has at most k non-zero coordinates. Now, for a given pair (m, n) we want to understand what is the biggest sparsity $k(m, n)$ such that there exists a set of vectors $\varphi_j \in \mathbb{R}^m$, $j = 1, \ldots, n$ and economical algorithm A mapping y into $\mathbb{R}^m$ in such a way that, for any u of sparsity $k(m, n)$, one would have an exact recovery $A(u) = u$. In other words, we want to describe matrices Φ with rows $\varphi_j \in \mathbb{R}^m$, $j = 1, \ldots, n$, such that there exists an economical algorithm of solving the following sparse recovery problem.

The sparse recovery problem can be stated as the problem of finding the sparsest vector $u^0 := u_\Phi^0(y) \in \mathbb{R}^m$:

$$\min \|v\|_0 \quad \text{subject to } \Phi v = y, \tag{P_0}$$

where $\|v\|_0 := |\mathrm{supp}(v)|$. D. Donoho and co-authors (see, for instance, Chen, Donoho and Saunders (2001), Donoho *et al.* (2006) and the history therein) have suggested an economical algorithm (Basis Pursuit) and have begun a systematic study of the following question. For which measurement matrices Φ should the highly non-convex combinatorial optimization problem (P_0) be equivalent to its convex relaxation problem

$$\min \|v\|_1 \quad \text{subject to} \quad \Phi v = y, \qquad (P_1)$$

where $\|v\|_1$ denotes the ℓ_1-norm of the vector $v \in \mathbb{R}^m$? Denote the solution to (P_1) by $A_\Phi(y)$. It is known that the problem (P_1) can be solved by linear programming techniques. The ℓ_1-minimization algorithm A_Φ from (P_1) is an economical algorithm that we consider in this section. It is known (see, for instance, Donoho et $al.$ (2006)) that for M-coherent matrices Φ we have $u_\Phi^0(\Phi u) = A_\Phi(\Phi u) = u$, provided u is k-sparse with $k < (1 + 1/M)/2$. This allows us to build rather simple deterministic matrices Φ with $k(m, n) \asymp n^{1/2}$ and recover A_Φ from (P_1) with the ℓ_1-minimization algorithm.

Recent progress (see surveys by Candès (2006) and DeVore (2006)) in compressed sensing has resulted in proving the existence of matrices Φ with $k(m, n) \asymp n/\log(m/n)$, which is substantially larger than $n^{1/2}$. We proceed to a detailed discussion of these recent results.

We begin with results from Donoho (2006). Donoho formulated the following three properties of matrices Φ with ℓ_2-normalized columns, and proved the existence of matrices satisfying these conditions. Let T be a subset of indices from $[1, m]$. Let Φ_T denote a matrix consisting of columns of Φ with indices from T.

CS1 The minimal singular value of Φ_T is $\geq \eta_1 > 0$ uniformly in T, satisfying $|T| \leq \rho n/\log m$.

CS2 Let W_T denote the range of Φ_T. Assume that for any T satisfying $|T| \leq \rho n/\log m$, we have

$$\|w\|_1 \geq \eta_2 n^{1/2}\|w\|_2, \quad \forall w \in W_T, \quad \eta_2 > 0.$$

CS3 Denote $T^c := \{j\}_{j=1}^m \setminus T$. For any T, $|T| \leq \rho n/\log m$ and for any $w \in W_T$, we have for any v satisfying $\Phi_{T^c}v = w$

$$\|v\|_{\ell_1(T^c)} \geq \eta_3 (\log(m/n))^{-1/2}\|w\|_1, \quad \eta_3 > 0.$$

It is proved in Donoho (2006) that if Φ satisfies CS1–CS3, then there exists $\rho_0 > 0$ such that $u_\Phi^0(\Phi u) = A_\Phi(\Phi u) = u$ provided $|\operatorname{supp} u| \leq \rho_0 n/\log m$. Analysis in Donoho (2006) relates the compressed sensing problem to the problem of estimating the Kolmogorov widths and their dual, the Gel'fand widths.

We give the corresponding definitions. For a compact $F \subset \mathbb{R}^m$, the Kolmogorov width is given by

$$d_n(F, \ell_p) := \inf_{L_n:\dim L_n \leq n} \sup_{f \in F} \inf_{a \in L_n} \|f - a\|_p,$$

where L_n is a linear subspace of $\mathbb{R}^m$ and $\|\cdot\|_p$ denotes the ℓ_p-norm. The

Gel'fand width is defined by

$$d^n(F, \ell_p) := \inf_{V_n} \sup_{f \in F \cap V_n} \|f\|_p,$$

where the infimum is taken over linear subspaces V_n with dimension $\geq m-n$. It is well known that the Kolmogorov and the Gel'fand widths are related by the duality formula. For instance, when $F = B_p^m$ is a unit ℓ_p-ball in $\mathbb{R}^m$ and $1 \leq q, p \leq \infty$, we have

$$d_n(B_p^m, \ell_q) = d^n(B_{q'}^m, \ell_{p'}), \quad p' := p/(p-1). \tag{2.9.1}$$

In the particular case $p = 2$, $q = \infty$ of our interest, (2.9.1) gives

$$d_n(B_2^m, \ell_\infty) = d^n(B_1^m, \ell_2). \tag{2.9.2}$$

It has been established in approximation theory (see Kashin (1977) and Garnaev and Gluskin (1984)) that

$$d_n(B_2^m, \ell_\infty) \leq C((1 + \log(m/n))/n)^{1/2}. \tag{2.9.3}$$

In other words, it was proved (see (2.9.3) and (2.9.2)) that for any pair (m, n) there exists a subspace V_n, $\dim V_n \geq m - n$ such that, for any $x \in V_n$, we have

$$\|x\|_2 \leq C((1 + \log(m/n))/n)^{1/2} \|x\|_1. \tag{2.9.4}$$

It was understood in Donoho (2006) that properties of the null space $\mathcal{N}(\Phi) := \{x : \Phi x = 0\}$ of a measurement matrix Φ play an important role in the compressed sensing problem. Donoho (2006) introduced the following two characteristics of Φ formulated in terms of $\mathcal{N}(\Phi)$:

$$w(\Phi, F) := \sup_{x \in F \cap \mathcal{N}(\Phi)} \|x\|_2$$

and

$$\nu(\Phi, T) := \sup_{x \in \mathcal{N}(\Phi)} \|x_T\|_1/\|x\|_1,$$

where x_T is a restriction of x onto T: $(x_T)_j = x_j$ for $j \in T$ and $(x_T)_j = 0$ otherwise. He proved that if Φ obeys the following two conditions,

$$\nu(\Phi, T) \leq \eta_1, \quad |T| \leq \rho_1 n/\log m, \tag{A1}$$

$$w(\Phi, B_1^m) \leq \eta_2((\log m)/n)^{1/2}, \tag{A2}$$

then for any $u \in B_1^m$ we have

$$\|u - A_\Phi(\Phi u)\|_2 \leq C((\log m)/n)^{1/2}.$$

We now proceed to the contribution of E. Candès, J. Romberg and T. Tao published in a series of papers (see Candès and Tao (2005)). They intro-

duced the following Restricted Isometry Property (RIP) of a sensing ma-
trix Φ: $\delta_S < 1$ is the S-restricted isometry constant of Φ if it is the smallest
quantity such that

$$(1 - \delta_S)\|c\|_2^2 \leq \|\Phi_T c\|_2^2 \leq (1 + \delta_S)\|c\|_2^2$$

for all subsets T with $|T| \leq S$ and all coefficient sequences $\{c_j\}_{j \in T}$. Candès
and Tao (2005) proved that if $\delta_{2S} + \delta_{3S} < 1$, then for S-sparse u we have
$A_\Phi(\Phi u) = u$ (recovery by ℓ_1-minimization is exact). They also proved
existence of sensing matrices Φ obeying the condition $\delta_{2S} + \delta_{3S} < 1$ for
large values of sparsity $S \asymp n/\log(m/n)$. For a positive number a denote

$$\sigma_a(v)_1 := \min_{w \in \mathbb{R}^m : |\operatorname{supp}(w)| \leq a} \|v - w\|_1.$$

Candès, Romberg and Tao (2006) proved that if $\delta_{3S} + 3\delta_{4S} < 2$, then

$$\|u - A_\Phi(\Phi u)\|_2 \leq C S^{-1/2} \sigma_S(u)_1. \tag{2.9.5}$$

We note that properties of the RIP-type matrices have already been employed in Kashin (1977) (see Kashin and Temlyakov (2007) for further discussion) for the widths estimation. The inequality (2.9.3) with an extra
factor $(1 + \log m/n)$ was established in Kashin (1977). The proof in Kashin
(1977) is based on properties of a random matrix Φ with elements $\pm 1/\sqrt{n}$.

Further investigation of the compressed sensing problem was conducted
by Cohen, Dahmen and DeVore (2007). They proved that if Φ satisfies the
RIP of order $2k$ with $\delta_{2k} < \delta < 1/3$, then

$$\|u - A_\Phi(\Phi u)\|_1 \leq \frac{2 + 2\delta}{1 - 3\delta} \sigma_k(u)_1. \tag{2.9.6}$$

The above inequality is the Lebesgue-type inequality (see Section 2.6) for the
approximation method $u \to A_\Phi(\Phi u)$. In Cohen *et al.* (2007) the inequality
(2.9.6) was called *instance optimality*. In the proof of (2.9.6) the authors
used the following property (null space property) of matrices Φ satisfying
the RIP of order $3k/2$: for any $x \in \mathcal{N}(\Phi)$ and any T with $|T| \leq k$, we have

$$\|x\|_1 \leq C\|x_{T^c}\|_1. \tag{2.9.7}$$

The null space property (2.9.7) is closely related to the property (A1) from
Donoho (2006). The proof of (2.9.6) from Cohen *et al.* (2007) gives an
inequality similar to (2.9.6) under the assumption that Φ has the null space
property (2.9.7) with $C < 2$.

We now discuss results of Kashin and Temlyakov (2007). We say that a
measurement matrix Φ has a Strong Compressed Sensing Property (SCSP)
if, for any $u \in \mathbb{R}^m$, we have

$$\|u - A_\Phi(\Phi u)\|_2 \leq C k^{-1/2} \sigma_k(u)_1, \tag{2.9.8}$$

for $k \asymp n/\log(m/n)$. We define a Weak Compressed Sensing Property (WCSP) by replacing (2.9.8) by the weaker inequality

$$\|u - A_\Phi(\Phi u)\|_2 \le C k^{-1/2} \|u\|_1. \qquad (2.9.9)$$

We say that Φ satisfies the Width Property (WP) if (2.9.4) holds for the null space $\mathcal{N}(\Phi)$. The main result of the paper Kashin and Temlyakov (2007) states that the above three properties of Φ are equivalent. We proceed to a detailed discussion of results from Kashin and Temlyakov (2007).

We mentioned above that it is known that, for any pair (m, n), $n < m$, there exists a subspace $\Gamma \subset \mathbb{R}^m$ with $\dim \Gamma \ge m - n$ such that

$$\|x\|_2 \le C n^{-1/2} (\ln(em/n))^{1/2} \|x\|_1, \quad \forall x \in \Gamma. \qquad (2.9.10)$$

We will discuss some properties of subspaces Γ satisfying (2.9.10) that are useful in compressed sensing. Let

$$S := S(m, n) := C^{-2} n (\ln(em/n))^{-1}.$$

For $x = (x_1, \ldots, x_m) \in \mathbb{R}^m$, define $\operatorname{supp}(x) := \{j : x_j \ne 0\}$.

Lemma 2.9.1. Let Γ satisfy (2.9.10) and $x \in \Gamma$. Then either $x = 0$ or $|\operatorname{supp}(x)| \ge S(m, n)$.

Proof. Assume $x \ne 0$. Then $\|x\|_1 > 0$. Denote $\Lambda := \operatorname{supp}(x)$. We have

$$\|x\|_1 = \sum_{j \in \Lambda} |x_j| \le |\Lambda|^{1/2} \left(\sum_{j \in \Lambda} |x_j|^2 \right)^{1/2} \le |\Lambda|^{1/2} \|x\|_2. \qquad (2.9.11)$$

Using (2.9.10), we get from (2.9.11)

$$\|x\|_1 \le |\Lambda|^{1/2} S(m, n)^{-1/2} \|x\|_1.$$

Thus

$$|\Lambda| \ge S(m, n). \qquad \square$$

Lemma 2.9.2. Let Γ satisfy (2.9.10) and let $x \ne 0$, $x \in \Gamma$. Then, for any Λ such that $|\Lambda| < S(m, n)/4$,

$$\sum_{j \in \Lambda} |x_j| < \|x\|_1/2.$$

Proof. As in (2.9.11),

$$\sum_{j \in \Lambda} |x_j| \le |\Lambda|^{1/2} S(m, n)^{-1/2} \|x\|_1 < \|x\|_1/2. \qquad \square$$

Lemma 2.9.3. Let Γ satisfy (2.9.10). Suppose $u \in \mathbb{R}^m$ is sparse with $|\operatorname{supp}(u)| < S(m, n)/4$. Then, for any $v = u + x$, $x \in \Gamma$, $x \ne 0$,

$$\|v\|_1 > \|u\|_1.$$

Proof. Let $\Lambda := \text{supp}(u)$. Then

$$\|v\|_1 = \sum_{j \in [1,m]} |v_j| = \sum_{j \in \Lambda} |u_j + x_j| + \sum_{j \notin \Lambda} |x_j|$$

$$\geq \sum_{j \in \Lambda} |u_j| - \sum_{j \in \Lambda} |x_j| + \sum_{j \notin \Lambda} |x_j| = \|u\|_1 + \|x\|_1 - 2 \sum_{j \in \Lambda} |x_j|.$$

By Lemma 2.9.2,

$$\|x\|_1 - 2 \sum_{j \in \Lambda} |x_j| > 0. \qquad \square$$

Lemma 2.9.3 guarantees that the following algorithm, known as the Basis Pursuit (see A_Φ defined above), will find a sparse u exactly, provided $|\text{supp}(u)| < S(m,n)/4$:

$$u_\Gamma := u + \arg\min_{x \in \Gamma} \|u + x\|_1.$$

Theorem 2.9.4. Let Γ satisfy (2.9.10). Then, for any $u \in \mathbb{R}^m$ and u' such that $\|u'\|_1 \leq \|u\|_1$, $u - u' \in \Gamma$,

$$\|u - u'\|_1 \leq 4\sigma_{S/16}(u)_1, \qquad (2.9.12)$$

$$\|u - u'\|_2 \leq (S/16)^{-1/2}\sigma_{S/16}(u)_1. \qquad (2.9.13)$$

Proof. It is given that $u - u' \in \Gamma$. Thus, (2.9.13) follows from (2.9.12) and (2.9.10). We now prove (2.9.12). Let Λ, $|\Lambda| = [S/16]$, be the set of indices of coordinates of u that are largest in absolute value. Let u_Λ denote the restriction of u onto this set, i.e., $(u_\Lambda)_j = u_j$ for $j \in \Lambda$ and $(u_\Lambda)_j = 0$ for $j \notin \Lambda$, and let $u^\Lambda := u - u_\Lambda$. Then

$$\sigma_{S/16}(u)_1 = \sigma_{|\Lambda|}(u)_1 = \|u - u_\Lambda\|_1 = \|u^\Lambda\|_1. \qquad (2.9.14)$$

We have

$$\|u - u'\|_1 \leq \|(u - u')_\Lambda\|_1 + \|(u - u')^\Lambda\|_1.$$

Next,

$$\|(u - u')^\Lambda\|_1 \leq \|u^\Lambda\|_1 + \|(u')^\Lambda\|_1.$$

Using $\|u'\|_1 \leq \|u\|_1$, we obtain

$$\|(u')^\Lambda\|_1 - \|u^\Lambda\|_1 = \|u'\|_1 - \|u\|_1 - \|u'_\Lambda\|_1 + \|u_\Lambda\|_1 \leq \|(u - u')_\Lambda\|_1.$$

Therefore,

$$\|(u')^\Lambda\|_1 \leq \|u^\Lambda\|_1 + \|(u - u')_\Lambda\|_1$$

and

$$\|u - u'\|_1 \le 2\|(u - u')_\Lambda\|_1 + 2\|u^\Lambda\|_1. \qquad (2.9.15)$$

Using the fact $u - u' \in \Gamma$, we estimate

$$\|(u - u')_\Lambda\|_1 \le |\Lambda|^{1/2}\|(u - u')_\Lambda\|_2 \le |\Lambda|^{1/2}\|u - u'\|_2$$
$$\le |\Lambda|^{1/2} S^{-1/2}\|u - u'\|_1. \qquad (2.9.16)$$

Our assumption on $|\Lambda|$ guarantees that $|\Lambda|^{1/2} S^{-1/2} \le 1/4$. Using this and substituting (2.9.16) into (2.9.15), we obtain

$$\|u - u'\|_1 \le \|u - u'\|_1/2 + 2\|u^\Lambda\|_1,$$

which gives (2.9.12):

$$\|u - u'\|_1 \le 4\|u^\Lambda\|_1. \qquad \square$$

Corollary 2.9.5. Let Γ satisfy (2.9.10). Then, for any $u \in \mathbb{R}^m$,

$$\|u - u_\Gamma\|_1 \le 4\sigma_{S/16}(u)_1, \qquad (2.9.17)$$

$$\|u - u_\Gamma\|_2 \le (S/16)^{-1/2}\sigma_{S/16}(u)_1. \qquad (2.9.18)$$

Proposition 2.9.6. Let Γ be such that (2.9.9) holds with u_Γ instead of $A_\Phi(\Phi u)$ and $k = n/\ln(em/n)$. Then Γ satisfies (2.9.10).

Proof. Let $u \in \Gamma$. Then $u_\Gamma = 0$, and we get from (2.9.9)

$$\|u\|_2 \le C(n/\ln(em/n))^{-1/2}\|u\|_1. \qquad \square$$

Theorem 2.9.7. The following three properties of Φ are equivalent: the Strong Compressed Sensing Property, the Weak Compressed Sensing Property, and the Width Property.

Proof. It is obvious that SCSP $\Rightarrow$ WCSP. Corollary 2.9.5 with $\Gamma = \mathcal{N}(\Phi)$ implies that WP $\Rightarrow$ SCSP. Proposition 2.9.6 with $\Gamma = \mathcal{N}(\Phi)$ implies that WCSP $\Rightarrow$ WP. Thus the three properties are equivalent. $\qquad \square$

The result (2.9.5) of Candès *et al.* (2006) states that the RIP with $S \asymp n/\log(m/n)$ implies the SCSP. Therefore, by Theorem 2.9.7 it implies the WP.

We note that there are very interesting results on greedy approximation in compressed sensing. We do not discuss these results here, and refer the reader to two of them: Tropp and Gilbert (2005) and Needell and Vershynin (2007).

CHAPTER THREE
Greedy approximation with respect to dictionaries: Banach spaces

3.1. Introduction

In this chapter we make a step from Hilbert spaces to more general Banach spaces. Let X be a Banach space with norm $\| \cdot \|$. We say that a set of elements (functions) $\mathcal{D}$ from X is a dictionary, respectively, symmetric dictionary, if each $g \in \mathcal{D}$ has norm bounded by one ($\|g\| \leq 1$),

$$g \in \mathcal{D} \quad \text{implies} \quad -g \in \mathcal{D},$$

and the closure of span $\mathcal{D}$ is X. We denote the closure (in X) of the convex hull of $\mathcal{D}$ by $A_1(\mathcal{D})$. We introduce a new norm, associated with a dictionary $\mathcal{D}$, in the dual space X^* by the formula

$$\|F\|_{\mathcal{D}} := \sup_{g \in \mathcal{D}} F(g), \quad F \in X^*.$$

In this chapter we will study greedy algorithms with regard to $\mathcal{D}$. For a non-zero element $f \in X$ we let F_f denote a norming (peak) functional for f:

$$\|F_f\| = 1, \qquad F_f(f) = \|f\|.$$

The existence of such a functional is guaranteed by Hahn–Banach theorem.

We begin with a generalization of the Pure Greedy Algorithm. The greedy step of the PGA can be interpreted in two ways. First, we look at the mth step for an element $\varphi_m \in \mathcal{D}$ and a number λ_m satisfying

$$\|f_{m-1} - \lambda_m \varphi_m\|_H = \inf_{g \in \mathcal{D}, \lambda} \|f_{m-1} - \lambda g\|_H. \tag{3.1.1}$$

Second, we look for an element $\varphi_m \in \mathcal{D}$ such that

$$\langle f_{m-1}, \varphi_m \rangle = \sup_{g \in \mathcal{D}} \langle f_{m-1}, g \rangle. \tag{3.1.2}$$

In a Hilbert space both versions (3.1.1) and (3.1.2) resulted in the same PGA. In a general Banach space the corresponding versions of (3.1.1) and (3.1.2) lead to different greedy algorithms. The Banach space version of (3.1.1) is straightforward: instead of the Hilbert norm $\| \cdot \|_H$ in (3.1.1) we use the Banach norm $\| \cdot \|_X$. This results in the following greedy algorithm (see Temlyakov (2003a)).

X-Greedy Algorithm (XGA). We define $f_0 := f$, $G_0 := 0$. Then, for each $m \geq 1$ we have the following inductive definition.

(1) $\varphi_m \in \mathcal{D}$, $\lambda_m \in \mathbb{R}$ are such that (we assume existence)

$$\|f_{m-1} - \lambda_m \varphi_m\|_X = \inf_{g \in \mathcal{D}, \lambda} \|f_{m-1} - \lambda g\|_X. \tag{3.1.3}$$

(2) Define

$$f_m := f_{m-1} - \lambda_m \varphi_m, \qquad G_m := G_{m-1} + \lambda_m \varphi_m.$$

The second version of the PGA in a Banach space is based on the concept of a norming (peak) functional. We note that in a Hilbert space a norming functional F_f acts as follows:

$$F_f(g) = \langle f/\|f\|, g \rangle.$$

Thus, (3.1.2) can be rewritten in terms of the norming functional $F_{f_{m-1}}$ as

$$F_{f_{m-1}}(\varphi_m) = \sup_{g \in \mathcal{D}} F_{f_{m-1}}(g). \tag{3.1.4}$$

This observation leads to the class of dual greedy algorithms. We define the Weak Dual Greedy Algorithm with weakness τ (WDGA(τ)) (see Dilworth, Kutzarova and Temlyakov (2002) and Temlyakov (2003a)) that is a generalization of the Weak Greedy Algorithm.

Weak Dual Greedy Algorithm (WDGA(τ)). Let $\tau := \{t_m\}_{m=1}^{\infty}$, $t_m \in [0,1]$, be a weakness sequence. We define $f_0 := f$. Then, for each $m \geq 1$ we have the following inductive definition.

(1) $\varphi_m \in \mathcal{D}$ is any element satisfying

$$F_{f_{m-1}}(\varphi_m) \geq t_m \|F_{f_{m-1}}\|_{\mathcal{D}}. \tag{3.1.5}$$

(2) Define a_m as

$$\|f_{m-1} - a_m \varphi_m\| = \min_{a \in \mathbb{R}} \|f_{m-1} - a \varphi_m\|.$$

(3) Let

$$f_m := f_{m-1} - a_m \varphi_m.$$

Let us make a remark that justifies the idea of the dual greedy algorithms in terms of real analysis. We consider here approximation in uniformly smooth Banach spaces. For a Banach space X we define the modulus of smoothness

$$\rho(u) := \sup_{\|x\|=\|y\|=1} \left(\frac{1}{2}(\|x + uy\| + \|x - uy\|) - 1 \right).$$

The uniformly smooth Banach space is the one with the property

$$\lim_{u \to 0} \rho(u)/u = 0.$$

It is easy to see that for any Banach space X its modulus of smoothness $\rho(u)$ is an even convex function satisfying the inequalities

$$\max(0, u - 1) \leq \rho(u) \leq u, \quad u \in (0, \infty).$$

We note that from the definition of modulus of smoothness we get the following inequality.

Lemma 3.1.1. Let $x \neq 0$. Then

$$0 \leq \|x + uy\| - \|x\| - uF_x(y) \leq 2\|x\|\rho(u\|y\|/\|x\|). \qquad (3.1.6)$$

Proof. We have

$$\|x + uy\| \geq F_x(x + uy) = \|x\| + uF_x(y).$$

This proves the first inequality. Next, from the definition of modulus of smoothness it follows that

$$\|x + uy\| + \|x - uy\| \leq 2\|x\|(1 + \rho(u\|y\|/\|x\|)). \qquad (3.1.7)$$

Also,

$$\|x - uy\| \geq F_x(x - uy) = \|x\| - uF_x(y). \qquad (3.1.8)$$

Combining (3.1.7) and (3.1.8), we obtain

$$\|x + uy\| \leq \|x\| + uF_x(y) + 2\|x\|\rho(u\|y\|/\|x\|).$$

This proves the second inequality. $\qquad \square$

Proposition 3.1.2. Let X be a uniformly smooth Banach space. Then, for any $x \neq 0$ and y we have

$$F_x(y) = \left(\frac{d}{du}\|x + uy\|\right)(0) = \lim_{u \to 0}(\|x + uy\| - \|x\|)/u. \qquad (3.1.9)$$

Proof. The equality (3.1.9) follows from (3.1.6) and the property that, for a uniformly smooth Banach space, $\lim_{u \to 0} \rho(u)/u = 0$. $\qquad \square$

Proposition 3.1.2 shows that in the WDGA we are looking for an element $\varphi_m \in \mathcal{D}$ that provides a big derivative of the quantity $\|f_{m-1} + ug\|$. Thus, we have two classes of greedy algorithms in Banach spaces. The first one is based on a greedy step of the form (3.1.3). We call this class the class of X-greedy algorithms. The second one is based on a greedy step of the form (3.1.5). We call this class the class of dual greedy algorithms. A very important feature of the dual greedy algorithms is that they can be modified into a weak form. The term 'weak' in the definition of the WDGA means that, at the greedy step (3.1.5), we do not aim for the optimal element of the dictionary which realizes the corresponding supremum but are satisfied with a weaker property than being optimal. The obvious reason for this is that we do not know, in general, that the optimal one exists. Another, practical reason is that the weaker the assumption, the easier it is satisfied and, therefore, it is easier to realize in practice.

The greedy algorithms defined above (XGA, WDGA) are the generalizations of the PGA and the WGA, studied in Chapter 2, to the case of Banach spaces. The results of Chapter 2 show that the PGA is not the most efficient greedy algorithm for approximation of elements of $A_1(\mathcal{D})$. It was mentioned in Chapter 2 (see Livshitz and Temlyakov (2003) for the proof) that there exist a dictionary $\mathcal{D}$, a positive constant C, and an element $f \in A_1(\mathcal{D})$ such that, for the PGA,

$$\|f_m\| \geq Cm^{-0.27}. \tag{3.1.10}$$

We note that even before the lower estimate (3.1.10) was proved, researchers began looking for other greedy algorithms that provide a good rate of approximation of functions from $A_1(\mathcal{D})$. Two different ideas have been used at this step. The first idea was that of relaxation: see Jones (1992), Barron (1993), DeVore and Temlyakov (1996) and Temlyakov (2000b). The corresponding algorithms (for example, the WRGA, studied in Chapter 2) were designed for approximation of functions from $A_1(\mathcal{D})$. These algorithms do not provide an expansion into a series but they have other good features. It was established (see Theorem 2.3.8) for the WRGA with $\tau = \{1\}$ in a Hilbert space that, for $f \in A_1(\mathcal{D})$,

$$\|f_m\| \leq Cm^{-1/2}.$$

Also, for the WRGA we always have $G_m \in A_1(\mathcal{D})$. The latter property, clearly, limits the applicability of the WRGA to the $A_1(\mathcal{D})$.

The second idea was the idea of building the best approximant from the $\mathrm{span}(\varphi_1, \ldots, \varphi_m)$ instead of the use of only one element φ_m for an update of the approximant. This idea was realized in the Weak Orthogonal Greedy Algorithm (see Chapter 2) in the case of a Hilbert space and in the Weak Chebyshev Greedy Algorithm (WCGA) (see Temlyakov (2001b)) in the case of a Banach space.

The realization of both ideas resulted in the construction of algorithms (the WRGA and WCGA) that are good for approximation of functions from $A_1(\mathcal{D})$. We present results on the WCGA in Section 3.2 and results on the WRGA in Section 3.3. The WCGA has the following advantage over the WRGA. It will be proved in Section 3.2 that the WCGA (under some assumptions on the weakness sequence τ) converges for each $f \in X$ in any uniformly smooth Banach space. The WRGA is simpler than the WCGA in the sense of computational complexity. However, the WRGA has limited applicability. It converges only for elements of the closure of the convex hull of a dictionary. In Sections 3.4 and 3.5 we study algorithms that combine good features of both algorithms the WRGA and the WCGA. In the construction of such algorithms we use different forms of relaxation.

The Weak Greedy Algorithm with Free Relaxation (WGAFR) (Temlyakov 2006c), studied in Section 3.4, is the most powerful of the versions considered here. We prove convergence of the WGAFR in Theorem 3.4.3. This theorem is the same as the corresponding convergence result for the

WCGA (see Theorem 3.2.4). The results on the rate of convergence for the WGAFR and the WCGA are also the same (see Theorem 3.4.4 and Theorem 3.2.12). Thus, the WGAFR performs in the same way as the WCGA from the point of view of convergence and rate of convergence, and outperforms the WCGA in terms of computational complexity.

In the WGAFR we are optimizing over two parameters w and λ at each step of the algorithm. In other words we are looking for the best approximation from a two-dimensional linear subspace at each step. In the other version of the weak relaxed greedy algorithms (see the GAWR), considered in Section 3.5, we approximate from a one-dimensional linear subspace at each step of the algorithm. This makes computational complexity of these algorithms very close to that of the PGA. The analysis of the GAWR version turns out to be more complicated than the analysis of the WGAFR. Also, the results obtained for the GAWR are not as general as in the case of the WGAFR. For instance, we present results on the GAWR only in the case $\tau = \{t\}$, when the weakness parameter t is the same for all steps.

The XGA and WDGA have a good feature that distinguishes them from all relaxed greedy algorithms, and also from the WCGA. For an element $f \in X$ they provide an expansion into a series,

$$f \sim \sum_{j=1}^{\infty} c_j(f)g_j(f), \quad g_j(f) \in \mathcal{D}, \quad c_j(f) > 0, \quad j = 1, 2, \ldots, \quad (3.1.11)$$

such that

$$G_m = \sum_{j=1}^{m} c_j(f)g_j(f), \quad f_m = f - G_m.$$

In Section 3.7 we discuss other greedy algorithms that provide the expansion (3.1.11).

All the algorithms studied in Sections 3.2–3.7 belong to the class of dual greedy algorithms. Results obtained in Sections 3.2–3.7 confirm that dual greedy algorithms provide powerful methods of nonlinear approximation. In Section 3.8 we present some results on the X-greedy algorithms. These results are similar to those for the dual greedy algorithms.

The algorithms studied in Sections 3.2–3.8 are very general approximation methods that work well in an arbitrary uniformly smooth Banach space X for any dictionary $\mathcal{D}$. This motivates an attempt, made in Section 3.9, to modify these theoretical approximation methods in a direction of practical applicability. In Section 3.9 we illustrate this idea by modifying the WCGA. We note that Section 3.6 is also devoted to modification of greedy algorithms in order to make them more practically feasible. The main idea of Section 3.6 is to replace the most difficult (expensive) step of an algorithm, namely the greedy step, by a thresholding step.

In Section 3.10 we give an example of how the greedy algorithms can be used in constructing deterministic cubature formulas with error estimates similar to those for the Monte Carlo Method.

As a typical example of a uniformly smooth Banach space we will use a space L_p, $1 < p < \infty$. It is well known (see, for instance, Donahue *et al.* (1997, Lemma B.1)) that in the case $X = L_p$, $1 \leq p < \infty$ we have

$$\rho(u) \leq u^p/p \quad \text{if} \quad 1 \leq p \leq 2 \quad \text{and} \quad \rho(u) \leq (p-1)u^2/2 \quad \text{if} \quad 2 \leq p < \infty.$$
$$(3.1.12)$$

It is also known (see Lindenstrauss and Tzafriri (1977, p. 63)) that, for any X with $\dim X = \infty$, we have

$$\rho(u) \geq (1 + u^2)^{1/2} - 1,$$

and for every X, $\dim X \geq 2$,

$$\rho(u) \geq Cu^2, \quad C > 0.$$

This limits the power-type modulus of smoothness of non-trivial Banach spaces to the case $1 \leq q \leq 2$.

3.2. The Weak Chebyshev Greedy Algorithm

Let $\tau := \{t_k\}_{k=1}^{\infty}$ be a given weakness sequence of non-negative numbers $t_k \leq 1$, $k = 1, \ldots$. We define first the Weak Chebyshev Greedy Algorithm (WCGA) (see Temlyakov (2001*b*)) that is a generalization for Banach spaces of the Weak Orthogonal Greedy Algorithm.

Weak Chebyshev Greedy Algorithm (WCGA). We define $f_0^c := f_0^{c,\tau} := f$. Then, for each $m \geq 1$ we have the following inductive definition.

(1) $\varphi_m^c := \varphi_m^{c,\tau} \in \mathcal{D}$ is any element satisfying

$$F_{f_{m-1}^c}(\varphi_m^c) \geq t_m \|F_{f_{m-1}^c}\|_{\mathcal{D}}.$$

(2) Define

$$\Phi_m := \Phi_m^{\tau} := \operatorname{span}\{\varphi_j^c\}_{j=1}^{m},$$

and define $G_m^c := G_m^{c,\tau}$ to be the best approximant to f from Φ_m.

(3) Let

$$f_m^c := f_m^{c,\tau} := f - G_m^c.$$

Remark 3.2.1. It follows from the definition of the WCGA that the sequence $\{\|f_m^c\|\}$ is a non-increasing sequence.

We proceed to a theorem on convergence of the WCGA. In the formulation of this theorem we need a special sequence which is defined for a given modulus of smoothness $\rho(u)$ and a given $\tau = \{t_k\}_{k=1}^{\infty}$.

Definition 3.2.2. Let $\rho(u)$ be an even convex function on $(-\infty, \infty)$ with the property: $\rho(2) \geq 1$ and

$$\lim_{u \to 0} \rho(u)/u = 0.$$

For any $\tau = \{t_k\}_{k=1}^{\infty}$, $0 < t_k \leq 1$, and $0 < \theta \leq 1/2$ we define $\xi_m := \xi_m(\rho, \tau, \theta)$ as a number u satisfying the equation

$$\rho(u) = \theta t_m u. \tag{3.2.1}$$

Remark 3.2.3. Assumptions on $\rho(u)$ imply that the function

$$s(u) := \rho(u)/u, \quad u \neq 0, \quad s(0) = 0,$$

is a continuous increasing function on $[0, \infty)$ with $s(2) \geq 1/2$. Thus (3.2.1) has a unique solution $\xi_m = s^{-1}(\theta t_m)$ such that $0 < \xi_m \leq 2$.

The following theorem from Temlyakov (2001b) gives a sufficient condition for convergence of the WCGA.

Theorem 3.2.4. Let X be a uniformly smooth Banach space with modulus of smoothness $\rho(u)$. Assume that a sequence $\tau := \{t_k\}_{k=1}^{\infty}$ satisfies the condition: for any $\theta > 0$ we have

$$\sum_{m=1}^{\infty} t_m \xi_m(\rho, \tau, \theta) = \infty.$$

Then, for any $f \in X$ we have

$$\lim_{m \to \infty} \|f_m^{c,\tau}\| = 0.$$

Corollary 3.2.5. Let a Banach space X have modulus of smoothness $\rho(u)$ of power type $1 < q \leq 2$, that is, $\rho(u) \leq \gamma u^q$. Assume that

$$\sum_{m=1}^{\infty} t_m^p = \infty, \quad p = \frac{q}{q-1}. \tag{3.2.2}$$

Then the WCGA converges for any $f \in X$.

Proof. Denote $\rho^q(u) := \gamma u^q$. Then

$$\rho(u)/u \leq \rho^q(u)/u,$$

and therefore for any $\theta > 0$ we have

$$\xi_m(\rho, \tau, \theta) \geq \xi_m(\rho^q, \tau, \theta).$$

For ρ^q we get from the definition of ξ_m that

$$\xi_m(\rho^q, \tau, \theta) = (\theta t_m/\gamma)^{\frac{1}{q-1}}.$$

Thus (3.2.2) implies that

$$\sum_{m=1}^{\infty} t_m \xi_m(\rho, \tau, \theta) \geq \sum_{m=1}^{\infty} t_m \xi_m(\rho^q, \tau, \theta) \asymp \sum_{m=1}^{\infty} t_m^p = \infty.$$

It remains to apply Theorem 3.2.4.   □

The following theorem from Temlyakov (2001b) gives the rate of convergence of the WCGA for f in $A_1(\mathcal{D})$.

Theorem 3.2.6. Let X be a uniformly smooth Banach space with modulus of smoothness $\rho(u) \leq \gamma u^q$, $1 < q \leq 2$. Then, for a sequence $\tau := \{t_k\}_{k=1}^{\infty}$, $t_k \leq 1$, $k = 1, 2, \ldots$, we have for any $f \in A_1(\mathcal{D})$ that

$$\|f_m^{c,\tau}\| \leq C(q, \gamma)\left(1 + \sum_{k=1}^{m} t_k^p\right)^{-1/p}, \qquad p := \frac{q}{q-1},$$

with a constant $C(q, \gamma)$ which may depend only on q and γ.

We will use the following two simple and well-known lemmas in the proof of the above two theorems.

Lemma 3.2.7. Let X be a uniformly smooth Banach space and let L be a finite-dimensional subspace of X. For any $f \in X \setminus L$, let f_L denote the best approximant of f from L. Then we have

$$F_{f-f_L}(\phi) = 0$$

for any $\phi \in L$.

Proof. Let us assume the contrary: there is a $\phi \in L$ such that $\|\phi\| = 1$ and

$$F_{f-f_L}(\phi) = \beta > 0.$$

For any λ we have from the definition of $\rho(u)$ that

$$\|f - f_L - \lambda\phi\| + \|f - f_L + \lambda\phi\| \leq 2\|f - f_L\|\left(1 + \rho\left(\frac{\lambda}{\|f - f_L\|}\right)\right). \quad (3.2.3)$$

Next

$$\|f - f_L + \lambda\phi\| \geq F_{f-f_L}(f - f_L + \lambda\phi) = \|f - f_L\| + \lambda\beta. \quad (3.2.4)$$

Combining (3.2.3) and (3.2.4) we get

$$\|f - f_L - \lambda\phi\| \leq \|f - f_L\|\left(1 - \frac{\lambda\beta}{\|f - f_L\|} + 2\rho\left(\frac{\lambda}{\|f - f_L\|}\right)\right). \quad (3.2.5)$$

Taking into account that $\rho(u) = o(u)$, we find $\lambda' > 0$ such that

$$\left(1 - \frac{\lambda'\beta}{\|f - f_L\|} + 2\rho\left(\frac{\lambda'}{\|f - f_L\|}\right)\right) < 1.$$

Then (3.2.5) gives

$$\|f - f_L - \lambda'\phi\| < \|f - f_L\|,$$

which contradicts the assumption that $f_L \in L$ is the best approximant of f.

$\square$

Lemma 3.2.8. For any bounded linear functional F and any dictionary $\mathcal{D}$, we have

$$\|F\|_{\mathcal{D}} := \sup_{g \in \mathcal{D}} F(g) = \sup_{f \in A_1(\mathcal{D})} F(f).$$

Proof. The inequality

$$\sup_{g \in \mathcal{D}} F(g) \leq \sup_{f \in A_1(\mathcal{D})} F(f)$$

is obvious. We prove the opposite inequality. Take any $f \in A_1(\mathcal{D})$. Then, for any $\epsilon > 0$ there exist $g_1^\epsilon, \ldots, g_N^\epsilon \in \mathcal{D}$ and numbers $a_1^\epsilon, \ldots, a_N^\epsilon$ such that $a_i^\epsilon > 0$, $a_1^\epsilon + \cdots + a_N^\epsilon \leq 1$ and

$$\left\| f - \sum_{i=1}^N a_i^\epsilon g_i^\epsilon \right\| \leq \epsilon.$$

Thus

$$F(f) \leq \|F\|\epsilon + F\left(\sum_{i=1}^N a_i^\epsilon g_i^\epsilon \right) \leq \epsilon\|F\| + \sup_{g \in \mathcal{D}} F(g),$$

which proves Lemma 3.2.8.

$\square$

We will also need one more lemma from Temlyakov (2001b).

Lemma 3.2.9. Let X be a uniformly smooth Banach space with modulus of smoothness $\rho(u)$. Take a number $\epsilon \geq 0$ and two elements f, f^ϵ from X such that

$$\|f - f^\epsilon\| \leq \epsilon, \quad f^\epsilon / A(\epsilon) \in A_1(\mathcal{D}),$$

with some number $A(\epsilon) > 0$. Then we have

$$\|f_m^{c,\tau}\| \leq \|f_{m-1}^{c,\tau}\| \inf_{\lambda \geq 0} \left(1 - \lambda t_m A(\epsilon)^{-1} \left(1 - \frac{\epsilon}{\|f_{m-1}^{c,\tau}\|} \right) + 2\rho\left(\frac{\lambda}{\|f_{m-1}^{c,\tau}\|} \right) \right),$$

for $m = 1, 2, \ldots$.

Proof. We have for any λ

$$\|f_{m-1}^c - \lambda\varphi_m^c\| + \|f_{m-1}^c + \lambda\varphi_m^c\| \leq 2\|f_{m-1}^c\| \left(1 + \rho\left(\frac{\lambda}{\|f_{m-1}^c\|} \right) \right), \quad (3.2.6)$$

and by (1) from the definition of the WCGA and Lemma 3.2.8 we get

$$F_{f_{m-1}^c}(\varphi_m^c) \geq t_m \sup_{g \in \mathcal{D}} F_{f_{m-1}^c}(g)$$

$$= t_m \sup_{\phi \in A_1(\mathcal{D})} F_{f_{m-1}^c}(\phi) \geq t_m A(\epsilon)^{-1} F_{f_{m-1}^c}(f^\epsilon).$$

By Lemma 3.2.7 we obtain

$$F_{f_{m-1}^c}(f^\epsilon) = F_{f_{m-1}^c}(f + f^\epsilon - f) \geq F_{f_{m-1}^c}(f) - \epsilon$$

$$= F_{f_{m-1}^c}(f_{m-1}^c) - \epsilon = \|f_{m-1}^c\| - \epsilon.$$

Thus, as in (3.2.5) we get from (3.2.6)

$$\|f_m^c\| \leq \inf_{\lambda \geq 0} \|f_{m-1}^c - \lambda \varphi_m^c\| \tag{3.2.7}$$

$$\leq \|f_{m-1}^c\| \inf_{\lambda \geq 0} \left(1 - \lambda t_m A(\epsilon)^{-1}\left(1 - \frac{\epsilon}{\|f_{m-1}^c\|}\right) + 2\rho\left(\frac{\lambda}{\|f_{m-1}^c\|}\right)\right),$$

which proves the lemma. $\square$

Proof of Theorem 3.2.4. The definition of the WCGA implies that $\{\|f_m^c\|\}$ is a non-increasing sequence. Therefore we have

$$\lim_{m \to \infty} \|f_m^c\| = \alpha.$$

We prove that $\alpha = 0$ by contradiction. Assume to the contrary that $\alpha > 0$. Then, for any m we have

$$\|f_m^c\| \geq \alpha.$$

We set $\epsilon = \alpha/2$ and find f^ϵ such that

$$\|f - f^\epsilon\| \leq \epsilon \quad \text{and} \quad f^\epsilon/A(\epsilon) \in A_1(\mathcal{D}),$$

with some $A(\epsilon)$. Then, by Lemma 3.2.9 we get

$$\|f_m^c\| \leq \|f_{m-1}^c\| \inf_\lambda (1 - \lambda t_m A(\epsilon)^{-1}/2 + 2\rho(\lambda/\alpha)).$$

Let us specify $\theta := \frac{\alpha}{8A(\epsilon)}$ and take $\lambda = \alpha \xi_m(\rho, \tau, \theta)$. Then we obtain

$$\|f_m^c\| \leq \|f_{m-1}^c\|(1 - 2\theta t_m \xi_m).$$

The assumption

$$\sum_{m=1}^{\infty} t_m \xi_m = \infty$$

implies that

$$\|f_m^c\| \to 0 \quad \text{as } m \to \infty.$$

We have a contradiction, which proves the theorem. $\square$

Proof of Theorem 3.2.6. By Lemma 3.2.9 with $\epsilon = 0$ and $A(\epsilon) = 1$, we have for $f \in A_1(\mathcal{D})$ that

$$\|f_m^c\| \leq \|f_{m-1}^c\| \inf_{\lambda \geq 0} \left(1 - \lambda t_m + 2\gamma \left(\frac{\lambda}{\|f_{m-1}^c\|}\right)^q\right). \qquad (3.2.8)$$

Choose λ from the equation

$$\frac{1}{2} \lambda t_m = 2\gamma \left(\frac{\lambda}{\|f_{m-1}^c\|}\right)^q,$$

which implies that

$$\lambda = \|f_{m-1}^c\|^{\frac{q}{q-1}} (4\gamma)^{-\frac{1}{q-1}} t_m^{\frac{1}{q-1}}.$$

Let

$$A_q := 2(4\gamma)^{\frac{1}{q-1}}.$$

Using the notation $p := \frac{q}{q-1}$, we get from (3.2.8)

$$\|f_m^c\| \leq \|f_{m-1}^c\| \left(1 - \frac{1}{2} \lambda t_m\right) = \|f_{m-1}^c\|(1 - t_m^p \|f_{m-1}^c\|^p / A_q).$$

Raising both sides of this inequality to the power p and taking into account the inequality $x^r \leq x$ for $r \geq 1$, $0 \leq x \leq 1$, we obtain

$$\|f_m^c\|^p \leq \|f_{m-1}^c\|^p (1 - t_m^p \|f_{m-1}^c\|^p / A_q).$$

By an analogue of Lemma 2.3.3 (see Temlyakov (2000b, Lemma 3.1)), using the estimate $\|f\|^p \leq 1 < A_q$ we get

$$\|f_m^c\|^p \leq A_q \left(1 + \sum_{n=1}^{m} t_n^p\right)^{-1}$$

which implies

$$\|f_m^c\| \leq C(q, \gamma) \left(1 + \sum_{n=1}^{m} t_n^p\right)^{-1/p}.$$

Theorem 3.2.6 is now proved. $\qquad\qquad\qquad\qquad\qquad\qquad\qquad\qquad$ □

Remark 3.2.10. Theorem 3.2.6 also holds for a slightly modified version of the WCGA, the WCGA(1), for which at step (1) we require

$$F_{f_{m-1}^{c(1)}} (\varphi_m^{c(1)}) \geq t_m \|f_{m-1}^{c(1)}\|. \qquad (3.2.9)$$

This statement follows from the fact that, in the proof of Theorem 3.2.6, the relation

$$F_{f_{m-1}^c}(\varphi_m^c) \geq t_m \sup_{g \in \mathcal{D}} F_{f_{m-1}^c}(g)$$

was used only to get (3.2.9).

Proposition 3.2.11. Condition (3.2.2) in Corollary 3.2.5 is sharp.

Proof. Let $1 < q \leq 2$. Consider $X = \ell_q$. It is known (Lindenstrauss and Tzafriri 1977, p. 67) that ℓ_q, $1 < q \leq 2$, is a uniformly smooth Banach space with modulus of smoothness $\rho(u)$ of power type q. Denote $p := \frac{q}{q-1}$ and take any $\{t_k\}_{k=1}^\infty$, $0 < t_k \leq 1$, such that

$$\sum_{k=1}^{\infty} t_k^p < \infty. \tag{3.2.10}$$

Choose $\mathcal{D}$ as a standard basis $\{e_j\}_{j=1}^\infty$, $e_j := (0, \ldots, 0, 1, 0, \ldots)$, for ℓ_q. Consider the following realization of the WCGA for

$$f := \left(1, t_1^{\frac{1}{q-1}}, t_2^{\frac{1}{q-1}}, \ldots\right).$$

First of all, (3.2.10) guarantees that $f \in \ell_q$. Next, it is well known that F_f can be identified as

$$F_f = (1, t_1, t_2, \ldots) / \left(1 + \sum_{k=1}^{\infty} t_k^p\right)^{1/p} \in \ell_p.$$

At the first step of the WCGA we pick $\varphi_1 = e_2$ and get

$$f_1^c = \left(1, 0, t_2^{\frac{1}{q-1}}, \ldots\right).$$

We continue with f replaced by f_1 and so on. After m steps we get

$$f_m^c = \left(1, 0, \ldots, 0, t_{m+1}^{\frac{1}{q-1}}, \ldots\right).$$

It is clear that for all m we have $\|f_m^c\|_{\ell_q} \geq 1$. $\square$

The following variant of Theorem 3.2.6 (see Temlyakov (2006c)) follows from Lemma 3.2.9.

Theorem 3.2.12. Let X be a uniformly smooth Banach space with modulus of smoothness $\rho(u) \leq \gamma u^q$, $1 < q \leq 2$. Take a number $\epsilon \geq 0$ and two elements f, f^ϵ from X such that

$$\|f - f^\epsilon\| \leq \epsilon, \quad f^\epsilon / A(\epsilon) \in A_1(\mathcal{D}),$$

with some number $A(\epsilon) > 0$. Then we have $(p := q/(q-1))$

$$\|f_m^{c,\tau}\| \leq \max\left(2\epsilon, C(q,\gamma)(A(\epsilon)+\epsilon)\left(1 + \sum_{k=1}^{m} t_k^p\right)^{-1/p}\right). \qquad (3.2.11)$$

3.3. Relaxation; co-convex approximation

In this section we study a generalization for Banach spaces of relaxed greedy algorithms considered in Chapter 2. We present here results from Temlyakov (2001b). Let $\tau := \{t_k\}_{k=1}^{\infty}$ be a given weakness sequence of numbers $t_k \in [0,1]$, $k = 1, \ldots$.

Weak Relaxed Greedy Algorithm (WRGA). We define $f_0^r := f_0^{r,\tau} := f$ and $G_0^r := G_0^{r,\tau} := 0$. Then, for each $m \geq 1$ we have the following inductive definition.

(1) $\varphi_m^r := \varphi_m^{r,\tau} \in \mathcal{D}$ is any element satisfying

$$F_{f_{m-1}^r}(\varphi_m^r - G_{m-1}^r) \geq t_m \sup_{g \in \mathcal{D}} F_{f_{m-1}^r}(g - G_{m-1}^r).$$

(2) Find $0 \leq \lambda_m \leq 1$ such that

$$\|f - ((1-\lambda_m)G_{m-1}^r + \lambda_m\varphi_m^r)\| = \inf_{0 \leq \lambda \leq 1} \|f - ((1-\lambda)G_{m-1}^r + \lambda\varphi_m^r)\|$$

and define

$$G_m^r := G_m^{r,\tau} := (1-\lambda_m)G_{m-1}^r + \lambda_m\varphi_m^r.$$

(3) Let

$$f_m^r := f_m^{r,\tau} := f - G_m^r.$$

Remark 3.3.1. It follows from the definition of the WRGA that the sequence $\{\|f_m^r\|\}$ is a non-increasing sequence.

We call the WRGA *relaxed* because at the mth step of the algorithm we use a linear combination (convex combination) of the previous approximant G_{m-1}^r and a new element φ_m^r. The relaxation parameter λ_m in the WRGA is chosen at the mth step depending on f. We prove here the analogues of Theorems 3.2.4 and 3.2.6 for the Weak Relaxed Greedy Algorithm.

Theorem 3.3.2. Let X be a uniformly smooth Banach space with modulus of smoothness $\rho(u)$. Assume that a sequence $\tau := \{t_k\}_{k=1}^{\infty}$ satisfies the condition: for any $\theta > 0$ we have

$$\sum_{m=1}^{\infty} t_m\xi_m(\rho,\tau,\theta) = \infty.$$

Then, for any $f \in A_1(\mathcal{D})$ we have

$$\lim_{m\to\infty} \|f_m^{r,\tau}\| = 0.$$

Theorem 3.3.3. Let X be a uniformly smooth Banach space with modulus of smoothness $\rho(u) \leq \gamma u^q$, $1 < q \leq 2$. Then, for a sequence $\tau := \{t_k\}_{k=1}^{\infty}$, $t_k \leq 1$, $k = 1, 2, \ldots$, we have for any $f \in A_1(\mathcal{D})$ that

$$\|f_m^{r,\tau}\| \leq C_1(q, \gamma) \left(1 + \sum_{k=1}^{m} t_k^p\right)^{-1/p}, \quad p := \frac{q}{q-1},$$

with a constant $C_1(q, \gamma)$ which may depend only on q and γ.

Proof of Theorems 3.3.2 and 3.3.3. This proof is similar to the proof of Theorems 3.2.4 and 3.2.6. Instead of Lemma 3.2.9 we use the following lemma.

Lemma 3.3.4. Let X be a uniformly smooth Banach space with modulus of smoothness $\rho(u)$. Then, for any $f \in A_1(\mathcal{D})$ we have

$$\|f_m^{r,\tau}\| \leq \|f_{m-1}^{r,\tau}\| \inf_{0 \leq \lambda \leq 1} \left(1 - \lambda t_m + 2\rho\left(\frac{2\lambda}{\|f_{m-1}^{r,\tau}\|}\right)\right), \quad m = 1, 2, \ldots.$$

Proof. We have

$$f_m^r := f - ((1 - \lambda_m)G_{m-1}^r + \lambda_m \varphi_m^r) = f_{m-1}^r - \lambda_m(\varphi_m^r - G_{m-1}^r)$$

and

$$\|f_m^r\| = \inf_{0 \leq \lambda \leq 1} \|f_{m-1}^r - \lambda(\varphi_m^r - G_{m-1}^r)\|.$$

As for (3.2.6), we have for any λ

$$\|f_{m-1}^r - \lambda(\varphi_m^r - G_{m-1}^r)\| + \|f_{m-1}^r + \lambda(\varphi_m^r - G_{m-1}^r)\|$$
$$\leq 2\|f_{m-1}^r\| \left(1 + \rho\left(\frac{\lambda\|\varphi_m^r - G_{m-1}^r\|}{\|f_{m-1}^r\|}\right)\right). \quad (3.3.1)$$

Next we get for $\lambda \geq 0$

$$\|f_{m-1}^r + \lambda(\varphi_m^r - G_{m-1}^r)\|$$
$$\geq F_{f_{m-1}^r}(f_{m-1}^r + \lambda(\varphi_m^r - G_{m-1}^r))$$
$$= \|f_{m-1}^r\| + \lambda F_{f_{m-1}^r}(\varphi_m^r - G_{m-1}^r) \geq \|f_{m-1}^r\| + \lambda t_m \sup_{g \in \mathcal{D}} F_{f_{m-1}^r}(g - G_{m-1}^r)$$
$$= \|f_{m-1}^r\| + \lambda t_m \sup_{\phi \in A_1(\mathcal{D}} F_{f_{m-1}^r}(\phi - G_{m-1}^r) \geq \|f_{m-1}^r\| + \lambda t_m\|f_{m-1}^r\|,$$

applying Lemma 3.2.8 for the last inequality. Using the trivial estimate $\|\varphi_m^r - G_{m-1}^r\| \leq 2$, we obtain

$$\|f_{m-1}^r - \lambda(\varphi_m^r - G_{m-1}^r)\| \leq \|f_{m-1}^r\| \left(1 - \lambda t_m + 2\rho\left(\frac{2\lambda}{\|f_{m-1}^r\|}\right)\right), \quad (3.3.2)$$

from (3.3.1), which proves Lemma 3.3.4. $\qquad \square$

The remaining part of the proof uses inequality (3.3.2) in the same way relation (3.2.7) was used in the proof of Theorems 3.2.4 and 3.2.6. The only additional difficulty here is that we are optimizing over $0 \le \lambda \le 1$. However, it is easy to check that the corresponding λ chosen in a similar way always satisfies the restriction $0 \le \lambda \le 1$. In the proof of Theorem 3.3.2 we choose $\theta = \alpha/8$ and $\lambda = \alpha \xi_m(\rho, \tau, \theta)/2$, and in the proof of Theorem 3.3.3 we choose λ from the equation

$$\frac{1}{2}\lambda t_m = 2\gamma(2\lambda)^q \|f_{m-1}^r\|^{-q}. \qquad \square$$

Remark 3.3.5. Theorems 3.3.2 and 3.3.3 hold for a slightly modified version of the WRGA, the WRGA(1), for which at step (1) we require

$$F_{f_{m-1}^{r(1)}}\left(\varphi_m^{r(1)} - G_{m-1}^{r(1)}\right) \ge t_m \|f_{m-1}^{r(1)}\|. \qquad (3.3.3)$$

This follows from the observation that in the proof of Lemma 3.3.4 we used the inequality from step (1) of the WRGA only to derive (3.3.3). It is clear from Lemma 3.2.8 that in the case of approximation of $f \in A_1(\mathcal{D})$, the requirement (3.3.3) is weaker and easier to check than step (1) of the WRGA.

3.4. Free relaxation

Both of the above algorithms, the WCGA and the WRGA, use the functional $F_{f_{m-1}}$ in a search for the mth element φ_m from the dictionary to be used in approximation. The construction of the approximant in the WRGA is different from the construction in the WCGA. In the WCGA we build the approximant G_m^c so as to maximally use the approximation power of the elements $\varphi_1, \ldots, \varphi_m$. The WRGA by its definition is designed for approximation of functions from $A_1(\mathcal{D})$. In building the approximant in the WRGA we keep the property $G_m^r \in A_1(\mathcal{D})$. As we mentioned in Section 3.3 the relaxation parameter λ_m in the WRGA is chosen at the mth step depending on f. The following modification of the above idea of relaxation in greedy approximation will be studied in this section (see Temlyakov (2006c)).

Weak Greedy Algorithm with Free Relaxation (WGAFR). Let $\tau := \{t_m\}_{m=1}^\infty$, $t_m \in [0, 1]$, be a weakness sequence. We define $f_0 := f$ and $G_0 := 0$. Then, for each $m \ge 1$ we have the following inductive definition.

(1) $\varphi_m \in \mathcal{D}$ is any element satisfying

$$F_{f_{m-1}}(\varphi_m) \ge t_m \|F_{f_{m-1}}\|_{\mathcal{D}}.$$

(2) Find w_m and λ_m such that

$$\|f - ((1 - w_m)G_{m-1} + \lambda_m \varphi_m)\| = \inf_{\lambda, w} \|f - ((1 - w)G_{m-1} + \lambda \varphi_m)\|$$

and define
$$G_m := (1 - w_m)G_{m-1} + \lambda_m \varphi_m.$$
(3) Let
$$f_m := f - G_m.$$
We begin with the following analogue of Lemma 3.2.9.

Lemma 3.4.1. Let X be a uniformly smooth Banach space with modulus of smoothness $\rho(u)$. Take a number $\epsilon \geq 0$ and two elements f, f^ϵ from X such that
$$\|f - f^\epsilon\| \leq \epsilon, \quad f^\epsilon/A(\epsilon) \in A_1(\mathcal{D}),$$
with some number $A(\epsilon) \geq \epsilon$. Then we have for the WGAFR
$$\|f_m\| \leq \|f_{m-1}\| \inf_{\lambda \geq 0} \left(1 - \lambda t_m A(\epsilon)^{-1} \left(1 - \frac{\epsilon}{\|f_{m-1}\|} \right) + 2\rho \left(\frac{5\lambda}{\|f_{m-1}\|} \right) \right),$$
for $m = 1, 2, \ldots$.

Proof. By the definition of f_m,
$$\|f_m\| \leq \inf_{\lambda \geq 0, w} \|f_{m-1} + wG_{m-1} - \lambda\varphi_m\|.$$

As in the arguments in the proof of Lemma 3.2.9, we use the inequality
$$\|f_{m-1} + wG_{m-1} - \lambda\varphi_m\| + \|f_{m-1} - wG_{m-1} + \lambda\varphi_m\| \qquad (3.4.1)$$
$$\leq 2\|f_{m-1}\|(1 + \rho(\|wG_{m-1} - \lambda\varphi_m\|/\|f_{m-1}\|)),$$
and estimate for $\lambda \geq 0$
$$\|f_{m-1} - wG_{m-1} + \lambda\varphi_m\| \geq F_{f_{m-1}}(f_{m-1} - wG_{m-1} + \lambda\varphi_m)$$
$$\geq \|f_{m-1}\| - F_{f_{m-1}}(wG_{m-1}) + \lambda t_m \sup_{g \in \mathcal{D}} F_{f_{m-1}}(g).$$

By Lemma 3.2.8, we continue:
$$= \|f_{m-1}\| - F_{f_{m-1}}(wG_{m-1}) + \lambda t_m \sup_{\phi \in A_1(\mathcal{D})} F_{f_{m-1}}(\phi)$$
$$\geq \|f_{m-1}\| - F_{f_{m-1}}(wG_{m-1}) + \lambda t_m A(\epsilon)^{-1} F_{f_{m-1}}(f^\epsilon)$$
$$\geq \|f_{m-1}\| - F_{f_{m-1}}(wG_{m-1}) + \lambda t_m A(\epsilon)^{-1}(F_{f_{m-1}}(f) - \epsilon).$$
We set $w^* := \lambda t_m A(\epsilon)^{-1}$ and obtain
$$\|f_{m-1} - w^*G_{m-1} + \lambda\varphi_m\| \geq \|f_{m-1}\| + \lambda t_m A(\epsilon)^{-1}(\|f_{m-1}\| - \epsilon). \quad (3.4.2)$$
Combining (3.4.1) and (3.4.2) we get
$$\|f_m\| \leq \|f_{m-1}\| \inf_{\lambda \geq 0}(1 - \lambda t_m A(\epsilon)^{-1}(1 - \epsilon/\|f_{m-1}\|)$$
$$+ 2\rho(\|w^*G_{m-1} - \lambda\varphi_m\|/\|f_{m-1}\|)).$$

We now estimate

$$\|w^*G_{m-1} - \lambda\varphi_m\| \le w^*\|G_{m-1}\| + \lambda.$$

Next,

$$\|G_{m-1}\| = \|f - f_{m-1}\| \le 2\|f\| \le 2(\|f^\epsilon\| + \epsilon) \le 2(A(\epsilon) + \epsilon).$$

Thus, under assumption $A(\epsilon) \ge \epsilon$ we get

$$w^*\|G_{m-1}\| \le 2\lambda t_m(A(\epsilon) + \epsilon)/A(\epsilon) \le 4\lambda.$$

Finally,

$$\|w^*G_{m-1} - \lambda\varphi_m\| \le 5\lambda.$$

This completes the proof of Lemma 3.4.1. □

Remark 3.4.2. It follows from the definition of the WGAFR that the sequence $\{\|f_m\|\}$ is a non-increasing sequence.

We now prove a convergence theorem for an arbitrary uniformly smooth Banach space. Modulus of smoothness $\rho(u)$ of a uniformly smooth Banach space is an even convex function such that $\rho(0) = 0$ and $\lim_{u\to 0} \rho(u)/u = 0$. The function $s(u) := \rho(u)/u$, $s(0) := 0$, associated with $\rho(u)$, is a continuous increasing function on $[0, \infty)$. Therefore, the inverse function $s^{-1}(\cdot)$ is well defined.

Theorem 3.4.3. Let X be a uniformly smooth Banach space with modulus of smoothness $\rho(u)$. Assume that a sequence $\tau := \{t_k\}_{k=1}^\infty$ satisfies the following condition. For any $\theta > 0$ we have

$$\sum_{m=1}^{\infty} t_m s^{-1}(\theta t_m) = \infty. \tag{3.4.3}$$

Then, for any $f \in X$ we have for the WGAFR

$$\lim_{m\to\infty} \|f_m\| = 0.$$

Proof. By Remark 3.4.2, $\{\|f_m\|\}$ is a non-increasing sequence. Therefore we have

$$\lim_{m\to\infty} \|f_m\| = \beta.$$

We prove that $\beta = 0$ by contradiction. Assume the contrary, that $\beta > 0$. Then, for any m we have

$$\|f_m\| \ge \beta.$$

We set $\epsilon = \beta/2$ and find f^ϵ such that

$$\|f - f^\epsilon\| \le \epsilon \quad \text{and} \quad f^\epsilon/A(\epsilon) \in A_1(\mathcal{D}),$$

with some $A(\epsilon) \geq \epsilon$. Then, by Lemma 3.4.1 we get

$$\|f_m\| \leq \|f_{m-1}\| \inf_{\lambda \geq 0}(1 - \lambda t_m A(\epsilon)^{-1}/2 + 2\rho(5\lambda/\beta)).$$

Let us specify $\theta := \beta/(40A(\epsilon))$ and take $\lambda = \beta s^{-1}(\theta t_m)/5$. Then we obtain

$$\|f_m\| \leq \|f_{m-1}\|(1 - 2\theta t_m s^{-1}(\theta t_m)).$$

The assumption

$$\sum_{m=1}^{\infty} t_m s^{-1}(\theta t_m) = \infty$$

implies that

$$\|f_m\| \to 0 \quad \text{as } m \to \infty.$$

We have a contradiction, which proves the theorem. $\qquad \square$

Theorem 3.4.4. Let X be a uniformly smooth Banach space with modulus of smoothness $\rho(u) \leq \gamma u^q$, $1 < q \leq 2$. Take a number $\epsilon \geq 0$ and two elements f, f^ϵ from X such that

$$\|f - f^\epsilon\| \leq \epsilon, \quad f^\epsilon/A(\epsilon) \in A_1(\mathcal{D}),$$

with some number $A(\epsilon) > 0$. Then we have for the WGAFR

$$\|f_m\| \leq \max\left(2\epsilon, C(q, \gamma)(A(\epsilon) + \epsilon)\left(1 + \sum_{k=1}^{m} t_k^p\right)^{-1/p}\right), \quad p := q/(q-1).$$

Proof. It is clear that it suffices to consider the case $A(\epsilon) \geq \epsilon$. Otherwise, $\|f_m\| \leq \|f\| \leq \|f^\epsilon\| + \epsilon \leq 2\epsilon$. Also, assume $\|f_m\| > 2\epsilon$ (otherwise Theorem 3.4.4 trivially holds). Then, by Remark 3.4.2, we have for all $k = 0, 1, \ldots, m$ that $\|f_k\| > 2\epsilon$. By Lemma 3.4.1 we obtain

$$\|f_k\| \leq \|f_{k-1}\| \inf_{\lambda \geq 0}\left(1 - \lambda t_k A(\epsilon)^{-1}/2 + 2\gamma\left(\frac{5\lambda}{\|f_{k-1}\|}\right)^q\right). \qquad (3.4.4)$$

Choose λ from the equation

$$\frac{\lambda t_k}{4A(\epsilon)} = 2\gamma\left(\frac{5\lambda}{\|f_{k-1}\|}\right)^q,$$

which implies that

$$\lambda = \|f_{k-1}\|^{\frac{q}{q-1}} 5^{-\frac{q}{q-1}} (8\gamma A(\epsilon))^{-\frac{1}{q-1}} t_k^{\frac{1}{q-1}}.$$

Define

$$A_q := 4(8\gamma)^{\frac{1}{q-1}} 5^{\frac{q}{q-1}}.$$

Using the notation $p := \frac{q}{q-1}$, we get from (3.4.4)

$$\|f_k\| \le \|f_{k-1}\| \left(1 - \frac{1}{4}\frac{\lambda t_k}{A(\epsilon)}\right) = \|f_{k-1}\| \left(1 - \frac{t_k^p \|f_{k-1}\|^p}{A_q A(\epsilon)^p}\right).$$

Raising both sides of this inequality to the power p and taking into account the inequality $x^r \le x$ for $r \ge 1$, $0 \le x \le 1$, we obtain

$$\|f_k\|^p \le \|f_{k-1}\|^p \left(1 - \frac{t_k^p \|f_{k-1}\|^p}{A_q A(\epsilon)^p}\right).$$

By an analogue of Lemma 2.3.3 (see Temlyakov (2000b, Lemma 3.1)), using the estimates $\|f\| \le A(\epsilon) + \epsilon$ and $A_q > 1$, we get

$$\|f_m\|^p \le A_q (A(\epsilon) + \epsilon)^p \left(1 + \sum_{k=1}^{m} t_k^p\right)^{-1},$$

which implies

$$\|f_m\| \le C(q, \gamma)(A(\epsilon) + \epsilon) \left(1 + \sum_{k=1}^{m} t_k^p\right)^{-1/p}.$$

Theorem 3.4.4 is proved.   □

3.5. Fixed relaxation

In this section we consider a relaxed greedy algorithm with relaxation prescribed in advance. Let a sequence $\mathbf{r} := \{r_k\}_{k=1}^{\infty}$, $r_k \in [0, 1)$, of relaxation parameters be given. Then, at each step of our new algorithm we build the mth approximant of the form $G_m = (1 - r_m)G_{m-1} + \lambda \varphi_m$. With an approximant of this form we are not limited to approximation of functions from $A_1(\mathcal{D})$ as in the WRGA. In this section we study the Greedy Algorithm with Weakness parameter t and Relaxation $\mathbf{r}$ (GAWR($t, \mathbf{r}$)). In addition to the acronym GAWR($t, \mathbf{r}$) we will use the abbreviated acronym GAWR for the name of this algorithm. We give a general definition of the algorithm in the case of a weakness sequence τ. We present in this section results from Temlyakov (2006c).

GAWR($\tau, \mathbf{r}$). Let $\tau := \{t_m\}_{m=1}^{\infty}$, $t_m \in (0, 1]$, be a weakness sequence and let $\mathbf{r} := \{r_m\}_{m=1}^{\infty}$, $r_m \in [0, 1)$, be a relaxation sequence. We define $f_0 := f$ and $G_0 := 0$. Then, for each $m \ge 1$ we have the following inductive definition.

(1) $\varphi_m \in \mathcal{D}$ is any element satisfying

$$F_{f_{m-1}}(\varphi_m) \ge t_m \|F_{f_{m-1}}\|_{\mathcal{D}}.$$

(2) Find $\lambda_m \geq 0$ such that

$$\|f - ((1 - r_m)G_{m-1} + \lambda_m \varphi_m)\| = \inf_{\lambda \geq 0} \|f - ((1 - r_m)G_{m-1} + \lambda \varphi_m)\|$$

and define

$$G_m := (1 - r_m)G_{m-1} + \lambda_m \varphi_m.$$

(3) Let

$$f_m := f - G_m.$$

In the case $\tau = \{t\}$ we write t instead of τ in the notation. We note that in the case $r_k = 0$, $k = 1, \ldots$, when there is no relaxation the GAWR$(\tau, \mathbf{0})$ coincides with the Weak Dual Greedy Algorithm. We now proceed to the GAWR. We begin with an analogue of Lemma 3.2.9.

Lemma 3.5.1. Let X be a uniformly smooth Banach space with modulus of smoothness $\rho(u)$. Take a number $\epsilon \geq 0$ and two elements f, f^ϵ from X such that

$$\|f - f^\epsilon\| \leq \epsilon, \quad f^\epsilon / A(\epsilon) \in A_1(\mathcal{D}),$$

with some number $A(\epsilon) > 0$. Then we have for the GAWR$(t, \mathbf{r})$

$$\|f_m\| \leq \|f_{m-1}\|(1 - r_m(1 - \epsilon/\|f_{m-1}\|))$$
$$+ 2\rho((r_m(\|f\| + A(\epsilon)/t))/((1 - r_m)\|f_{m-1}\|)), \quad m = 1, 2, \ldots.$$

Theorem 3.5.2. Let a sequence $\mathbf{r}$ satisfy the conditions

$$\sum_{k=1}^{\infty} r_k = \infty, \quad r_k \to 0 \quad \text{as } k \to \infty.$$

Then the GAWR$(t, \mathbf{r})$ converges in any uniformly smooth Banach space for each $f \in X$ and for all dictionaries $\mathcal{D}$.

Proof. We prove this theorem in two steps.

I First, we prove that $\liminf_{m \to \infty} \|f_m\| = 0$. The proof goes by contradiction. We want to prove that $\liminf_{m \to \infty} \|f_m\| = 0$. Assume the contrary. Then there exists K and $\beta > 0$ such that we have for all $k \geq K$ that $\|f_k\| \geq \beta$. By Lemma 3.5.1, for $m > K$

$$\|f_m\| \leq \|f_{m-1}\|\left(1 - r_m\left(1 - \frac{\epsilon}{\beta}\right) + 2\rho\left(\frac{r_m(\|f\| + A(\epsilon)/t)}{(1 - r_m)\beta}\right)\right).$$

We choose $\epsilon := \beta/2$. Using the assumption that X is uniformly smooth and the assumption $r_k \to 0$ as $k \to \infty$, we find $N \geq K$ such that for $m \geq N$ we have

$$2\rho\left(\frac{r_m(\|f\| + A(\epsilon)/t)}{(1 - r_m)\beta}\right) \leq r_m/4.$$

Then, for $m > N$,

$$\|f_m\| \le \|f_{m-1}\|(1 - r_m/4).$$

The assumption $\sum_{m=1}^{\infty} r_m = \infty$ implies that $\|f_m\| \to 0$ as $m \to \infty$. The obtained contradiction to the assumption $\beta > 0$ completes the proof of part I.

II Secondly, we prove that $\lim_{m \to \infty} \|f_m\| = 0$. Using the assumption $r_k \to 0$ as $k \to \infty$, we find N_1 such that for $k \ge N_1$ we have $r_k \le 1/2$. For such k we obtain from Lemma 3.5.1

$$\|f_k\| - \epsilon \le (1 - r_k)(\|f_{k-1}\| - \epsilon) + 2\|f_{k-1}\|\rho\left(\frac{Br_k}{\|f_{k-1}\|}\right), \qquad (3.5.1)$$

with $B := 2(\|f\| + A(\epsilon)/t)$. Denote $a_k := \|f_{k-1}\| - \epsilon$. We note that from the definition of f_k it follows that

$$a_{k+1} \le a_k + r_k\|f\|. \qquad (3.5.2)$$

Using the fact that the function $\rho(u)/u$ is monotone increasing on $[0, \infty)$, we obtain from (3.5.1) for $a_k > 0$

$$a_{k+1} \le a_k\left(1 - r_k + 2\frac{\|f_{k-1}\|}{a_k}\rho\left(\frac{Br_k}{\|f_{k-1}\|}\right)\right)$$

$$\le a_k\left(1 - r_k + 2\rho\left(\frac{Br_k}{a_k}\right)\right). \qquad (3.5.3)$$

We now introduce an auxiliary sequence $\{b_k\}$ of positive numbers that is defined by the equation

$$2\rho(Br_k/b_k) = r_k.$$

The property $\rho(u)/u \to 0$ as $u \to 0$ implies $b_k \to 0$ as $k \to \infty$. Inequality (3.5.3) guarantees that for $k \ge N_1$ such that $a_k \ge b_k$, we have $a_{k+1} \le a_k$.
    Let

$$U := \{k : k \ge N_1, \quad a_k \ge b_k\}.$$

If the set U is finite then we get

$$\limsup_{k \to \infty} a_k \le \lim_{k \to \infty} b_k = 0.$$

This implies

$$\limsup_{m \to \infty} \|f_m\| \le \epsilon.$$

    Consider the case when U is infinite. We note that part I of the proof implies that there is a subsequence $\{k_j\}$ such that $a_{k_j} \le 0$, $j = 1, 2, \ldots$. This means that

$$U = \cup_{j=1}^{\infty}[l_j, n_j],$$

with the property $n_{j-1} < l_j - 1$. For $k \notin U$, $k \geq N_1$ we have

$$a_k < b_k. \tag{3.5.4}$$

For $k \in [l_j, n_j]$, we have by (3.5.2) and the monotonicity property of a_k, when $k \in [l_j, n_j]$, that

$$a_k \leq a_{l_j} \leq a_{l_j-1} + r_{l_j-1}\|f\| \leq b_{l_j-1} + r_{l_j-1}\|f\|. \tag{3.5.5}$$

By (3.5.4) and (3.5.5) we obtain

$$\limsup_{k\to\infty} a_k \leq 0 \;\Rightarrow\; \limsup_{m\to\infty} \|f_m\| \leq \epsilon.$$

Taking into account that $\epsilon > 0$ is arbitrary, we complete the proof.   □

We now proceed to results on the rate of approximation. We will need the following technical lemma (see Temlyakov (1999, 2006c)).

Lemma 3.5.3. Let a sequence $\{a_n\}_{n=1}^{\infty}$ have the following property. For, given positive numbers $\alpha < \gamma \leq 1$, $A > a_1$, we have, for all $n \geq 2$,

$$a_n \leq a_{n-1} + A(n-1)^{-\alpha}. \tag{3.5.6}$$

If for some $\nu \geq 2$ we have

$$a_\nu \geq A\nu^{-\alpha},$$

then

$$a_{\nu+1} \leq a_\nu(1 - \gamma/\nu). \tag{3.5.7}$$

Then there exists a constant $C(\alpha, \gamma)$ such that, for all $n = 1, 2, \ldots$, we have

$$a_n \leq C(\alpha, \gamma)An^{-\alpha}.$$

Theorem 3.5.4. Let X be a uniformly smooth Banach space with modulus of smoothness $\rho(u) \leq \gamma u^q$, $1 < q \leq 2$. Let $\mathbf{r} := \{2/(k+2)\}_{k=1}^{\infty}$. Consider the GAWR$(t, \mathbf{r})$. For a pair of functions f, f^ϵ, satisfying

$$\|f - f^\epsilon\| \leq \epsilon, \quad f^\epsilon/A(\epsilon) \in A_1(\mathcal{D}),$$

we have

$$\|f_m\| \leq \epsilon + C(q, \gamma)(\|f\| + A(\epsilon)/t)m^{-1+1/q}.$$

Proof. By Lemma 3.5.1 we obtain

$$\|f_k\| - \epsilon \leq (1 - r_k)(\|f_{k-1}\| - \epsilon) + C\gamma\|f_{k-1}\|\left(\frac{r_k(\|f\| + A(\epsilon)/t)}{\|f_{k-1}\|}\right)^q. \tag{3.5.8}$$

Consider, as in the proof of Theorem 3.5.2, the sequence $a_n := \|f_{n-1}\| - \epsilon$. We plan to apply Lemma 3.5.3 to the sequence $\{a_n\}$. We set $\alpha := 1 - 1/q \leq 1/2$. The parameters $\gamma \in (\alpha, 1]$ and A will be chosen later. We note that

$$\|f_m\| \leq \|f_{m-1}\| + r_m\|f\|.$$

Therefore, condition (3.5.6) of Lemma 3.5.3 is satisfied with $A \geq 2\|f\|$. Let $a_k \geq Ak^{-\alpha}$. Then, by (3.5.8) we get

$$a_{k+1} \leq a_k(1 - r_k + C\gamma(r_k(\|f\| + A(\epsilon)/t)/a_k)^q$$

$$\leq a_k\left(1 - \frac{2}{k+2} + \frac{C\gamma(\|f\| + A(\epsilon)/t)^q 2^q}{A^q} \frac{k^{\alpha q}}{(k+2)^q}\right).$$

Setting $A := \max(2\|f\|, 2(2C\gamma)^{1/q}(\|f\| + A(\epsilon)/t))$, we obtain

$$a_{k+1} \leq a_k\left(1 - \frac{3}{2(k+2)}\right).$$

Thus condition (3.5.7) of Lemma 3.5.3 is satisfied with $\gamma = 3/4$. Applying Lemma 3.5.3 we obtain

$$\|f_m\| \leq \epsilon + C(q, \gamma)(\|f\| + A(\epsilon)/t)m^{-1+1/q}. \qquad \square$$

We conclude this section by the following remark. The algorithms GAWR and WGAFR are both dual-type greedy algorithms. The first steps are similar for both algorithms: we use the norming functional $F_{f_{m-1}}$ in the search for an element φ_m. The WGAFR provides more freedom than the GAWR does in choosing good coefficients w_m and λ_m. This results in more flexibility in choosing the weakness sequence $\tau = \{t_m\}$. For instance, condition (3.4.3) of Theorem 3.4.3 is satisfied if $\tau = \{t\}$, $t \in (0,1]$ for any uniformly smooth Banach space. In the case $\rho(u) \leq \gamma u^q$, $1 < q \leq 2$, condition (3.4.3) is satisfied if

$$\sum_{m=1}^{\infty} t_m^p = \infty, \quad p := q/(q-1).$$

3.6. Thresholding algorithms

We begin with a remark on computational complexity of greedy algorithms. The main point of Section 3.4 is in proving that relaxation allows us to build greedy algorithms (see the WGAFR) that are computationally simpler than the WCGA and perform as well as the WCGA. We note that the WCGA and the WGAFR differ in the second step of the algorithm. However, the most computationally involved step of all greedy algorithms is the greedy step (the first step of the algorithm). One of the goals of relaxation was to get rid of the assumption $f \in A_1(\mathcal{D})$ (as in the WRGA). All relaxed greedy algorithms from Sections 3.4 and 3.5 are applicable to (and converge for) any $f \in X$. We want to point out that the information $f \in A_1(\mathcal{D})$ allows us to simplify substantially the greedy step of the algorithm. It is remarked in Section 3.2 (see Remark 3.2.10) that we can replace the first step of the WCGA by the following search criterion:

$$F_{f_{m-1}}(\varphi_m) \geq t_m\|f_{m-1}\|. \tag{3.6.1}$$

A similar remark (see Section 3.3, Remark 3.3.5) holds for the WRGA. The requirement (3.6.1) is weaker than the requirement of the greedy step of the WCGA. However, Theorem 3.2.6 holds for this modification of the WCGA. Relation (3.6.1) is a threshold-type inequality and can be checked more easily than the greedy inequality.

We now consider two algorithms defined and studied in Temlyakov (2006c) with a different type of thresholding. These algorithms work for any $f \in X$. We begin with the Dual Greedy Algorithm with Relaxation and Thresholding (DGART).

DGART. We define $f_0 := f$ and $G_0 := 0$. Then, for a given parameter $\delta \in (0, 1/2]$ we have the following inductive definition for $m \geq 1$.

(1) $\varphi_m \in \mathcal{D}$ is any element satisfying

$$F_{f_{m-1}}(\varphi_m) \geq \delta. \qquad (3.6.2)$$

If there is no $\varphi_m \in \mathcal{D}$ satisfying (3.6.2) then we stop.

(2) Find w_m and λ_m such that

$$\|f - ((1 - w_m)G_{m-1} + \lambda_m\varphi_m)\| = \inf_{\lambda,w} \|f - ((1 - w)G_{m-1} + \lambda\varphi_m)\|$$

and define

$$G_m := (1 - w_m)G_{m-1} + \lambda_m\varphi_m.$$

(3) Let

$$f_m := f - G_m.$$

If $\|f_m\| \leq \delta\|f\|$ then we stop, otherwise we proceed to the $(m + 1)$th iteration.

The following algorithm is a thresholding-type modification of the WCGA. This modification can be applied to any $f \in X$.

Chebyshev Greedy Algorithm with Thresholding (CGAT). For a given parameter $\delta \in (0, 1/2]$, we conduct instead of the greedy step of the WCGA the following thresholding step: find $\varphi_m \in \mathcal{D}$ such that $F_{f_{m-1}}(\varphi_m) \geq \delta$. Choosing such a φ_m, if one exists, we apply steps (2) and (3) of the WCGA. If such φ_m does not exist, then we stop. We also stop if $\|f_m\| \leq \delta\|f\|$.

Theorem 3.6.1. Let X be a uniformly smooth Banach space with modulus of smoothness $\rho(u) \leq \gamma u^q$, $1 < q \leq 2$. Take a number $\epsilon \geq 0$ and two elements f, f^ϵ from X such that

$$\|f - f^\epsilon\| \leq \epsilon, \quad f^\epsilon/A(\epsilon) \in A_1(\mathcal{D}),$$

with some number $A(\epsilon) > 0$. Then the DGART (CGAT) will stop after $m \leq C(\gamma)\delta^{-p}\ln(1/\delta)$, $p := q/(q-1)$, iterations with

$$\|f_m\| \leq \epsilon + \delta A(\epsilon).$$

Proof. We begin with the error bound. For both algorithms, the DGART and the CGAT, our stopping criterion guarantees that either $\|F_{f_m}\|_{\mathcal{D}} \leq \delta$ or $\|f_m\| \leq \delta\|f\|$. In the latter case the required bound follows from simple inequalities:

$$\|f\| \leq \epsilon + \|f^\epsilon\| \leq \epsilon + A(\epsilon).$$

Thus, assume that $\|F_{f_m}\|_{\mathcal{D}} \leq \delta$ holds. In the case of the CGAT we apply Lemma 3.2.7 with $L = \text{span}(\varphi_1, \ldots, \varphi_m)$ and obtain

$$\|f_m\| = F_{f_m}(f_m) = F_{f_m}(f) \leq \epsilon + F_{f_m}(f^\epsilon) \leq \epsilon + \|F_{f_m}\|_{\mathcal{D}}A(\epsilon) \leq \epsilon + \delta A(\epsilon).$$

For the DGART we apply Lemma 3.2.7 with f_{m-1} and $L = \text{span}(G_{m-1}, \varphi_m)$, and get

$$\|f_m\| = F_{f_m}(f_m) = F_{f_m}(f_{m-1}) = F_{f_m}(f)$$
$$\leq \epsilon + F_{f_m}(f^\epsilon) \leq \epsilon + \|F_{f_m}\|_{\mathcal{D}}A(\epsilon) \leq \epsilon + \delta A(\epsilon).$$

This proves the required bound.

We now proceed to the bound of m. We prove the bound for both algorithms simultaneously. We note that for the DGART

$$\|f_k\| = \inf_{\lambda,w} \|f_{k-1} + wG_{k-1} - \lambda\varphi_k\| \leq \inf_{\lambda \geq 0} \|f_{k-1} - \lambda\varphi_k\|.$$

We write for all $k \leq m$, $\lambda \geq 0$

$$\|f_{k-1} - \lambda\varphi_k\| + \|f_{k-1} + \lambda\varphi_k\| \leq 2\|f_{k-1}\|(1 + \rho(\lambda/\|f_{k-1}\|)). \qquad (3.6.3)$$

Next,

$$\|f_{k-1} + \lambda\varphi_k\| \geq F_{f_{k-1}}(f_{k-1} + \lambda\varphi_k) \geq \|f_{k-1}\| + \lambda\delta. \qquad (3.6.4)$$

Combining (3.6.3) with (3.6.4), we obtain

$$\|f_k\| \leq \inf_{\lambda \geq 0} \|f_{k-1} - \lambda\varphi_k\| \leq \inf_{\lambda \geq 0} \left(\|f_{k-1}\| - \lambda\delta + 2\|f_{k-1}\|\gamma(\lambda/\|f_{k-1}\|)^q\right).$$
$$(3.6.5)$$

Solving the equation $\delta x/2 = 2\gamma x^q$ we get $x_1 = (\delta/(4\gamma))^{1/(q-1)}$. Setting $\lambda := x_1\|f_{k-1}\|$ we obtain

$$\|f_k\| \leq \|f_{k-1}\|(1 - \delta x_1/2) = \|f_{k-1}\|(1 - c(\gamma)\delta^p).$$

Thus,

$$\|f_k\| \leq \|f\|(1 - c(\gamma)\delta^p)^k.$$

By the stopping condition $\|f_m\| \leq \delta\|f\|$, we deduce that $m \leq n$, where n is the smallest integer for which

$$(1 - c(\gamma)\delta^p)^n \leq \delta.$$

This implies

$$m \leq C(\gamma)\delta^{-p} \ln(1/\delta). \qquad \square$$

We proceed to one more thresholding-type algorithm (see Temlyakov (2005a)). Keeping in mind possible applications of this algorithm, we do not assume that a dictionary $\mathcal{D}$ is symmetric: $g \in \mathcal{D}$ implies $-g \in \mathcal{D}$. To indicate this we use the notation $\mathcal{D}^+$ for such a dictionary. We do not assume that elements of a dictionary $\mathcal{D}^+$ are normalized ($\|g\| = 1$ if $g \in \mathcal{D}^+$) and assume only that $\|g\| \leq 1$ if $g \in \mathcal{D}^+$. By $A_1(\mathcal{D}^+)$ we denote the closure of the convex hull of $\mathcal{D}^+$. Let $\epsilon = \{\epsilon_n\}_{n=1}^{\infty}$, $\epsilon_n > 0$, $n = 1, 2, \ldots$.

Incremental Algorithm with schedule ϵ (IA(ϵ)). Let $f \in A_1(\mathcal{D}^+)$. Denote $f_0^{i,\epsilon} := f$ and $G_0^{i,\epsilon} := 0$. Then, for each $m \geq 1$ we have the following inductive definition.

(1) $\varphi_m^{i,\epsilon} \in \mathcal{D}^+$ is any element satisfying

$$F_{f_{m-1}^{i,\epsilon}}(\varphi_m^{i,\epsilon} - f) \geq -\epsilon_m.$$

(2) Define

$$G_m^{i,\epsilon} := (1 - 1/m)G_{m-1}^{i,\epsilon} + \varphi_m^{i,\epsilon}/m.$$

(3) Let

$$f_m^{i,\epsilon} := f - G_m^{i,\epsilon}.$$

We note that, as in Lemma 3.2.8, we have for any bounded linear functional F and any $\mathcal{D}^+$

$$\sup_{g \in \mathcal{D}^+} F(g) = \sup_{f \in A_1(\mathcal{D}^+)} F(f).$$

Therefore, for any F and any $f \in A_1(\mathcal{D}^+)$,

$$\sup_{g \in \mathcal{D}^+} F(g) \geq F(f).$$

This guarantees existence of $\varphi_m^{i,\epsilon}$.

Theorem 3.6.2. Let X be a uniformly smooth Banach space with modulus of smoothness $\rho(u) \leq \gamma u^q$, $1 < q \leq 2$. Define

$$\epsilon_n := K_1 \gamma^{1/q} n^{-1/p}, \quad p = \frac{q}{q - 1}, \quad n = 1, 2, \ldots.$$

Then, for any $f \in A_1(\mathcal{D}^+)$ we have

$$\|f_m^{i,\epsilon}\| \leq C(K_1)\gamma^{1/q} m^{-1/p}, \quad m = 1, 2 \ldots.$$

Proof. We will use the abbreviated notation $f_m := f_m^{i,\epsilon}$, $\varphi_m := \varphi_m^{i,\epsilon}$, $G_m := G_m^{i,\epsilon}$. Writing

$$f_m = f_{m-1} - (\varphi_m - G_{m-1})/m,$$

we immediately obtain the trivial estimate

$$\|f_m\| \leq \|f_{m-1}\| + 2/m. \tag{3.6.6}$$

Since

$$f_m = (1 - 1/m)f_{m-1} - (\varphi_m - f)/m$$
$$= (1 - 1/m)(f_{m-1} - (\varphi_m - f)/(m - 1)), \tag{3.6.7}$$

we obtain

$$\|f_{m-1} - (\varphi_m - f)/(m - 1)\| \tag{3.6.8}$$
$$\leq \|f_{m-1}\|(1 + 2\rho(2((m - 1)\|f_{m-1}\|)^{-1})) + \epsilon_m(m - 1)^{-1},$$

in a similar way to (3.6.5). Using the definition of ϵ_m and the assumption $\rho(u) \leq \gamma u^q$, we make the following observation. There exists a constant $C(K_1)$ such that, if

$$\|f_{m-1}\| \geq C(K_1)\gamma^{1/q}(m - 1)^{-1/p}, \tag{3.6.9}$$

then

$$2\rho(2((m - 1)\|f_{m-1}\|)^{-1}) + \epsilon_m((m - 1)\|f_{m-1}\|)^{-1} \leq 1/(4m), \tag{3.6.10}$$

and therefore, by (3.6.7) and (3.6.8),

$$\|f_m\| \leq (1 - 3/(4m))\|f_{m-1}\|. \tag{3.6.11}$$

Taking into account (3.6.6), we apply Lemma 3.5.3 to the sequence $a_n = \|f_n\|$, $n = 1, 2, \ldots$ with $\alpha = 1/p$, $\beta = 3/4$, and complete the proof of Theorem 3.6.2. $\qquad\square$

3.7. Greedy expansions

3.7.1. Introduction

From the definition of a dictionary it follows that any element $f \in X$ can be approximated arbitrarily well by finite linear combinations of the dictionary elements. The primary goal of this section is to study representations of an element $f \in X$ by a series

$$f \sim \sum_{j=1}^{\infty} c_j(f)g_j(f), \quad g_j(f) \in \mathcal{D}, \quad c_j(f) > 0, \quad j = 1, 2, \ldots. \tag{3.7.1}$$

In building the representation (3.7.1) we should construct two sequences: $\{g_j(f)\}_{j=1}^{\infty}$ and $\{c_j(f)\}_{j=1}^{\infty}$. In this section the construction of $\{g_j(f)\}_{j=1}^{\infty}$

will be based on ideas used in greedy-type nonlinear approximation (greedy-type algorithms). This justifies the use of the term *greedy expansion* for (3.7.1) considered in the section. The construction of $\{g_j(f)\}_{j=1}^{\infty}$ is, clearly, the most important and difficult part in building the representation (3.7.1). On the basis of the contemporary theory of nonlinear approximation with respect to redundant dictionaries, we may conclude that the method of using a norming functional in greedy steps of an algorithm is the most productive in approximation in Banach spaces. This method was utilized in the Weak Chebyshev Greedy Algorithm and in the Weak Dual Greedy Algorithm. We use this same method in new algorithms considered in this section. A new qualitative result of this section establishes that we have a lot of flexibility in constructing a sequence of coefficients $\{c_j(f)\}_{j=1}^{\infty}$.

Denote

$$r_{\mathcal{D}}(f) := \sup_{F_f} \|F_f\|_{\mathcal{D}} := \sup_{F_f} \sup_{g \in \mathcal{D}} F_f(g).$$

We note that, in general, a norming functional F_f is not unique. This is why we take $\sup_{F_f}$ over all norming functionals of f in the definition of $r_{\mathcal{D}}(f)$. It is known that in the case of uniformly smooth Banach spaces (our primary object here) the norming functional F_f is unique. In such a case we do not need $\sup_{F_f}$ in the definition of $r_{\mathcal{D}}(f)$: we have $r_{\mathcal{D}}(f) = \|F_f\|_{\mathcal{D}}$.

We begin with a description of a general scheme that provides an expansion for a given element f. Later, specifying this general scheme, we will obtain different methods of expansion.

Dual-Based Expansion (DBE). Let $t \in (0, 1]$ and $f \neq 0$. Denote $f_0 := f$. Assume $\{f_j\}_{j=0}^{m-1} \subset X$, $\{\varphi_j\}_{j=1}^{m-1} \subset \mathcal{D}$ and a set of coefficients $\{c_j\}_{j=1}^{m-1}$ of expansion have already been constructed. If $f_{m-1} = 0$ then we stop (set $c_j = 0$, $j = m, m+1, \ldots$ in the expansion) and get $f = \sum_{j=1}^{m-1} c_j \varphi_j$. If $f_{m-1} \neq 0$ then we conduct the following two steps.

(1) Choose $\varphi_m \in \mathcal{D}$ such that

$$\sup_{F_{f_{m-1}}} F_{f_{m-1}}(\varphi_m) \geq t r_{\mathcal{D}}(f_{m-1}).$$

(2) Define

$$f_m := f_{m-1} - c_m \varphi_m,$$

where $c_m > 0$ is a coefficient either prescribed in advance or chosen from a concrete approximation procedure.

We call the series

$$f \sim \sum_{j=1}^{\infty} c_j \varphi_j \tag{3.7.2}$$

the Dual-Based Expansion of f with coefficients $c_j(f) := c_j$, $j = 1, 2, \ldots$ with respect to $\mathcal{D}$.

Denote

$$S_m(f, \mathcal{D}) := \sum_{j=1}^{m} c_j \varphi_j.$$

Then it is clear that

$$f_m = f - S_m(f, \mathcal{D}).$$

We prove some convergence results for the DBE in Sections 3.7.2 and 3.7.3. In Section 3.7.3 we consider a variant of the Dual-Based Expansion with coefficients chosen by a certain simple rule. The rule depends on two numerical parameters, $t \in (0, 1]$ (the weakness parameter from the definition of the DBE) and $b \in (0, 1)$ (the tuning parameter of the approximation method). The rule also depends on a majorant μ of the modulus of smoothness of the Banach space X.

Dual Greedy Algorithm with parameters (t, b, μ) (DGA(t, b, μ)). Let X be a uniformly smooth Banach space with modulus of smoothness $\rho(u)$, and let $\mu(u)$ be a continuous majorant of $\rho(u)$: $\rho(u) \leq \mu(u)$, $u \in [0, \infty)$. For parameters $t \in (0, 1]$, $b \in (0, 1]$ we define sequences $\{f_m\}_{m=0}^{\infty}$, $\{\varphi_m\}_{m=1}^{\infty}$, $\{c_m\}_{m=1}^{\infty}$ inductively. Let $f_0 := f$. If for $m \geq 1$ $f_{m-1} = 0$ then we set $f_j = 0$ for $j \geq m$ and stop. If $f_{m-1} \neq 0$ then we conduct the following three steps.

(1) Take any $\varphi_m \in \mathcal{D}$ such that

$$F_{f_{m-1}}(\varphi_m) \geq t r_{\mathcal{D}}(f_{m-1}). \tag{3.7.3}$$

(2) Choose $c_m > 0$ from the equation

$$\|f_{m-1}\| \mu(c_m/\|f_{m-1}\|) = \frac{tb}{2} c_m r_{\mathcal{D}}(f_{m-1}). \tag{3.7.4}$$

(3) Define

$$f_m := f_{m-1} - c_m \varphi_m. \tag{3.7.5}$$

In Section 3.7.3 we prove the following convergence result.

Theorem 3.7.1. Let X be a uniformly smooth Banach space with the modulus of smoothness $\rho(u)$ and let $\mu(u)$ be a continuous majorant of $\rho(u)$ with the property $\mu(u)/u \downarrow 0$ as $u \to +0$. Then, for any $t \in (0, 1]$ and $b \in (0, 1)$, the DGA(t, b, μ) converges for each dictionary $\mathcal{D}$ and all $f \in X$.

The following result from Section 3.7.3 gives the rate of convergence.

Theorem 3.7.2. Assume X has a modulus of smoothness $\rho(u) \leq \gamma u^q$, $q \in (1, 2]$. Denote $\mu(u) = \gamma u^q$. Then, for any dictionary $\mathcal{D}$ and any $f \in A_1(\mathcal{D})$, the rate of convergence of the $\mathrm{DGA}(t, b, \mu)$ is given by

$$\|f_m\| \leq C(t, b, \gamma, q) m^{-\frac{t(1-b)}{p(1+t(1-b))}}, \quad p := \frac{q}{q-1}.$$

3.7.2. Convergence of the Dual-Based Expansion

We begin with the following lemma.

Lemma 3.7.3. Let $f \in X$. Assume that the coefficients $\{c_j\}_{j=1}^{\infty}$ of the expansion

$$f \sim \sum_{j=1}^{\infty} c_j \varphi_j, \qquad f_m := f - \sum_{j=1}^{m} c_j \varphi_j$$

are non-negative and satisfy the following two conditions:

$$\sum_{j=1}^{\infty} c_j r_{\mathcal{D}}(f_{j-1}) < \infty, \tag{3.7.6}$$

$$\sum_{j=1}^{\infty} c_j = \infty. \tag{3.7.7}$$

Then

$$\liminf_{m \to \infty} \|f_m\| = 0. \tag{3.7.8}$$

Proof. The proof of this lemma is similar to the proof of Lemma 1 from Ganichev and Kalton (2003). Denote $s_n := \sum_{j=1}^{n} c_j$. Then (3.7.7) implies (see Bary (1961, p. 904)) that

$$\sum_{n=1}^{\infty} c_n/s_n = \infty. \tag{3.7.9}$$

Using (3.7.6), we get

$$\sum_{n=1}^{\infty} s_n r_{\mathcal{D}}(f_{n-1}) c_n / s_n = \sum_{n=1}^{\infty} c_n r_{\mathcal{D}}(f_{n-1}) < \infty.$$

Thus, by (3.7.9),

$$\liminf_{n \to \infty} s_n r_{\mathcal{D}}(f_{n-1}) = 0,$$

and also $(s_{n-1} \leq s_n)$

$$\liminf_{n \to \infty} s_n r_{\mathcal{D}}(f_n) = 0.$$

Let

$$\lim_{k\to\infty} s_{n_k} r_{\mathcal{D}}(f_{n_k}) = 0. \tag{3.7.10}$$

Consider $\{F_{f_{n_k}}\}$. The unit sphere in the dual X^* is weakly* compact (see Habala, Hájek and Zizler (1996, p. 45)). Let $\{F_i\}_{i=1}^{\infty}$, $F_i := F_{f_{n_{k_i}}}$ be a w^*-convergent subsequence. Denote

$$F := w^* - \lim_{i\to\infty} F_i.$$

We will complete the proof of Lemma 3.7.3 by contradiction. We assume that (3.7.8) does not hold, that is, there exist $\alpha > 0$ and $N \in \mathbb{N}$ such that

$$\|f_m\| \geq \alpha, \quad m \geq N, \tag{3.7.11}$$

and will thence derive a contradiction.

We begin by deducing from (3.7.11) that $F \neq 0$. Indeed, we have

$$F(f) = \lim_{i\to\infty} F_i(f), \tag{3.7.12}$$

and

$$F_i(f) = F_i\left(f_{n_{k_i}} + \sum_{j=1}^{n_{k_i}} c_j\varphi_j\right) = \|f_{n_{k_i}}\| + \sum_{j=1}^{n_{k_i}} c_j F_i(\varphi_j) \geq \alpha - s_{n_{k_i}} r_{\mathcal{D}}(f_{n_{k_i}}), \tag{3.7.13}$$

for big i. Relations (3.7.12), (3.7.13) and (3.7.10) imply that $F(f) \geq \alpha$, and hence $F \neq 0$. This implies that there exist $g \in \mathcal{D}$ for which $F(g) > 0$. However,

$$F(g) = \lim_{i\to\infty} F_i(g) \leq \lim_{i\to\infty} r_{\mathcal{D}}(f_{n_{k_i}}) = 0.$$

We have a contradiction, which completes the proof of Lemma 3.7.3. □

In the paper Temlyakov (2007b) we pushed to the extreme the flexibility of choice of the coefficients $c_j(f)$ in (3.7.1). We made these coefficients independent of an element $f \in X$. Surprisingly, for properly chosen coefficients we obtained results for the corresponding dual greedy expansion similar to the above Theorems 3.7.1 and 3.7.2. Even more surprisingly, we obtained similar results for the corresponding X-greedy expansions. We proceed to the formulation of these results. Let $\mathcal{C} := \{c_m\}_{m=1}^{\infty}$ be a fixed sequence of positive numbers. We restrict ourselves to positive numbers because of the symmetry of the dictionary $\mathcal{D}$.

X-Greedy Algorithm with coefficients $\mathcal{C}$ (XGA($\mathcal{C}$)). We define $f_0 := f$, $G_0 := 0$. Then, for each $m \geq 1$ we have the following inductive definition.

(1) $\varphi_m \in \mathcal{D}$ is such that (assuming existence)

$$\|f_{m-1} - c_m\varphi_m\|_X = \inf_{g\in\mathcal{D}} \|f_{m-1} - c_m g\|_X.$$

(2) Let

$$f_m := f_{m-1} - c_m \varphi_m, \qquad G_m := G_{m-1} + c_m \varphi_m.$$

Dual Greedy Algorithm, weakness τ, coefficients $\mathcal{C}$ (DGA($\tau, \mathcal{C}$)).
Let $\tau := \{t_m\}_{m=1}^{\infty}$, $t_m \in [0, 1]$, be a weakness sequence. We define $f_0 := f$,
$G_0 := 0$. Then, for each $m \geq 1$ we have the following inductive definition.

(1) $\varphi_m \in \mathcal{D}$ is any element satisfying

$$F_{f_{m-1}}(\varphi_m) \geq t_m \|F_{f_{m-1}}\|_{\mathcal{D}}.$$

(2) Define

$$f_m := f_{m-1} - c_m \varphi_m, \qquad G_m := G_{m-1} + c_m \varphi_m.$$

In the case $\tau = \{t\}$, $t \in (0, 1]$, we write t instead of τ in the notation.
The first result on convergence properties of the DGA($t, \mathcal{C}$) was obtained in
Temlyakov (2007a). We prove it here.

Theorem 3.7.4. Let X be a uniformly smooth Banach space with the
modulus of smoothness $\rho(u)$. Assume $\mathcal{C} = \{c_j\}_{j=1}^{\infty}$ is such that $c_j \geq 0$,
$j = 1, 2, \ldots$,

$$\sum_{j=1}^{\infty} c_j = \infty,$$

and for any $y > 0$,

$$\sum_{j=1}^{\infty} \rho(y c_j) < \infty. \qquad (3.7.14)$$

Then, for the DGA($t, \mathcal{C}$) we have

$$\liminf_{m \to \infty} \|f_m\| = 0. \qquad (3.7.15)$$

Proof. The proof is by contradiction. Assume (3.7.15) does not hold. Then
$\exists \alpha > 0$ and $\exists N \in \mathbb{N}$ such that, for all $m \geq N$,

$$\|f_m\| \geq \alpha > 0.$$

From the definition of the modulus of smoothness we have

$$\|f_{n-1} - c_n \varphi_n\| + \|f_{n-1} + c_n \varphi_n\| \leq 2\|f_{n-1}\|(1 + \rho(c_n / \|f_{n-1}\|)). \qquad (3.7.16)$$

Using the definition of φ_n,

$$F_{f_{n-1}}(\varphi_n) \geq t r_{\mathcal{D}}(f_{n-1}), \qquad (3.7.17)$$

we get

$$\|f_{n-1} + c_n \varphi_n\| \geq F_{f_{n-1}}(f_{n-1} + c_n \varphi_n) \qquad (3.7.18)$$

$$= \|f_{n-1}\| + c_n F_{f_{n-1}}(\varphi_n) \geq \|f_{n-1}\| + c_n t r_{\mathcal{D}}(f_{n-1}).$$

Combining (3.7.16) and (3.7.18), we get

$$\|f_n\| = \|f_{n-1} - c_n \varphi_n\| \le \|f_{n-1}\|(1 + 2\rho(c_n/\|f_{n-1}\|)) - c_n tr_\mathcal{D}(f_{n-1}). \quad (3.7.19)$$

We note that by Remark 3.2.3

$$\|f_{n-1}\|\rho(c_n/\|f_{n-1}\|) \le \alpha\rho(c_n/\alpha), \quad n > N.$$

Therefore, by the assumption (3.7.14)

$$\sum_{n=1}^{\infty} \|f_{n-1}\|\rho(c_n/\|f_{n-1}\|) < \infty. \quad (3.7.20)$$

This and (3.7.19) imply

$$\sum_{n=1}^{\infty} c_n r_\mathcal{D}(f_{n-1}) \le t^{-1}\left(\|f\| + 2\sum_{n=1}^{\infty} \|f_{n-1}\|\rho(c_n/\|f_{n-1}\|)\right) < \infty.$$

It remains to apply Lemma 3.7.3 to complete the proof.   □

In Temlyakov (2007b) we proved an analogue of Theorem 3.7.4 for the XGA($\mathcal{C}$) and improved upon the convergence in Theorem 3.7.4 in the case of uniformly smooth Banach spaces with power-type modulus of smoothness. Under an extra assumption on $\mathcal{C}$ we replaced $\liminf$ by $\lim$. Here is the corresponding result from Temlyakov (2007b).

Theorem 3.7.5. Let $\mathcal{C} \in \ell_q \setminus \ell_1$ be a monotone sequence. Then the DGA($t, \mathcal{C}$) and the XGA($\mathcal{C}$) converge for each dictionary and all $f \in X$ in any uniformly smooth Banach space X with modulus of smoothness $\rho(u) \le \gamma u^q$, $q \in (1, 2]$.

In Temlyakov (2007b) we also addressed a question of rate of approximation for $f \in A_1(\mathcal{D})$. We proved the following theorem.

Theorem 3.7.6. Let X be a uniformly smooth Banach space with modulus of smoothness $\rho(u) \le \gamma u^q$, $q \in (1, 2]$. We set $s := (1 + 1/q)/2$ and $\mathcal{C}_s := \{k^{-s}\}_{k=1}^{\infty}$. Then the DGA($t, \mathcal{C}_s$) and XGA($\mathcal{C}_s$) (for this algorithm $t = 1$) converge for $f \in A_1(\mathcal{D})$ with the following rate: for any $r \in (0, t(1 - s))$,

$$\|f_m\| \le C(r, t, q, \gamma)m^{-r}.$$

In the case $t = 1$, Theorem 3.7.6 provides the rate of convergence m^{-r} for $f \in A_1(\mathcal{D})$ with r arbitrarily close to $(1 - 1/q)/2$. Theorem 3.7.2 provides a similar rate of convergence. It would be interesting to know if the rate $m^{-(1-1/q)/2}$ is the best that can be achieved in greedy expansions (for each $\mathcal{D}$, any $f \in A_1(\mathcal{D})$, and any X with $\rho(u) \le \gamma u^q$, $q \in (1, 2]$). We note that there are greedy approximation methods that provide an error bound of the order $m^{1/q-1}$ for $f \in A_1(\mathcal{D})$ (see Temlyakov (2003a, 2006c) for recent results). However, these approximation methods do not provide an expansion.

3.7.3. A modification of the Weak Dual Greedy Algorithm

We begin this subsection with a proof of Theorem 3.7.1. Here we give a definition of the DGA(τ, b, μ), $\tau = \{t_k\}_{k=1}^{\infty}$, $t_k \in (0, 1]$ that coincides with the definition of the DGA(t, b, μ) from Section 3.7.1 in the case $\tau = \{t\}$.

Dual Greedy Algorithm with parameters (τ, b, μ) (DGA(τ, b, μ)).
Let X be a uniformly smooth Banach space with modulus of smoothness $\rho(u)$, and let $\mu(u)$ be a continuous majorant of $\rho(u)$: $\rho(u) \leq \mu(u)$, $u \in [0, \infty)$. For a sequence $\tau = \{t_k\}_{k=1}^{\infty}$, $t_k \in (0, 1]$ and a parameter $b \in (0, 1]$, we define sequences $\{f_m\}_{m=0}^{\infty}$, $\{\varphi_m\}_{m=1}^{\infty}$, $\{c_m\}_{m=1}^{\infty}$ inductively. Let $f_0 := f$. If $f_{m-1} = 0$ for some $m \geq 1$, then we set $f_j = 0$ for $j \geq m$ and stop. If $f_{m-1} \neq 0$ then we conduct the following three steps.

(1) Take any $\varphi_m \in \mathcal{D}$ such that

$$F_{f_{m-1}}(\varphi_m) \geq t_m r_{\mathcal{D}}(f_{m-1}). \qquad (3.7.21)$$

(2) Choose $c_m > 0$ from the equation

$$\|f_{m-1}\| \mu(c_m / \|f_{m-1}\|) = \frac{t_m b}{2} c_m r_{\mathcal{D}}(f_{m-1}). \qquad (3.7.22)$$

(3) Define

$$f_m := f_{m-1} - c_m \varphi_m. \qquad (3.7.23)$$

Proof of Theorem 3.7.1. In this case $\tau = \{t\}$, $t \in (0, 1]$. We have by (3.7.19)

$$\|f_m\| = \|f_{m-1} - c_m \varphi_m\| \leq \|f_{m-1}\|(1 + 2\rho(c_m / \|f_{m-1}\|)) - c_m t r_{\mathcal{D}}(f_{m-1}). \qquad (3.7.24)$$

Using the choice of c_m, we find

$$\|f_m\| \leq \|f_{m-1}\| - t(1 - b)c_m r_{\mathcal{D}}(f_{m-1}). \qquad (3.7.25)$$

In particular, (3.7.25) implies that $\{\|f_m\|\}$ is a monotone decreasing sequence and

$$t(1 - b)c_m r_{\mathcal{D}}(f_{m-1}) \leq \|f_{m-1}\| - \|f_m\|.$$

Thus

$$\sum_{m=1}^{\infty} c_m r_{\mathcal{D}}(f_{m-1}) < \infty. \qquad (3.7.26)$$

We have the following two cases:

$$\text{(I)} \quad \sum_{m=1}^{\infty} c_m = \infty, \qquad \text{(II)} \quad \sum_{m=1}^{\infty} c_m < \infty.$$

In case (I), by Lemma 3.7.3 we obtain

$$\liminf_{m \to \infty} \|f_m\| = 0 \;\; \Rightarrow \;\; \lim_{m \to \infty} \|f_m\| = 0.$$

It remains to consider case (II). We prove convergence in this case by contradiction. Assume

$$\lim_{m \to \infty} \|f_m\| = \alpha > 0. \tag{3.7.27}$$

By (II) we have $f_m \to f_\infty \neq 0$ as $m \to \infty$. We note that by uniform smoothness of X we get

$$\lim_{m \to \infty} \|F_{f_m} - F_{f_\infty}\| = 0.$$

We have $F_{f_\infty} \neq 0$, and therefore there is a $g \in \mathcal{D}$ such that $F_{f_\infty}(g) > 0$. However,

$$F_{f_\infty}(g) = \lim_{m \to \infty} F_{f_m}(g) \leq \lim_{m \to \infty} r_\mathcal{D}(f_m) = 0. \tag{3.7.28}$$

Indeed, by (3.7.22) and (3.7.27) we get

$$r_\mathcal{D}(f_{m-1}) \leq \alpha c_m^{-1} \mu(c_m/\alpha) \frac{2}{tb} \to 0,$$

as $m \to \infty$.

Theorem 3.7.1 is proved. $\qquad\qquad\qquad\qquad\qquad\qquad\qquad\qquad\qquad\qquad\square$

Remark 3.7.7. It is clear from the above proof that Theorem 3.7.1 holds for an algorithm obtained from the $\mathrm{DGA}(\tau, b, \mu)$, by replacing (3.7.22) by

$$\|f_{m-1}\| \mu(c_m/\|f_{m-1}\|) = \frac{b}{2} c_m F_{f_{m-1}}(\varphi_m). \tag{3.7.29}$$

Also, a parameter b in (3.7.22) and (3.7.29) can be replaced by varying parameters $b_m \in (a, b) \subset (0, 1)$.

We proceed to study the rate of convergence of the $\mathrm{DGA}(\tau, b, \mu)$ in the uniformly smooth Banach spaces with the power-type majorant of modulus of smoothness: $\rho(u) \leq \mu(u) = \gamma u^q$, $1 < q \leq 2$. We now prove a statement more general than Theorem 3.7.2.

Theorem 3.7.8. Let $\tau := \{t_k\}_{k=1}^\infty$ be a non-increasing sequence $1 \geq t_1 \geq t_2 \cdots > 0$ and $b \in (0, 1)$. Assume X has a modulus of smoothness $\rho(u) \leq \gamma u^q$, $q \in (1, 2]$. Denote $\mu(u) = \gamma u^q$. Then, for any dictionary $\mathcal{D}$ and any $f \in A_1(\mathcal{D})$, the rate of convergence of the $\mathrm{DGA}(\tau, b, \mu)$ is given by

$$\|f_m\| \leq C(b, \gamma, q) \left(1 + \sum_{k=1}^m t_k^p \right)^{-\frac{t_m(1-b)}{p(1 + t_m(1-b))}}, \qquad p := \frac{q}{q-1}.$$

Proof. As in (3.7.25), we get

$$\|f_m\| \leq \|f_{m-1}\| - t_m(1-b) c_m r_\mathcal{D}(f_{m-1}). \tag{3.7.30}$$

Thus we need to estimate $c_m r_\mathcal{D}(f_{m-1})$ from below. It is clear that

$$\|f_{m-1}\|_{A_1(\mathcal{D})} = \|f - \sum_{j=1}^{m-1} c_j \varphi_j\|_{A_1(\mathcal{D})} \leq \|f\|_{A_1(\mathcal{D})} + \sum_{j=1}^{m-1} c_j. \qquad (3.7.31)$$

Denote $b_n := 1 + \sum_{j=1}^{n} c_j$. Then, by (3.7.31) we get

$$\|f_{m-1}\|_{A_1(\mathcal{D})} \leq b_{m-1}.$$

Next, by Lemma 3.2.8 we get

$$r_\mathcal{D}(f_{m-1}) = \sup_{g \in \mathcal{D}} F_{f_{m-1}}(g) = \sup_{\varphi \in A_1(\mathcal{D})} F_{f_{m-1}}(\varphi)$$

$$\geq \|f_{m-1}\|_{A_1(\mathcal{D})}^{-1} F_{f_{m-1}}(f_{m-1}) \geq \|f_{m-1}\|/b_{m-1}. \qquad (3.7.32)$$

Substituting (3.7.32) into (3.7.30), we get

$$\|f_m\| \leq \|f_{m-1}\|(1 - t_m(1-b)c_m/b_{m-1}). \qquad (3.7.33)$$

From the definition of b_m we find

$$b_m = b_{m-1} + c_m = b_{m-1}(1 + c_m/b_{m-1}).$$

Using the inequality

$$(1+x)^\alpha \leq 1 + \alpha x, \quad 0 \leq \alpha \leq 1, \quad x \geq 0,$$

we obtain

$$b_m^{t_m(1-b)} \leq b_{m-1}^{t_m(1-b)}(1 + t_m(1-b)c_m/b_{m-1}). \qquad (3.7.34)$$

Multiplying (3.7.33) and (3.7.34), and using that $t_m \leq t_{m-1}$, we get

$$\|f_m\| b_m^{t_m(1-b)} \leq \|f_{m-1}\| b_{m-1}^{t_{m-1}(1-b)} \leq \|f\| \leq 1. \qquad (3.7.35)$$

The function $\mu(u)/u = \gamma u^{q-1}$ is increasing on $[0, \infty)$. Therefore the c_m from (3.7.22) is greater than or equal to c_m' from (see (3.7.32))

$$\gamma \|f_{m-1}\| (c_m'/\|f_{m-1}\|)^q = \frac{t_m b}{2} c_m' \|f_{m-1}\|/b_{m-1}, \qquad (3.7.36)$$

$$c_m' = \left(\frac{t_m b}{2\gamma}\right)^{\frac{1}{q-1}} \frac{\|f_{m-1}\|^{\frac{q}{q-1}}}{b_{m-1}^{\frac{1}{q-1}}}. \qquad (3.7.37)$$

Setting

$$p := \frac{q}{q-1}, \qquad A^{-1} := (1-b)\left(\frac{b}{2\gamma}\right)^{\frac{1}{q-1}} \leq 1/2,$$

we obtain

$$\|f_m\| \leq \|f_{m-1}\|\left(1 - \frac{t_m^p}{A} \frac{\|f_{m-1}\|^p}{b_{m-1}^p}\right) \qquad (3.7.38)$$

from (3.7.30), (3.7.32) and (3.7.37). Noting that $b_m \geq b_{m-1}$, we infer from (3.7.38) that

$$(\|f_m\|/b_m)^p \leq (\|f_{m-1}\|/b_{m-1})^p(1 - A^{-1}t_m^p(\|f_{m-1}\|/b_{m-1})^p). \quad (3.7.39)$$

Taking into account that $\|f\| \leq 1 < A$, we obtain from (3.7.39) by an analogue of Lemma 2.3.3 (see Temlyakov (2000b, Lemma 3.1))

$$(\|f_m\|/b_m)^p \leq A\left(1 + \sum_{k=1}^{m} t_k^p\right)^{-1}. \quad (3.7.40)$$

Combining (3.7.35) and (3.7.40), we get

$$\|f_m\| \leq C(b,\gamma,q)\left(1 + \sum_{k=1}^{m} t_k^p\right)^{-\frac{t_m(1-b)}{p(1+t_m(1-b))}}, \quad p := \frac{q}{q-1}.$$

This completes the proof of Theorem 3.7.8. $\square$

In the case $\tau = \{t\}$, $t \in (0,1]$, we get Theorem 3.7.2 from Theorem 3.7.8.

Remark 3.7.9. Theorem 3.7.8 holds for an algorithm obtained from the DGA(τ, b, μ) by replacing (3.7.22) by (3.7.29).

It follows from the proof of Theorem 3.7.8 that it holds for a modification of the DGA(τ, b, μ) when we replace the quantity $r_{\mathcal{D}}(f_{m-1})$ in the definition by its lower estimate (see (3.7.32)) $\|f_{m-1}\|/b_{m-1}$, with $b_{m-1} := 1 + \sum_{j=1}^{m-1} c_j$. Clearly, this modification is more suitable for practical implementation than the DGA(τ, b, μ). We formulate the above remark as a separate result.

Modified Dual Greedy Algorithm (τ, b, μ) (MDGA(τ, b, μ)). Let X be a uniformly smooth Banach space with modulus of smoothness $\rho(u)$ and let $\mu(u)$ be a continuous majorant of $\rho(u)$: $\rho(u) \leq \mu(u)$, $u \in [0,\infty)$. For a sequence $\tau = \{t_k\}_{k=1}^{\infty}$, $t_k \in (0,1]$ and a parameter $b \in (0,1)$, we define for $f \in A_1(\mathcal{D})$ sequences $\{f_m\}_{m=0}^{\infty}$, $\{\varphi_m\}_{m=1}^{\infty}$, $\{c_m\}_{m=1}^{\infty}$ inductively. Let $f_0 := f$. If for $m \geq 1$ $f_{m-1} = 0$, then we set $f_j = 0$ for $j \geq m$ and stop. If $f_{m-1} \neq 0$ then we conduct the following three steps.

(1) Take any $\varphi_m \in \mathcal{D}$ such that

$$F_{f_{m-1}}(\varphi_m) \geq t_m\|f_{m-1}\|\left(1 + \sum_{j=1}^{m-1} c_j\right)^{-1}.$$

(2) Choose $c_m > 0$ from the equation

$$\mu(c_m/\|f_{m-1}\|) = \frac{t_m b}{2}c_m\left(1 + \sum_{j=1}^{m-1} c_j\right)^{-1}.$$

(3) Define

$$f_m := f_{m-1} - c_m \varphi_m.$$

Theorem 3.7.10. Let $\tau := \{t_k\}_{k=1}^{\infty}$ be a non-increasing sequence $1 \geq t_1 \geq t_2 \cdots > 0$ and $b \in (0,1)$. Assume X has a modulus of smoothness $\rho(u) \leq \gamma u^q$, $q \in (1,2]$. Denote $\mu(u) = \gamma u^q$. Then, for any dictionary $\mathcal{D}$ and any $f \in A_1(\mathcal{D})$, the rate of convergence of the MDGA(τ, b, μ) is given by

$$\|f_m\| \leq C(b, \gamma, q) \left(1 + \sum_{k=1}^{m} t_k^p \right)^{-\frac{t_m(1-b)}{p(1+t_m(1-b))}}, \qquad p := \frac{q}{q-1}.$$

Let us discuss an application of Theorem 3.7.2 in the case of a Hilbert space. It is well known and easy to check that, for a Hilbert space H,

$$\rho(u) \leq (1 + u^2)^{1/2} - 1 \leq u^2/2.$$

Therefore, by Theorem 3.7.2 with $\mu(u) = u^2/2$, the DGA(t, b, μ) provides the following error estimate:

$$\|f_m\| \leq C(t,b) m^{-\frac{t(1-b)}{2(1+t(1-b))}} \quad \text{for } f \in A_1(\mathcal{D}). \qquad (3.7.41)$$

The estimate (3.7.41) with $t = 1$ gives

$$\|f_m\| \leq C(b) m^{-\frac{1-b}{2(2-b)}} \quad \text{for } f \in A_1(\mathcal{D}). \qquad (3.7.42)$$

The exponent $(1 - b)/(2(2 - b))$ in this estimate tends to $1/4$ when b tends to 0. Comparing (3.7.42) with the upper estimate for the PGA (see Section 2.3), we observe that the DGA$(1, b, u^2/2)$ with small b has a better upper estimate for the rate of convergence than the known estimates for the PGA. We note also that inequality (2.3.21) indicates that the exponent in the power rate of decay of error for the PGA is less than 0.1898.

Let us figure out how the DGA$(1, b, u^2/2)$ works in Hilbert space. Consider its mth step. Let $\varphi_m \in \mathcal{D}$ be from (3.7.3). Then it is clear that φ_m maximizes $\langle f_{m-1}, g \rangle$ over the dictionary $\mathcal{D}$ and

$$\langle f_{m-1}, \varphi_m \rangle = \|f_{m-1}\| r_{\mathcal{D}}(f_{m-1}).$$

The PGA would use φ_m with the coefficient $\langle f_{m-1}, \varphi_m \rangle$ at this step. The DGA$(1, b, u^2/2)$ uses the same φ_m and only a fraction of $\langle f_{m-1}, \varphi_m \rangle$:

$$c_m = b \|f_{m-1}\| r_{\mathcal{D}}(f_{m-1}). \qquad (3.7.43)$$

Thus the choice $b = 1$ in (3.7.43) corresponds to the PGA. However, it is clear from the above considerations that our technique, designed for general Banach spaces, does not work in the case $b = 1$. The above discussion brings us the following surprising observation. The use of a small fraction $(c_m = b \langle f_{m-1}, g \rangle)$ of an optimal coefficient results in an improvement of the upper estimate for the rate of convergence.

3.7.4. Convergence of the WDGA

We now study convergence of the Weak Dual Greedy Algorithm (WDGA) defined in the Introduction of this chapter. We present in this subsection results from Ganichev and Kalton (2003). We will prove the convergence result under an extra assumption on a Banach space X.

Definition 3.7.11. (Property Γ)   A uniformly smooth Banach space has property Γ if there is a constant $\beta > 0$ such that, for any $x, y \in X$ satisfying $F_x(y) = 0$, we have
$$\|x + y\| \geq \|x\| + \beta F_{x+y}(y).$$

Property Γ in the above form was introduced in Ganichev and Kalton (2003). This condition (formulated somewhat differently) was considered previously in the context of greedy approximation in Livshitz (2003).

Theorem 3.7.12.   Let X be a uniformly smooth Banach space with property Γ. Then the WDGA(τ) with $\tau = \{t\}$, $t \in (0, 1]$, converges for each dictionary and all $f \in X$.

Proof.   Let $\{f_m\}_{m=0}^\infty$ be a sequence generated by the WDGA(t). Then
$$f_{m-1} = f_m + a_m \varphi_m, \quad F_{f_m}(\varphi_m) = 0. \tag{3.7.44}$$
We use property Γ with $x := f_m$ and $y := a_m \varphi_m$ and obtain
$$\|f_{m-1}\| \geq \|f_m\| + \beta a_m F_{f_{m-1}}(\varphi_m). \tag{3.7.45}$$
This inequality, and monotonicity of the sequence $\{\|f_m\|\}$, imply that
$$\sum_{m=1}^\infty a_m F_{f_{m-1}}(\varphi_m) < \infty \Rightarrow \sum_{m=1}^\infty a_m r_D(f_{m-1}) < \infty. \tag{3.7.46}$$
As in the proof of Theorem 3.7.1, we consider separately two cases:
$$\text{(I)} \quad \sum_{m=1}^\infty a_m = \infty, \quad \text{(II)} \quad \sum_{m=1}^\infty a_m < \infty.$$
In case (I), by (3.7.46) and Lemma 3.7.3 we obtain
$$\liminf_{m\to\infty} \|f_m\| = 0 \Rightarrow \lim_{m\to\infty} \|f_m\| = 0.$$
In case (II) we argue by contradiction. Assume
$$\lim_{m\to\infty} \|f_m\| = \alpha > 0.$$
Then, by (II) we have $f_m \to f_\infty \neq 0$ as $m \to \infty$. By uniform smoothness of X we get
$$\lim_{m\to\infty} \|F_{f_m} - F_{f_\infty}\| = 0, \quad \lim_{m\to\infty} \|F_{f_m} - F_{f_{m-1}}\| = 0. \tag{3.7.47}$$

In particular, (3.7.44) and (3.7.47) imply that

$$\lim_{m\to\infty} F_{f_{m-1}}(\varphi_m) = 0 \;\Rightarrow\; \lim_{m\to\infty} r_{\mathcal{D}}(f_m) = 0. \qquad (3.7.48)$$

We have $F_{f_\infty} \neq 0$, and therefore there is a $g \in \mathcal{D}$ such that $F_{f_\infty}(g) > 0$. However, by (3.7.47) and (3.7.48),

$$F_{f_\infty}(g) = \lim_{m\to\infty} F_{f_m}(g) \le \lim_{m\to\infty} r_{\mathcal{D}}(f_m) = 0.$$

The obtained contradiction completes the proof.

We now give a direct proof in case (I) that does not use Lemma 3.7.3. By property Γ we get

$$\|f_m\| \le \|f_{m-1}\| - \beta a_m F_{f_{m-1}}(\varphi_m) \le \|f_{m-1}\| - t\beta a_m \|F_{f_{m-1}}\|_{\mathcal{D}}. \quad (3.7.49)$$

Let $\epsilon > 0$, $A(\epsilon) > 0$, and f^ϵ be such that

$$\|f - f^\epsilon\| \le \epsilon, \quad f^\epsilon / A(\epsilon) \in A_1(\mathcal{D}).$$

Then

$$\|f_{m-1}\| = F_{f_{m-1}}(f_{m-1}) = F_{f_{m-1}}(f - f^\epsilon + f^\epsilon - G_{m-1})$$
$$\le \epsilon + \|F_{f_{m-1}}\|_{\mathcal{D}}(A(\epsilon) + b_m),$$

where $b_m := \sum_{k=1}^{m-1} a_k$. Therefore,

$$\|F_{f_{m-1}}\|_{\mathcal{D}} \ge (\|f_{m-1}\| - \epsilon)/(A(\epsilon) + b_m). \qquad (3.7.50)$$

We complete the proof by obtaining a contradiction. If $\lim_{m\to\infty} \|f_m\| = \alpha > 0$, and $\epsilon := \alpha/2$, then (3.7.49) and (3.7.50) imply

$$\|f_m\| \le \|f_{m-1}\|\left(1 - \frac{t\beta a_m}{2(A(\epsilon) + b_m)}\right).$$

Assumption (I) implies

$$\sum_{m=1}^{\infty} \frac{a_m}{A(\epsilon) + b_m} = \infty \;\Rightarrow\; \|f_m\| \to 0. \qquad \square$$

We now turn to the L_p-spaces. The following results, Proposition 3.7.13 and Theorem 3.7.14, are from Ganichev and Kalton (2003).

Proposition 3.7.13. The L_p-space with $1 < p < \infty$ has property Γ.

Proof. Let $p \in (1, \infty)$. Consider the following function:

$$\phi_p(u) := \frac{u|1 + u|^{p-2}(1 + u) - u}{|1 + u|^p - pu - 1}, \quad u \neq 0, \quad \phi_p(0) := 2/p.$$

We note that $|1 + u|^p - pu - 1 > 0$ for $u \neq 0$. Indeed, it is sufficient to check the inequality for $u \ge -1/p$. In this case $|1 + u|^p = (1 + u)^p > 1 + pu$, $u \neq 0$.

It is easy to check that

$$\lim_{u \to 0} \phi_p(u) = 2/p.$$

Thus, $\phi_p(u)$ is continuous on $(-\infty, \infty)$. This and

$$\lim_{u \to -\infty} \phi_p(u) = \lim_{u \to \infty} \phi_p(u) = 1$$

imply that $\phi_p(u) \le C_p$.

We now proceed to property Γ. For any two real functions $x(s)$, $y(s)$, the inequality $\phi_p(u) \le C_p$ implies

$$|x(s) + y(s)|^{p-2}(x(s) + y(s))y(s) - |x(s)|^{p-2}x(s)y(s) \qquad (3.7.51)$$

$$\le C_p(|x(s) + y(s)|^p - p|x(s)|^{p-2}x(s)y(s) - |x(s)|^p).$$

Suppose that $F_x(y) = 0$. This means that

$$\int |x(s)|^{p-2}x(s)y(s)\,\mathrm{d}s = 0. \qquad (3.7.52)$$

Integrating inequality (3.7.51) and taking into account (3.7.52), we get

$$\|x + y\|^{p-1}F_{x+y}(y) \le C_p(\|x + y\|^p - \|x\|^p). \qquad (3.7.53)$$

Next,

$$\|x\| = F_x(x) = F_x(x + y) \le \|x + y\|.$$

Therefore, (3.7.53) implies

$$F_{x+y}(y) \le pC_p(\|x + y\| - \|x\|). \qquad (3.7.54)$$

It remains to note that (3.7.54) is equivalent to property Γ with $\beta = (pC_p)^{-1}$. $\qquad \square$

Combining Theorem 3.7.12 with Proposition 3.7.13 we obtain the following result.

Theorem 3.7.14. Let $p \in (1, \infty)$. Then the WDGA(τ) with $\tau = \{t\}$, $t \subset (0, 1]$, converges for each dictionary and all $f \in L_p$.

3.8. Relaxation; X-greedy algorithms

In Sections 3.2–3.7 we studied dual greedy algorithms. In this section we define some generalizations of the X-Greedy Algorithm using the idea of relaxation. We begin with an analogue of the WGAFR.

X-Greedy Algorithm with Free Relaxation (XGAFR). We define $f_0 := f$ and $G_0 := 0$. Then, for each $m \ge 1$ we have the following inductive definition.

(1) $\varphi_m \in \mathcal{D}$ and $\lambda_m \geq 0$, w_m are such that

$$\|f - ((1 - w_m)G_{m-1} + \lambda_m \varphi_m)\| = \inf_{g \in \mathcal{D}, \lambda \geq 0, w} \|f - ((1 - w)G_{m-1} + \lambda g)\|$$

and

$$G_m := (1 - w_m)G_{m-1} + \lambda_m \varphi_m.$$

(2) Let

$$f_m := f - G_m.$$

Using this definition, we obtain that for any $t \in (0, 1]$

$$\|f_m\| \leq \inf_{\lambda \geq 0, w} \|f - ((1 - w)G_{m-1} + \lambda \varphi_m^t)\|,$$

where the $\varphi_m^t \in \mathcal{D}$ is an element satisfying

$$F_{f_{m-1}}(\varphi_m^t) \geq t\|F_{f_{m-1}}\|_{\mathcal{D}}.$$

Setting $t = 1$ we obtain a version of Lemma 3.4.1 for the XGAFR.

Lemma 3.8.1. Let X be a uniformly smooth Banach space with modulus of smoothness $\rho(u)$. Take a number $\epsilon \geq 0$ and two elements f, f^ϵ from X such that

$$\|f - f^\epsilon\| \leq \epsilon, \quad f^\epsilon/A(\epsilon) \in A_1(\mathcal{D}),$$

with some number $A(\epsilon) \geq \epsilon$. Then we have for the XGAFR

$$\|f_m\| \leq \|f_{m-1}\| \inf_{\lambda \geq 0} \left(1 - \lambda A(\epsilon)^{-1}\left(1 - \frac{\epsilon}{\|f_{m-1}\|}\right) + 2\rho\left(\frac{5\lambda}{\|f_{m-1}\|}\right)\right),$$

for $m = 1, 2, \dots$.

Theorems 3.4.3 and 3.4.4 were derived from Lemma 3.4.1. In the same way we derive from Lemma 3.8.1 the following analogues of Theorems 3.4.3 and 3.4.4 for the XGAFR.

Theorem 3.8.2. Let X be a uniformly smooth Banach space with modulus of smoothness $\rho(u)$. Then, for any $f \in X$ we have for the XGAFR

$$\lim_{m \to \infty} \|f_m\| = 0.$$

Theorem 3.8.3. Let X be a uniformly smooth Banach space with modulus of smoothness $\rho(u) \leq \gamma u^q$, $1 < q \leq 2$. Take a number $\epsilon \geq 0$ and two elements f, f^ϵ from X such that

$$\|f - f^\epsilon\| \leq \epsilon, \quad f^\epsilon/A(\epsilon) \in A_1(\mathcal{D}),$$

with some number $A(\epsilon) > 0$. Then we have for the XGAFR

$$\|f_m\| \leq \max\left(2\epsilon, C(q, \gamma)(A(\epsilon) + \epsilon)(1 + m)^{-1/p}\right), \quad p := q/(q - 1).$$

We now proceed to an analogue of the GAWR.

X-Greedy Algorithm with Relaxation r (XGAR(r)). Given a relaxation sequence $\mathbf{r} := \{r_m\}_{m=1}^{\infty}$, $r_m \in [0,1)$, we define $f_0 := f$ and $G_0 := 0$. Then, for each $m \geq 1$ we have the following inductive definition.

(1) $\varphi_m \in \mathcal{D}$ and $\lambda_m \geq 0$ are such that

$$\|f - ((1 - r_m)G_{m-1} + \lambda_m \varphi_m)\| = \inf_{g \in \mathcal{D}, \lambda \geq 0} \|f - ((1 - r_m)G_{m-1} + \lambda g)\|$$

and

$$G_m := (1 - r_m)G_{m-1} + \lambda_m \varphi_m.$$

(2) Let

$$f_m := f - G_m.$$

We note that in the case $r_k = 0$, $k = 1, \ldots$, when there is no relaxation, the XGAR(0) coincides with the X-Greedy Algorithm. Practically nothing is known about convergence and rate of convergence of the X-Greedy Algorithm. However, relaxation helps to prove convergence results for the XGAR(r). Here are analogues of the corresponding results for the GAWR.

Lemma 3.8.4. Let X be a uniformly smooth Banach space with modulus of smoothness $\rho(u)$. Take a number $\epsilon \geq 0$ and two elements f, f^ϵ from X such that

$$\|f - f^\epsilon\| \leq \epsilon, \quad f^\epsilon / A(\epsilon) \in A_1(\mathcal{D}),$$

with some number $A(\epsilon) > 0$. Then we have for the XGAR(r)

$$\|f_m\| \leq \|f_{m-1}\| \left(1 - r_m\left(1 - \frac{\epsilon}{\|f_{m-1}\|}\right) + 2\rho\left(\frac{r_m(\|f\| + A(\epsilon))}{(1 - r_m)\|f_{m-1}\|}\right)\right),$$

for $m = 1, 2, \ldots$.

Theorem 3.8.5. Let a sequence $\mathbf{r} := \{r_k\}_{k=1}^{\infty}$, $r_k \in [0,1)$, satisfy the conditions

$$\sum_{k=1}^{\infty} r_k = \infty, \quad \text{and} \quad r_k \to 0 \quad \text{as } k \to \infty.$$

Then the XGAR(r) converges in any uniformly smooth Banach space for each $f \in X$ and for all dictionaries $\mathcal{D}$.

Theorem 3.8.6. Let X be a uniformly smooth Banach space with modulus of smoothness $\rho(u) \leq \gamma u^q$, $1 < q \leq 2$. Let $\mathbf{r} := \{2/(k+2)\}_{k=1}^{\infty}$. Consider the XGAR(r). For a pair of functions f, f^ϵ satisfying

$$\|f - f^\epsilon\| \leq \epsilon, \quad f^\epsilon / A(\epsilon) \in A_1(\mathcal{D}),$$

we have

$$\|f_m\| \leq \epsilon + C(q, \gamma)(\|f\| + A(\epsilon))m^{-1+1/q}.$$

3.9. Greedy algorithms with approximate evaluations and restricted search

In this section we study a modification of the WCGA that is motivated by numerical applications. In this modification, we allow steps of the WCGA to be performed approximately with some error control. We show that the modified version of the WCGA performs as well as the WCGA. We develop the theory of the Approximate Weak Chebyshev Greedy Algorithm in a general setting: X is an arbitrary uniformly smooth Banach space and $\mathcal{D}$ is any dictionary. We begin with some remarks on the WCGA. It is clear that in the case of an infinite dictionary $\mathcal{D}$ there is no direct computationally feasible way to evaluate $\sup_{g \in \mathcal{D}} F_{f_{m-1}^c}(g)$. This makes the greedy step, even in a weak version, very difficult to realize in practice. At the second step of the WCGA we are looking for the best approximant of f from Φ_m. We know that such an approximant exists. However, in practice we cannot find it exactly: we can only find it approximately.

The above observations motivated us to consider a variant of the WCGA with an eye towards practically implementable algorithms. We note that Approximate Weak Greedy Algorithms in Hilbert spaces were studied in Gribonval and Nielsen (2001 a) and Galatenko and Livshitz (2003, 2005).

In Temlyakov (2005 a) we studied the following modification of the WCGA. Let three sequences $\tau = \{t_k\}_{k=1}^{\infty}$, $\delta = \{\delta_k\}_{k=0}^{\infty}$, $\eta = \{\eta_k\}_{k=1}^{\infty}$ of numbers from $[0, 1]$ be given.

Approximate Weak Chebyshev Greedy Algorithm (AWCGA). We define $f_0 := f_0^{\tau,\delta,\eta} := f$. Then, for each $m \geq 1$ we have the following inductive definition.

(1) F_{m-1} is a functional with properties

$$\|F_{m-1}\| \leq 1, \qquad F_{m-1}(f_{m-1}) \geq \|f_{m-1}\|(1 - \delta_{m-1});$$

and $\varphi_m := \varphi_m^{\tau,\delta,\eta} \in \mathcal{D}$ is any element satisfying

$$F_{m-1}(\varphi_m) \geq t_m \sup_{g \in \mathcal{D}} F_{m-1}(g).$$

(2) Define

$$\Phi_m := \operatorname{span}\{\varphi_j\}_{j=1}^m,$$

and let

$$E_m(f) := \inf_{\varphi \in \Phi_m} \|f - \varphi\|.$$

Let $G_m \in \Phi_m$ be such that

$$\|f - G_m\| \leq E_m(f)(1 + \eta_m).$$

(3) Let

$$f_m := f_m^{\tau,\delta,\eta} := f - G_m.$$

The term *approximate* in this definition means that we use a functional F_{m-1} that is an approximation to the norming (peak) functional $F_{f_{m-1}}$ and also that we use an approximant $G_m \in \Phi_m$ which satisfies a weaker assumption than being a best approximant to f from Φ_m. Thus, in the *approximate* version of the WCGA, we have addressed the issue of non-exact evaluation of the norming functional and the best approximant. We did not address the issue of finding the $\sup_{g \in \mathcal{D}} F_{f_{m-1}^c}(g)$. In the paper Temlyakov (2005*b*) we addressed this issue. We did it in two steps. First we considered the corresponding modification of the WCGA, and then the modification of the AWCGA. These modifications are done in the style of the concept of *depth search* from Donoho (2001).

We now consider a countable dictionary $\mathcal{D} = \{\pm\psi_j\}_{j=1}^{\infty}$. We denote $\mathcal{D}(N) := \{\pm\psi_j\}_{j=1}^{N}$. Let $\mathcal{N} := \{N_j\}_{j=1}^{\infty}$ be a sequence of natural numbers.

Restricted Weak Chebyshev Greedy Algorithm (RWCGA). We define $f_0 := f_0^{c,\tau,\mathcal{N}} := f$. Then, for each $m \geq 1$ we have the following inductive definition.

(1) $\varphi_m := \varphi_m^{c,\tau,\mathcal{N}} \in \mathcal{D}(N_m)$ is any element satisfying

$$F_{f_{m-1}}(\varphi_m) \geq t_m \sup_{g \in \mathcal{D}(N_m)} F_{f_{m-1}}(g).$$

(2) Define

$$\Phi_m := \Phi_m^{\tau,\mathcal{N}} := \mathrm{span}\{\varphi_j\}_{j=1}^{m},$$

and define $G_m := G_m^{c,\tau,\mathcal{N}}$ to be the best approximant to f from Φ_m.

(3) Let

$$f_m := f_m^{c,\tau,\mathcal{N}} := f - G_m.$$

We formulate some results from Temlyakov (2005*a*, 2005*b*) in a particular case of a uniformly smooth Banach space with modulus of smoothness of power type (see Temlyakov (2005*a*, 2005*b*) for the general case). The following theorem was proved in Temlyakov (2005*a*).

Theorem 3.9.1. Let a Banach space X have modulus of smoothness $\rho(u)$ of power type $1 < q \leq 2$, that is, $\rho(u) \leq \gamma u^q$. Assume that

$$\sum_{m=1}^{\infty} t_m^p = \infty, \quad p = \frac{q}{q-1},$$

and

$$\delta_m = o(t_m^p), \qquad \eta_m = o(t_m^p).$$

Then the AWCGA converges for any $f \in X$.

We now give two theorems from Temlyakov (2005b) on greedy algorithms with restricted search.

Theorem 3.9.2. Let a Banach space X have modulus of smoothness $\rho(u)$ of power type $1 < q \leq 2$, that is, $\rho(u) \leq \gamma u^q$. Assume that $\lim_{m \to \infty} N_m = \infty$ and

$$\sum_{m=1}^{\infty} t_m^p = \infty, \quad p = \frac{q}{q-1}.$$

Then the RWCGA converges for any $f \in X$.

For $b > 0$, $K > 0$, we define the class

$$A_1^b(K, \mathcal{D}) := \{f : d(f, A_1(\mathcal{D}(n))) \leq K n^{-b}, \quad n = 1, 2, \ldots\}.$$

Here, $A_1(\mathcal{D}(n))$ is a convex hull of $\{\pm \psi_j\}_{j=1}^n$, and for a compact set F

$$d(f, F) := \inf_{\phi \in F} \|f - \phi\|.$$

Theorem 3.9.3. Let X be a uniformly smooth Banach space with modulus of smoothness $\rho(u) \leq \gamma u^q$, $1 < q \leq 2$. Then, for $t \in (0, 1]$ there exist $C_1(t, \gamma, q, K)$, $C_2(t, \gamma, q, K)$ such that, for $\mathcal{N}$ with $N_m \geq C_1(t, \gamma, q, K) m^{r/b}$, $m = 1, 2, \ldots$, we have for any $f \in A_1^b(K, \mathcal{D})$

$$\|f_m^{c,\tau,\mathcal{N}}\| \leq C_2(t, \gamma, q, K) m^{-r}, \quad \tau = \{t\}, \quad r := 1 - 1/q.$$

We note that we can choose an algorithm from Theorem 3.9.3 that satisfies the *polynomial depth search* condition $N_m \leq C m^a$ from Donoho (2001).

We proceed to an algorithm that combines approximate evaluations with restricted search. Let three sequences $\tau = \{t_k\}_{k=1}^{\infty}$, $\delta = \{\delta_k\}_{k=0}^{\infty}$, $\eta = \{\eta_k\}_{k=1}^{\infty}$ of numbers from $[0, 1]$ be given. Let $\mathcal{N} := \{N_j\}_{j=1}^{\infty}$ be a sequence of natural numbers.

Restricted Approximate Weak Chebyshev Greedy Algorithm (RAWCGA). We define $f_0 := f_0^{\tau,\delta,\eta,\mathcal{N}} := f$. Then, for each $m \geq 1$ we have the following inductive definition.

(1) F_{m-1} is a functional with properties

$$\|F_{m-1}\| \leq 1, \quad F_{m-1}(f_{m-1}) \geq \|f_{m-1}\|(1 - \delta_{m-1}),$$

and $\varphi_m := \varphi_m^{\tau,\delta,\eta,\mathcal{N}} \in \mathcal{D}(N_m)$ is any element satisfying

$$F_{m-1}(\varphi_m) \geq t_m \sup_{g \in \mathcal{D}(N_m)} F_{m-1}(g).$$

(2) Define

$$\Phi_m := \mathrm{span}\{\varphi_j\}_{j=1}^m,$$

and let

$$E_m(f) := \inf_{\varphi \in \Phi_m} \|f - \varphi\|.$$

Let $G_m \in \Phi_m$ be such that

$$\|f - G_m\| \leq E_m(f)(1 + \eta_m).$$

(3) Let

$$f_m := f_m^{\tau,\delta,\eta,\mathcal{N}} := f - G_m.$$

Theorem 3.9.4. Let a Banach space X have modulus of smoothness $\rho(u)$ of power type $1 < q \leq 2$, that is, $\rho(u) \leq \gamma u^q$. Assume $\lim_{m \to \infty} N_m = \infty$,

$$\sum_{m=1}^{\infty} t_m^p = \infty, \quad p = \frac{q}{q-1},$$

and

$$\delta_m = o(t_m^p), \qquad \eta_m = o(t_m^p).$$

Then the RAWCGA converges for any $f \in X$.

We now make some general remarks on m-term approximation with the depth search constraint. The depth search constraint means that for a given m we restrict ourselves to systems of elements (subdictionaries) containing at most $N := N(m)$ elements. Let X be a linear metric space and for a set $\mathcal{D} \subset X$, let $\mathcal{L}_m(\mathcal{D})$ denote the collection of all linear subspaces spanned by m elements of $\mathcal{D}$. For a linear subspace $L \subset X$, the ϵ-neighbourhood $U_\epsilon(L)$ of L is the set of all $x \in X$ which are at a distance not exceeding ϵ from L (*i.e.*, those $x \in X$ which can be approximated to an error not exceeding ϵ by the elements of L). For any compact set $F \subset X$ and any integers $N, m \geq 1$, we define the (N, m)-entropy numbers (see Temlyakov (2003a, p. 94))

$$\epsilon_{N,m}(F, X) := \inf_{\#\mathcal{D}=N} \inf\{\epsilon : F \subset \cup_{L \in \mathcal{L}_m(\mathcal{D})} U_\epsilon(L)\}.$$

We let $\Sigma_m(\mathcal{D})$ denote the collection of all functions (elements) in X which can be expressed as a linear combination of at most m elements of $\mathcal{D}$. Thus each function $s \in \Sigma_m(\mathcal{D})$ can be written in the form

$$s = \sum_{g \in \Lambda} c_g g, \quad \Lambda \subset \mathcal{D}, \quad \#\Lambda \leq m,$$

where the c_g are real or complex numbers. For a function $f \in X$ we define its best m-term approximation error

$$\sigma_m(f) := \sigma_m(f, \mathcal{D}) := \inf_{s \in \Sigma_m(\mathcal{D})} \|f - s\|.$$

For a function class $F \subset X$ we define

$$\sigma_m(F) := \sigma_m(F, \mathcal{D}) := \sup_{f \in F} \sigma_m(f, \mathcal{D}).$$

We can express $\sigma_m(F, \mathcal{D})$ as

$$\sigma_m(F, \mathcal{D}) = \inf\{\epsilon : F \subset \cup_{L \in \mathcal{L}_m(\mathcal{D})} U_\epsilon(L)\}.$$

It follows therefore that

$$\inf_{\#\mathcal{D}=N} \sigma_m(F, \mathcal{D}) = \epsilon_{N,m}(F, X).$$

In other words, finding best dictionaries consisting of N elements for m-term approximation of F is the same as finding sets $\mathcal{D}$ which attain the (N, m)-entropy numbers $\epsilon_{N,m}(F, X)$. It is easy to see that $\epsilon_{m,m}(F, X) = d_m(F, X)$, where $d_m(F, X)$ is the Kolmogorov width of F in X. This establishes a connection between (N, m)-entropy numbers and the Kolmogorov widths. One can find a further discussion on the nonlinear Kolmogorov (N, m)-widths and the entropy numbers in Temlyakov (2003a).

3.10. An application of greedy algorithms for the discrepancy estimates

Let $1 \leq p < \infty$. We recall the definition of the L_p-discrepancy of points $\{\xi^1, \ldots, \xi^m\} \subset \Omega_d := [0,1]^d$. Let $\chi_{[a,b]}(\cdot)$ be the characteristic function of the interval $[a, b]$. For $x, y \in \Omega_d$, let

$$B(x, y) := \prod_{j=1}^{d} \chi_{[0,x_j]}(y_j).$$

Then the L_p-discrepancy of $\xi := \{\xi^1, \ldots, \xi^m\} \subset \Omega_d$ is defined by

$$D(\xi, m, d)_p := \left\| \int_{\Omega_d} B(x, y) \, dy - \frac{1}{m} \sum_{\mu=1}^{m} B(x, \xi^\mu) \right\|_{L_p(\Omega_d)}.$$

It will be convenient for us to study a slight modification of $D(\xi, m, d)_p$. For $a, t \in [0, 1]$, let

$$H(a, t) := \chi_{[0,a]}(t) - \chi_{[a,1]}(t),$$

and for $x, y \in \Omega_d$

$$H(x, y) := \prod_{j=1}^{d} H(x_j, y_j).$$

We define the symmetrized L_p-discrepancy by

$$D^s(\xi, m, d)_p := \left\| \int_{\Omega_d} H(x, y) \, dy - \frac{1}{m} \sum_{\mu=1}^{m} H(x, \xi^\mu) \right\|_{L_p(\Omega_d)}.$$

The L_∞-discrepancies $D(\xi, m, d)_\infty$ and $D^s(\xi, m, d)_\infty$ are defined in the same way, with the L_p-norm replaced by the L_∞-norm.

Using the identity

$$\chi_{[0,x_j]}(y_j) = \frac{1}{2}(H(1,y_j) + H(x_j,y_j)),$$

we get a simple inequality,

$$D(\xi, m, d)_\infty \leq D^s(\xi, m, d)_\infty. \tag{3.10.1}$$

We are interested in ξ with small discrepancy. Consider

$$D(m, d)_p := \inf_\xi D(\xi, m, d)_p, \qquad D^s(m, d)_p := \inf_\xi D^s(\xi, m, d)_p.$$

For $1 < p < \infty$ the following relation is known,

$$D(m, d)_p \asymp m^{-1}(\ln m)^{(d-1)/2} \tag{3.10.2}$$

(see Beck and Chen (1987, p. 5)), with constants in $\asymp$ depending on p and d. The correct order of $D(m, d)_p$, $p = 1, \infty$, for $d \geq 3$ is unknown. The following estimate was obtained in Heinrich, Novak, Wasilkowski and Wozniakowski (2001):

$$D(m, d)_\infty \leq C d^{1/2} m^{-1/2}. \tag{3.10.3}$$

It is pointed out in Heinrich *et al.* (2001) that (3.10.3) is only an existence theorem and even the constant C in (3.10.3) is unknown. Their proof is a probabilistic one. There are also some other estimates in Heinrich *et al.* (2001) with explicit constants. We mention one of them,

$$D(m, d)_\infty \leq C(d \ln d)^{1/2}((\ln m)/m)^{1/2}, \tag{3.10.4}$$

with an explicit constant C. The proof of (3.10.4) is also probabilistic.

In this section we apply greedy-type algorithms to obtain upper estimates of $D(m, d)_p$, $1 \leq p \leq \infty$ in the style of (3.10.3) and (3.10.4). The important feature of our proof is that it is deterministic, and moreover it is constructive. Formally, the optimization problem

$$D(m, d)_p = \inf_\xi D(\xi, m, d)_p$$

is deterministic: one needs to minimize over $\{\xi^1, \ldots, \xi^m\} \subset \Omega_d$. However, minimization by itself does not provide any upper estimate. It is known (see Davis *et al.* (1997)) that simultaneous optimization over many parameters ($\{\xi^1, \ldots, \xi^m\}$ in our case) is a very difficult problem. We note that

$$D(m, d)_p = \sigma_m^e(J, \mathcal{B})_p := \inf_{g_1, \ldots, g_m \in \mathcal{B}} \left\| J(\cdot) - \frac{1}{m}\sum_{\mu=1}^m g_\mu \right\|_{L_p(\Omega_d)},$$

where

$$J(x) = \int_{\Omega_d} B(x, y)\, \mathrm{d}y$$

and

$$\mathcal{B} = \{B(x,y) : y \in \Omega_d\}.$$

It was proved in Davis *et al.* (1997) that if an algorithm finds the best m-term approximation for each $f \in \mathbb{R}^N$ for every dictionary $\mathcal{D}$, with the number of elements of order N^k, $k \geq 1$, then this algorithm solves an NP-hard problem. Thus, in nonlinear m-term approximation we look for methods (algorithms) which provide approximation close to best m-term approximation, and at each step solve an optimization problem over only one parameter (ξ^μ in our case). In this section we will provide such an algorithm for estimating $\sigma_m^e(J, \mathcal{B})_p$. We call this algorithm 'constructive' because it provides an explicit construction with feasible one-parameter optimization steps.

We proceed to the construction. We will use in our construction the IA(ϵ) which was studied in Section 3.6. We will use the following corollaries of Theorem 3.6.2.

Corollary 3.10.1. We apply Theorem 3.6.2 for $X = L_p(\Omega_d)$, $p \in [2, \infty)$, $\mathcal{D}^+ = \{H(x,y) : y \in \Omega_d\}$, $f = J^s(x)$, where

$$J^s(x) = \int_{\Omega_d} H(x,y) \, dy \in A_1(\mathcal{D}^+).$$

Using (3.1.12), we get by Theorem 3.6.2 a constructive set $\xi^1, \ldots, \xi^m$, such that

$$D^s(\xi, m, d)_p = \|(J^s)_m^{i,\epsilon}\|_{L_p(\Omega_d)} \leq Cp^{1/2}m^{-1/2},$$

with absolute constant C.

Corollary 3.10.2. We apply Theorem 3.6.2 for $X = L_p(\Omega_d)$, $p \in [2, \infty)$, $\mathcal{D}^+ = \{B(x,y) : y \in \Omega_d\}$, $f = J(x)$, where

$$J(x) = \int_{\Omega_d} B(x,y) \, dy \in A_1(\mathcal{D}^+).$$

Using (3.1.12), we get by Theorem 3.6.2 a constructive set $\xi^1, \ldots, \xi^m$, such that

$$D(\xi, m, d)_p = \|J_m^{i,\epsilon}\|_{L_p(\Omega_d)} \leq Cp^{1/2}m^{-1/2},$$

with absolute constant C.

Corollary 3.10.3. We apply Theorem 3.6.2 for $X = L_p(\Omega_d)$, $p \in [2, \infty)$, $\mathcal{D}^+ = \{B(x,y)/\|B(\cdot,y)\|_{L_p(\Omega_d)} : y \in \Omega_d\}$, $f = J(x)$. Using (3.1.12), we get by Theorem 3.6.2 a constructive set $\xi^1, \ldots, \xi^m$ such that

$$\left\| \int_{\Omega_d} B(x,y)\,\mathrm{d}y - \frac{1}{m} \sum_{\mu=1}^{m} \left(\frac{p}{p+1} \right)^d \left(\prod_{j=1}^{d} (1 - \xi_j^\mu)^{-1/p} \right) B(x, \xi^\mu) \right\|_{L_p(\Omega_d)}$$

$$\leq C \left(\frac{p}{p+1} \right)^d p^{1/2} m^{-1/2},$$

with absolute constant C.

We note that in the case $X = L_p(\Omega_d)$, $p \in [2, \infty)$, $\mathcal{D}^+ = \{H(x,y) : x \in \Omega_d\}$, $f = J^s(y)$, the implementation of the $\mathrm{IA}(\epsilon)$ is a sequence of maximization steps when we maximize functions of d variables. An important advantage of the L_p-spaces is a simple and explicit form of the norming functional F_f of a function $f \in L_p(\Omega_d)$. The F_f acts as (for real L_p-spaces)

$$F_f(g) = \int_{\Omega_d} \|f\|_p^{1-p} |f|^{p-2} f g \, \mathrm{d}y.$$

Thus the $\mathrm{IA}(\epsilon)$ should find at a step m an approximate solution to the following optimization problem (over $y \in \Omega_d$)

$$\int_{\Omega_d} |f_{m-1}^{i,\epsilon}(x)|^{p-2} f_{m-1}^{i,\epsilon}(x) H(x,y) \, \mathrm{d}x \longrightarrow \max.$$

Let us discuss one possible application of the WRGA instead of the $\mathrm{IA}(\epsilon)$. An obvious change is that instead of the cubature formula

$$\frac{1}{m} \sum_{\mu=1}^{m} H(x, \xi^\mu),$$

in the case of $\mathrm{IA}(\epsilon)$, we have the cubature formula

$$\sum_{\mu=1}^{m} w_\mu^m H(x, \xi^\mu), \quad \sum_{\mu=1}^{m} |w_\mu^m| \leq 1,$$

in the case of the WRGA. It is a disadvantage of the WRGA. An advantage of the WRGA is that we are more flexible in selecting an element φ_m^r satisfying

$$F_{f_{m-1}^r}^r (\varphi_m^r - G_{m-1}^r) \geq t_m \sup_{g \in \mathcal{D}} F_{f_{m-1}^r}^r (g - G_{m-1}^r)$$

than an element $\varphi_m^{i,\epsilon}$ satisfying

$$F_{f_{m-1}^{i,\epsilon}} (\varphi_m^{i,\epsilon} - f) \geq -\epsilon_m.$$

We will now derive an estimate for $D(m, d)_\infty$ from Corollary 3.10.2.

Proposition 3.10.4. For any m there exists a constructive set

$$\xi = \{\xi^1, \ldots, \xi^m\} \subset \Omega_d$$

such that

$$D(\xi, m, d)_\infty \leq Cd^{3/2}(\max(\ln d, \ln m))^{1/2}m^{-1/2}, \quad d, m \geq 2 \qquad (3.10.5)$$

with an effective absolute constant C.

Proof. We use the inequality

$$D(\xi, m, d)_\infty \leq c(d, p)d(3d + 4)D(\xi, m, d)_p^{p/(p+d)}, \qquad (3.10.6)$$

from Niederreiter, Tichy and Turnwald (1990), and the estimate for $c(d, p)$

$$c(d, p) \leq 3^{1/3}d^{-1+2/(1+p/d)}, \qquad (3.10.7)$$

from Heinrich *et al.* (2001). Specifying $p = d\max(\ln d, \ln m)$ and using Corollary 3.10.2 we get (3.10.5) from (3.10.6) and (3.10.7). □

REFERENCES

A. R. Barron (1993), 'Universal approximation bounds for superposition of n sigmoidal functions', *IEEE Trans. Inform. Theory* **39**, 930–945.

A. Barron, A. Cohen, W. Dahmen and R. DeVore (2005), Approximation and learning by greedy algorithms. Manuscript.

B. M. Baishanski (1983), 'Approximation by polynomials of given length', *Illinois J. Math.* **27**, 449–458.

N. K. Bary (1961), *Trigonometric Series*, Nauka, Moscow (in Russian). English translation: Pergamon Press, Oxford (1964).

J. Beck and W. Chen (1987), *Irregularities of Distribution*, Cambridge University Press.

W. Bednorz (2006), Greedy bases are best for m-term approximation. Manuscript.

M. S. Birman and M. Z. Solomyak (1977), 'Estimates of singular numbers of integral operators', *Uspekhi Mat. Nauk* **32**, 17–84. English translation in *Russian Math. Surveys* **32** (1977).

J. Bourgain (1992), 'A remark on the behaviour of L^p-multipliers and the range of operators acting on L^p-spaces', *Israel J. Math.* **79**, 193–206.

E. Candès (2006), Compressive sampling. In *Proc. International Congress of Mathematics* (Madrid 2006), Vol. 3, pp. 1433–1452.

E. Candès, J. Romberg and T. Tao (2006), 'Stable signal recovery from incomplete and inaccurate measurements', *Comm. Pure Appl. Math.* **59**, 1207–1223.

E. Candès and T. Tao (2005), 'Decoding by linear programming', *IEEE Trans. Inform. Theory* **51**, 4203–4215.

B. Carl (1981), 'Entropy numbers, s-numbers, and eigenvalue problems', *J. Funct. Anal.* **41**, 290–306.

S. S. Chen, D. L. Donoho and M. A. Saunders (2001), 'Atomic decomposition by basis pursuit', *SIAM Review* **43**, 129–159.

J. A. Cochran (1977), 'Composite integral operators and nuclearity', *Ark. Mat.* **15**, 215–222.

A. Cohen, R. A. DeVore and R. Hochmuth (2000), 'Restricted nonlinear approximation', *Constr. Approx.* **16**, 85–113.

A. Cohen, W. Dahmen and R. DeVore (2007), Compressed sensing and k-term approximation. Manuscript.

R. R. Coifman and M. V. Wickerhauser (1992), 'Entropy-based algorithms for best-basis selection', *IEEE Trans. Inform. Theory* **38**, 713–718.

A. Cordoba and P. Fernandez (1998), 'Convergence and divergence of decreasing rearranged Fourier series', *SIAM J. Math. Anal.* **29**, 1129–1139.

F. Cucker and S. Smale (2001), 'On the mathematical foundations of learning', *Bull. Amer. Math. Soc.* **39**, 1–49.

G. Davis, S. Mallat and M. Avellaneda (1997), 'Adaptive greedy approximations', *Constr. Approx.* **13**, 57–98.

R. A. DeVore (1998), Nonlinear approximation. In *Acta Numerica*, Vol. 7, Cambridge University Press, pp. 51–150.

R. A. DeVore (2006), Optimal computation. In *Proc. International Congress of Mathematics* (Madrid 2006), Vol. 1, pp. 187–215.

R. DeVore, B. Jawerth and V. Popov (1992), 'Compression of wavelet decompositions', *Amer. J. Math.* **114**, 737–785.

R. DeVore, G. Kerkyacharian, D. Picard and V. Temlyakov (2004), On mathematical methods of learning. IMI Preprint 10, Department of Mathematics, University of South Carolina.

R. DeVore, G. Kerkyacharian, D. Picard and V. Temlyakov (2006), 'Mathematical methods for supervised learning', *Found. Comput. Math.* **6**, 3–58.

R. A. DeVore, S. V. Konyagin and V. N. Temlyakov (1998), 'Hyperbolic wavelet approximation', *Constr. Approx.* **14**, 1–26.

R. A. DeVore and G. G. Lorenz (1993), *Constructive Approximation*, Springer, Berlin.

R. DeVore, G. Petrova and V. N. Temlyakov (2003), 'Best basis selection for approximation in L_p', *Found. Comput. Math.* **3**, 161–185.

R. A. DeVore and V. A. Popov (1988), Interpolation spaces and non-linear approximation. In *Function Spaces and Approximation*, Vol. 1302 of *Lecture Notes in Mathematics*, Springer, pp. 191–205.

R. A. DeVore and V. N. Temlyakov (1995), 'Nonlinear approximation by trigonometric sums', *J. Fourier Anal. Appl.* **2**, 29–48.

R. A. DeVore and V. N. Temlyakov (1996), 'Some remarks on greedy algorithms', *Adv. Comput. Math.* **5**, 173–187.

R. A. DeVore and V. N. Temlyakov (1997), 'Nonlinear approximation in finite-dimensional spaces', *J. Complexity* **13**, 489–508.

S. J. Dilworth, N. J. Kalton and D. Kutzarova (2003), 'On the existence of almost greedy bases in Banach spaces', *Studia Math.* **158**, 67–101.

S. J. Dilworth, N. J. Kalton, D. Kutzarova and V. N. Temlyakov (2003), 'The thresholding greedy algorithm, greedy bases, and duality', *Constr. Approx.* **19**, 575–597.

S. Dilworth, D. Kutzarova and V. Temlyakov (2002), 'Convergence of some greedy algorithms in Banach spaces', *J. Fourier Anal. Appl.* **8**, 489–505.

S. J. Dilworth, D. Kutzarova and P. Wojtaszczyk (2002), 'On approximate ℓ_1 systems in Banach spaces', *J. Approx. Theory* **114**, 214–241.

M. Donahue, L. Gurvits, C. Darken and E. Sontag (1997), 'Rate of convex approximation in non-Hilbert spaces', *Constr. Approx.* **13**, 187–220.

D. L. Donoho (1993), 'Unconditional bases are optimal bases for data compression and for statistical estimation', *Appl. Comput. Harmon. Anal.* **1**, 100–115.

D. L. Donoho (1997), 'CART and best-ortho-basis: A connection', *Ann. Statist.* **25**, 1870–1911.

D. L. Donoho (2001), 'Sparse components of images and optimal atomic decompositions', *Constr. Approx.* **17**, 353–382.

D. Donoho (2006), 'Compressed sensing', *IEEE Trans. Inform. Theory* **52**, 1289–1306.

D. Donoho, M. Elad and V. N. Temlyakov (2006), 'Stable recovery of sparse overcomplete representations in the presence of noise', *IEEE Trans. Inform. Theory* **52**, 6–18.

D. Donoho, M. Elad and V. N. Temlyakov (2007), 'On the Lebesgue type inequalities for greedy approximation', *J. Approx. Theory* **147**, 185–195.

D. Donoho and I. Johnstone (1994), 'Ideal spatial adaptation via wavelet shrinkage', *Biometrica* **81**, 425–455.

V. V. Dubinin (1997), Greedy algorithms and applications. PhD Thesis, University of South Carolina.

C. Fefferman and E. Stein (1972), 'H^p spaces of several variables', *Acta Math.* **129**, 137–193.

T. Figiel, W. B. Johnson and G. Schechtman (1988), 'Factorization of natural embeddings of ℓ_p^n into L_r I', *Studia Math.* **89**, 79–103.

M. Frazier and B. Jawerth (1990), 'A discrete transform and decomposition of distribution spaces', *J. Funct. Anal.* **93**, 34–170.

I. Fredholm (1903), 'Sur une classe d'équations fonctionelles', *Acta Math.* **27**, 365–390.

J. H. Friedman and W. Stuetzle (1981), 'Projection pursuit regression', *J. Amer. Statist. Assoc.* **76**, 817–823.

V. V. Galatenko and E. D. Livshitz (2003), 'On convergence of approximate weak greedy algorithms', *East J. Approx.* **9**, 43–49.

V. V. Galatenko and E. D. Livshitz (2005), 'Generalized approximate weak greedy algorithms', *Math. Notes* **78**, 170–184.

M. Ganichev and N. J. Kalton (2003), 'Convergence of the weak dual greedy algorithm in L_p-spaces', *J. Approx. Theory* **124**, 89–95.

A. Garnaev and E. Gluskin (1984), 'The widths of a Euclidean ball', *Dokl. Akad. Nauk USSR* **277**, 1048–1052. English translation in *Soviet Math. Dokl.* **30**, 200–204.

A. C. Gilbert, S. Muthukrishnan and M. J. Strauss (2003), Approximation of functions over redundant dictionaries using coherence. In *Proc. 14th Annual ACM–SIAM Symposium on Discrete Algorithms*, pp. 243–252.

S. Gogyan (2005), 'Greedy algorithm with regard to Haar subsystems', *East J. Approx.* **11**, 221–236.

S. Gogyan (2006), On convergence of weak thresholding greedy algorithm in $L^1(0,1)$. Manuscript.

R. Gribonval and M. Nielsen (2001a), 'Approximate weak greedy algorithms', *Adv. Comput. Math.* **14**, 361–368.

R. Gribonval and M. Nielsen (2001b), 'Some remarks on non-linear approximation with Schauder bases', *East J. Approx.* **7**, 267–285.

P. Habala, P. Hájek and V. Zizler (1996), *Introduction to Banach Spaces*, Vol. I, Matfyzpress, Univerzity Karlovy.

S. Heinrich, E. Novak, G. Wasilkowski and H. Wozniakowski (2001), 'The inverse of the star-discrepancy depends linearly on the dimension', *Acta Arith.* **96**, 279–302.

E. Hille and J. D. Tamarkin (1931), 'On the characteristic values of linear integral equations', *Acta Math.* **57**, 1–76.

P. J. Huber (1985), 'Projection pursuit', *Ann. Statist.* **13**, 435–475.

L. Jones (1987), 'On a conjecture of Huber concerning the convergence of projection pursuit regression', *Ann. Statist.* **15**, 880–882.

L. Jones (1992), 'A simple lemma on greedy approximation in Hilbert space and convergence rates for projection pursuit regression and neural network training', *Ann. Statist.* **20**, 608–613.

N. J. Kalton, N. T. Beck and J. W. Roberts (1984), *An F-Space Sampler*, Vol. 5 of *London Math. Soc. Lecture Notes*, Cambridge University Press, Cambridge.

A. Kamont and V. N. Temlyakov (2004), 'Greedy approximation and the multivariate Haar system', *Studia Math.* **161**, 199–223.

B. S. Kashin (1977), 'Widths of certain finite-dimensional sets and classes of smooth functions', *Izv. Akad. Nauk SSSR, Ser. Mat.* **41**, 334–351. English translation in *Math. USSR IZV.* **11**.

B. S. Kashin (1985), 'On approximation properties of complete orthonormal systems', *Tr. Mat. Inst. Steklova* **172**, 187–191. English translation in *Proc. Steklov Inst. Math.* **3**, 207–211.

B. S. Kashin and V. N. Temlyakov (2007), A remark on compressed sensing. Manuscript.

G. Kerkyacharian and D. Picard (2004), 'Entropy, universal coding, approximation, and bases properties', *Constr. Approx.* **20**, 1–37.

G. Kerkyacharian, D. Picard and V. N. Temlyakov (2006), 'Some inequalities for the tensor product of greedy bases and weight-greedy bases', *East J. Approx.* **12**, 103–118.

S. V. Konyagin and M. A. Skopina (2001), 'Comparison of the L_1-norms of total and truncated exponential sums', *Mat. Zametki* **69**, 699–707.

S. V. Konyagin and V. N. Temlyakov (1999a), 'A remark on greedy approximation in Banach spaces', *East J. Approx.* **5**, 1–15.

S. V. Konyagin and V. N. Temlyakov (1999b), 'Rate of convergence of pure greedy algorithms', *East J. Approx.* **5**, 493–499.

S. V. Konyagin and V. N. Temlyakov (2002), 'Greedy approximation with regard to bases and general minimal systems', *Serdica Math. J.* **28**, 305–328.

S. V. Konyagin and V. N. Temlyakov (2003a), 'Convergence of greedy approximation I: General systems', *Studia Math.* **159**, 143–160.

S. V. Konyagin and V. N. Temlyakov (2003b), 'Convergence of greedy approximation II: The trigonometric system', *Studia Math.* **159**, 161–184.

S. V. Konyagin and V. N. Temlyakov (2004), Some error estimates in learning theory. In *Approximation Theory: A Volume Dedicated to Borislav Bojanov*, Marin Drinov Academy Publishing House, Sofia, pp. 126–144.

S. V. Konyagin and V. N. Temlyakov (2005), 'Convergence of greedy approximation for the trigonometric system', *Anal. Math.* **31**, 85–115.

S. V. Konyagin and V. N. Temlyakov (2007), 'The entropy in learning theory: Error estimates', *Constr. Approx.* **25**, 1–27.

T. W. Körner (1996), 'Divergence of decreasing rearranged Fourier series', *Ann. of Math.* **144**, 167–180.

T. W. Körner (1999), 'Decreasing rearranged Fourier series', *J. Fourier Anal. Appl.* **5**, 1–19.

H. Lebesgue (1909), 'Sur les intégrales singulières', *Ann. Fac. Sci. Univ. Toulouse* (3) **1**, 25–117.

W. S. Lee, P. L. Bartlett and R. C. Williamson (1996), 'Efficient agnostic learning of neural networks with bounded fan-in', *IEEE Trans. Inform. Theory* **42**, 2118–2132.

D. Leviatan and V. N. Temlyakov (2005), 'Simultaneous greedy approximation in Banach spaces', *J. Complexity* **21**, 275–293.

D. Leviatan and V. N. Temlyakov (2006), 'Simultaneous approximation by greedy algorithms', *Adv. Comput. Math.* **25**, 73–90.

J. Lindenstrauss and L. Tzafriri (1977), *Classical Banach Spaces*, Vol. I, Springer, Berlin.

E. D. Livshitz (2003), 'Convergence of greedy algorithms in Banach spaces', *Math. Notes* **73**, 342–368.

E. D. Livshitz (2006), 'On the recursive greedy algorithm', *Izv. RAN. Ser. Mat.* **70**, 95–116.

E. D. Livshitz (2007a), 'Optimality of the greedy algorithm for some function classes', *Mat. Sb.* **198**, 95–114.

E. D. Livshitz (2007b), On lower estimates of rate of convergence of greedy algorithms. Manuscript.

E. D. Livshitz and V. N. Temlyakov (2001), 'On the convergence of weak greedy algorithms', *Tr. Mat. Inst. Steklova* **232**, 236–247.

E. D. Livshitz and V. N. Temlyakov (2003), 'Two lower estimates in greedy approximation', *Constr. Approx.* **19**, 509–523.

A. Lutoborski and V. N. Temlyakov (2003), 'Vector greedy algorithms', *J. Complexity* **19**, 458–473.

V. E. Maiorov, K. I. Oskolkov and V. N. Temlyakov (2002), Gridge approximations and Radon compass. In *Approximation Theory: A Volume Dedicated to Blagovest Sendov*, DARBA, Sofia, pp. 284–309.

S. Mallat and Z. Zhang (1993), 'Matching pursuit in a time-frequency dictionary', *IEEE Trans. Signal Proc.* **41**, 3397–3415.

D. Needell and R. Vershynin (2007), Uniform uncertainty principle and signal recovery via regularized orthogonal matching pursuit. Manuscript.

H. Niederreiter, R. F. Tichy and G. Turnwald (1990), 'An inequality for differences of distribution functions', *Arch. Math.* **54**, 166–172.

M. Nielsen (2006), Trigonometric quasi-greedy bases for $L_p(\mathbb{T}; w)$. Manuscript.

S. N. Nikol'skii(1975), *Approximation of Functions of Several Variables and Embedding Theorems*, Springer.

P. Oswald (2001), 'Greedy algorithms and best m-term approximation with respect to biorthogonal systems', *J. Fourier Anal. Appl.* **7**, 325–341.

P. Petrushev (1988), Direct and converse theorems for spline and rational approximation and Besov spaces. In *Function Spaces and Applications*, Vol. 1302 of *Lecture Notes in Mathematics*, pp. 363–377.

T. Poggio and S. Smale (2003), 'The mathematics of learning: Dealing with data', *Notices Amer. Math. Soc.*, **50**, 537–544.

E. Schmidt (1906), 'Zur Theorie der linearen und nichtlinearen Integralgleichungen I', *Math. Annalen* **63**, 433–476.

A. V. Sil'nichenko (2004), 'Rate of convergence of greedy algorithms', *Mat. Zametki* **76**, 628–632.

F. Smithies (1937), 'The eigen-values and singular values of integral equations', *Proc. London Math. Soc.* (2) **43**, 255–279.

T. Stromberg and R. Heath Jr. (2003), 'Grassmannian frames with applications to coding and communications', *Appl. Comput. Harm. Anal.* **14**, 257–275.

V. N. Temlyakov (1988), 'Approximation by elements of a finite dimensional subspace of functions from various Sobolev or Nikol'skii spaces', *Matem. Zametki* **43**, 770–786. English translation in *Math. Notes* **43**, 444–454.

V. N. Temlyakov (1989*a*), 'Approximation of functions with bounded mixed derivative', *Proc. Steklov Inst.* **1**, 1–122.

V. N. Temlyakov (1989*b*) 'Estimates of the best bilinear approximations of functions of two variables and some of their applications', *Math. USSR Sb.* **62**, 95–109.

V. N. Temlyakov (1990), 'Bilinear approximation and applications', *Proc. Steklov Inst. Math.* **3**, 221–248.

V. N. Temlyakov (1992*a*), 'On estimates of approximation numbers and best bilinear approximation', *Constr. Approx.* **8**, 23–33.

V. N. Temlyakov (1992*b*), 'Estimates of best bilinear approximations of functions and approximation numbers of integral operators', *Math. Notes* **51**, 510–517.

V. N. Temlyakov (1993*b*), 'Bilinear approximation and related questions', *Proc. Steklov Inst. Math.* **4**, 245–265.

V. N. Temlyakov (1998*a*), 'The best m-term approximation and greedy algorithms', *Adv. Comput. Math.* **8**, 249–265.

V. N. Temlyakov (1998*b*), 'Nonlinear m-term approximation with regard to the multivariate Haar system', *East J. Approx.* **4**, 87–106.

V. N. Temlyakov (1998*c*), 'Greedy algorithm and m-term trigonometric approximation', *Constr. Approx.* **14**, 569–587.

V. N. Temlyakov (1999), 'Greedy algorithms and m-term approximation with regard to redundant dictionaries', *J. Approx. Theory* **98**, 117–145.

V. N. Temlyakov (2000*a*), 'Greedy algorithms with regard to multivariate systems with special structure', *Constr. Approx.* **16**, 399–425.

V. N. Temlyakov (2000*b*), 'Weak greedy algorithms', *Adv. Comput. Math.* **12**, 213–227.

V. N. Temlyakov (2001*a*), Lecture notes on approximation theory, Chapter I, University of South Carolina, pp. 1–20.

V. N. Temlyakov (2001*b*), 'Greedy algorithms in Banach spaces', *Adv. Comput. Math.* **14**, 277–292.

V. N. Temlyakov (2002*a*), 'Universal bases and greedy algorithms for anisotropic function classes', *Constr. Approx.* **18**, 529–550.

V. N. Temlyakov (2002*b*), 'A criterion for convergence of weak greedy algorithms', *Adv. Comput. Math.* **17**, 269–280.

V. N. Temlyakov (2002*c*), Nonlinear approximation with regard to bases. In *Approximation Theory X*, Vanderbilt University Press, Nashville, TN, pp. 373–402.

V. N. Temlyakov (2003*a*), 'Nonlinear methods of approximation', *Found. Comput. Math.* **3**, 33–107.

V. N. Temlyakov (2003*b*), 'Cubature formulas, discrepancy, and nonlinear approximation', *J. Complexity* **19**, 352–391.

V. N. Temlyakov (2004), 'A remark on simultaneous greedy approximation', *East J. Approx.* **10**, 17–25.

V. N. Temlyakov (2005*a*), 'Greedy type algorithms in Banach spaces and applications', *Constr. Approx.* **21**, 257–292.

V. N. Temlyakov (2005*b*), 'Greedy algorithms with restricted depth search', *Proc. Steklov Inst. Math.* **248**, 255–267.

V. N. Temlyakov (2005*c*), Approximation in learning theory. IMI Preprint 05, Department of Mathematics, University of South Carolina.

V. N. Temlyakov (2005*d*), On universal estimators in learning theory. IMI Preprint 17, Department of Mathematics, University of South Carolina.

V. N. Temlyakov (2006*a*), Greedy approximations. In *Foundations of Computational Mathematics: Santander 2005*, Vol. 331 of *London Mathematical Society Lecture Notes*, Cambridge University Press, pp. 371–394.

V. N. Temlyakov (2006*b*), Greedy approximations with regard to bases. In *Proc. International Congress of Mathematics* (Madrid 2006), Vol. 2, pp. 1479–1504.

V. N. Temlyakov (2006*c*), Relaxation in greedy approximation. IMI Preprint 03, Department of Mathematics, University of South Carolina.

V. N. Temlyakov (2006*d*), 'Optimal estimators in learning theory', *Approximation and Probability, Banach Center Publications*, **72**, 341–366.

V. N. Temlyakov (2007*a*), 'Greedy expansions in Banach spaces', *Adv. Comput. Math.* **26**, 431–449.

V. N. Temlyakov (2007*b*), 'Greedy algorithms with prescribed coefficients', *J. Fourier Anal. Appl.* **13**, 71–86.

J. F. Traub, G. W. Wasilkowski and H. Wozniakowski (1988), *Information-Based Complexity*, Academic Press, New York.

J. A. Tropp (2004), 'Greed is good: Algorithmic results for sparse approximation', *IEEE Trans. Inform. Theory* **50**, 2231–2242.

J. A. Tropp and A. C. Gilbert (2005), Signal recovery from partial information via orthogonal matching pursuit. Preprint, University of Michigan.

P. Wojtaszczyk (1997), 'On unconditional polynomial bases in L_p and Bergman spaces', *Constr. Approx.* **13**, 1–15.

P. Wojtaszczyk (2000), 'Greedy algorithms for general systems', *J. Approx. Theory* **107**, 293–314.

P. Wojtaszczyk (2002*a*), Greedy type bases in Banach spaces. In *Constructive Function Theory*, DARBA, Sofia, pp. 1–20.

P. Wojtaszczyk (2002*b*), 'Existence of best *m*-term approximation', *Functiones et Approximatio* **XXX**, 127–133.

P. Wojtaszczyk (2006), 'Greediness of the Haar system in rearrangement invariant spaces', *Banach Center Publications* **72**, 385–395.

H. Weyl (1911), 'Das asymptotische Verteilungsgesetz der Eigenwerte linearer partieller Differentialgleichungen', *Math. Ann.* **71**, 441–479.

A. Zygmund (1959), *Trigonometric Series*, Cambridge University Press.